THEORY OF PLASMAS

TEORIYA PLAZMY

ТЕОРИЯ ПЛАЗМЫ

The Lebedev Physics Institute Series

Editors: Academicians D. V. Skobel'tsyn and N. G. Basov

P. N. Lebedev Physics Institute, Academy of Sciences of the USSR

Volume 30	Physical Optics
Volume 31	Quantum Electronics in Lasers and Masers, Part 1
Volume 32	Plasma Physics
Volume 33	Studies of Nuclear Reactions
Volume 34	Photomesic and Photonuclear Processes
Volume 35	Electronic and Vibrational Spectra of Molecules
Volume 36	Photodisintegration of Nuclei in the Giant Resonance Region
Volume 37	Electrical and Optical Properties of Semiconductors
Volume 38	Wideband Cruciform Radio Telescope Research
Volume 39	Optical Studies in Liquids and Solids
Volume 40	Experimental Physics: Methods and Apparatus
Volume 41	The Nucleon Compton Effect at Low and Medium Energies
Volume 42	Electronics in Experimental Physics
Volume 43	Nonlinear Optics
Volume 44	Nuclear Physics and Interaction of Particles with Matter
Volume 45	Programming and Computer Techniques in Experimental Physics
Volume 46	Cosmic Rays and Nuclear Interactions at High Energies
Volume 47	Radio Astronomy: Instruments and Observations
Volume 48	Surface Properties of Semiconductors and Dynamics of Ionic Crystals
Volume 49	Quantum Electronics and Paramagnetic Resonance
Volume 50	Electroluminescence
Volume 51	Physics of Atomic Collisions
Volume 52	Quantum Electronics in Lasers and Masers, Part 2
Volume 53	Studies in Nuclear Physics
Volume 54	Photomesic and Photonuclear Reactions and Investigation Methods with Synchrotrons
Volume 55	Optical Properties of Metals and Intermolecular Interactions
Volume 56	Physical Processes in Lasers
Volume 57	Theory of Interaction of Elementary Particles at High Energies
Volume 58	Investigations in Nonlinear Optics and Hyperacoustics
Volume 59	Luminescence and Nonlinear Optics
Volume 60	Spectroscopy of Laser Crystals with Ionic Structure
Volume 61	Theory of Plasmas
Volume 62	Methods in Stellar Atmosphere and Interplanetary Plasma Research
Volume 63	Nuclear Reactions and Interaction of Neutrons and Matter
Volume 65	Stellarators
Volume 67	Physical Investigations in Strong Magnetic Fields
Volume 70	Group-Theoretical Methods in Physics

In preparation

Volume 64	Primary Cosmic Radiation
Volume 66	Theory of Collective Particle Acceleration and Relativistic Electron Beam Emission
Volume 68	Radiative Recombination in Semiconducting Crystals
Volume 69	Nuclear Reactions and Accelerators of Charged Particles
Volume 71	Photonuclear and Photomesic Processes
Volume 72	Physical Acoustics and Optics: Molecular Scattering of Light; Propagation of Hypersound; Metal Optics
Volume 73	Microwave–Plasma Interactions
Volume 74	Neutral Current Layers in a Plasma
Volume 75	Optical Properties of Semiconductors

Proceedings (Trudy) of the P. N. Lebedev Physics Institute

Volume 61

THEORY OF PLASMAS

Edited by
Academician D. V. Skobel'tsyn
Director, P. N. Lebedev Physics Institute
Academy of Sciences of the USSR, Moscow

Translated from Russian by
Julian B. Barbour

CONSULTANTS BUREAU
NEW YORK AND LONDON

Library of Congress Cataloging in Publication Data

Main entry under title:

Theory of plasmas.

(Proceedings (Trudy) of the P. N. Lebedev Physics Institute; v. 61)
Translation of Teoriiâ plazmy.
Includes bibliographical references.
CONTENTS: Kikvidze, R. R. and Rukhadze, A. A. Theory of oscillations and stability of a semiconductor plasma with low carrier density in a strong electric field. — Pustovalov, V. V. and Silin, V. P. Non-linear theory of the interaction of waves in a plasma.
1. Plasma (Ionized gases) 2. Plasma waves. 3. Semiconductors, Effect of radiation on. I. Skobel'tŝyn, Dmitrii Vladimirovich, 1892- ed. II. Series: Akademiiâ nauk SSSR. Fizicheskiĭ institut. Proceedings; v. 61.
QCl.A4114 vol. 61 [QC718] 530 .08s [530.4'4]
ISBN 978-1-4757-0132-6 ISBN 978-1-4757-0130-2 (eBook) 74-34269
DOI 10.1007/978-1-4757-0130-2

The original Russian text was published by Nauka Press in Moscow in 1972 for the Academy of Sciences of the USSR as Volume 61 of the Proceedings of the P. N. Lebedev Physics Institute. This translation is published under an agreement with the Copyright Agency of the USSR (VAAP).

A Division of Plenum Publishing Corporation
227 West 17th Street, New York, N.Y. 10011

United Kingdom edition published by Consultants Bureau, London
A Division of Plenum Publishing Company, Ltd.
Davis House (4th Floor), 8 Scrubs Lane, Harlesden, London, NW10 6SE, England

PREFACE

This volume contains two papers that review certain theoretical problems that have been studied in the Laboratory of Plasma Accelerators and Plasma Physics of the P. N. Lebedev Physics Institute of the Academy of Sciences of the USSR.

The review of R. R. Kikvidze and A. A. Rukhadze, "Theory of oscillations and stability of a semiconductor plasma with low carrier density in a strong electric field," is devoted to a solid-state plasma. The main attention is devoted to the fact that in such a plasma electromagnetic waves are effectively generated if there is a negative current–voltage characteristic in the carrier current; this effect can compete in importance with the well-known Gunn effect.

In their fundamental review paper "Nonlinear theory of the interaction of waves in a plasma," V. V. Pustovalov and V. P. Silin set forth the fundamentals of the theory of nonlinear interaction of waves in a hot rarefied plasma. Besides a systematic exposition of the procedure for deriving the equations that describe the nonlinear interaction of waves in an isotropic or an anisotropic (magnetized) plasma, they study many concrete examples relating to the interaction of definite types of waves under different conditions.

CONTENTS

THEORY OF OSCILLATIONS AND STABILITY OF A SEMICONDUCTOR PLASMA WITH LOW CARRIER DENSITY IN A STRONG ELECTRIC FIELD

R. R. Kikvidze and A. A. Rukhadze

Introduction 1

Chapter I. Homogeneous Semiconductor Plasma in a Strong Electric Field 5
- 1.1. Physical Meaning of the Instability Associated with a Negative Current-Voltage Characteristic 5
- 1.2. Basic Equations Describing a Seminconductor Plasma with Low Carrier Density 8
- 1.3. Oscillation Spectra of a Homogeneous Semiconductor Plasma in a Strong Electric Field 11
- 1.4. Effect of Diffusion on the Instability of a Semiconductor Plasma with Negative Current–Voltage Characteristic 20

Chapter II. Spatially Inhomogeneous Solid-State Plasma in External Fields 25
- 2.1. Equation of Small Oscillations of an Inhomogeneous Plasma with Low Carrier Density 25
- 2.2. Spectrum of Short-Wave Oscillations of an Inhomogeneous Solid-State Plasma 28
- 2.3. Surface Waves in a Solid-State Plasma in the Absence of a Magnetic Field 30
- 2.4. Surface Waves in a Bounded Magnetized Plasma with Current 33

Literature Cited 35

NONLINEAR THEORY OF THE INTERACTION OF WAVES IN A PLASMA

V. V. Pustovalov and V. P. Silin

Preface 37
Notation 38
Introduction 41

Chapter I. Fundamentals of the Nonlinear Electrodynamics of a Medium with Time and Spatial Dispersion 42
- 1. The Electromagnetic Field Equations and the Material Equation 42
- 2. Multi-Index Permittivity Tensors and Truncated Equations of the Electromagnetic Field 45
- 3. Averaging of the Truncated Equations of the Electromagnetic Field with Respect to the Statistical Ensemble 49
- 4. Discussion of Results. Equation for the Frequencies of Interacting Electromagnetic Waves 55

Chapter II. Multi-Index Permittivity Tensors of a Homogeneous Isotropic Plasma 57
5. Solution of the Kinetic Equation for an Isotropic Plasma as a Power Series in the Self-Consistent Field 59
6. Permittivity Tensor of an Isotropic Plasma as the Simplest Example of the Multi-Index Tensor $\varepsilon_{ij1...jn}$ for n = 1. Spectra and Polarization of the Characteristic Oscillations of an Isotropic Plasma 64
7. Three-Index Tensor $S_{ijs}(\omega, \mathbf{k}; \omega', \mathbf{k}')$ in an Isotropic Plasma 74
8. The Four-Index Tensor $V_{iajb}(\omega, \mathbf{k}; \omega', \mathbf{k}')$ in an Isotropic Plasma 84

Chapter III. Examples of Nonlinear Processes in an Isotropic Plasma 89
9. Nonlinear Interaction of Longitudinal Oscillations of an Isotropic Plasma 91
10. Nonlinear Interaction of Longitudinal and Transverse Oscillations in an Isotropic Plasma 108
11. Nonlinear Interaction of Transverse Characteristic Oscillations of an Isotropic Plasma 116
12. Effective Cross Sections of Various Scattering Processes in an Isotropic Plasma 121
13. Corrections to the Spectra of the Characteristic Oscillations of an Isotropic Plasma due to Nonlinear Interaction 127
14. Nonlinear Interaction of Characteristic Oscillations of a Plasma and the Theory of the Scattering of Electromagnetic Waves on Fluctuations of the Plasma 133

Chapter IV. Multi-Index Tensors of the Complex Permittivity of a Homogeneous Magnetoactive Plasma 140
15. Solution of the Transport Equation for a Magnetoactive Plasma in the Form of a Power Series in the Self-Consistent Electric Field 142
16. Permittivity Tensor of a Magnetoactive Plasma as the Simplest Example of the Multi-Index Tensor for n = 1 149
17. The Inverse Tensor (3.17) and the Spectral Function of the Electric Fields of the Characteristic Oscillations of a Magnetoactive Plasma 158
18. Quasilongitudinal Oscillations of a Magnetoactive Plasma. Spectra and Spectral Functions 166
19. Three-Index Tensor $S_{ijs}(\omega, \mathbf{k}; \omega', \mathbf{k}')$ in a Magnetoactive Plasma 176
20. Four-Index Tensor $V_{iajb}(\omega, \mathbf{k}; \omega', \mathbf{k}')$ in a Magnetoactive Plasma 197
21. Conservation Laws in the Nonlinear Interaction of Plasma Oscillations as a Consequence of the Symmetry of the Multi-Index Tensors 203

Chapter V. Specific Nonlinear Processes in a Magnetoactive Plasma 209
22. Decay of Quasilongitudinal Characteristic Oscillations of an Isothermal Magnetoactive Plasma 210
23. Decay of Quasilongitudinal Characteristic Oscillations of a Nonisothermal Magnetoactive Plasma 215
24. Induced Combination Scattering of High-Frequency Transverse Waves in a Cold Magnetoactive Plasma 220
25. Induced Scattering of Low-Frequency Quasilongitudinal Characteristic Oscillations of Magnetoactive Plasma on the Plasma Particles 222
26. Induced Scattering of Quasilongitudinal Characteristic Cyclotron Oscillations of a Magnetoactive Plasma on the Plasma Particles 233

Conclusions 240
Appendix. Multi-Index Complex Permittivity Tensors of a Cold Plasma in the Framework of the Hydrodynamic Description 241
Literature Cited 252

THEORY OF OSCILLATIONS AND STABILITY OF A SEMICONDUCTOR PLASMA WITH LOW CARRIER DENSITY IN A STRONG ELECTRIC FIELD

R. R. Kikvidze and A. A. Rukhadze

INTRODUCTION

Interest in plasma effects in solids has been greatly stimulated by the wide-scale introduction of semiconductor materials in recent years into radio engineering and electronics. The possibility of amplifying and generating electromagnetic waves has attracted particular attention. Very many papers on this subject have been published. By and large, the overwhelming majority of the papers have followed lines laid down by analogous investigations of a gas-discharge plasma; the specific features characteristic of a solid-state plasma have not really been taken into account. Among such papers we may mention the pioneering [1, 2], which first pointed out the possibility of the existence in metals and semiconductors of low-frequency slow waves, or helicons, which are well known in a gas plasma as helical waves [3]. These papers, which were confirmed experimentally in a short space of time [4], laid the foundation for many investigations of plasma effects in solids.

At about the same time [5], Pines and Schrieffer considered the problem of the excitation of electromagnetic waves when beams of charged particles interact with a solid-state plasma. This investigation, too, was similar to the well-known work of Akhiezer and Fainberg [6] and Bohm and Gross [7], who discovered the two-stream instability in a gas plasma.

The idea of exciting electromagnetic waves in semiconductors by a beam of charged particles appeared very attractive but difficult to realize in practice. It was found to be more realistic to excite electromagnetic waves by carrier drift when an electric field is applied to a semiconductor. In a gas plasma, investigations of the analogous problem led to very hopeful results (see [8], which contains a detailed bibliography). Many of the results of the theory of the oscillations of a gas plasma in an external electric field were generalized to the case of a solid-state plasma [9-11].

However, it was only quite recently that the foundations were laid of a truly comprehensive and consistent theory of the oscillations and the stability of a solid-state plasma in an external electric field. Above all, we must here mention [12-14], in which the theory of the oscillations and the stability of a solid-state plasma in an external electric field is constructed on the basis of the transport equation for the carriers with collision integral in Davydov's form [11] (see also [3, 16]). In these papers, the spectra of the oscillations of a solid-state plasma when there is electric drift of the carriers were analyzed in considerable detail and criteria were obtained for the excitation of different modes. It was shown that low-frequency oscillations of the solid-state plasma can be excited at relatively low velocities of the electric drift of the carriers, lower than their thermal velocity. However, the excitation of oscillations investigated in these papers is possible only in a plasma with two carrier species. Such a conclusion is a result of the circumstance that in [12-14] it was assumed that the energy and momentum relaxation times (or, in other words, the collision frequency) and also the effective mass do not depend on the carrier velocity. For this reason, the theory developed in the cited papers resembled the theory of oscillations of a weakly ionized plasma [17-20] and was applicable rather to such a plasma than to a solid-state plasma.

Important progress in the theory of the oscillations and the stability of a solid-state plasma was made in [21-30], which predicted and investigated the instability when there is negative differential conductivity in the carrier current. Instability due to a negative current–voltage characteristic is more characteristic of a solid-state plasma than a gas plasma, although it should be pointed out that negative differential conductivity is also possible in a gas plasma, especially in a weakly ionized one [31] (see also [16]). Interest in this instability quickened in 1963 after Gunn had discovered experimentally the phenomenon of amplification and spontaneous generation of high-frequency electromagnetic waves in gallium arsenide when a sufficiently strong electric field is applied [32]. This effect, which became known as the Gunn effect, agreed well in its qualitative features with the theoretical predictions developed in [21, 22] on the basis of the concept of negative differential conductivity.

Obviously, a medium can have a negative current – voltage characteristic only if the linear Ohm's law is violated. There are many reasons why Ohm's law should break down. As long ago as 1937, Davydov [15] noted that there can be a departure from Ohm's law in a semiconductor in even relatively weak electric fields. As we shall see from what follows, light carriers are effectively heated in a semiconductor in an electric field; their temperature becomes greater than the lattice temperature (or, as one says, "hot electrons" appear). The characteristic feature of such a plasma is the possibility that the energy and momentum relaxation times and also the mass of the carriers, and therefore the conductivity, depend on the carrier velocity, i.e., on the electric field. If the conductivity is a decreasing function of the electric field, the current – voltage characteristic exhibits a section with negative slope, and this can give rise to an instability, or, and this is the same thing, the spontaneous excitation of electromagnetic oscillations in the medium.

The most widespread reason for the appearance of a negative current–voltage characteristic may be a dependence of the energy and momentum relaxation times of the carriers on their velocity. It was this cause of instability in a solid-state plasma that was investigated in [24-29] and is taken as the basis of the present review. There are other mechanisms that can give rise to negative differential conductivity. For example, in extrinsic semiconductors a section of negative current–voltage characteristic can arise because of a decrease in the carrier density due to an increase in the probability of carrier capture at impurities when the carrier energy increases in the electric field [33]. In the Gunn effect, the negative differential conductivity in a two-valley semiconductor like gallium arsenide is due to heated carriers being transferred from one conduction band to another, this being accompanied by an increase in their effective mass, which reduces the mobility. However, these are rather special reasons and can be operative only in semiconductors of a certain type, whereas the relaxation times can depend on the carrier velocity in almost any semiconductor.

In a theoretical investigation of the oscillation spectra of a solid-state plasma one must distinguish two limiting cases: plasmas with high and low carrier density, respectively. If the energy and momentum relaxation times in a semiconductor are entirely determined by carrier scattering on vibrations of the crystal lattice and the interaction of the carriers with one another can be ignored, one says that the solid-state plasma has a low carrier density. This condition is satisfied if [34]

$$n \ll n_{cr} \approx \frac{1}{4\pi} \frac{\varepsilon^{3/2}\sqrt{m}}{e^{*4}\tau_\varepsilon}. \tag{1}$$

Here, n is the carrier density, ε is the mean carrier kinetic energy; m is the effective mass; e^* is the effective charge; $e^{*2} = (e^2/\varepsilon)L$ (e is the electron charge, L is the Coulomb logarithm; ε_0 is the permittivity of the crystal); and τ_ε is the energy relaxation time for carrier scattering on lattice vibrations. For the majority of semiconductors, $n_{cr} \approx 10^{14}$-10^{15} cm^{-3}. In the opposite limit, when $n \gg n_{cr}$, the interaction of the carriers with one another plays an important role in relaxation processes; in such cases, one speaks of a semiconductor plasma with a high carrier density.

The behavior of a solid-state plasma with high carrier density in an external electric field was investigated in [24-27]. The effective-temperature approximation [3, 16] can be used to describe such a plasma, since the energy distribution function of the carriers can be made Maxwellian with an effective temperature that depends on the electric field, by scattering on lattice vibrations. Such an approximation is equivalent to the hydrodynamic description of the motion of the carriers (one-or two-fluid hydrodynamics), in which the momentum and energy relaxation times are averaged over the Maxwellian distribution with effective temperature.

We have a different situation in a plasma with low carrier density, for which the condition (1) holds. In general, the carrier velocity distribution in such a plasma is not Maxwellian when an external electric field is applied. To describe such a plasma, one cannot apply hydrodynamic equations and one must use a transport equation with collision integral in Davydov's form [15]. The non-Maxwellian distribution function gives rise to a large number of effects that make a semiconductor plasma with low carrier density very different from a plasma with high carrier density. It is for this reason that we are here studying a solid-state plasma with low carrier density in an external electric field.

The first part of the review is devoted to the properties of a spatially homogeneous and unbounded plasma in an electric field. We derive the basic equations that describe a solid-state plasma with low carrier density and we set forth methods of solving these equations approximately at low frequencies of the oscillations. On the basis of the solutions obtained in § 1.1-1.3, we investigate the oscillations and stability of a solid-state plasma in an electric field, but under conditions when spatial dispersion can be ignored in the permittivity tensor of the medium. This last assumption means that the effect of carrier diffusion in the oscillation process is ignored. The momentum and energy relaxation times are assumed to depend on the carrier velocity in the form [24]

$$\frac{1}{\tau} = \nu = \nu_0\left(\frac{v}{v_0}\right)^{-2q}, \quad \frac{1}{\tau_\varepsilon} = \delta\nu = \delta_0\nu_0\left(\frac{v}{v_0}\right)^{2(r-1)}. \tag{2}$$

Here ν_0 and δ_0 are the values of ν and δ (δ characterizes the fraction of energy transferred to the lattice in carrier scattering; for typical semiconductors, $\delta \sim 10^{-3}$-10^{-2} at $v = v_0 = (T/m)^{1/2}$, where T is the lattice temperature; r and q are certain numbers that characterize the actual carrier scattering mechanism. The values of r and q for different scattering mechanisms are given in [24].

For $r = 1$ and $q = 0$, i.e., when the relaxation times do not depend on the carrier velocity, the current–voltage characteristic in a solid-state plasma is always positive, irrespective of any carrier heating. Electromagnetic oscillations can be excited in such a plasma only if two carrier species are present, and then in the two-stream manner. But if $r \neq 1$ and $q \neq 0$, then if the carriers are heated and certain conditions hold, the current–voltage characteristic may include a section with negative slope. In a plasma with low carrier density the characteristic can actually have only an N-shaped form (we restrict ourselves to confining heating mechanisms [35], i.e., we impose a restriction on the numbers r and q, requiring the distribution function to be normalizable). The analysis of the stability of small oscillations shows that the appearance of a negative current–voltage characteristic in a semiconductor plasma with low carrier density is accompanied by the excitation of essentially electrostatic (longitudinal) electromagnetic oscillations, since the growth rates of transverse waves are relatively small.

The effect of diffusion on the excitation of oscillations in a semiconductor plasma with low carrier density in an external electric field is considered in §1.4 of Chapter I of the review. Here, we explain a method of solving Davydov's equation with allowance for diffusion and heat conduction processes, this being similar to the well-known Chapman–Enskog method [36]. It is shown that carrier diffusion stabilizes oscillations whose wavelength is less than the Debye radius.

Chapter II of the review is devoted to an investigation of a spatially inhomogeneous and bounded solid-state plasma in an external electric field. We may mention here that one of the papers that stimulated the intensive investigations of plasma effects in a solid was [37], in which it is shown that helical waves can be excited by an electric current in a gas plasma with inhomogeneous conductivity. This paper was followed by a great many papers, both theoretical and experimental, which attempted to explain the observed excitation of oscillations in semiconductors by the development of a helical instability in a solid-state plasma. All these investigations are fairly fully reviewed in [9, 10]. Since then, the theory of the oscillations and stability of a spatially inhomogeneous gas plasma has made considerable advances. It has been shown that in an inhomogeneous plasma with a current there can develop drift and drift-dissipative instabilities due to inhomogeneity of the density and temperature of the charged particles, current-convective and slipping instabilities due to inhomogeneity of the velocity of electric drift of the electrons, and others. The reader can become acquainted with the theory of oscillations and stability of an inhomogeneous gas plasma in the reviews [38-41] (see also [19]), in which there is a detailed bibligoraphy on this subject. These advances in the physics of a gas plasma have not yet been matched in the theory of solid-state plasmas. The papers [42-44], which are devoted to the drift-temperature and current-convective instabilities of a solid-state plasma, like the cited papers on the helical instability (see the reviews [9, 10]), do not take into account all the features of a solid-state plasma. In particular, they do not consider effects related to the dependence of the relaxation times on the carrier velocity, so that they do not allow for the deviation of the carrier velocity distribution from a Maxwellian distribution. To establish the importance of these effects, we consider in the present review, following [45], the stability of a spatially inhomogeneous solid-state plasma with low carrier density when there is a negative current–voltage characteristic. We show that in a strongly magnetized solid-state plasma to which an electric field is applied along the direction of the magnetic field a current-convective instability can arise in both an inhomogeneous unbounded plasma (§ 2.1 and § 2.2) as well as in a homogeneous but bounded plasma (§ 2.3 and § 2.4). Under certain conditions, this instability may be dominant, and the frequency and growth rate of the oscillations in this case are at least an order of magnitude greater than the corresponding quantities when there is an instability due to a negative current–voltage characteristic in a spatially homogeneous and unbounded plasma.

In this paper we consider the excitation of oscillations in a solid-state plasma with one carrier species. This applies to both the instability due to the negative current–voltage characteristic as well as the current-convective instability. It is not difficult to generalize the results to an intrinsic semiconductor with two carrier species. In this review, we make such a generalization for one special example. It must, however, be borne in mind that in an intrinsic semiconductor one can also have the excitation of two-stream instabilities; the conditions of excitation and the growth rates of these instabilities are also given in the review.*

* For the physical effects in semiconductors with negative differential conductivity and the oscillations that are then excited, see the review [46], which is primarily concerned with a semiconductor plasma with high carrier density.

CHAPTER I

HOMOGENEOUS SEMICONDUCTOR PLASMA IN A STRONG ELECTRIC FIELD

§ 1.1. Physical Meaning of the Instability Associated with a Negative Current – Voltage Characteristic

To bring out the physical nature of the electrostatic instability of a solid-state plasma associated with a negative current–voltage characteristic, we shall consider this instability from a more general point of view. For simplicity, we shall restrict ourselves to a solid-state plasma with one carrier species whose density does not change (i.e., we do not allow for carrier creation and recombination); the effective carrier mass is also assumed constant. An arbitrary electrostatic field in such a medium is described by the equations

$$\operatorname{div} \mathbf{E} = 4\pi e (n - n_0), \qquad \frac{\partial en}{\partial t} + \operatorname{div} \mathbf{j} = 0. \tag{1.1}$$

In a stationary and homogeneous equilibrium state of the medium, $n = n_0 = \text{const}$ and, therefore, $\operatorname{div} \mathbf{j}_0 = 0$ and $\operatorname{div} \mathbf{E}_0 = 0$, i.e., if there is no excess charge in the medium, it is only possible for there to be homogeneous electric fields $\mathbf{E}_0$ and currents $\mathbf{j}_0$ in the medium. If an excess charge arises as a result of a perturbation of the density, $n = n_0 + n_1$, there may appear in the medium perturbed inhomogeneous fields $\mathbf{E} = \mathbf{E}_0 + \mathbf{E}_1$ and currents $\mathbf{j} = \mathbf{j}_0 + \mathbf{j}_1$ and, if the state of the medium is unstable, these may increase with the time.

To follow the time evolution of the process, we must augment the system (1.1) with the material equation. Without particularizing the model of the medium, we can write the material equation in the general form

$$\mathbf{j} = en\mathbf{F}(\mathbf{E}), \tag{1.2}$$

where $\mathbf{F}(\mathbf{E})$ takes into account the arbitrary dependence of the carrier mobility on the electric field.

From the system (1.1) and with allowance for the material equation (1.2), we find

$$\frac{\partial}{\partial t} \operatorname{div} \mathbf{E} + \operatorname{div} \{\mathbf{F}(\mathbf{E}) [\operatorname{div} \mathbf{E} + 4\pi e n_0]\} = 0. \tag{1.3}$$

Linearizing this equation with respect to the small perturbation $\mathbf{E}_1$ of the field, which depends on the time and the coordinates as a plane electromagnetic wave, $\exp(-i\omega t + i\mathbf{kr})$, we obtain

$$\omega (\mathbf{kE}_1) - \mathbf{kF}(\mathbf{E}_0)(\mathbf{kE}_1) + 4\pi i e n_0 \frac{\partial (\mathbf{kF})}{\partial \mathbf{E}_0} \mathbf{E}_1 = 0. \tag{1.4}$$

Hence, remembering that the field is electrostatic, $\mathbf{E}_1 = \nabla \Phi$, we find the spectrum of small oscillations, this describing the time evolution of the electrostatic perturbations:

$$\omega = \mathbf{kF}(\mathbf{E}_0) - 4\pi i e n_0 \frac{k_i k_j}{k^2} \frac{\partial F_i (\mathbf{E}_0)}{\partial E_{0j}}. \tag{1.5}$$

In the case of an isotropic medium, when $\mathbf{F}(\mathbf{E}_0) = \mu(\mathbf{E}_0)\mathbf{E}_0$, we then obtain

$$\omega = \mu(\mathbf{E}_0)(\mathbf{kE}_0) - 4\pi i e n_0 \mu(\mathbf{E}_0)\left[1 + \cos^2 \varphi \frac{\partial \ln \mu (\mathbf{E}_0)}{\partial \ln E_0}\right], \tag{1.6}$$

where φ is the angle between the vectors $\mathbf{k}$ and $\mathbf{E}_0$. This expression shows that it is most probable for waves to be excited that propagate along the external field and that excitation is possible only if

$$1+\frac{\partial \ln \mu(\mathbf{E}_0)}{\partial \ln \mathbf{E}_0}<0. \tag{1.7}$$

This inequality is identical with the condition for the appearance of negative differential conductivity in the carrier current:

$$\frac{\partial j_0}{\partial E_0} \sim \left(1+\frac{\partial \ln \mu(\mathbf{E}_0)}{\partial \ln E_0}\right)<0. \tag{1.8}$$

The electrostatic instability associated with a negative current–voltage characteristic can be explained as follows. Suppose that in a certain bounded region of the medium an excess charge arises in a fluctuation (the carriers flow in from the neighboring region), this violating the neutrality of the medium. This excess charge increases the electric field in this region of space and decreases it in the neighboring region. Because of the law of charge conservation (the continuity equation) an excess current must arise in the medium to equalize the charge by flowing into the neighboring region. However, because of the negative current–voltage characteristic, the current in the medium decreases with increasing electric field. This, in its turn, means that the current from the region of excess charge decreases while that from the depleted region increases. As a result, the charge continues to accumulate, and the electric field in the region of excess charge grows – the medium is unstable.

A negative current–voltage characteristic in the carrier current may be due to various factors. For example, if we write down the current in a semiconductor plasma with one carrier species in the generally adopted form

$$\mathbf{j}=en(\mathbf{E}_0)\,\mu(\mathbf{E}_0)\,\mathbf{E}_0, \tag{1.9}$$

we can readily see that a negative current–voltage characteristic can be due to a decrease of either the carrier density $n(E_0)$ or the carrier mobility $\mu(E_0)$ with increasing electric field.

Mechanisms leading to a negative current–voltage characteristic because of a decrease in the number of carriers with increasing field are considered in [33]. These mechanisms include carrier recombination at charged impurities if they are present in the semiconductor and the probability of recombination increases with increasing electric field or the capture of carriers at trapping levels increases, etc. These mechanisms are too specific and can be manifested only in a small class of semiconductors of a certain type. For this reason we do not consider a dependence of the carrier density on the field in the present review.

A dependence of the mobility on the electric field is a more widespread reason for the appearance of a negative current–voltage characteristic in a semiconductor plasma. Two mechanisms are associated with a dependence of the mobility on the external electric field:

$$\mu(\mathbf{E}_0)=\frac{e}{m(\mathbf{E}_0)\,\bar{\nu}(\mathbf{E}_0)}, \tag{1.10}$$

where m is the effective mass and $\bar{\nu}$ is the averaged collision frequency of the carriers. The first of them is possible only in a many-valley semiconductor and is due to the transfer of carriers from one conduction band to another when the electric field is increased. If the effective carrier mass is increased on the transfer, their mobility decreases, which leads to the appearance of a section with negative current–voltage characteristic. A mechanism of this kind has been investigated theoretically in [21, 22], in which it was predicted that electrostatic oscillations could be excited in such semiconductors. And, indeed, the excitation of oscillations in a two-valley semiconductor (for example, gallium arsenide) in an external electric field was observed by Gunn [32], and became known as the Gunn effect. From the general standpoint adopted at the beginning of this section, one can also include such a mechanism of negative current–voltage characteristic. However, in what follows we shall not consider this mechanism, regarding it too as specific.

A second mechanism leading to a decrease in the mobility consists of an increase in the effective collision frequency of the carriers with increasing energy due to heating in an external electric field. This mechanism leading to a section of negative current–voltage characteristic in a solid has been called the overheating mechanism. In [24-26, 47, 48] a study is made of the instability of a solid-state plasma when a negative current–voltage characteristic is made possible by this mechanism under conditions when the carrier density is high, i.e., the opposite condition to (1) is satisfied, and the effective-temperature approximation therefore applies. In the present review we consider the opposite limit of a semiconductor with low carrier density. However, the general arguments at the beginning of this section apply to both limiting cases provided $\bar{\nu}$ is understood as an energy-dependent frequency of collisions of carriers with the lattice averaged over the carrier equilibrium distribution function.

Denoting the mean carrier density by ε, which must not however be generally identified with the temperature since the carrier velocity distribution is not Maxwellian, we find that because of the dependence of the conductivity σ on the carrier energy,

$$\mathbf{j}_0 = \sigma(\varepsilon)\,\mathbf{E}_0. \tag{1.11}$$

On the other hand, in a stationary state of the medium the ohmic heat $\mathbf{j}_0\mathbf{E}_0$ liberated in the carrier gas when a current flows is passed to the lattice, which constitutes a reservoir with infinite heat capacity, as a result of which the lattice is not heated. The heat $P(\varepsilon)$ transferred to the lattice is a function of the carrier energy. Thus,

$$P(\varepsilon) = j_0 E_0 = \sigma(\varepsilon)\,E_0^2. \tag{1.12}$$

From the relations (1.11) and (1.12), we can readily obtain an expression for the differential conductivity of the medium [48]

$$\sigma_{\text{diff}} = \frac{\partial j_0}{\partial E_0} = \frac{\left(\frac{\partial P}{\partial \varepsilon} + \frac{\partial \sigma}{\partial \varepsilon} E_0^2\right)}{\left(\frac{\partial P}{\partial \varepsilon} - \frac{\partial \sigma}{\partial \varepsilon} E_0^2\right)}\,\sigma(\varepsilon). \tag{1.13}$$

In weak electric fields, when the carrier heating can be ignored [i.e., when one can ignore the second terms in the numerator and denominator of (1.13)], $\sigma_{\text{diff}} = \sigma = \text{const}$. With increasing field there is a departure from the linear Ohm's law, because of the carrier heating. If at the same time

$$\frac{\partial P}{\partial \varepsilon} > 0 \quad \left(\frac{\partial P}{\partial \varepsilon} < 0\right), \qquad \frac{\partial \sigma}{\partial \varepsilon} > 0 \quad \left(\frac{\partial \sigma}{\partial \varepsilon} < 0\right), \tag{1.14}$$

then the denominator of the expression (1.13) may vanish when the electric field increases. Then the differential conductivity increases unboundedly. In such cases one says that the current–voltage characteristic has an S-shaped form (Fig. 1, curve 1). But if

$$\frac{\partial P}{\partial \varepsilon} > 0 \quad \left(\frac{\partial P}{\partial \varepsilon} < 0\right), \qquad \frac{\partial \sigma}{\partial \varepsilon} < 0 \quad \left(\frac{\partial \sigma}{\partial \varepsilon} > 0\right), \tag{1.15}$$

then the numerator of (1.13) may vanish with increasing electric field, and then the differential conductivity of the medium vanishes as well. In this case, the current–voltage characteristic has an N-shaped form (Fig. 1, curve 2).

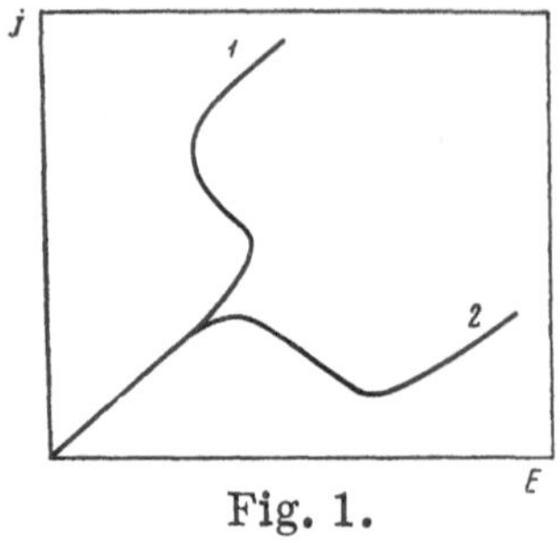

Fig. 1.

§ 1.2. Basic Equations Describing a Semiconductor Plasma with Low Carrier Density

Before we turn to the investigation of the oscillations and the stability of a solid-state plasma in an external electric field, let us formulate the basic equations that describe a plasma with low carrier density. As usual, to investigate the spectra of small electromagnetic oscillations of a solid-state plasma as a material medium, we shall proceed from Maxwell's equations in the form [49]

$$\operatorname{div} \mathbf{D} = 4\pi\rho_0, \qquad \operatorname{div} \mathbf{B} = 0,$$
$$\operatorname{rot} \mathbf{E} = -\frac{1}{c}\frac{\partial \mathbf{B}}{\partial t}, \qquad \operatorname{rot} \mathbf{B} = \frac{1}{c}\frac{\partial \mathbf{D}}{\partial t} + \frac{4\pi}{c}\mathbf{j}_0. \tag{1.16}$$

Here, ρ_0 and $\mathbf{j}_0$ are, respectively, the charge density and the current of the external sources of the field in the medium, **E** and **B** are the electric field and the magnetic induction, and **D** is the electric displacement. The vector function **D** contains all the information on the electromagnetic properties of the medium and is related to the electric field by the material equation $\mathbf{D} = \mathbf{D}(\mathbf{E})$. This arbitrary functional relation becomes linear in the case of weak fields **E** and **D**. For media that are homogeneous in space and time, when the field can be represented in the form of a plane monochromatic wave $\mathbf{E} = \mathbf{E}(\omega, \mathbf{k}) \exp(i\omega t + i\mathbf{kr})$ (the fields **D** and **B** are represented similarly), the linear material relation in the general case can be expressed as

$$D_i(\omega, \mathbf{k}) = \varepsilon_{ij}(\omega, \mathbf{k}) E_j(\omega, \mathbf{k}). \tag{1.17}$$

The permittivity tensor $\varepsilon_{ij}(\omega, \mathbf{k})$ characterizes the electromagnetic properties of the medium. The calculation of this tensor is the problem for the particular model of the medium. It is this tensor that we calculate in the present review for a semiconductor plasma with low carrier density in an external electric field.

Knowing the tensor $\varepsilon_{ij}(\omega, \mathbf{k})$, we can readily determine the spectrum of characteristic oscillations of the medium. Indeed, from Maxwell's equation (1.16) with allowance for the material relation (1.17) and in the absence of external field sources ($\rho_0 = 0$, $\mathbf{j}_0 = 0$), we obtain a system of homogeneous algebraic equations for the components of the field **E**:

$$\left\{k^2\delta_{ij} - k_i k_j - \frac{\omega^2}{c^2}\varepsilon_{ij}(\omega, \mathbf{k})\right\} E_j = 0. \tag{1.18}$$

The solvability condition of this system,

$$\left|k^2\delta_{ij} - k_i k_j - \frac{\omega^2}{c^2}\varepsilon_{ij}(\omega, \mathbf{k})\right| = 0, \tag{1.19}$$

is the dispersion relation for determining the spectrum of characteristic oscillations of the medium: $\omega = \omega(\mathbf{k})$.

If all the roots of Eq. (1.19) $\omega_s(\mathbf{k})$ (where s = 1, 2, ...) are such that Im $\omega_s < 0$, then any perturbations that arise as a result of a fluctuation in the medium are damped with time. In such cases one says that the medium is stable. But if the roots of this equation include some for which Im $\omega_s(\mathbf{k}) > 0$, initial perturbations in the medium grow with time. One then says that the medium is unstable; electromagnetic oscillations can be excited in it.

Under conditions when one is concerned with low-frequency and short-wavelength oscillations, so that $\omega \ll kc$, there is no need to investigate the general dispersion equation (1.19). Such a field in the medium is electrostatic to a high degree of accuracy, i.e.,

$$\mathbf{E} = -\nabla\Phi, \quad \mathbf{B} = 0. \tag{1.20}$$

Instead of the system of Maxwell's equations (1.16) we can then restrict ourselves to Poisson's equation:

$$\Delta\Phi = 4\pi(\rho + \rho_0). \tag{1.21}$$

Here, ρ is the density of the charge induced in the medium by the field **E**, and it is related to the density of the induced current **j** by the continuity equation:

$$\frac{\partial\rho}{\partial t} + \operatorname{div}\mathbf{j} = 0. \tag{1.22}$$

For media that are homogeneous in space and in time and in the absence of external field sources ($\rho_0 = 0$, $\mathbf{j}_0 = 0$), Eqs. (1.21) and (1.22) with allowance for the material equation (1.17) lead to the following dispersion relation for the electrostatic (longitudinal) field oscillations:

$$\varepsilon(\omega, \mathbf{k}) = \frac{k_i k_j}{k^2}\varepsilon_{ij}(\omega, \mathbf{k}) = 0. \tag{1.23}$$

In the present review we shall be largely concerned with longitudinal field oscillations.

Now that we have written down the equations of the electromagnetic field, let us formulate the equations of motion of the medium. In other words, let us particularize our model of a solid-state plasma with low carrier density. We have pointed out above that the carrier velocity distribution in such a plasma may be far from Maxwellian. Therefore, even if we are interested in low-frequency oscillations and the case of frequent collisions between the carriers and the lattice, we cannot, in general, use a hydrodynamic description. We shall therefore proceed from the most general model — a transport equation with self-consistent interaction that takes into account carrier collisions:

$$\frac{\partial f}{\partial t} + \mathbf{v}\frac{\partial f}{\partial \mathbf{r}} + e\left\{\mathbf{E} + \frac{1}{c}[\mathbf{v}\times\mathbf{B}]\right\}\frac{\partial f}{\partial \mathbf{p}} = St_{ee}(f) + St(f). \tag{1.24}$$

The first term on the right-hand side takes into account collisions between the carriers and the second collisions between them and lattice vibrations. We point out once more that inelastic processes that change the number of particles in the solid (capture or generation) will not be considered in this review. Moreover, we shall assume that the carriers remain in one conduction band for the whole time, and we therefore ignore effects due to the transfer of carriers from one band to another, as is the case in the Gunn effect.

For a plasma with low carrier density, the first term on the right-hand side of Eq. (1.24) is small compared with the second, $\mathrm{St}_{ee}(f) \ll \mathrm{St}(f)$. If we note that $\mathrm{St}_{ee}(f) \sim \nu_{ee} f$, where ν_{ee} is the frequency of collisions of the charged carriers with one another and $\mathrm{St}(f) \sim \delta\nu f \sim f/\tau_\varepsilon$, where τ_ε is the energy relaxation time for carrier scattering on lattice vibrations, we arrive at the condition (1). A semiconductor with a high carrier density, in which the opposite inequality holds, is investigated in [24-26, 47, 48]. In what follows, we shall compare our results with those of these papers.

We shall be interested in relatively weak external electric fields $\mathbf{E}_0$, for which the drift velocity v_d of the carriers is less than their thermal velocity v_t, i.e.,

$$v_d \sim \frac{eE_0}{m}\tau \ll v_t. \tag{1.25}$$

Note that the fields can still be fairly strong and lead to an appreciable heating of the carriers; the carriers are heated appreciably under conditions when the electric drift velocity exceeds the velocity of sound in the medium, $v_s \sim \sqrt{\delta} v_0$, where v_0 is the thermal velocity of the unheated carriers. We shall show below that when carriers are heated the inequality (1.25) is satisfied automatically. It is important that this is also satisfied at the time when the field is switched on and the carriers are not yet heated. For typical semiconductors, $v_0 \sim 10^8$ cm/sec, and the condition (1.25) is well satisfied in fields up to a few keV/cm. Under these assumptions, the anisotropic part of the distribution function of the carriers, which characterizes their macroscopic motion – electric drift and diffusion – is small compared with the isotropic part. This enables us to represent the carrier distribution function as a series in Legendre polynomials and restrict ourselves to the first two terms of the expansion:

$$f = f_0 + \frac{\mathbf{v}}{v}\mathbf{f}_1. \tag{1.26}$$

The expansion (1.26) is like the one used in [15] (see also [3, 16]) for a spatially homogeneous plasma in an electric field. Since the electric drift of the carriers in the cases in which we are interested is always greater than the diffusion drift, the condition obtained in [15] for one to be able to ignore the higher terms of the expansion (1.26) remains in force for a spatially inhomogeneous plasma. This condition is identical with (1.25).

Substituting the expansion (1.26) into Eq. (1.24) and using the explicit form of the collision integral $\mathrm{St}(f)$ given in [15], we obtain a system of two coupled equations for the symmetric and the asymmetric parts of the distribution functions:

$$\begin{aligned} &\frac{\partial f_0}{\partial t} + \frac{v}{3}\operatorname{div}\mathbf{f}_1 + \frac{e}{3mv^2}\frac{\partial}{\partial v}(v^2\mathbf{E}\mathbf{f}_1) + S_0(f_0) = 0, \\ &\frac{\partial \mathbf{f}_1}{\partial t} + v\operatorname{grad} f_0 + \frac{e\mathbf{E}}{m}\frac{\partial f_0}{\partial v} + \frac{e}{mc}[\mathbf{B}\times\mathbf{f}_1] + \nu(v)\,\mathbf{f}_1 = 0, \end{aligned} \tag{1.27}$$

where

$$S_0(f_0) = -\frac{1}{2v^2}\frac{\partial}{\partial v}\left\{v^2\delta\nu\left(\frac{T}{m}\frac{\partial f_0}{\partial v} + vf_0\right)\right\}. \tag{1.28}$$

The functions $\nu(v)$ and $\tau(v)\nu(v)$, which characterize the momentum and energy relaxation times, $\tau = 1/\nu$ and $\tau_\varepsilon = 1/\delta\nu$, are functions of the carrier velocity in a semiconductor. Following [24], we shall choose them in the form (2).

All our above arguments concerning the validity of (1.26) and Eqs. (1.27) referred to the equilibrium state of the plasma. However, it is perfectly obvious that they also remain true for a nonequilibrium state if the characteristic times of variation of the distribution function (or the collision frequency) are greater than the relaxation times ($\omega \ll \delta\nu$) and the characteristic dimensions (wavelengths of the oscillations) are greater than the mean-free-path* ($\lambda \gg l = v_t/\nu$). Therefore, we shall use these equations to describe both the equilibrium state and small low-frequency oscillations of a semiconductor plasma in an external electric field.

* As is shown in [13], the system (1.27) holds for the description of plasma oscillations under conditions of weak spatial dispersion, i.e., when $|\omega + i\nu| \gg kv_t$. Therefore, in general, it is also valid when $\omega \gg \nu$ provided $\omega \gg kv_t$.

§ 1.3. Oscillation Spectra of a Homogeneous Semiconductor Plasma in a Strong Electric Field

Equilibrium Distribution Function of the Carriers. We consider a spatially homogeneous solid-state plasma in an external stationary electric field $\mathbf{E}_0$. Before we calculate the permittivity tensor of such a plasma, we determine the equilibrium distribution function of the carriers. Using the collision integral (1.28), we can readily find a solution of the system (1.27) and write it in the form

$$f_{00}=A\exp\left\{-\int_0^v mv\,dv\left[T+\frac{2e^2E_0^2\,(1+\Omega^2\cos^2\vartheta/\nu^2)}{3m\delta\,(\nu^2+\Omega^2)}\right]^{-1}\right\},$$
$$\mathbf{f}_{10}=-\mathbf{V}_0\frac{\partial f_{00}}{\partial v}, \tag{1.29}$$

where $\mathbf{V}_0$, the velocity of electric drift of the carriers in the fields $\mathbf{E}_0$ and $\mathbf{B}_0$, is

$$\mathbf{V}_0=\frac{e}{m\nu}\frac{\nu^2}{\Omega^2+\nu^2}\left\{\mathbf{E}_0+\frac{\Omega}{\nu}\left[\frac{[\mathbf{E}_0\times\mathbf{B}_0]}{B_0}+\frac{\Omega}{\nu}\frac{\mathbf{B}_0\,(\mathbf{B}_0\mathbf{E}_0)}{B_0^2}\right]\right\}. \tag{1.30}$$

Here $\Omega = eB_0/mc$ is the frequency of Larmor gyration of the carriers, θ is the angle between the fields $\mathbf{E}_0$ and $\mathbf{B}_0$, and A is the normalization constant, determined by the condition

$$\int dp f_{00}=n_0. \tag{1.31}$$

From formulas (1.29) we find an expression for the current density due to one carrier species:

$$\mathbf{j}_0=e\int dp\mathbf{v}f=\frac{4\pi m^3e}{3}\int v^3dv\mathbf{f}_{10}. \tag{1.32}$$

In the case of constant ν and δ, i.e., when q = 0 and r = 1, the distribution function f_{00} is identical with a Maxwellian distribution with temperature $T_e = T + \frac{2}{3}mV_0^2/\delta$; in this case the velocity of electric drift is constant for all carriers. Thus, in a semiconductor in an external electric field the carriers are heated, and the heating becomes important under conditions when $V_0 > v_S$. Note that $V_0 \sim (\delta T_e/m)^{1/2} \ll v_t$ and the condition (1.25) (which is written down for the case $\mathbf{E}_0 \| \mathbf{B}_0$) is automatically satisfied. The current–voltage characteristic, i.e., the dependence of $\mathbf{j}_0$ on $\mathbf{E}_0$, always has a positive slope in this case (for all orientations of $\mathbf{j}_0$ and $\mathbf{E}_0$).

A qualitatively different picture is obtained when one allows for the dependence of ν and δ on the carrier velocity, i.e., when $q \neq 0$ and $r \neq 1$. Then as we shall show, the carrier current can exhibit a section with negative slope in the current–voltage characteristic. It is obvious that this can only occur when there is carrier heating. We shall therefore assume straight away that the lattice temperature in the exponential in (1.29) can be ignored. To avoid cumbersome formulas, we consider two cases of orientation of the fields $\mathbf{E}_0$ and $\mathbf{B}_0$. If they are parallel, we obtain from Eqs. (1.27) and (1.28)

$$f_{00}=A\exp\{-\alpha_0v^{2(r-q)}\},\quad \alpha_0=\frac{3m^2\delta_0\nu_0^2}{4\,(r-q)\,e^2E_0^2v_0^{2(r-q-1)}}. \tag{1.33}$$

In general, this distribution is far from Maxwellian (only when r − q = 1 is it Maxwellian), and one cannot associate a temperature with the carriers. However, all our arguments concerning the carrier heating at constant ν and δ remain true in this case if T_e is understood as the mean carrier energy and ν and δ as the values of these quantities averaged over the distribution (1.33).

It should be noted that the function (1.33) can be normalized only if

$$r-q>0. \tag{1.34}$$

In this case the carrier distribution function decreases exponentially with increasing carrier velocity and one speaks of a confining heating mechanism. In the opposite case, the number of carriers with high velocities is exponentially large and one speaks of a carrier "runaway" [35]. We restrict ourselves to confining heating mechanisms, and therefore in the case of parallel fields, $\mathbf{E}_0$ and $\mathbf{B}_0$ we shall assume that inequality (1.34) holds. However, the results that are obtained below can be generalized with certain reservations to nonconfining carrier heating mechanisms. The point is that the dependences (2) that we have chosen are, strictly speaking, valid only in a certain narrow range of carrier energies. The requirement (1.34) is due to our extending these dependences to the entire energy range. In the case of the real $\nu(v)$ and $\delta(v)$ dependences, a nonconfining heating mechanism can go over into a confining mechanism when the carrier energy increases and the distribution function can then be normalized. If the confining heating mechanism lies in the region of very high energies, so that the contribution of this region to all physical quantities is negligible, it can be ignored and the results will hold for all r and q, irrespective of the condition (1.34).

Using (1.33) and (1.30), we can readily calculate the equilibrium current in a medium produced by one carrier species:

$$\mathbf{j}_0=\sigma_0\gamma_0(E_0)\,\mathbf{E}_0, \tag{1.35}$$

where $\sigma_0 = e^2n_0/m\nu_0$ is the conductivity in the absence of heating, and

$$\gamma_0(E_0)=\frac{2}{3}\frac{\Gamma\left(\frac{2r+3}{2r-2q}\right)}{\Gamma\left(\frac{3}{2r-2q}\right)}\frac{(r-q)}{v_0^{2q}\alpha_0^{q/r-q}}. \tag{1.36}$$

Since $\alpha_0 \sim E_0^2$, we have $\gamma_0(E_0) \sim E_0^{2q/r-q}$, and therefore $j_0 \sim E_0^{r+q/r-q}$. It follows that for

$$\frac{r+q}{r-q}<0 \tag{1.37}$$

the current in the medium decreases with increasing electric field, $dj_0/dE_0 < 0$, or as one says, the current–voltage characteristic of the current of this carrier species is negative. Taking into account the condition (1.34) (the condition that the distribution function can be normalized, or there is a confining heating mechanism), we write the inequality (1.37) in the form

$$r+q<0, \quad \text{or} \quad q<0 \tag{1.38}$$

In the case of crossed fields, $\mathbf{E}_0 \perp \mathbf{B}_0$, simple formulas can be obtained in the limits $\Omega^2 \gg \nu^2$ or $\nu^2 \gg \Omega^2$. When $\nu > \Omega$, the magnetic field can be ignored in formulas (1.29) and we obtain exactly the same picture as in the case when there is no magnetic field. If $\Omega > \nu$, we have [cf. (1.33)]

$$f_{00}=A\exp\{-\alpha_1 v^{2(r+q)}\}, \qquad \alpha_1=\frac{3m^2\delta_0\Omega^2 v_0^{-2(r+q-1)}}{4(r+q)e^2E_0^2}. \tag{1.39}$$

These formulas have the same kind of structure as formulas (1.33) and differ from them by the substitution $q \to -q$. The condition of normalizability of the function (1.39) can therefore be written in the form [cf (1.34)]

$$r+q>0. \tag{1.40}$$

From formulas (1.30) and (1.39) we find the current in the medium created by one carrier species in crossed fields:

$$\mathbf{j}_0=\sigma_0\left\{g_0(E_0)\frac{\nu_0^2}{\Omega^2}\mathbf{E}_0+\frac{\nu_0}{\Omega}\frac{[\mathbf{E}_0\times\mathbf{B}_0]}{B_0}\right\}, \tag{1.41}$$

where

$$g_0(E_0)=\frac{2}{3}\frac{\Gamma\left(\frac{2r+3}{2r+2q}\right)}{\Gamma\left(\frac{3}{2r+2q}\right)}\frac{(r+q)}{\alpha_1^{-\frac{q}{r+q}}v_0^{-2q}}. \qquad (1.42)$$

This expression also differs from (1.36) by the substitution $q \to -q$. As a result, the condition for a section of negative current–voltage characteristic to arise in a semiconductor plasma in crossed fields when the carriers are heated can be written in the form [cf. (1.38)]*

$$r-q<0 \quad \text{or} \quad q>0. \qquad (1.43)$$

In the case we are considering, for which the dependence of the momentum and energy relaxation times is given by formulas (2),

$$\sigma\sim\frac{1}{\nu}\sim\varepsilon^{q}, \qquad P\sim\varepsilon\delta\nu\sim\varepsilon^{r}. \qquad (1.44)$$

As a result, we obtain [24]

$$\sigma_{\text{diff}}=\frac{r+q}{r-q}\sigma. \qquad (1.45)$$

In the absence of an external magnetic field or for parallel electric and magnetic fields, the condition for a negative current–voltage characteristic to appear has the form $r+q<0$ (i.e., $q<0$), and the normalization condition for the carrier distribution function is $r-q>0$. This means that $d\sigma/d\varepsilon \sim q<0$, and $dP/d\varepsilon \sim r$ and the expression (1.13) may, in general, have any sign. However, the treatment of only a completely confining heating mechanism (normalizability of the distribution function) leads to the requirement

$$\frac{\partial P}{\partial\varepsilon}-\frac{\partial\sigma}{\partial\varepsilon}E_0^2>0. \qquad (1.46)$$

Therefore, as can be seen from the relations (1.13) and (1.45), in a semiconductor plasma with low carrier density in an external electric field the current–voltage characteristic must have an N-shaped form. Similar arguments show that the current–voltage characteristic must be N-shaped in crossed fields as well.

Thus, in a semiconductor plasma with low carrier density in an external electric field and when there is a confining carrier heating mechanism, one can only have an N-shaped current–voltage characteristic, irrespective of the orientation of the external magnetic field. In a plasma with a high carrier density, the distribution function is Maxwellian and can therefore be normalized irrespective of the sign of the difference $r-q$. As a result, the sign of q can be arbitrary when there is a negative current–voltage characteristic, which may be either N- or S-shaped when there is carrier heating.

Oscillations of a Solid-State Plasma in the Absence of an External Magnetic Field. Having found the carrier distribution function in an equilibrium state, we can turn to the study of the stability of this state. As we have already pointed out, the investigation of the stability of a spatially homogeneous medium reduces to an analysis of the spectra of small oscillations and this, in its turn, is completely determined by the permittivity tensor. To calculate the permittivity tensor, we must, as is usually done (see, for example, [49]), find the small deviation δf of the distribution function from the equilibrium function under the

* We mean here the current–voltage characteristic of the component of the current parallel to the electric field [the second term in (1.41) is the Hall current].

influence of the fields $\mathbf{E}_1$ and $\mathbf{B}_1$ of the perturbation by solving the linearized system of equations (1.27). However, if we are interested in low-frequency oscillations under conditions of weak spatial dispersion, when

$$\omega \ll \delta\nu, \quad k^2 v_t^2 \ll \omega\nu, \tag{1.47}$$

there is no need to solve the system (1.27). The point is that the solutions (1.29) and (1.30) that we have found are suitable not only for stationary and homogeneous fields, but also for nonstationary and inhomogeneous fields provided the characteristic frequencies and wavelengths satisfy the conditions (1.47). Therefore, to calculate the permittivity tensor, we can use these solutions, omitting in them the subscript 0 of the density n and the fields $\mathbf{E}$ and $\mathbf{B}$ and representing them in the form

$$n = n_0 + n_1, \quad \mathbf{E} = \mathbf{E}_0 + \mathbf{E}_1, \quad \mathbf{B} = \mathbf{B}_0 + \mathbf{B}_1, \tag{1.48}$$

where n_1, $\mathbf{E}_1$, and $\mathbf{B}_1$ are, respectively, the small perturbations of the equilibrium density and the equilibrium fields in the medium. Then, linearizing these expressions, we can find the perturbation of the equilibrium function:

$$f_0 = f_{00} + \delta f_0, \quad \mathbf{f}_1 = \mathbf{f}_{10} + \delta\mathbf{f}_1, \tag{1.49}$$

where δf_0 is the isotropic and $\delta\mathbf{f}_1$ the anisotropic part of the perturbed distribution function. Then the perturbed current in the plasma produced by one carrier species is given by

$$\delta j_i = \sigma_{ij}(\omega, \mathbf{k}) E_{1j} = e \int d\mathbf{p} v_i \left(\frac{\mathbf{v}}{v}\,\delta\mathbf{f}_1\right). \tag{1.50}$$

However, this is not the final expression, since it does not express δj solely in terms of $\mathbf{E}_1$; $\delta\mathbf{f}_1$ contains n_1 and $\mathbf{B}_1$. To eliminate these quantities, we must use the continuity equation

$$en_1\omega = \mathbf{k}\delta\mathbf{j} \tag{1.51}$$

and the field equation

$$\mathbf{k} \times \mathbf{E}_1 = \frac{\omega}{c}\,\mathbf{B}_1. \tag{1.52}$$

Substituting n_1 and $\mathbf{B}_1$ into (1.50), we obtain the relation

$$\delta j_i = \sigma_{ij}(\omega, \mathbf{k}) E_j. \tag{1.53}$$

Knowing the conductivity tensor $\sigma_{ij}(\omega, \mathbf{k})$, we find the permittivity tensor of carriers of one species in a semiconductor plasma:

$$\varepsilon_{ij}(\omega, \mathbf{k}) = \varepsilon_{ij}^{(0)} + \frac{4\pi i}{\omega}\,\sigma_{ij}(\omega, \mathbf{k}), \tag{1.54}$$

where $\varepsilon_{ij}^{(0)}$ is the lattice permittivity. In what follows, we shall for simplicity consider isotropic semiconductors or semiconductors with a crystal lattice of cubic symmetry, for which

$$\varepsilon_{ij}^{(0)} = \varepsilon_0 \delta_{ij}. \tag{1.55}$$

The generalization of our results to crystals of noncubic symmetry is not difficult.

If the semiconductor consists of carriers of different species, its conductivity is the sum of the conductivities of the carriers of each species. The expression (1.54) in this case must be written in the form

$$\varepsilon_{ij}(\omega, \mathbf{k}) = \varepsilon_{ij}^{(0)} + \frac{4\pi i}{\omega} \sum \sigma_{ij}(\omega, \mathbf{k}), \tag{1.56}$$

where the summation is over all carrier species.

This method of calculating the conductivity tensor and the permittivity tensor of a semiconductor plasma is very simple in conception but tedious in execution. Therefore, we shall not reproduce it here, but give only the final results in the sections devoted to the particular low-frequency spectra of the oscillations of a semiconductor plasma in an external electric field.

Following the method explained above, we obtain the following expression for the permittivity tensor of a solid-state plasma with one carrier species in an external electric field but in the absence of a magnetic field [29]:

$$\varepsilon_{ij}(\omega, \mathbf{k}) = \varepsilon_0\delta_{ij} + \frac{4\pi i}{\omega}\sigma_{ij}(\omega, \mathbf{k}) = \varepsilon_0\delta_{ij} + \frac{4\pi i}{\omega}\sigma_0\gamma_0(E_0)\left\{\left(1 - \beta\frac{\mathbf{ku}}{\omega}\right)\delta_{ij} + \right.$$
$$\left. + \beta\frac{k_i u_j}{\omega} + \frac{1}{\omega - \mathbf{ku}}\left[\frac{2q}{r-q}\omega\frac{E_{0i}E_{0j}}{E_0^2} + \frac{u_i k_j}{\omega}(\omega - \mathbf{ku}) + \beta\frac{k^2 u_i u_j}{\omega}\right]\right\}. \quad (1.57)$$

Here

$$\beta = \frac{\Gamma\left(1 + \frac{3}{2r-2q}\right)\Gamma\left(\frac{2r+2q+3}{2r-2q}\right)}{\Gamma^2\left(\frac{2r+3}{2r-2q}\right)}, \qquad \mathbf{u} = \frac{e\mathbf{E}_0}{m\nu_0}\gamma_0(E_0). \quad (1.58)$$

The expression (1.57) has the same kind of structure as the expression obtained in [24] in the effective-temperature approximation and differs from it only by the explicit form of the quantities u, β, and γ_0. For $q = 0$ and $r = 1$ (i.e., when ν and δ are constant and $\beta = \gamma_0 = 1$), they go over into the expression for the permittivity tensor obtained in [13] by solving the system of transport equations (1.27). Note that in this last case all the carriers acquire the same electric drift velocity in the electric field and the permittivity tensor (1.57) can therefore be obtained by using the transformation formulas proposed in [50] and noting that in the intrinsic coordinate system the plasma conductivity is $\sigma_{ij} = \sigma_0\delta_{ij}$.

Using the dispersion relation (1.23) and also the permittivity tensor (1.57) we obtain the following expression for the spectrum of longitudinal oscillations of a solid-state plasma in the absence of a magnetic field:

$$\omega = \mathbf{ku} - i\frac{\omega_L^2}{\varepsilon_0\nu_0}\gamma_0(E_0)\left(1 + \frac{2q}{r-q}\cos^2\varphi\right). \quad (1.59)$$

Here $\omega_L = (4\pi e^2 n_0/m)^{1/2}$ is the plasma frequency of the carriers and φ is the angle between the direction of propagation of the wave and the external electric field. The condition for an instability to arise or, that is, longitudinal electromagnetic oscillations to be excited, in the solid-state plasma can then be written in the form

$$1 + \frac{2q}{r-q}\cos^2\varphi < 0 \quad \text{or} \quad r + q - 2q\sin^2\varphi < 0. \quad (1.60)$$

In writing down the last inequality we have taken into account the normalizability condition for the carrier equilibrium distribution function, which in this case has the form $r - q > 0$. Comparing the conditions (1.60) and (1.38), we see that the excitation of oscillations in a semiconductor plasma with one carrier species in an external electric field is possible only when there is a negative current–voltage characteristic. And the most favorable conditions for excitation correspond to waves propagating along the external field, i.e., for $\varphi = 0$; waves propagating at right angles to the electric field are always damped. The maximal growth rate is

$$\gamma = \operatorname{Im}\omega = \frac{\omega_L^2}{\varepsilon_0\nu_0}\gamma_0(E_0)\left|\frac{r+q}{r-q}\right|, \quad (1.61)$$

and the maximal frequency of the excited oscillations is determined by the condition of applicability of the treatment and may reach $\omega_{max} \lessgtr \delta\nu$.

This instability associated with a negative current–voltage characteristic is possible not only in semiconductors with one carrier species, but also in an intrinsic semiconductor with two carrier species. At the same time, it is very important that the current–voltage characteristic of the total current may be positive; for instability to develop it is sufficient for there to be negative differential conductivity in the current of one of the carrier species. This can be readily seen by writing down the dispersion relation of longitudinal oscillations for an intrinsic semiconductor:

$$\varepsilon_0 + i\sum \frac{\omega_L^2 \gamma_0(E_0)}{\nu_0(\omega - \mathbf{ku})}\left(1 + \frac{2q}{r-q}\cos^2\varphi\right) = 0. \tag{1.62}$$

The summation is over both carrier species, which may have not only different masses but also different energies because of different heating and therefore different scattering mechanisms, i.e., different r and q. If the current–voltage characteristic in the current of light carriers (and therefore of the total current) is negative, then, considering the frequency region $\omega \approx \mathbf{ku}$, we can ignore the contribution of heavy carriers in Eq. (1.62). And the entire picture of the excitation of oscillations in an intrinsic semiconductor is the same as in a plasma with one carrier species, whose role is played by the light carriers. If the current–voltage characteristic is negative for the current of the heavy carriers, then, considering the region of frequencies $\omega \approx \mathbf{ku}$, we find the following spectrum from Eq. (1.62):

$$\omega = \mathbf{ku}_+ - i\frac{\frac{\omega_{L+}^2}{\nu_{0+}}\gamma_{0+}(E_0)\left(1 + \frac{2q_+}{r_+ - q_+}\cos^2\varphi\right)}{\varepsilon_0 - i\frac{\omega_{L-}^2}{\mathbf{ku}_{rel}\nu_{0-}}\gamma_{0-}(E_0)\left(1 + \frac{2q_-}{r_- - q_-}\cos^2\varphi\right)}, \tag{1.63}$$

where $\mathbf{U}_{rel} = \mathbf{U}_- + \mathbf{U}_+$. The subscripts + and − refer, respectively, to the positive (heavy) and negative (light) carriers. It can be seen from the expression (1.63) that oscillations are excited when

$$1 + \frac{2q_+}{r_+ - q_+}\cos^2\varphi < 0, \quad \text{or} \quad r_+ + q_+ - 2q_+\sin^2\varphi < 0, \tag{1.64}$$

which can be satisfied only when there is a negative current–voltage characteristic in the current of the positive carriers when $r_+ + q_+ < 0$. The growth rate of the oscillations is then

$$\gamma = \mathrm{Im}\,\omega = \frac{\omega_{L+}^2}{\nu_{0+}}\varepsilon_0\gamma_{0+}(E_0)\frac{1 + \frac{2q_+}{r_+ - q_+}\cos^2\varphi}{\varepsilon_0^2 + \frac{\omega_{L-}^4\gamma_{0-}^2(E_0)}{(\mathbf{ku}_{rel})^2\nu_{0-}^2}\left(1 + \frac{2q_-}{r_- - q_-}\cos^2\varphi\right)}. \tag{1.65}$$

This formula, like (1.63), is applicable only if $\gamma < \omega \approx \mathbf{ku}_+ < \delta\nu$.

In what follows, in analyzing the oscillations excited in a plasma with one carrier species, we shall not generalize to the case of an intrinsic semiconductor. Such a generalization is along the lines indicated above and is not difficult.

Plasma Oscillations in External Electric and Magnetic Fields. If there is, in addition, a strong magnetic field ($\Omega^2 \gg \nu^2$) parallel to the electric field, the oscillation spectrum of a plasma with low carrier density is given by the expression [27]

$$\omega = \mathbf{ku} - i\frac{\omega_L^2}{\varepsilon_0\nu_0}\left[\gamma_0\frac{r+q}{r-q}\cos^2\varphi + \frac{\nu_0^2}{\Omega^2}\beta_1\gamma_0^{-1}(E_0)\sin^2\varphi\right]. \tag{1.66}$$

Comparing this expression with (1.59), we see that in the case of parallel fields in a strongly magnetized solid-state plasma, as in the case when no magnetic field is present, longitudinal oscillations can be unstable only if there is a negative current–voltage characteristic, when $r + q < 0$ (i.e., $q < 0$). The most favorable excitation conditions correspond to oscillations propagating along the magnetic field. In general, oscillations propagating at right angles to the magnetic field are not excited. In this sense, we see that the magnetic field has a stabilizing influence on the plasma instability associated with a negative current–voltage characteristic. However, this influence is too weak and is manifested only in a very narrow range of angles near $\varphi \approx \pi/2$, i.e., at right angles to the magnetic field, when $\cos^2\varphi \lessapprox \nu_0^2/\Omega^2 \ll 1$.

A magnetic field has a greater influence on a semiconductor plasma when $\mathbf{E}_0 \perp \mathbf{B}_0$. In this case, the spectrum of longitudinal oscillations is given by [27]

$$\omega = \mathrm{k}u - i\frac{\omega_L^2}{\varepsilon_0\nu_0}\left[g_0\frac{\nu_0^2}{\Omega^2}\left(1 - \frac{2q}{r+q}\cos^2\varphi\right) + \beta_2 g_0^{-1}\sin^2\varphi\right]. \tag{1.67}$$

From this it is clear that the excitation of oscillations in this case too can only occur when there is a negative current–voltage characteristic in the equilibrium current, when $r - q < 0$ (i.e., $q > 0$) (we recall that the condition of normalizability of the equilibrium carrier distribution in crossed fields is $r + q > 0$). As in the cases considered above, oscillations that propagate along the external electric field are excited most effectively, although the opening angle of the region of unstable oscillations is very small: $\varphi \lessapprox \nu_0^2/\Omega^2 \ll 1$.

Oscillations propagating at a large angle to the electric field are damped. In contrast to the foregoing cases, the growth rate of the oscillations is small in the case of crossed fields. Indeed, its maximal value,

$$\gamma_{\max} = \mathrm{Im}\,\omega = \frac{\omega_L^2}{\varepsilon_0\nu_0}\frac{\nu_0^2}{\Omega^2}g_0\left|\frac{r-q}{r+q}\right|, \tag{1.68}$$

is smaller by a factor Ω^2/ν^2 than that for the development of longitudinal oscillations in the case of parallel fields [or without magnetic fields, see (1.61)].

Nonelectrostatic Electromagnetic Waves in a Solid-State Plasma. It should be noted that in the case of a positive current–voltage characteristic (for example, for constant ν and δ) longitudinal oscillations of a magnetized plasma are always damped, both for a semiconductor with one carrier species and for an intrinsic semiconductor with two carrier species. This conclusion makes it particularly important to analyze nonelectrostatic waves, the more so since numerous investigations [10, 11, 13, 14, 51] indicate instability of a solid-state plasma in an external electric field even when ν and δ are constant.

On the other hand, it is well known that for an anisotropic medium (for example, a plasma in external electric and magnetic fields) longitudinal waves are not, in general, characteristic oscillations. It is only under certain conditions that characteristic oscillations of an anisotropic medium can be regarded as electrostatic with a high degree of accuracy. To establish these conditions and therefore determine the region of applicability of the above results, we must investigate arbitrary nonelectrostatic oscillations of a solid-state plasma in an external electric field.

In the absence of a magnetic field, we obtain, substituting the expression (1.57) for the permittivity tensor into (1.19), a fairly lengthy equation for determining the spectrum of arbitrary nonelectrostatic oscillations of a solid-state plasma in an external electric field.

In the absence of a magnetic field, we obtain, substituting the expression (1.57) for the permittivity tensor into (1.19), a fairly lengthy equation for determining the spectrum of arbitrary nonelectrostatic oscillations of a solid-state plasma with one carrier species. This equation can be readily analyzed for purely longitudinal ($\varphi = 0$) and purely transverse ($\varphi = \pi/2$) propagation of waves.

For waves that propagate along the electric field ($\varphi = 0$), the dispersion relation splits up into two equations, one of which describes transverse waves that are damped with time and the other corresponds to longitudinal oscillations with the spectrum (1.59). In this case, the longitudinal waves are strictly characteristic oscillations of the system.

In the case of transverse propagation of the waves ($\varphi = \pi/2$), the nature of the excitation of the oscillations is reversed. In this case the longitudinal waves are damped and the transverse may grow provided the current–voltage characteristic of the carriers is negative. Excitation of oscillations is possible only when $c^2k^2 \gg \omega^2 \varepsilon_0$, and their spectrum is given by [29]

$$\omega = -i\frac{k^2c^2\nu_0}{\omega_L^2\gamma_0}\frac{r-q}{r+q}. \tag{1.69}$$

In this case the instability is aperiodic. Comparing the growth rates for longitudinal and transverse waves, determined, respectively, by the expressions (1.61) and (1.69), we can readily see that when $k^2c^2 \gg \omega^2 \varepsilon_0$ the former is appreciably greater than the latter. This means that under real conditions only longitudinal oscillations are ever excited in a solid-state plasma. As we have already mentioned, the preferential excitation of longitudinal waves is a consequence of the current–voltage characteristics being N-shaped in a plasma with low carrier density.

It should be noted that in an intrinsic semiconductor with two carrier species in an external electric field transverse waves can also be excited when there is a positive current–voltage characteristic. Indeed, generalizing the expression (1.57) to the case of an intrinsic semiconductor (i.e., summing over the carrier species) and substituting it into Eq. (1.19), we obtain the following expression for waves propagating at right angles to the electric field ($\varphi = \pi/2$) in the limit $k^2c^2 > \omega^2\varepsilon_0$ [13, 14]:

$$\omega = i\frac{v_{rel}^2}{c^2}\frac{\omega_{L+}^2\omega_{L-}^2}{\omega_{L+}^2\nu_- + \omega_{L-}^2\nu_+}, \tag{1.70}$$

where v_{rel} is the velocity of relative drift of the carriers, i.e., $v_{rel} = v_{d-} - v_{d+}$. For simplicity we have here restricted ourselves to the case of constant ν and δ. This instability is of the two-stream type and, as is shown in [14], can only develop if $k^2c^2 < \omega_L^2 v_{rel}^2/v_t^2$. Physically, this condition means that the magnetic pressure of the field of the equilibrium current at distances $\sim 1/k$ exceeds the kinetic pressure of the free carriers of the solid. Such a situation must lead to a compression of the carrier plasma, i.e., to a pinch effect. The time $\tau \sim 1/\mathrm{Im}\,\omega$ determined by (1.70) is equal to the hydrodynamic time of compression of an electron-hole plasma in a viscous medium formed by a rigid crystal lattice.* Thus, we conclude that a semiconductor plasma with two carrier species is unstable against compression if the transverse dimensions are sufficiently large ($L_\perp \sim 1/k > (v_t/v_{rel})(c/\omega_L)$). The growth rate of such an instability is however smaller by a factor v_t^2/v_{rel}^2 than the growth rate of the electrostatic instability due to the negative current–voltage characteristic [see (1.61)]. Therefore, in a medium with negative differential conductivity one cannot expect a hydrodynamic compression of the plasma, since the electrostatic instability changes the state of the medium before the transverse instability can appear.

In a magnetized solid-state plasma with one carrier species in external parallel fields, the oscillations are of the same nature as when there is no magnetic field. In particular, in a mag-

* A consistent theory of the stability of the pinch effect in a solid-state plasma has been developed in [52, 53].

netized plasma with a low carrier density it is still true that there is a preferential excitation of electrostatic waves when there is a negative current–voltage characteristic in the carrier current, which is a consequence of this characteristic's being N-shaped. This conclusion applies equally to solid-state plasmas in crossed electric and magnetic fields.

We note finally that, in the approximation corresponding to our conclusion that an electrostatic instability can arise in a magnetized solid-state plasma, oscillations are always damped in the case of a positive current–voltage characteristic, irrespective of the number of carrier species. Assertions to the opposite effect in [11, 13, 14, 51] that helical, Alfvén, and drift waves can be excited are based on an analysis of the oscillations of a solid-state plasma in a higher approximation in ν/Ω. Calculating the permittivity tensor to terms of order ω/ν and ν/Ω for transverse waves propagating along the external fields in a plasma with two carrier species, we obtain the dispersion relation [51]

$$k^2c^2 = \omega^2\left[\varepsilon_0 \pm \sum_\alpha \frac{\omega_{L\alpha}^2(\omega - \mathbf{ku}_\alpha)}{\omega^2\Omega_\alpha}\left(1 \pm \frac{\omega - \mathbf{ku}_\alpha + i\nu_\alpha}{\Omega_\alpha}\right)\right]. \qquad (1.71)$$

Here it is assumed that ν and δ are constant. In an intrinsic semiconductor with equal number of carriers of opposite sign, when $\sum_\alpha \omega_{L\alpha}^2/\Omega_\alpha = 0$, we then obtain

$$\omega^2 = \frac{k^2c^2 \pm \sum_\alpha \omega_{L\alpha}^2 \mathbf{ku}_\alpha/\Omega_\alpha}{\varepsilon_0 + \sum_\alpha \omega_{L\alpha}^2/\Omega_\alpha}. \qquad (1.72)$$

It can be seen that if

$$k^2c^2 < \left|\sum_\alpha \frac{\omega_{L\alpha}^2 \mathbf{ku}_\alpha}{\Omega_\alpha}\right|$$

there arises an aperiodic instability ($\omega^2 < 0$) corresponding to the excitation of waves of Alfvén type [for $\mathbf{u}_\alpha = 0$, i.e., in the absence of an electric field, the spectrum (1.72) corresponds to Alfvén waves in a solid-state plasma]. But if the densities of the carriers of opposite sign are not equal and $\sum_\alpha \omega_{L\alpha}^2/\Omega_\alpha \neq 0$, we obtain from the dispersion relation (1.72) the spectrum of helical waves in the presence of electric drift of the carriers:

$$\begin{aligned} \operatorname{Re}\omega &= \left(\sum_\alpha \frac{\omega_{L\alpha}^2}{\Omega_\alpha}\right)^{-1}\left[k^2c^2 + \sum_\alpha \frac{\omega_{L\alpha}^2 \mathbf{ku}_\alpha}{\Omega_\alpha}\right], \\ \gamma = \operatorname{Im}\omega &= -\sum_\alpha \frac{\nu_\alpha}{\Omega_\alpha}(\operatorname{Re}\omega - \mathbf{ku}_\alpha). \end{aligned} \qquad (1.73)$$

When $\mathbf{ku}_\alpha > \operatorname{Re}\omega$, i.e., when the velocity of electric drift of the carriers of greater density exceeds the phase velocity of the wave, $\gamma > 0$, which corresponds to the excitation of helical waves (helicons) in the solid-state plasma.

It is easy to show that in a plasma with one carrier species and constant ν and δ neither Alfvén nor helical waves are excited. This is perfectly natural, since in this case all the carriers of one species have a single drift velocity, and by a simple transformation of the coordinate system (transition to the drifting system) it follows that the decay rate of the oscillations must have the same form as in the absence of an electric field; in this system, the oscillations are always damped.

It follows from the analysis in this section of the oscillation spectra of a solid-state plasma with low carrier density that longitudinal electrostatic waves will always be excited under real conditions when there is a negative current–voltage characteristic. Even in the cases when transverse waves can be excited because of the negative current–voltage characteristic, their growth rates are always less than the growth rate of the longitudinal oscillations. This means that the amplitudes of the electrostatic oscillations reach appreciable values and the plasma goes over into a new equilibrium state before the amplitudes of the transverse perturbations grow significantly. Such a state must be characterized by an inhomogeneous distribution of the carriers or, as one frequently says, a domain structure (waves of charge or an electrostatic field propagating along an external field are frequently called domains [46]). We have attributed the preferential excitation of longitudinal waves in a solid-state plasma with low carrier density to the fact that the negative current–voltage characteristic is N-shaped. For in the case of an N-shaped characteristic the electric field in the plasma is a many-valued function of the current – to one value of the current there correspond several (at least two) values of the field. Therefore, a perturbation produced by a fluctuation can carry the plasma from the state with smaller field strength into a state with a higher field strength without changing the current in the plasma, i.e., the perturbation leads to oscillations of the field, but not of the current; but these are electrostatic oscillations.

With regard to the transverse (nonelectrostatic) waves, they can be excited in practice only in a plasma with two carrier species and when there is a positive current–voltage characteristic (since they are otherwise suppressed by the stronger electrostatic instability of the plasma. In this case one can have excitation of helical waves (helicons) and Alfvén waves under conditions when the velocity of electric drift is greater than the phase velocity of these waves. In addition, if the transverse dimensions of the sample are sufficiently large, a solid-state plasma with a current in the absence of an external magnetic field can sustain an instability similar to the well-known pinch effect in a plasma, this leading to compression of the carrier plasma.

§ 1.4. Effect of Diffusion on the Instability of a Semiconductor Plasma with Negative Current – Voltage Characteristic

In the foregoing sections, we have investigated the low-frequency oscillations of a solid-state plasma ($\omega < \delta\nu$) in an external electric field. The analysis showed that in this region of frequencies the most favorable conditions for excitation by electric drift of the carriers correspond to longitudinal waves when the medium has a negative current–voltage characteristic; the growth rate for these waves is maximal. It is therefore of interest to establish the importance of the restrictions on which the treatment of this instability was based. In particular, in the foregoing section we completely ignored spatial dispersion in the derivation of the permittivity tensor, assuming

$$\omega\nu \gg k^2 v_t^2. \tag{1.74}$$

This means that we ignored dissipative processes like heat conduction and carrier diffusion. On the other hand, it is obvious physically that electrostatic oscillations in a plasma with one carrier species are possible, even if they are damped in time, only if the oscillation frequency is greater than the reciprocal time of diffusion of carriers over the wavelength of these oscillations. Otherwise, the carriers have sufficient time during one period of the oscillations to diffuse and rearrange themselves in space in such a way as to screen the longitudinal field; there will be ordinary Debye screening of the field in the medium.

If carrier diffusion is allowed for, one cannot neglect the terms grad f_0 and div f_1 in the system (1.27), as we did in §1.3. This greatly complicates the solution of the system of Davydov's equations. However, if one assumes that

$$k^2 v_t^2 \ll \delta\nu^2, \qquad \omega \ll \delta\nu, \tag{1.75}$$

then to solve this system one can develop a method of successive approximation along the lines of the Chapman–Enskog expansion [36], regarding ω/ν and λ/L, where λ is the wavelength and L is the characteristic scale of an inhomogeneity, as small parameters. A method of this kind was developed in [28] in order to allow for carrier diffusion. We represent the carrier distribution in the form

$$f = f_{00} + f_{01} + \frac{\mathbf{v}}{v}(\mathbf{f}_{10} + \mathbf{f}_{11}), \tag{1.76}$$

where f_{00} and $\mathbf{f}_{10}$ are the symmetric and antisymmetric parts of the distribution function in a homogeneous plasma, i.e., without allowance for the terms grad f_0 and div $\mathbf{f}_1$ (expressions for which have been given above) and f_{01} and $\mathbf{f}_{11}$ are the corrections due to these terms. Substituting the expression (1.76) into the transport equation (1.24), we obtain a system similar to (1.27), which we shall solve by successive approximation, regarding the terms with spatial and time derivatives as small, which is true if (1.75) is satisfied. In the zeroth approximation, we ignore both the time and the spatial derivatives, and the solution of the system of transport equations is given by formulas (1.29). In the next approximation, we take into account the small terms with spatial and time derivatives. As a result, we obtain a system of equations whose solution can be found without great difficulty and has the form [45]

$$f_{01} = e^{-\Psi(v)}\left\{\int_0^v dv_1\left[\frac{3m}{ev_1^2(\mathbf{EV}_0)}\int_\infty^{v_1} v_2^2 dv_2\left(\frac{\partial f_{00}}{\partial t} + \frac{v_2}{3}\operatorname{div}\mathbf{f}_{10}\right) - \frac{(\mathbf{E},\ \mathbf{F}_1 + \mathbf{F}_2)}{(\mathbf{EV}_0)}\right]e^{\Psi(v_1)} + A_1\right\},$$

$$\mathbf{f}_{11} = -\mathbf{V}_0\frac{\partial f_{01}}{\partial v} - \mathbf{F}_1 - \mathbf{F}_2. \tag{1.77}$$

Here, f_{00}, $\mathbf{f}_{10}$, and $\mathbf{V}_0$ are determined by the expressions (1.29) and (1.30) and

$$\Psi(v) = \int_0^v \frac{3m\delta\nu}{2e(\mathbf{EV}_0)}v_1 dv_1 = \int_0^v mv_1 dv_1\left[T + \frac{2e^2E^2(1 + \Omega^2\cos^2\vartheta/\nu^2)}{3m\delta(\nu^2 + \Omega^2)}\right]^{-1},$$

$$\mathbf{F}_1 = \frac{\nu}{\nu^2 + \Omega^2}\left\{\frac{\partial \mathbf{f}_{10}}{\partial t} + \frac{\Omega}{\nu}\frac{1}{B}\left[\frac{\partial \mathbf{f}_{10}}{\partial t}\times\mathbf{B}\right] + \frac{\Omega^2}{\nu^2}\frac{1}{B^2}\mathbf{B}\left(\frac{\partial \mathbf{f}_{10}}{\partial t}\mathbf{B}\right)\right\}, \tag{1.78}$$

$$\mathbf{F}_2 = \frac{v\nu}{\nu^2 + \Omega^2}\left\{\nabla f_{00} + \frac{\Omega}{\nu}\frac{1}{B}[\nabla f_{00}\times\mathbf{B}] + \frac{\Omega^2}{\nu^2}\frac{1}{B^2}\mathbf{B}(\mathbf{B}\nabla f_{00})\right\}.$$

In deriving these solutions, we have used the requirement that the distribution function be bounded as $v \to \infty$ (as we have already mentioned several times, we consider only confining heating mechanisms, when there is no runaway of the carriers inthe electric field). This enabled us to dispense with one of the two constants of integration that arise in the solution of a second-order differential equation. To this one must also add the normalization constant A contained in the function f_{00}. To determine these two constants, A and A_1, we have only one normalization condition. We can impose a second additional condition by requiring that the function f_{01} should not change the carrier density [36]. We then obtain the two relations

$$\int d\mathbf{p} f = \int d\mathbf{p} f_{00} = n, \quad \int d\mathbf{p} f_{01} = 0 \tag{1.79}$$

for finding the two constants A and A_1.

Thus, the problem of determining the carrier distribution function is solved. Because it is cumbersome, we shall not write out here the explicit form of this function. We mention only that the function is suitable for describing not only an equilibrium but also a nonequilibrium state provided the inequalities (1.74) are satisfied. Expressing the fields and carrier density as small deviations from the equilibrium values and linearing the solution (1.76) and the carrier current

$$\mathbf{j} = e\int d\mathbf{p}\mathbf{v}f = \frac{4\pi m^3 e}{3}\int_0^\infty v^3 dv\,(\mathbf{f}_{10} + \mathbf{f}_{11}) \tag{1.80}$$

in the small deviations from the equilibrium state, we can find the permittivity tensor and investigate the oscillation spectrum and the stability of the solid-state plasma in an external electric field.

Let us first consider the case of a plasma with constant ν and δ. Following the above method and ignoring terms of order ω/ν and $\mathbf{k}\mathbf{V}_0/\nu$, we can readily obtain an expression for the carrier current:

$$\mathbf{j} = en\mathbf{V}_0 - \frac{e\nu}{\nu^2 + \Omega^2}\left\{\frac{\partial}{\partial \mathbf{r}}(nv_t^2) + \frac{\Omega}{\nu}\frac{1}{B}\left[\frac{\partial}{\partial \mathbf{r}}(nv_t^2)\times\mathbf{B}\right] + \frac{\Omega^2}{\nu^2}\frac{1}{B^2}\mathbf{B}\left(\mathbf{B}\frac{\partial}{\partial \mathbf{r}}(nv_t^2)\right)\right\}, \tag{1.81}$$

where $\mathbf{V}_0$ is determined by the expression (1.30) and $v_t^2 = (T/m) + \frac{2}{3}V_0^2/\delta$. The absence in the expression (1.81) of terms with time derivatives is due to the fact that in this case the distribution function is Maxwellian (this can be seen most clearly in the hydrodynamic model).

Varying (1.81) with respect to the small perturbations, we find an expression for the permittivity tensor in the absence of an external field:

$$\varepsilon_{ij}(\omega,\,\mathbf{k}) = \varepsilon_0\delta_{ij} + i\frac{\omega_L^2}{\omega^2\nu}\left\{(\omega - \mathbf{k}\mathbf{u})\,\delta_{ij} + k_i k_j + \frac{u_i - ik_i v_t^2/\nu}{\omega - \mathbf{k}\mathbf{u} + ik^2 v_t^2/\nu}\left[(\omega - \mathbf{k}\mathbf{u})\,k_j + k^2 u_j\right]\right\}. \tag{1.82}$$

Here $\mathbf{u} = e\mathbf{E}_0/m\nu$ is the velocity of electric drift of the carriers in the absence of the magnetic field. In deriving this expression we have ignored terms proportional to $\mathbf{k}\mathbf{u}/\delta\nu$, since in a homogeneous plasma, in contrast to an inhomogeneous one (see Ch. II) they lead to small real and therefore unimportant, corrections of order $\omega_L^2/\varepsilon_0\,\delta\nu^2$ to the oscillation frequency.

Using the tensor (1.82), we obtain the following dispersion equation for longitudinal (electrostatic) oscillations of the plasma:

$$\varepsilon = \frac{k_i k_j}{k^2}\varepsilon_{ij} = \varepsilon_0 + i\frac{\omega_L^2}{\nu}\frac{1}{\omega - \mathbf{k}\mathbf{u} + ik^2 v_t^2/\nu} = 0. \tag{1.83}$$

If carrier diffusion is ignored, when $\omega \gg k^2 v_t^2/\nu$, the expression (1.82) goes over into (1.57) (in the case of constant ν and δ, i.e., $r = 1$, $q = 0$). The dispersion relation (1.83) has the well-known stable oscillation spectrum

$$\omega = \mathbf{k}\mathbf{u} - i\omega_L^2/\varepsilon_0\nu. \tag{1.84}$$

In the opposite limit, when $|\omega - \mathbf{k}\mathbf{u}| < k^2 v_t^2/\nu$, we obtain from the relation

$$\varepsilon_0 + \frac{\omega_L^2}{k^2 v_t^2} = 0, \tag{1.85}$$

which corresponds to Debye screening of the electrostatic field in the plasma; under these conditions, electrostatic oscillations in the plasma are impossible. And this is, in fact, the stabilizing influence of carrier diffusion on longitudinal plasma oscillations. Writing the general solution of Eq. (1.83) in the form

$$\omega = \mathbf{k}\mathbf{u} - i\left(\frac{k^2 v_t^2}{\nu} + \frac{\omega_L^2}{\varepsilon_0\nu}\right), \tag{1.86}$$

we see that diffusion can be ignored if $\varepsilon_0 k^2 v_t^2 < \omega_L^2$, i.e., for oscillations with wavelength greater than the Debye radius of the carriers. Therefore, the maximal frequency of the oscillations for which carrier diffusion is still unimportant is

$$\omega_{\max} = k_{\max} u \sim \frac{u}{v_t} \frac{\omega_L}{\sqrt{\varepsilon_0}}. \tag{1.87}$$

All the above arguments remain qualitatively true when allowance is made for the dependence of ν and δ on the carrier velocity. However, the expression for the permittivity tensor is then more cumbersome. Since we are here interested in only electrostatic oscillations, we give the final results of the analysis of the dispersion relation for longitudinal oscillations in the case $\mathbf{k} = (0, 0, |\mathbf{k}|)$ (as we have shown above, it is these oscillations that have the maximal growth rate). The spectrum of excited longitudinal oscillations with allowance for carrier diffusion is determined in this case by

$$\omega = \mathbf{k}\mathbf{u} - i\left[\frac{\omega_L^2}{\varepsilon_0 \nu_0} \frac{r+q}{r-q} \gamma_0(E_0) + b \frac{k^2 v_t^2}{\nu_0}\right], \tag{1.88}$$

where

$$b = b_1 - b_2,$$

$$b_1 = \frac{2}{3} \frac{3+2q}{2(r-q)} \frac{1}{\Gamma\left(\frac{3}{2(r-q)}\right)} \left(\frac{\delta_0 v_0^2}{v_d^2} \frac{3}{4(r-q)}\right)^{\frac{r-2q-1}{r-q}} \times$$

$$\times \left\{ I(3+2q) + (r-q)\left[\frac{\Gamma\left(\frac{3+2q}{2r-2q}\right)\Gamma\left(\frac{5}{2r-2q}\right)}{\Gamma\left(\frac{3}{2r-2q}\right)} - \frac{1+2q}{3+2q}\Gamma\left(\frac{5+2q}{2r-2q}\right)\right]\right\},$$

$$b_2 = \frac{3+2q}{2} \frac{\Gamma\left(\frac{3+2q}{2r-2q}\right)}{\Gamma\left(\frac{3}{2r-2q}\right)} \left(\frac{3}{4(r-q)} \frac{\delta_0 v_0^2}{v_d^2}\right)^{-\frac{q}{r-q}} a_1, \tag{1.89}$$

$$a_1 = \frac{3+2q}{3(r-q)} \frac{1}{\Gamma\left(\frac{3}{2r-2q}\right)} \left(\frac{3}{4(r-q)} \frac{\delta_0 v_0^2}{v_d^2}\right)^{\frac{r-q-1}{r-q}} I(3-2r+2q),$$

$$I(\alpha) = \int_0^\infty dx\, e^{-x} x^{\frac{1-2r+2q}{2r-2q}} \left(\frac{\Gamma\left(\frac{3+2q}{2r-2q}\right)}{\Gamma\left(\frac{3}{2r-2q}\right)} x^{\frac{1}{r-q}} - x^{\frac{1+q}{r-q}}\right) \int_0^x dx_1\, x_1^{-\frac{1+2r}{2r-2q}} e^{x_1} \int_\infty^{x_1} dx_2\, x_2^{\frac{\alpha}{2r-2q}} e^{-x_2},$$

and the expressions for u, v_d, v_t and $\gamma_0(E_0)$ are given above. In deriving formula (1.88), we ignored terms of order $kv_d/\nu_0\delta_0$ and $\omega/\delta_0\nu_0$, which give unimportant small contributions to the frequency and growth rate of the oscillations.

It is easy to show that the coefficient in front of the last term in the expression (1.88) is always greater than zero. Therefore, carrier diffusion has a stabilizing effect on the electrostatic instability of a solid-state plasma associated with a negative current–voltage characteristic in the carrier current. Oscillations are excited only if

$$\frac{\omega_L^2}{\varepsilon_0 \nu_0} \left|\frac{r+q}{r-q}\right| \gamma_0 > b \frac{k^2 v_t^2}{\nu_0}, \tag{1.90}$$

or, and this is the same thing, under conditions when the wavelength is greater than the Debye radius of the carriers and there is little diffusion. The maximal frequency of the excited oscillations is then of the order $\omega_{\max} \sim \frac{u}{v_t} \frac{\omega_L}{\sqrt{\varepsilon_0}}$.

All that we have said concerning the stabilizing effect of carrier diffusion on the plasma instability associated with a negative current–voltage characteristic remains true when an external magnetic field is present. Moreover, for $\Omega > \nu$, when terms of order ν^2/Ω^2 can be ignored (or, and this is the same thing, carrier diffusion at right angles to the magnetic field can be ignored), all the above formulas hold for a magnetized plasma as well.

To conclude the present section, we make some estimates of the plasma parameters for typical semiconductors with low carrier density in which the above instabilities can be manifested. Let us elucidate the conditions for the development of the various instabilities and estimate the frequency and wavelength of the oscillations that are excited. We note first that for typical semiconductors the energy relaxation time for carrier scattering on lattice vibrations is of order $\tau_\varepsilon = 1/\delta\nu \sim 10^8$–$10^{-9}$ sec. Therefore, the critical density below which the carrier density can be assumed low in a semiconductor when they are heated by an electric field to a mean energy $\varepsilon \sim 1$ eV is $n_{cr} \sim 10^{14}$–10^{15} cm^{-3}, in accordance with (1). It must be remembered that in certain semiconductors the energy relaxation time is anomalously short $\tau_\varepsilon \sim 10^{-10}$–$10^{-11}$ sec (for example, in gallium arsenide) and therefore n_{cr} may reach values of approximately 10^{17} cm^{-3}. If one bears in mind that the momentum relaxation time for the majority of semiconductors is of order $\tau = 1/\nu \sim 10^{-11}$–$10^{-12}$ sec, it is easy to estimate the electric field in the plasma that can guarantee carrier heating to an energy $\varepsilon \sim 1$ eV; it is $E_0 \sim 0.3$-1 kV/cm, which is readily obtained in real experiments. The velocities of electric drift of the carriers in such fields reach values of about $5 \cdot 10^6$–10^7 cm/sec, i.e., they exceed the velocity of sound by more than an order of magnitude. In such strong electric fields, electromagnetic oscillations must be excited quite efficiently in the solid-state plasma due to the development of instabilities. Our method of investigating the stability of a plasma in an external electric field, which is based on variation of the equilibrium carrier current, enables us to draw conclusions about the spectra of oscillations whose frequencies are less than the reciprocal energy relaxation time, i.e., $\omega < 1/\tau_\varepsilon = \delta\nu \sim 10^8$–$10^9$ sec^{-1}.

Of the various instabilities of a plasma with low carrier density that we have considered, the most interesting one is naturally the electrostatic instability associated with a negative current–voltage characteristic, since this has the maximal growth rate. When this instability develops in a solid-state plasma, a longitudinal wave is excited, and this travels with the velocity of electric drift of the carriers, $\omega = \mathbf{k}\mathbf{u}$, and with a growth rate of order $\gamma_{max} \sim \omega_L^2/\varepsilon_0\nu_0$. This means that the instability is convective, and in semiconductors of finite size it can actually develop only under conditions when $\gamma \sim \omega_L^2/\varepsilon_0\nu_0 \gg u/L_{11}$, where L_{11} is the longitudinal (along the drift) scale of the system. If $n \sim 10^{14}$ cm^{-3}, $\nu \sim 10^{12}$ sec^{-1}, and $\varepsilon_0 \sim 10$ and $u \sim 10^7$ cm/sec, it follows that $L_{11} \gg 10^{-4}$–10^{-3} cm, and the larger the system the more the wave amplitude grows; for $L_{11} \gtrsim 0.1$ cm, the wave amplification may reach 100 dB. As we have shown, carrier diffusion leads to stabilization of the electrostatic instability associated with a negative current–voltage characteristics under conditions when the wavelength is less than the Debye radius. Therefore, only long-wavelength oscillations will be excited: $\lambda_{min} > v_t(\varepsilon_0)^{1/2}/\omega_L$. If $v_t \sim 10^8$ cm/sec, this gives $\lambda_{min} > 10^{-3}$–10^{-4} cm and $\omega_{max} \sim (u\omega_L/v_t)\varepsilon_0^{-1/2} \lesssim 10^{10}$ sec^{-1}. Thus, one can have fairly effective excitation of high-voltage electromagnetic waves of the centimeter range (by this we mean the vacuum wavelength). Experimentally, this must be manifested in the form of isolated traveling charge waves, or domains, whose minimal dimensions are determined by λ_{min} and whose maximal dimensions are determined by the length of the system.*

Electrostatic instabilities do not develop in the plasma when there is a positive current–voltage characteristic. However, in a semiconductor with two carrier species (whose densities are not, in general, equal) transverse (nonelectrostatic) waves can be excited. Of the instabilities we have considered above, the most interesting one is the excitation of helical waves (helicons) in a plasma with unequal carrier densities. In accordance with formulas (1.73), the maximal frequency of the excited oscillations under conditions when the electron density in the

*See the review [46] for more detail on the size and form of the domains.

plasma exceeds the hole density may become of order $\omega_{max} \sim \omega_L^2 v_-^2/\Omega_- c^2$, and their growth rate $\gamma_{max} \sim \nu_- \omega_{max}/\Omega_-$. If $n_- \sim 10^{14}$–10^{15} cm^{-3} and $u \sim 5 \cdot 10^6$ cm/sec in magnetic fields not greater than 1 keV, we have $\omega_{max} \sim 10^7$–10^8 sec^{-1}. The condition $\Omega_- \gg \nu_-$ then requires that $\nu_- < 10^{10}$ sec^{-1}, i.e., experiments to detect drift excitation of helical waves in a plasma with low carrier density must be made at very low temperatures.

A further important practical conclusion follows from our theoretical analysis. In a plasma with low carrier density the frequencies of the excited oscillations increase with increasing electric drift. Therefore, to achieve the maximal oscillations frequencies one must make the experiments with the strongest possible electric fields. But there is then the danger of overheating and melting of the entire crystal. To avoid this, one must use pulsed electric fields in the experiments. In a plasma with low carrier density, the low conductivity means that the pulse duration is very long and the formation of the pulses does not present any difficulty. For example, in fields $E_0 \sim 1$ kV/cm when $n \sim 10^{15}$ cm^{-3} and $\nu \sim 10^{10}$ sec^{-1} damage to the crystal will be avoided if the pulse duration of the field is less than 10^{-5}–10^{-6} sec.

CHAPTER II

SPATIALLY INHOMOGENEOUS SOLID-STATE PLASMA IN EXTERNAL FIELDS

§ 2.1. Equation of Small Oscillations of an Inhomogeneous Plasma with Low Carrier Density

One of the first of the papers that stimulated the intensive investigations of plasma effects in semiconductors was [37]. This paper showed that in a gas plasma with a current in a sufficiently strong magnetic field the so-called helical instability can develop if the conductivity is inhomogeneous over the plasma cross section. This paper was followed by a great many investigations, both theoretical and experimental, in which the possible manifestation of a helical instability in the electron-hole plasma of a solid was investigated. These investigations are reviewed in [8-10].

In recent years, the theory of oscillations and the stability of a spatially inhomogeneous gas plasma has made great progress. There has been developed a general theory of the oscillations of a spatially inhomogeneous plasma based on the approximation of geometrical optics or, as it is known in quantum mechanics, the quasiclassical approximation [39] (the WKB method). Among the most interesting results of this theory we must mention the discovery in an inhomogeneous magnetized plasma of drift and drift-dissipative oscillations. Such plasma oscillations are of particular interest when an external longitudinal electric field (along the magnetic field) is applied to the plasma, when there is excitation and amplification of oscillations due to the electric drift of charged particles. The various forms of instabilities of a spatially inhomogeneous plasma in external electric and magnetic fields are called drift, beam-drift, and current-drift instabilities. A review of the investigations into the theory of oscillations and the stability of a spatially inhomogeneous gas plasma can be found in [38-41], which also contains a detailed bibiliography of the literature on this question.

The achievements in the theory of the oscillations and stability of a spatially inhomogeneous plasma have not yet been adequately matched in the theory of solid-state plasma. The oscillations spectra and the nature of their excitation in a semiconductor plasma in which there is a regular inhomogeneity have by no means been fully investigated as yet. Inhomogeneities in a semiconductor can arise for a great variety of reasons. For example, the finite dimensions of the semiconductor sample in space is a sign of its inhomogeneity. When one works with pulsed electric fields, the electric field distribution can also be inhomogeneous because of the skin effect in the semiconductor, and then the current and energy (temperature) of the carriers are also inhomogeneous.

Of particular interest from the point of view of a semiconductor plasma are oscillations of an inhomogeneous plasma that can be excited when there is only one carrier species. For a gas plasma such an instability (which can develop in a purely electron plasma) was investigated in [43, 53, 54] and was called the slipping instability; this is also a form of the current-convective instability. However, one cannot merely transfer the results of these investigations to the case of a solid-state plasma, because they do not allow for the specific features of a solid-state plasma, such as the dependence of the energy and momentum relaxation times on the carrier velocities, the heating of the carriers in an external electric field, and so forth. To establish the importance of these effects, we investigate in this, the second chapter of the paper, the problem of the excitation of longitudinal (electrostatic) oscillations in an inhomogeneous semiconductor plasma with low carrier density in parallel external electric and magnetic fields.

First, we obtain the equation of small electrostatic oscillations of the field in an inhomogeneous plasma. To do this, we shall follow the general method of solving the system of Davydov's equations set forth in the first chapter, the only difference being that for the variation of the total current of the carriers with respect to the perturbations we shall assume that the equilibrium quantities n_0, E_0, and B_0 depend on the coordinates. We shall demonstrate the method of deriving the field equation for the example of a plasma with constant ν and δ. Our point of departure is the expression for the current

$$j_i=\frac{e^2}{m\nu}\beta_{ij}\left(nE_j-\frac{m}{e}\frac{\partial}{\partial r_j}nv_{\rm t}^2\right), \tag{2.1}$$

where

$$\beta_{ij}=\frac{\nu^2}{\Omega^2+\nu^2}\left(\delta_{ij}+\frac{\Omega}{\nu}\frac{1}{B_0}e_{ijl}B_{0l}+\frac{\Omega^2}{\nu^2}\frac{1}{B_0^2}B_{0i}B_{0j}\right). \tag{2.2}$$

It should be noted that, because of the inhomogeneity of the plasma pressure in the equilibrium state, a current arises that is equalized by the inhomogeneity of the magnetic field:

$$\nabla\left(nmv_{\rm t}^2+\frac{1}{8\pi}B_0^2\right)=0. \tag{2.3}$$

For fields $B_0>10^2$ Oe, in a plasma with low carrier density the inequality $B_0^2/8\pi \gg nmv_{\rm t}^2$ is satisified,* and the inhomogeneity of the magnetic field is much less pronounced than the inhomogeneity of the density and the electric field. Therefore, we shall ignore the magnetic field inhomogeneity in what follows, except in the case of a plasma with constant ν and δ, when the entire effect of the current-convective instability in the approximation we are considering is attributable to this inhomogeneity.

Varying the expression (2.1) with respect to small perturbations $n=n_0+n_1$, $\mathbf{E}=\mathbf{E}_0+\mathbf{E}_1$ (the magnetic field is not perturbed because the electric field is electrostatic, $E_1=-\nabla\Phi$) and using the continuity equation and Poisson's equation, we can readily obtain the desired field equation in an inhomogeneous plasma:

$$\left(1-\frac{k_zV_0}{\omega}+i\frac{k_z^2v_{\rm t}^2}{\omega\nu}-\frac{k_yv_{\rm t}^2}{\omega\Omega}\frac{\partial\ln B_0}{\partial x}\right)\Delta\Phi-\frac{4\pi i\sigma_0}{\varepsilon_0\omega}k_z^2\Phi-\frac{4\pi\sigma_0}{\varepsilon_0\omega}\left(\frac{4}{3}\frac{k_zV_0}{\delta\nu}k_z^2+k_y\frac{\nu}{\Omega}\frac{\partial}{\partial x}\ln\frac{\sigma_0}{B_0}+i\frac{4}{3}\frac{k_zV_0}{\delta\Omega}k_y\frac{\partial\ln B_0}{\partial x}\right)\Phi=0. \tag{2.4}$$

Here we have restricted the treatment to a plasma that is inhomogeneous in one dimension in parallel fields along the OZ axis; the direction of the inhomogeneity of the plasma is assumed

* The oscillations excited under such conditions can always be assumed to be electrostatic, since a small perturbation of the plasma cannot change the strong magnetic field in the sample.

to coincide with the OX axis. All the perturbed quantities have been represented in the form $f(x)\exp(ik_y y + ik_z z - i\omega t)$. In addition, in deriving Eq. (2.4) we have completely ignored terms of order ν^2/Ω^2, or, thus, effects of transverse diffusion and carrier conductivity.

It should be noted that Eq. (2.4) is not the same as the equation investigated in [43] for $\omega \ll \nu$ and with neglect of diffusion. This is because the model of a weakly ionized plasma used in [43] makes no allowance at all for ohmic heating of the carriers nor a dependence of the plasma temperature on the electric field. If we also ignore heating in our model, then to achieve agreement with [43] we must take into account the terms of order ω/ν, which we have ignored above.

Similarly, we derive an equation for the potential of the field of perturbations when allowance is made for a dependence of ν and δ on the carriers velocity. Omitting the fairly lengthy calculations, we give only the final result:

$$\left(1 - \frac{k_z u}{\omega} + ib_1 \frac{k_z^2 v_t^2}{\omega \nu_0} - ia_1 \frac{k_z v_d}{\delta_0 \nu_0} - b_3 \frac{k_y v_t^2}{\omega \Omega} \frac{\partial \ln B_0}{\partial x}\right) \Delta\Phi - \frac{4\pi\sigma_0}{\varepsilon_0 \omega}\left[ik_z^2 \gamma_0 \frac{r+q}{r-q} + \right.$$

$$\left. + \frac{\nu_0}{\Omega} k_y \frac{\partial}{\partial x} \ln \frac{\sigma_0}{B_0} + k_z^2 a_2 \frac{\omega}{\delta_0 \nu_0} + \frac{4}{3} b_4 \frac{k_z v_d}{\delta_0 \nu_0} + ik_y \frac{k_z v_d}{\delta_0 \Omega}\left(b_5 \frac{\partial \ln n_0}{\partial x} + b_6 \frac{\partial \ln E_0}{\partial x} + \frac{4}{3} b_7 \frac{\partial \ln B_0}{\partial x}\right)\right]\Phi = 0. \qquad (2.5)$$

Here we have introduced the notation

$$b_3 = \frac{2(r-q)}{3}\Gamma\left(\frac{5}{2r-2q}\right)\Big/\Gamma\left(\frac{3}{2r-2q}\right),$$

$$b_4 = \frac{3+2q}{6}\left(\frac{3\delta_0 v_0^2}{4(r-q)v_d^2}\right)^{1-\frac{q+1}{r-q}}\left\{\frac{r+q}{r-q} I(3+2q) \frac{1}{\Gamma\left(\frac{3}{2r-2q}\right)} + \right.$$

$$\left. + \frac{6(r-q)-14}{2(r-q+1)} \frac{\Gamma\left(\frac{5}{2r-2q}\right)\Gamma\left(\frac{3+2q}{2r-2q}\right)}{\Gamma^2\left(\frac{3}{2r-2q}\right)} - \frac{6r-10q-14}{2(r-q+1)} + \frac{\Gamma\left(\frac{5+2q}{2r-2q}\right)}{\Gamma\left(\frac{3}{2r-2q}\right)} + \frac{4}{3}(1+q)\frac{\Gamma\left(\frac{5+2q}{2r-2q}\right)}{\Gamma\left(\frac{3}{2r-2q}\right)}\right\},$$

$$b_5 = -\frac{(3+2q)\,2}{9}\left(\frac{3\delta_0 v_0^2}{4(r-q)v_d^2}\right)^{1-\frac{1}{r-q}}\left\{I(3)\frac{1}{\Gamma\left(\frac{3}{2r-2q}\right)} + \frac{r-q}{1-q}\left[\frac{\Gamma\left(\frac{3+2q}{2r-2q}\right)\Gamma\left(\frac{5-2q}{2r-2q}\right)}{\Gamma^2\left(\frac{3}{2r-2q}\right)} - \frac{\Gamma\left(\frac{5}{2r-2q}\right)}{\Gamma\left(\frac{3}{2r-2q}\right)}\right]\right\}, \qquad (2.6)$$

$$b_6 = \frac{2}{3}\,\frac{3+2q}{1-q}\left(\frac{3\delta_0 v_0^2}{4(r-q)v_d^2}\right)^{1-\frac{1}{r-q}}\left\{\frac{\Gamma\left(\frac{3+2q}{2r-2q}\right)\Gamma\left(\frac{5-2q}{2r-2q}\right)}{\Gamma^2\left(\frac{3}{2r-2q}\right)} - \frac{\Gamma\left(\frac{5}{2r-2q}\right)}{\Gamma\left(\frac{3}{2r-2q}\right)}\right\},$$

$$b_z = \frac{3+2q}{6}\left(\frac{3\delta_0 v_0^2}{4(r-q)v_d^2}\right)^{1-\frac{1}{r-q}}\frac{I(3)}{\Gamma\left(\frac{3}{2r-2q}\right)} + \frac{2}{3(r-q)}\frac{\Gamma\left(\frac{5}{2r-2q}\right)}{\Gamma\left(\frac{3}{2r-2q}\right)},$$

$$a_2 = \frac{2(3+2q)}{3(1-q)}\left[\frac{\Gamma\left(\frac{3+2q}{2r-2q}\right)\Gamma\left(\frac{5-2q}{2r-2q}\right)}{\Gamma^2\left(\frac{3}{2r-2q}\right)} - \frac{\Gamma\left(\frac{5}{2r-2q}\right)}{\Gamma\left(\frac{3}{2r-2q}\right)}\right].$$

The quantities u, v_d, v_t, γ_0, b_1, b_2, a_1, and $I(\alpha)$ have been determined above. When ν and δ are constant (i.e., when $r = 1$ and $q = 0$) $b_1 = b_3 = \gamma_0 = b_4 = b_z = 1$ and $a_1 = a_2 = b_5 = b_6 = 0$, and Eq. (2.5) goes over into (2.4).

Now that we have derived field equations (2.4) and (2.5), the problem of the plasma stability reduces to analyzing the spectra of the eigenvalues of these equations. The field equations must be augmented by boundary conditions, which are usually introduced by integrating the field equations over a physically infinitesimally small layer near the surface of the solid-state plasma. Below, in § 2.3 and 2.4, such boundary conditions will be obtained in our investigation of the oscillations of a homogeneous but spatially bounded plasma.

§ 2.2. Spectrum of Short-Wave Oscillations of an Inhomogeneous Solid-State Plasma

In the general case of an arbitrary inhomogeneity of the medium, the solution of the differential equations (2.4) and (2.5) with variable coefficients and the determination of the eigenvalue spectra is impossible in practice. However, if one is interested in the higher oscillation modes, whose wavelength along the inhomogeneity is appreciably shorter than the characteristic inhomogeneity scale of the plasma, the problem simplifies appreciably; for in this case the WKB method can be used to solve the field equation and determine the eigenvalue spectra of higher order. We seek solutions of the field equations in the form

$$\Phi(x) = C \exp(i\psi(x)), \tag{2.7}$$

where $\psi(x)$ is an eikonal. In the case of a homogeneous medium, $\nabla\psi(x) = k_x = \text{const}$. The quantity k_x is called the component of the wave vector along the OX axis. But if the medium is inhomogeneous, then $k_x(x)$ is a function of the coordinate x with characteristic scale of variation equal to the inhomogeneity scale of the medium. Under conditions when

$$k_x L_0 = \frac{L_0}{\lambda_x} \gg 1, \tag{2.8}$$

where L_0 is the characteristic inhomogeneity scale of the plasma, $k_x(x)$ depends on x far less strongly than the field $\Phi(x)$ does (the dependence of the latter is determined by the wavelength $\lambda_x \sim 1/k_x$). Therefore, when the solution (2.7) is substituted into the field equations the derivatives of the wave vector $k_x(x)$ can be ignored in the zeroth approximation. As a result, the field equations are reduced to algebraic equations, whose solvability conditions (conditions for the existence of nontrivial solutions) are equations for finding $k_x(x)$ or, as one says, the eikonal equations. The dispersion relation that determines the frequency spectrum of such short-wavelength oscillations is [39]

$$\int_{x_\mu}^{x_\nu} dx k_x(\omega, x) = \pi n, \tag{2.9}$$

where n is an integer much larger than unity, x_μ and x_ν are the so-called turning points, at which $\text{Re}\, k_x^2(\omega, x) = 0$; the plasma oscillations are trapped between these points. In quantum mechanics, relations of the type (2.9) are called Bohr-Sommerfeld phase integrals, or quasi-classical quantization rules. This relation also remains true when the plasma is everywhere transparent, i.e., everywhere $\text{Re}\, k_x^2(\omega, x) \geq 0$; by x_μ and x_ν one must then understand the natural boundaries of the solid. The relation (2.9) is not altered for oscillations trapped between the surface of the semiconductor and a turning point within it (these questions are discussed in more detail in the reviews [39], which also contain a special bibliography).

Finally, we note that to obtain the condition of instability of the oscillations and to estimate their growth rate it is sufficient to analyze the integrand in (2.9); for noting that the integration is over the transparency region of the plasma, we can estimate the complex frequency of the oscillations using the mean-value theorem:

$$k_x(\omega, \bar{x}) = \pi n / L, \tag{2.10}$$

where $\bar{x}$ is some mean point in the transparency region and L is the linear dimension of this region. Obviously, the oscillation frequency ω defined in this manner is equal to the root of the local eikonal equation solved for $\omega(\bar{x})$, where $\bar{x}$ is the same mean point in the transparency region of the plasma. However, such an estimate of the oscillation frequency and the growth rate is not very accurate quantitatively, but it does reflect correctly the qualitative nature of the oscillations of the inhomogeneous medium. Moreover, in the case of a weakly inhomogeneous medium, this estimate can give a good quantitative description of the spectra of short-wavelength oscillations. With regard to the instability condition obtained in this manner, it is a necessary but not sufficient condition of instability of an inhomogeneous plasma. On the other hand, fulfillment of this condition in the whole of the transparency region of the plasma is certainly a sufficient condition of instability.

Bearing in mind the simplicity of the analysis of the local oscillation spectra of an inhomogeneous plasma and the great mathematical difficulties that arise when one uses the exact dispersion relation (2.9) (not to mention the fact that the use of this relation requires knowledge of the actual form of the plasma inhomogeneity), we shall restrict ourselves below to a local method of analysis of the spectra, taking the eikonal equation as our point of departure. We begin our analysis with the simplest case of a plasma with constant ν and δ, whose oscillations are described by the second-order differential equation (2.4). The corresponding local oscillation spectrum is given by

$$\omega = k_z V_0 - \frac{4\pi\sigma_0}{\varepsilon_0\Omega} \nu k_y \frac{1}{k^2} \frac{\partial}{\partial x} \ln \frac{\sigma_0}{B_0} + \frac{k_y v_t^2}{\Omega} \frac{\partial \ln B_0}{\partial x} - i \left\{ \frac{k_z^2 v_t^2}{\nu} + \frac{4\pi\sigma_0}{\varepsilon_0} \frac{k_z^2}{k^2} \left(1 + \frac{4}{3} \frac{k_y}{k_z} \frac{V_0}{\delta\Omega} \frac{\partial \ln B_0}{\partial x} \right) \right\}. \qquad (2.11)$$

In writing down this solution we have ignored the small unimportant correction to the real part of the frequency of order $\omega_L^2/\varepsilon_0 \delta \nu^2 \ll 1$. It can be seen that the imaginary part of the frequency consists of two terms, the first of which takes into account carrier diffusion and the second their electrical conductivity. Carrier diffusion always leads to damping of the oscillations, although its role is decisive only in the short-wavelength limit, when the wavelength is shorter than the Debye radius, $k > \omega_L/\varepsilon_0^{1/2} v_t$. In the opposite limit, the main contribution to the imaginary part of the frequency is due to the plasma conductivity. And whereas in a homogeneous plasma a finite conductivity always leads to damping of the oscillations (of course, only for constant ν and δ when the current–voltage characteristic is positive), in an inhomogeneous plasma with current and when

$$\eta_B = \frac{4}{3} \frac{k_y}{k_z} \frac{V_0}{\delta\Omega} \frac{\partial \ln B_0}{\partial x} > 1 \qquad (2.12)$$

it plays a destabilizing role, tending to excite oscillations. The nature of the instability that then arises is similar to the well-known current-convective instability studied in detail for a gas plasma in [37, 41, 43, 53, 54]. However, the instability condition (2.12) differs from the conditions obtained in the cited papers by the large factor of order δ^{-1} on the left-hand side of the inequality, which is a consequence of our allowing for ohmic heating of the carriers in the external electric field.

The maximal growth rate of the instability for sufficiently large η_B may be very large:

$$\gamma_{\max} \approx \frac{\omega_L^2}{\varepsilon_0 \nu} \frac{k_z^2}{k^2} \eta_B. \qquad (2.13)$$

The order of magnitude of this expression exceeds by a factor η_B the growth rate of the electrostatic instability associated with a negative current–voltage characteristic in the medium. The frequency of the excited oscillations is increased by the same amount. This current-convective instability at constant ν and δ develops because of the inhomogeneity of the magnetic field and is therefore of little interest, since it is difficult to produce such a field in real semiconductors.

Let us investigate Eq. (2.5), which takes into account the dependence of ν and δ on the carrier velocity. The solution of the eikonal equation in the form of the local spectrum is

$$\omega = k_z u + b_3 \frac{k_y v_t^2}{\Omega} \frac{\partial \ln B_0}{\partial x} - \frac{4\pi\sigma_0}{\varepsilon_0 k^2} \frac{\nu_0}{\Omega} k_y \frac{\partial}{\partial x} \ln \frac{\sigma_0}{B_0} - i \left\{ b \frac{k_z^2 v_T^2}{\nu_0} + \frac{4\pi\sigma_0}{\varepsilon_0} \frac{k_z^2}{k^2} \left[\frac{r+q}{r-q} \gamma_0 + \right.\right.$$
$$\left.\left. + \frac{k_y}{k_z} \frac{v_d}{\delta_0 \Omega} \left(b_5 \frac{\partial \ln n_0}{\partial x} + b_6 \frac{\partial \ln E_0}{\partial x} + b_7 \frac{\partial \ln B_0}{\partial x} \right) \right] \right\}. \qquad (2.14)$$

In writing down this expression, we have ignored, as above, terms proportional to $\omega/\delta_0\nu_0$ and $k_z v_d/\delta_0\nu_0$, which make small unimportant corrections of order $\omega_L^2/\varepsilon_0\delta_0\nu_0^2$ to the real and the imaginary part of the oscillation frequency. It can be seen from our result that carrier diffusion always stabilizes the oscillations (we recall that $b > 0$, see §1.4), whereas the conductivity can be a cause of excitation. Under conditions when the wavelength is greater than the Debye radius, $k < \omega_L/v_t(\varepsilon_0)^{1/2}$, carrier diffusion can be ignored and the nature of the amplification (or damping) of oscillations is completely determined by the plasma conductivity. Two types of instability must be distinguished. If

$$\frac{r+q}{r-q}\gamma_0 > \frac{k_y}{k_z}\frac{v_d}{\delta_0\Omega}\left(b_5\frac{\partial \ln n_0}{\partial x} + b_6\frac{\partial \ln E_0}{\partial x} + b_7\frac{\partial \ln B_0}{\partial x}\right), \tag{2.15}$$

the excitation of oscillations in the plasma is possible only when there is a negative current–voltage characteristic in the carrier current, when $r + q < 0$ (in accordance with the normalization condition, $r - q > 0$). This instability also occurs in a homogeneous plasma and was investigated in detail in the first chapter of the review. But if the opposite inequality to (2.15) holds, oscillations can also be excited when there is a positive current–voltage characteristic and they are then entirely due to the plasma inhomogeneity. The instability is of current-convective type and is similar to that considered above for a plasma with constant ν and δ. An essentially new feature in the allowance for the dependence of ν and δ on the carrier velocity is the fact that the current-convective instability arises not only because of the inhomogeneity of the magnetic field but also (and this is very important) because of the inhomogeneity of the carrier number density and the external electric field. For constant ν and δ, we have $b_5 = b_6 = 0$, and the inhomogeneity of the carrier density and the electric field do not contribute to the oscillation growth rate. Bearing in mind that in a magnetized solid-state plasma the inhomogeneity of the carrier density is almost always much stronger than the magnetic field inhomogeneity, we conclude that in a plasma with nonconstant ν and δ the current-convective instability must be manifested much more strongly.

We note finally that the order of magnitude of the current-convective instability is greater than the instability associated with negative current–voltage characteristic if

$$\frac{k_z}{k_y} < \frac{v_d}{\delta_0\Omega}\frac{1}{L_0}, \tag{2.16}$$

where L_0 is the characteristic inhomogeneity scale of the plasma determined by the transverse dimension of the sample. For a plasma with $v_d \sim (0.5\text{-}1)\cdot 10^7$ cm/sec, $\Omega \sim (1\text{-}5)\cdot 10^{11}$ sec^{-1}, $\delta_0 \sim 10^{-3}\text{-}10^{-2}$, $L_0 \sim 10^{-1}$ cm, the inequality (2.16) is satisfied when $k_z/k_y < 10^{-1}$. This means that even when $k_y \sim 1/L_0$ (the first azimuthal mode) the current-convective instability is decisive for wavelengths $\lambda_z \sim 1/k_z > 10L_0 \sim 1$ cm. We therefore regard this instability as the most probable for long-wavelength oscillations in a magnetized solid-state plasma.

§ 2.3. Surface Waves in a Solid-State Plasma in the Absence of a Magnetic Field

Hitherto, we have considered the oscillations and stability of a plasma that is in effect spatially unbounded, in other words, we have been considering oscillations whose wavelength is appreciably shorter than the sample dimensions. Even in the cases when we have investigated an inhomogeneous plasma, we have restricted the treatment to short-wavelength oscillations with wavelength shorter than the characteristic inhomogeneity scale. Thus, we have assumed everywhere above that

$$\lambda_x \ll L, \tag{2.17}$$

where L is either the linear dimension of the sample (as a rule, in experiments the transverse dimension of the crystal is minimal and under this condition λ_x must be understood as the wavelength in the transverse direction) or the characteristic inhomogeneity scale of the plasma. When there is such a restriction, we lose sight of the fundamental modes, whose wavelength is greater than or comparable with the sample dimensions. On the other hand, such long-wavelength oscillations do have their interest, since they are the ones that are most readily excited in a semiconductor plasma under the influence of electric drift of the carriers. This already follows from the fact that the shorter the wavelength of the perturbation, the greater the role of carrier diffusion, which, as we have already seen, tends to stabilize the instability. However, as we shall show below, the growth rates of such long-wavelength oscillations are always less than the corresponding growth rates of the short-wavelength volume oscillations.

To investigate the excitation of oscillations whose transverse wavelength is greater than or of order of the sample dimensions, $\lambda_x \gg L$, we must consider a spatially bounded semiconductor plasma. In addition, in real cases a semiconductor always has finite dimensions, so that such a problem is more realistic. By definition, a spatially bounded plasma is inhomogeneous. Therefore, to investigate the excitation of oscillations in such a medium we must from the very beginning consider oscillations (and the stability) of an inhomogeneous plasma and only at the end can we go over to the homogeneous limit within the sample. The equation of small oscillations of an inhomogeneous strongly magnetized semiconductor plasma was obtained in § 2.1 [see (2.5)], although we only analyzed the eigenvalue spectrum of this equation for short-wavelength oscillations described in the approximation of geometrical optics.

To establish the effect of the finite dimensions of the medium on the spectrum of excited oscillations, it is sufficient to consider a plane-parallel sample of homogeneous plasma bounded in one direction x. In the directions y and z, the semiconductor plasma (see Fig. 2, II) is assumed unbounded, and in the direction x it is bounded by vacuum (Fig. 2, I, III). We know from the results of the foregoing sections, that in a semiconductor plasma that has a negative current–voltage characteristic longitudinal oscillations are preferentially excited. Therefore, in this review we shall restrict the investigation to electrostatic surface waves in a solid-state plasma with low carrier density. In addition, allowance for diffusion always leads, as we have seen, to damping (in time) of such oscillations in both homogeneous and inhomogeneous media. In this connection, we shall restrict ourselves below to the frequency region

$$\omega \gg \frac{k^2 v_T^2}{\nu}, \quad \frac{v_T^2}{L^2 \nu}, \tag{2.18}$$

when the effects of diffusion and heat conduction of the carriers can be ignored.

We shall first consider the bounded semiconductor plasma in the absence of an external magnetic field. To obtain the equation of small oscillations of the field in such a plasma we use the continuity equation

$$-i\omega e n_1 + \frac{\partial \delta j_x}{\partial x} + ik_y \delta j_y + ik_z \delta j_z = 0. \tag{2.19}$$

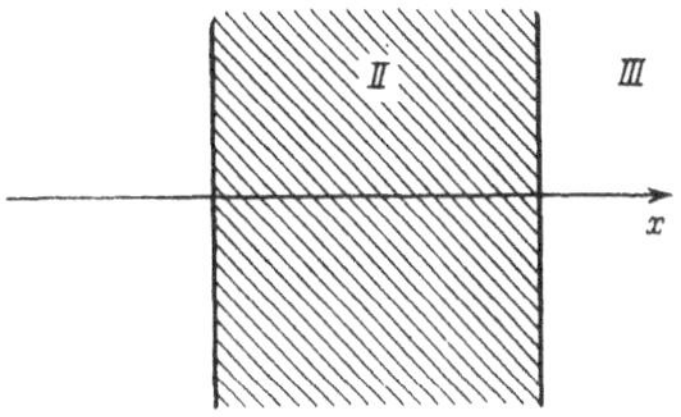

Fig. 2

Substituting the expression (1.50) for δj_x, which depends on x through the equilibrium concentration $n_0 = n_0(x)$ (for simplicity, we assume that the external field is homogeneous), and determining en_1 from Poisson's equation, we finally obtain

$$\omega_1 \frac{\partial^2\Phi}{\partial x^2} + \frac{\partial}{\partial x}\left(i \frac{\omega_L^2}{\varepsilon_0 \nu_0} \gamma_0 \frac{\partial \Phi}{\partial x}\right) - \left\{k_y^2\left(\omega_1 + i \frac{\omega_L^2}{\varepsilon_0\nu_0}\gamma_0\right) + k_z^2\left(\omega_1 + i\frac{r+q}{r-q}\frac{\omega_L^2}{\varepsilon_0\nu_0}\gamma_0\right)\right\}\Phi = 0, \tag{2.20}$$

where $\omega_1 = \omega - k_z u$, and u and γ_0 have been determined above (see Ch. I).

The equations for the potential of the field in the bounded medium must be augmented by boundary conditions. Following [55, 56], we shall obtain them by direct integration of Eq. (2.20) over a narrow layer at the plasma–vacuum interface. We then obtain

$$\{\Phi\} = 0,$$
$$\left\{\left(\omega_1 + i\frac{\omega_L^2}{\varepsilon_0\nu_0}\gamma_0\right)\frac{\partial\Phi}{\partial x}\right\} = 0. \tag{2.21}$$

We can now simplify Eq. (2.20) by assuming that the plasma is homogeneous within the sample Solving this equation by the usual method as an equation with constant coefficients and satisfying the boundary conditions (2.21), we obtain the desired dispersion relation for the oscillation spectrum of the bounded solid-state plasma:

$$\tan k_x a = \frac{2k_x k_\perp \omega_1 (\omega_1 + i\omega_L^2\gamma_0/\varepsilon_0\nu_0)}{k_x^2(\omega_1 + i\omega_L^2\gamma_0/\varepsilon_0\nu_0)^2 - k_\perp^2\omega_1^2}, \tag{2.22}$$

where $k_\perp^2 = k_y^2 + k_z^2$, and k_x is determined from the characteristic equation

$$\left(\omega_1 + i\frac{\omega_L^2}{\varepsilon_0\nu_0}\gamma_0\right)(k_x^2 + k_y^2) + k_z^2\left(\omega_1 + i\frac{r+q}{r-q}\frac{\omega_L^2}{\varepsilon_0\nu_0}\gamma_0\right) = 0. \tag{2.23}$$

For the analysis of the dispersion equation (2.22), we must distinguish two cases: a) the quantity k_x is real, and the waves are then called volume waves; b) the quantity k_x is purely imaginary and the waves are then surface waves (the waves propagate on the surface of the sample and are exponentially damped inside it).

For short-wavelength volume oscillations, when $k_x a \gg 1$, we find the relation $k_x a = \pi(n + 1/2)$, from Eq. (2.22); this relation is identical with the dispersion relation for an unbounded homogeneous semiconductor plasma analyzed in Ch. I of the review. In the same limiting case, $k_x a \gg 1$, for surface short-wavelength oscillations (k_x purely imaginary) $\tan k_x a \to i$ and from Eq. (2.22) we obtain the following dispersion equation:

$$k_x\left(\omega_1 + i\frac{\omega_L^2}{\varepsilon_0\nu_0}\gamma_0\right) - ik_\perp\omega_1 = 0. \tag{2.24}$$

The solution of this equation in the most interesting case of axisymmetric waves ($k_y = 0$) has the form

$$\omega = k_z u - i\frac{r+q}{2r}\frac{\omega_L^2}{\varepsilon_0\nu_0}\gamma_0. \tag{2.25}$$

As one would expect, the spectrum of such oscillations does not depend on the sample dimensions. It can be seen that axisymmetric surface short-wavelength oscillations in a bounded sample can be excited only if there is a negative current–voltage characteristic, when $r + q < 0$, $r > 0$, $q < 0$.

Let us consider long-wavelength oscillations, when $k_x a \ll 1$. The dispersion relation (2.22) in this limit for both volume and surface long-wavelength oscillations reduces to

$$\omega_1^2 + i\frac{\omega_L^2}{\varepsilon_0 \nu_0}\gamma_0 \omega_1 - \frac{a}{2}\frac{\omega_L^4 \gamma_0^2}{\varepsilon_0^2 \nu_0^2 k_\perp}\left(k_y^2 + k_z^2 \frac{r+q}{r-q}\right) = 0. \tag{2.26}$$

From the solution of this equation,

$$\omega = k_z u - i\frac{\omega_L^2 \gamma_0}{\varepsilon_0 \nu_0}\frac{a}{2k_\perp}\left(k_y^2 + k_z^2 \frac{r+q}{r-q}\right), \tag{2.27}$$

it can be seen that long-wavelength oscillations can be excited only when there is a negative current–voltage characteristic in the carrier current, and that axisymmetric modes ($k_y = 0$) are excited preferentially.

Thus, the excitation of electrostatic oscillations in a bounded sample of semiconductor plasma with low carrier density in the absence of an external magnetic field has the same nature as in the case of an unbounded plasma, i.e., it can arise only when there is a negative current–voltage characteristic in the carrier current.

Note that the situation is just the same for a solid-state plasma with high carrier density, for which the effective-temperature approximation applies [47, 48].

§ 2.4. Surface Waves in a Bounded Magnetized Plasma with Current

We consider the excitation of surface and also long-wavelength ($\lambda_x \gtrsim L$) oscillations in a bounded region of solid-state plasma in external parallel electric and magnetic fields. In § 2.1 we have already obtained Eq. (2.5) for the potential of the field of the oscillations in an inhomogeneous unbounded semiconductor plasma in external parallel electric and magnetic fields. However, using this equation we have investigated only the spectra of short-wavelength volume oscillations of the plasma described in the approximation of geometrical optics. We shall now use this equation to investigate the spectra of long-wavelength oscillations of a bounded plasma. For simplicity, we shall assume that only the carrier density $n_0(x)$ depends on the coordinates; we shall assume that the fields $\mathbf{E}_0$ and $\mathbf{B}_0$ are homogeneous. From Eq. (2.5) we then obtain

$$\omega_1 \frac{\partial^2 \Phi}{\partial x^2} - \left(k_\perp^2 \omega_1 + \frac{\omega_L^2}{\varepsilon_0 \Omega} k_y \frac{\partial \ln n_0}{\partial x} + i\frac{\omega_L^2}{\varepsilon_0 \nu_0}\frac{r+q}{r-q}\gamma_0 k_z^2 - i\frac{\omega_L^2}{\varepsilon_0 \nu_0} b_5 \frac{k_y k_z v_d}{\delta_0 \Omega}\frac{\partial \ln n_0}{\partial x}\right)\Phi = 0. \tag{2.28}$$

In writing down this equation we have also ignored carrier diffusion, assuming that $\omega \gg k_z^2 v_t^2/\nu$, and we have omitted the small terms of order $\omega/\delta\nu$ and $k_z v_t/\delta\nu$, which lead to unimportant corrections to the oscillation frequency ω.

We note an important difference between a magnetized plasma and a semiconductor plasma in the absence of an external magnetic field. In a magnetized plasma, the carriers are confined by the strong magnetic field ($\Omega^2 \gg \nu^2$) and the transverse carrier drift due to the pressure gradient is small, of order ν^2/Ω^2. This circumstance allows us to remove all restrictions on the time of development of the instability in such a medium.

Integrating the field equation (2.28) over a narrow layer at the plasma-vacuum boundary, we obtain the boundary conditions

$$\{\Phi\} = 0,$$

$$\left\{\omega_1 \frac{\partial \Phi}{\partial x} - \left(1 - ib_5 \frac{k_z v_d}{\delta_0 \nu_0}\right) k_y \frac{\omega_L^2}{\varepsilon_0 \Omega}\Phi\right\} = 0. \tag{2.29}$$

Assuming that the plasma within the sample is homogeneous, we obtain from Eq. (2.28) with the boundary conditions (2.29) a dispersion relation for the oscillation spectrum of the bounded magnetized solid-state plasma:

$$\tan k_x a = \frac{2k_x k_\perp \omega_1^2}{(k_x^2 - k_\perp^2)\,\omega_1^2 + \left(1 - i b_5 \frac{k_z v_d}{\delta_0 \nu_0}\right) k_y^2 \frac{\omega_L^4}{\varepsilon_0^2 \Omega^2}}, \tag{2.30}$$

where $k_\perp^2 = k_y^2 + k_z^2$, and k_x is determined from the characteristic equation

$$(k_\perp^2 + k_x^2)\,\omega_1 + i \frac{\omega_L^2}{\varepsilon_0 \nu_0} \gamma_0 k_z^2 \frac{r+q}{r-q} = 0. \tag{2.31}$$

In the limiting case of short-wavelength volume oscillations (i.e., $k_x a \gg 1$ and k_x real), Eq. (2.30) goes over into the dispersion relation for the spectrum of volume oscillations of a homogeneous unbounded solid-state plasma investigated in Ch. I. For short-wavelength surface oscillations, when $k_x a \gg 1$ and k_x is purely imaginary, the dispersion relation (2.30) takes the form (we remember that in this case $\tan k_x a \to i$)

$$(k_x + ik_\perp)^2 \omega_1^2 + \left(1 - i \frac{k_z v_d}{\delta_0 \nu_0} b_5\right)^2 \frac{k_y^2 \omega_L^2}{\varepsilon_0^2 \Omega^2} = 0. \tag{2.32}$$

Substituting k_x defined by (2.31) into Eq. (2.32), we readily see that in a magnetized medium asymmetric surface waves alone can be amplified (i.e., $k_y \neq 0$, $k_z \neq 0$) only when there is a negative current–voltage characteristic. The frequency and the growth rate are determined by

$$\omega = k_z u - i \left[4k_\perp^2 k_y^2 + \frac{\Omega^2}{\nu_0^2} \left(\frac{r+q}{r-q} \gamma_0 k_z^2\right)^2\right]^{-1} \frac{\omega_L^2 \gamma_0}{\varepsilon_0 \nu_0} \frac{r+q}{r-q} k_y^2 k_\perp^2. \tag{2.33}$$

The maximal growth rate of such oscillations is $\gamma_{max} \sim \omega_L^2/\varepsilon_0 \Omega$ and is attained when $k_y/k_z \sim \Omega/\nu_0$; it is less than the characteristic growth rate of the short-wavelength volume oscillations associated with a negative current–voltage characteristic by a factor Ω/ν_0.

In the limit of long waves, when $k_x a \ll 1$, the dispersion equation (2.30) for both volume and surface oscillations reduces to the form

$$\omega_1^2 + i\omega_1 \frac{\omega_L^2 \gamma_0}{\varepsilon_0 \nu_0} \frac{r+q}{r-q} \frac{k_z^2 a}{2k_\perp} - \left(1 - 2i \frac{k_z v_d}{\delta_0 \nu_0} b_5\right) \frac{\omega_L^4}{\varepsilon_0^2 \Omega^2} \frac{k_y^2 a}{2k_\perp} = 0. \tag{2.34}$$

This equation describes two types of instability; an instability due to the negative current–voltage characteristic and a current-convective instability; for if

$$\frac{k_y}{k_z} \frac{v_d}{\delta_0 \Omega} \sqrt{\frac{k_\perp}{a}} < 1, \tag{2.35}$$

the last term on the left-hand side of Eq. (2.34) can be ignored, and we then obtain

$$\omega = k_z u - i \frac{\omega_L^2 \gamma_0}{\varepsilon_0 \nu_0} \frac{k_z^2 a}{2k_\perp} \frac{r+q}{r-q}. \tag{2.36}$$

It follows from this expression that if (2.35) is satisfied long-wavelength oscillations are unstable only if there is a negative current–voltage characteristic in the carrier current. But if the opposite inequality to (2.35) holds, we obtain from Eq. (2.34)

$$\omega = k_z u - i \frac{\omega_L^2}{\varepsilon_0 \nu_0} \frac{k_z v_d}{\delta_0 \Omega} b_5 \frac{k_y}{2} \sqrt{\frac{a}{k_\perp}}. \tag{2.37}$$

The excitation of oscillations in this case does not depend on the form of the current–voltage characteristic and is due to current convection. Note that the condition (2.35) is similar to the corresponding condition that determines the nature of the excitation of volume oscillations in unbounded and inhomogeneous plasmas [see (2.15) and (2.16)].

Thus, in the case of a magnetized plasma, the nature of the excitation of long-wavelength surface oscillations in a bounded semiconductor sample is the same as of volume oscillations in an unbounded inhomogeneous plasma. We note only that $k_{\perp}a \ll 1$ for long-wavelength oscillations, and therefore the effect of current convection in bounded semiconductor samples is somewhat less than in an unbounded plasma.

LITERATURE CITED

1. O. V. Konstantinov and V. I. Perel', Zh. Eksp. Teor. Fiz., 38:161 (1960).
2. P. Aigrain, Proc. Fifth Intern. Conf. on Physics of Semiconductors, Prague, 1960; Academic Press, New York (1961), p. 224.
3. V. L. Ginzburg, The Propagation of Electromagnetic Waves in Plasmas, Pergamon Press, Oxford (1964).
4. R. Bowers, G. Legendy, and F. Rose Phys. Rev. Lett., 7:339 (1961).
5. D. Pines and J. R. Schrieffer, Phys. Rev., 124:1387 (1961).
6. A. I. Akhiezer and Ya. B. Fainberg, Dokl. Nauk SSSR, 64:555 (1949); Zh. Eksp. Teor. Fiz., 21:1262 (1951).
7. D. Bohm and E. Gross, Phys. Rev., 75:1851 (1949).
8. A. I. Akhiezer et al., Collective Oscillations in a Plasma, Oxford (1967).
9. A. A. Vedenov, Usp. Fiz. Nauk, 84:533 (1964).
10. B. Anker-Johnson, IEEE, I:43 (1966); Phys. Rev., 164:1050 (1967).
11. J. Bok and P. Nozieres, J. Phys. Chem. Solids, 24:709 (1963).
12. V. I. Pustovoit, Zh. Eksp. Teor. Fiz., 43:2281 (1962); Fiz. Tverd. Tela, 5:2490 (1963).
13. S. P. Bakanov and A. A. Rukhadze, Zh. Eksp. Teor. Fiz., 48:1656 (1965); ZLFT, 36:7 (1966).
14. S. P. Bakanov and A. A. Rukhadze, Fiz. Tverd. Tela, 10:482 (1968).
15. B. I. Davydov, Zh. Eksp. Teor. Fiz., 7:1069 (1937).
16. V. L. Ginzburg and A. V. Gurevich, Usp. Fiz. Nauk, 70:201, 393 (1960).
17. A. I. Akhiezer and A. G. Sitenko, Zh. Eksp. Teor. Fiz., 30:216 (1956).
18. K. N. Stepanov and V. S. Tkalich, Zh. Eksp. Teor. Fiz., 28:1789 (1958).
19. A. B. Mikhailovskii and O. P. Pogutse, Zh. Tekh. Fiz., 36:205 (1966).
20. B. Milich and A. A. Rukhadze, Zh. Tekh. Fiz., 38:229 (1968).
21. B. K. Ridley and T. B. Watkins, Proc. Phys. Soc., 78:293 (1961).
22. B. K. Ridley, Proc. Phys. Soc., 82:954 (1963).
23. V. L. Bonch–Bruevich, Fiz. Tverd. Tela, 6:2047 (1964); 7:21, 47 (1965).
24. F. G. Bass, Zh. Eksp. Teor. Fiz., 48:275 (1965).
25. F. G. Bass, S. I. Khankina, and V. M. Yakovenko, Zh. Eksp. Teor. Fiz., 50:102 (1966).
26. S. I. Khankina and V. M. Yakovenko, Radiofizika, 9:207 (1966).
27. R. R. Kikvidze, A. A. Rukhadze, and E. P. Fetisov, Fiz. Tverd. Tela, 9:1349 (1967).
28. R. R. Kikvidze, Fiz. Tverd. Tela, 10:276 (1968).
29. R. R. Kikvidze, and E. P. Fetisov, Fiz. Tverd. Tela, 10:1147 (1968).
30. A. F. Volkov and Sh. M. Kogan, Zh. Eksp. Teor. Fiz., 52:1647 (1967).
31. A. V. Gurevich, Zh. Eksp. Teor. Fiz., 36:624 (1959).
32. J. Gunn, Usp. Fiz. Nauk, 89:147 (1965) (Russian translation).
33. V. L. Bonch–Bruevich and S. G. Kalashnikov, Fiz. Tverd. Tela, 7:750 (1965).
34. H. Fröhlich and B. Paranjape, Proc. Phys. Soc., B69:21 (1956).

35. I. B. Levinson, Fiz. Tverd. Tela, 6:2113 (1964).
36. S. Chapman and T. G. Cowling, Mathematical Theory of Non-Uniform Gases, Cambridge (1952).
37. B. B. Kadmotsev and A. V. Nedospasov, J. Nucl. Phys., 1:230 (1960).
38. A. B. Mikhailovskii, in: Reviews of Plasma Physics, Vol. 3, Consultants Bureau, New York (1967), p. 159.
39. A. A. Rukhadze and V. P. Silin, Usp. Fiz. Nauk, 82:499 (1964); 96:87 (1968).
40. B. B. Kadomtsev, Plasma Turbulence, Academic Press, London (1965).
41. L. S. Bogdankevich, B. Milich, and A. A. Rukhadze, Yadernyii Sintez., 7:199 (1967).
42. L. É. Gurevich and I. V. Ioffe, Fiz. Tverd. Tela, 6:2926 (1964); 8:1661, 2887 (1966).
43. E. E. Lovetskii, Zh. Tekh. Fiz., 36:1017 (1966).
44. L. D. Tsendin, Fiz. Tekh. Poluprov., 2:807 (1968).
45. R. R. Kikvidze, A. A. Rukhadze, and E. P. Fetisov, Fiz. Tverd. Tela, 11:731 (1969).
46. A. F. Volkov and Sh. M. Kogan, Usp. Fiz. Nauk, 96:633 (1963).
47. S. I. Khankina and V. M. Yakovenko, Fiz. Tverd, Tela, 9:578, 2943 (1967).
48. S. I. Khankina and V. M. Yakovenko, Radiofizika, 11:1259 (1968).
49. V. P. Silin and A. A. Rukhadze, Electromagnetic Properties of Plasmas and Similar Media [in Russian], Atomizdat, Moscow (1961).
50. A. A. Rukhadze, Zh. Tekh. Fiz., 32:669 (1962); A. A. Rukhadze and V. P. Silin, Usp. Fiz. Nauk, 76:79 (1962).
51. V. G. Veselago, M. A. Glushkov, and A. A. Rukhadze, Fiz. Tverd. Tela, 8:24 (1966).
52. Yu. L. Igitkhanov, Zh. Eksp. Teor. Fiz., 56:1619 (1969).
53. L. S. Bogdankevich, E. E. Lovetskii, and A. A. Rukhadze, Nucl. Fusion, 6:9, 176 (1966).
54. A. B. Mikhailovskii and A. A. Rukhadze, Zh. Tekh. Fiz., 35:1841 (1965); I. I. Zhelyazkov and A. A. Rukhadze, Zh. Tekh. Fiz., 40:259 (1970).
55. A. B. Mikhailovskii and É. A. Pashitskii, Zh. Tekh. Fiz., 35:1960 (1965).
56. S. P. Bakanov, L. S. Bogdankevich, and A. A. Rukhadze, Zh. Tekh. Fiz., 36:1639 (1966).

NONLINEAR THEORY OF THE INTERACTION OF WAVES IN A PLASMA

V. V. Pustovalov and V. P. Silin

PREFACE

In this review we set forth the general theory of the nonlinear interaction of electromagnetic waves in a plasma, and we consider numerous concrete consequences of the theory. The basis of this theory, which forms one of the branches of macroscopic nonlinear electrodynamics, consists of the equations of the electromagnetic field,

$$\operatorname{div} \mathbf{D}' = 0, \qquad \operatorname{rot} \mathbf{E} = -\frac{1}{c}\frac{\partial \mathbf{B}}{\partial t},$$

$$\operatorname{rot} \mathbf{B} = \frac{1}{c}\frac{\partial \mathbf{D}'}{\partial t}, \qquad \operatorname{div} \mathbf{B} = 0$$

with the material equation in the form of a power series

$$D_i'(\mathbf{r}, t) = \sum_{n=1}^{\infty} \int_{-\infty}^{t} dt_1 \int d\mathbf{r}_1 \ldots \int_{-\infty}^{t_{n-1}} dt_n \int d\mathbf{r}_n \times$$

$$\times \varepsilon_{ij(1)\ldots j(n)}(t - t_1, \mathbf{r} - \mathbf{r}_1; \ldots; t_{n-1} - t_n, \mathbf{r}_{n-1} - \mathbf{r}_n) E_{j(1)}(t_1, \mathbf{r}_1) \ldots E_{j(n)}(t_n, \mathbf{r}_n).$$

These relations, which hold not only in a plasma but also for a large number of material media, will be particularized for the example of a homogeneous unbounded collisionless plasma described by the system of equations with self-consistent field

$$\frac{\partial f}{\partial t} + \mathbf{v}\frac{\partial f}{\partial \mathbf{r}} + \frac{e}{c}[\mathbf{v}\mathbf{B}_0]\frac{\partial f}{\partial \mathbf{p}} + e\left\{\mathbf{E} + \frac{1}{c}[\mathbf{v}\mathbf{B}]\right\}\frac{\partial f}{\partial \mathbf{p}} = 0,$$

$$\operatorname{div} \mathbf{E} = 4\pi e \int d\mathbf{p} f, \qquad \operatorname{rot} \mathbf{E} = -\frac{1}{c}\frac{\partial \mathbf{B}}{\partial t},$$

$$\operatorname{rot} \mathbf{B} = \frac{1}{c}\frac{\partial \mathbf{E}}{\partial t} + \frac{4\pi}{c} e \int d\mathbf{p}\mathbf{v} f, \qquad \operatorname{div} \mathbf{B} = 0.$$

We consider both an isotropic ($\mathbf{B}_0 = 0$) and magnetoactive ($\mathbf{B}_0 = \text{const} \neq 0$) plasma.

NOTATION

Basically, the notation corresponds to that adopted in the book [1].

Scalar Quantities

e and m are the charge and mass of the electron ($e < 0$).

e_i and M are the charge and mass of the ion.

N_e and N_i are the number of electrons and ions in a cubic centimeter.

T_e and T_i are the electron and ion temperatures in degrees Kelvin.

$\varkappa$ is Boltzmann's constant.

c is the velocity of light in vacuum.

B_0 is the absolute magnitude of the constant external and homogeneous magnetic field.

$v_{Te} = (\varkappa T_e/m)^{1/2}$ and $v_{Ti} = (\varkappa T_i/M)^{1/2}$ are the thermal velocities of the electrons and ions.

$\omega_{Le} = (4\pi N_e e^2/m)^{1/2}$ and $\omega_{Li} = (4\pi N_i e_i^2/M)^{1/2}$ are the electron and ion plasma frequencies.

$\Omega_e = (eB_0/mc)$ and $\Omega_i = (e_i B_0/Mc)$ are the gyroscopic frequencies of gyration of the electrons and ions ($\Omega_e < 0$).

$r_{De} = (\varkappa T_e/4\pi N_e e^2)^{1/2}$ and $r_{Di} = (\varkappa T_i/4\pi N_i e_i^2)^{1/2}$ are the Debye radii of the electrons and ions.

$\rho_e = (v_{Te}/\Omega_e)$ and $\rho_i = (v_{Ti}/\Omega_i)$ are the Larmor radii of the electrons and ions ($\rho_e < 0$).

ω, ω', ω'' and k, k', k'' are the frequencies and wave numbers of interacting oscillations, $\omega = \omega' + \omega''$.

(x, y, z) is a right-handed rectangular coordinate system with z axis along the external magnetic field $\mathbf{B}_0$.

k_z, k'_z, k''_z are the projections of the wave vectors **k**, **k'**, **k''** of the oscillations onto the z axis.

$k_\perp$, $k'_\perp$, $k''_\perp$ are the absolute values of the projections of the wave vectors **k**, **k'**, **k''** of the oscillations onto the plane (x, y) perpendicular to the external magnetic field $\mathbf{B}_0$.

v_z, $v_\perp$, φ are the velocity components of a plasma particle in a right-handed cylindrical system with z axis along $\mathbf{B}_0$.

ϑ, ϑ', ϑ'' are the angles between the external magnetic field $\mathbf{B}_0$ and the wave vectors **k**, **k'**, **k''**, respectively.

θ, θ', θ'' are the angles between the components $\mathbf{k}_\perp$, $\mathbf{k}'_\perp$, $\mathbf{k}''_\perp$ of the wave vectors **k**, **k'**, **k''** in the plane (x, y) and the x axis.

$z_{e,i} \equiv (k_\perp \rho_{e,i})^2$, $z'_{e,i} \equiv (k'_\perp \rho_{e,i})^2$, $z''_{e,i} \equiv (k''_\perp \rho_{e,i})^2$.

$J_n(z)$ is a Bessel function with integral index $n \gtrless 0$ and real argument z.

$I_n(z)$ is a modified Bessel function; $I_n(z) = i^{-n} J_n(iz)$.

n and l are integers; the letter l is also used as an index to characterize longitudinal oscillations, in particular, plasma (Langmuir) oscillations.

$J_+(\beta) \equiv \beta \exp(-\beta^2/2) \int\limits_{i\infty}^{\beta} d\tau \exp(\tau^2/2)$ (see [1], p. 91).

$\beta_e \equiv (\omega/kv_{T_e})$, $\beta_i \equiv (\omega/kv_{T_i})$ are the arguments of the function $J_+(\beta)$ in an isotropic plasma for the electrons and ions; in a magnetoactive plasma and also in the case of a one-dimensional interaction in an isotropic plasma the same notation is used for the combinations $(\omega/|k_z|v_{T_e}) = \beta_e$, $(\omega/|k_z|v_{T_i}) = \beta_i$.

β, β', β'' are the same quantities with wave vectors $\mathbf{k}$, $\mathbf{k'}$, $\mathbf{k''}$, respectively.

$\beta_n^{e,i} \equiv (\omega - n\Omega_{e,i})/|k_z|v_{T_{e,i}}$ are the arguments of the function J_+ in a magnetoactive plasma.

ε_1, ε_2, g are the nonvanishing components of the permittivity tensor of a cold magnetoactive plasma (see [1], p. 140, (20.4)).

$n^2 = c^2k^2/\omega^2$ is the square of the refractive index.

W, M, C, S are the Stokes parameters determining the polarization of a transverse electromagnetic wave.

$W(\mathbf{k})$ is the energy density of a characteristic plasma oscillation in the six-dimensional phase space $(\mathbf{k}, \mathbf{r})$; $[W(\mathbf{k})] = $ erg.

$W_l(\mathbf{k})$ and $W_{tr}(\mathbf{k})$ are the energy densities of longitudinal (l) and transverse (tr) oscillations.

$N_\mathbf{k}$ is the number density of plasma oscillations in the six-dimensional place space $(\mathbf{k}, \mathbf{r})$; $N_\mathbf{k} = W(\mathbf{k})/\hbar\omega(\mathbf{k})$; $\hbar$ is Planck's constant.

$S(\mathbf{r}, t) = \frac{1}{(2\pi)^3} \int d\mathbf{k} N_\mathbf{k}(\mathbf{r}, t)$ is the entropy of the plasma oscillations.

$\omega(\mathbf{k})$ is the spectrum of an oscillation with wave vector $\mathbf{k}$.

$\gamma(\mathbf{k})$ is the damping constant of a plasma oscillation with wave vector $\mathbf{k}$.

$\rho(\mathbf{r}, t)$ is the density of induced charge in the plasma.

$d\sigma$ is the differential scattering cross section.

Δ is the determinant of the matrix $\left\| \varepsilon_{ij}(\omega, \mathbf{k}) - \frac{c^2k^2}{\omega^2}\left(\delta_{ij} - \frac{k_ik_j}{k^2}\right) \right\|$ in the dispersion equation

$$\Delta = 0; \quad \Delta \equiv \left| \varepsilon_{ij}(\omega, \mathbf{k}) - \frac{c^2k^2}{\omega^2}\left(\delta_{ij} - \frac{k_ik_j}{k^2}\right) \right|.$$

t is the time, τ is the characteristic time of a nonlinear interaction process.

τ_0, τ_1, ... τ_n, ... are variables of integration in the interval $(-\infty, 0)$.

f_0 is the particle distribution function of a plasma in a state unperturbed by the electromagnetic fields of oscillations.

$f_1, f_2, ..., f_n, ...$ are the nonequilibrium terms in the distribution function

$$f(\mathbf{p}, \mathbf{r}, t) = f_0 + f_1 + ... + f_n + ...$$

* denotes the complex conjugate value.

Im and Re are the signs of the imaginary and real parts, which are sometimes indicated by primes, for example, $\varepsilon' = $ Re ε, $\varepsilon'' = $ Im ε, $\Delta' = $ Re Δ, $\Delta'' = $ Im Δ.

P is the sign of the principal part and $\delta(\omega)$ is the Dirac function; all singularities of the form $1/\omega$ are understood in the sense of the δ_+ function:

$$(1/\omega) = (P/\omega) - i\pi\delta(\omega).$$

In a number for formulas we understand summation over particle species; this will be clear from the context. The summation sign Σ and the index of the particle species in quantities like e, m, N, T, v_T, ω_L, Ω, r_D, ρ, z, ... will be omitted.

Vectors

$\mathbf{B}_0$ is the external constant and homogeneous magnetic field applied to a plasma.

$\mathbf{h}$ is a unit vector along the external magnetic field, $\mathbf{h} \equiv \mathbf{B}_0/B_0$.

$\mathbf{v}$ is the velocity vector of a plasma particle; $\mathbf{p}$ is the particle momentum; $\mathbf{r}$ is the radius vector.

$\mathbf{k}, \mathbf{k}', \mathbf{k}''$ are wave vectors of the oscillations $\mathbf{k} = \mathbf{k}' + \mathbf{k}''$.

$\varkappa, \varkappa', \varkappa''$ are unit vectors along the wave vectors $\mathbf{k}, \mathbf{k}', \mathbf{k}''$, respectively.

$\mathbf{e}, \mathbf{e}', \mathbf{e}''$ are unit vectors in the planes perpendicular to the wave vectors $\mathbf{k}, \mathbf{k}', \mathbf{k}''$, respectively; they define the polarization of the oscillations.

$[\varkappa\mathbf{h}]$ is the vector product of the vectors $\varkappa$ and $\mathbf{h}$; $(\varkappa\mathbf{h})$ is their scalar product.

$E_i(\mathbf{r}, t)$, $B_i(\mathbf{r}, t)$, and $E_i(\omega, \mathbf{k})$, $B_i(\omega, \mathbf{k})$ are the electric and magnetic fields and their Fourier transforms; the expansion in a Fourier integral is with respect to plane waves of the form $\exp\{-i\omega t + i\mathbf{kr}\}$, and the original and the transform are denoted by the same letter, for example,

$$E_i(\mathbf{r}, t) = \int d\omega d\mathbf{k} \exp\{-i\omega t + i\mathbf{kr}\} E_i(\omega, \mathbf{k}).$$

$\mathbf{D}'(\mathbf{r}, t)$ is the vector of the electric displacement.

$\mathbf{j}(\mathbf{r}, t)$ is the density of the induced current.

$\mathbf{v}(\tau, \mathbf{v}) \equiv (\mathbf{vh})\mathbf{h} + [\mathbf{vh}]\sin\Omega\tau + [\mathbf{h}[\mathbf{vh}]]\cos\Omega\tau$ is the velocity of a charged particle at the time τ gyrating in the magnetic field $\mathbf{B}_0 = \mathbf{h}B_0$ with gyroscopic frequency Ω; $\mathbf{v} \equiv \mathbf{v}(0, \mathbf{v})$ is the velocity of a particle at $\tau = 0$.

$$\delta\mathbf{R}(\tau, \mathbf{v}) \equiv (\mathbf{vh})\mathbf{h}\tau + \frac{1-\cos\Omega\tau}{\Omega}[\mathbf{vh}] + \frac{\sin\Omega\tau}{\Omega}[\mathbf{h}[\mathbf{vh}]] = \int_0^\tau d\tau' \mathbf{v}(\tau', \mathbf{v}).$$

Tensors

δ_{ij} is the Kronecker delta.

e_{isj} is the completely antisymmetric unit pseudotensor of third rank.

$a_{ij} \equiv \delta_{ij} - h_i h_j + ei_{isj} h_s$ is a tensor at right angles to the direction of the external magnetic field $\mathbf{B}_0$.

$$\Gamma_{ij}(\omega) \equiv \frac{1}{\omega} h_i h_j + \frac{1}{2} a_{ij} \frac{1}{\omega + \Omega} + \frac{1}{2} a_{ij} \frac{1}{\omega - \Omega}.$$

$$W_{ij}(\tau) \equiv h_i h_j - \sin\Omega\tau e_{isj} h_s + \cos\Omega\tau(\delta_{ij} - h_i h_j), \quad v_i(\tau, \mathbf{v}) = W_{ij}(\tau) v_j.$$

$$\rho_{ij}(\tau) \equiv h_i h_j \tau + e_{isj} h_s (\cos\Omega\tau - 1)/\Omega + (\sin\Omega\tau/\Omega)(\delta_{ij} - h_i h_j),$$

$$\delta R_i(\tau, \mathbf{v}) = \rho_{ij}(\tau) v_j.$$

$$\alpha_{ij} \equiv \alpha_{ij}(\omega, \mathbf{k}, \mathbf{v}) \equiv \frac{1}{\omega}\{k_i v_j + \delta_{ij}(\omega - \mathbf{kv})\}; \quad \alpha_{ij}(\omega, \mathbf{k}; \mathbf{v}(\tau, \mathbf{v})) = \alpha_{ij}(\tau).$$

$$\beta_{ij} \equiv \beta_{ij}(\omega, \mathbf{k}, \mathbf{v}) \equiv \delta_{ij} + \frac{k_i v_j}{\omega - \mathbf{kv}}.$$

$\varepsilon_{ij}(\omega, \mathbf{k})$ is the permittivity tensor of the plasma; $\delta\varepsilon_{ij}(\omega, \mathbf{k}) = \varepsilon_{ij}(\omega, \mathbf{k}) - \delta_{ij}$;

$\varepsilon_{ij}^{H}(\omega, \mathbf{k})$ is the Hermitian part of the permittivity tensor;

$\varepsilon_{ij}^{a}(\omega, \mathbf{k})$ is the anti-Hermitian part of the permittivity tensor.

$\varepsilon_{ij(1)\cdots j(n)}$ is the multi-index tensor of $(n + 1)$-rank that generalizes $\varepsilon_{ij}(\omega, \mathbf{k})$ in nonlinear theory.

$S_{ijs} \equiv \varepsilon_{ijs} + \varepsilon_{isj}$ is the first nonlinear tensor.

$V_{ijsr} \equiv \varepsilon_{ijsr} + \varepsilon_{ijrs}$.

$M_{ij}(\omega, \mathbf{k}) \equiv \varepsilon_{ij}(\omega, \mathbf{k}) - \frac{c^2k^2}{\omega^2}\left(\delta_{ij} - \frac{k_ik_j}{k^2}\right)$ is the Maxwell tensor.

$A_{ij}(\omega, \mathbf{k})$ is the inverse of the Maxwell tensor (called the inverse tensor); $A_{ij}(\omega, \mathbf{k}) \equiv M_{ij}^{-1}(\omega, \mathbf{k})$.

$(E_jE_i)_{\omega,k}$ is the spectral function of the field.

Contractions of tensors are denoted by the same letters as the tensors but without indices. For example, $\varepsilon(\omega, \mathbf{k}) = \varkappa_i \varepsilon_{ij}(\omega, \mathbf{k})\varkappa_j$ is the longitudinal permittivity; $S(\omega, \mathbf{k}; \omega', \mathbf{k}') = S_{ijs}(\omega, \mathbf{k}; \omega', \mathbf{k}') \varkappa_i \varkappa_j'' \varkappa_s'$. Summation over repeated indices is understood; for example, $\delta\varepsilon_{ij}k_j'' \equiv \delta\varepsilon_{ix}k_x'' + \delta\varepsilon_{iy}k_y'' + \delta\varepsilon_{iz}k_z''$. The letter i is used in three ways: as tensor index ($\delta\varepsilon_{ij}$; $i = x, y, z$), as the subscript of ion quantities (T_i is the temperature of the plasma ions), and as the imaginary unit: $i^2 = -1$.

INTRODUCTION

Systematic study of the effects of nonlinear interaction of waves in plasma was started long after the first theoretical papers had been published on this subject more than ten years ago. Nevertheless, the continuing interest of theoreticians and experimentalists in nonlinear processes in plasma resulted in the accumulation of extensive material, some of which has been set forth already in a number of review papers and monographs [2-11].

In the present review, we study from a unified point of view a large class of questions of the theory of nonlinear interaction of waves in an isotropic and a magnetoactive plasma. The exposition is deductive. In Ch. 1 we derive and discuss the basic general relationships of statistical nonlinear electrodynamics. The following four chapters are devoted to the analysis of special cases of these relations in isotropic plasmas (Chapters I and III) and magnetoactive plasmas (Chapters IV and V). In Chapters II and IV we discuss more general properties of the equations of the nonlinear interaction of waves (the structure of the multi-index tensors of the complex permittivity and the procedure for calculating them, conservation laws, etc), while in Chapters III and V we describe the evolution of the spectral functions of the electric fields of the plasma oscillations in specific nonlinear processes of induced combination scattering and induced wave scattering on particles. In the Appendix that follows Chapter V we derive the three- and four-index tensors of the complex permittivity of a cold isotropic or magnetoactive plasma in the model of two-fluid hydrodynamics.

This review is based on materials of investigations* that the authors made in conjunction with L. M. Gorbunov, A. P. Kropotkin, A. M. Timerbulatov, and N. V. Sholokhov. We make no claim to discuss all the results and the methodological directions in the nonlinear theory of the interaction of waves in a plasma. Our choice of material is also reflected in the bibliography, which contains less than a tenth of the papers on nonlinear theory of the interaction of waves in a plasma. Nevertheless, the results set forth here can serve, in our view, as a useful addition to the already existing reviews on this subject.

---

* Apart from journal papers, we have made wide use of the dissertations of L. M. Gorbunov, A. P. Kropotkin, and V. V. Pustovalov at the P. N. Lebedev Physics Institute under the direction of one of the authors (V. P. Silin).

CHAPTER I

FUNDAMENTALS OF THE NONLINEAR ELECTRODYNAMICS OF A MEDIUM WITH TIME AND SPATIAL DISPERSION

§ 1. The Electromagnetic Field Equations and the Material Equation

We shall regard the interaction of electromagnetic waves in a plasma as one of the problems of nonlinear electrodynamics. Therefore, we must in the first place consider the equations of the electromagnetic field [1]. It must also be emphasized that the fundamentals of nonlinear electrodynamics set forth in this chapter apply not only to plasmas but also to a much larger class of material media.

We note first that the electrodynamics of real media is, in fact, nonlinear only because of the nonlinearity of the material equation. We do not take into account the comparatively small nonlinear effects that can arise in a vacuum without allowance for the medium. In addition, we restrict ourselves to effects that are associated with only weak nonlinearity of the medium, for which the material equation can be represented as an expansion in powers of the electromagnetic fields and one need only consider a finite number of terms of this series [see formula (1.8)] below].

In recent years, nonlinear electrodynamics has developed most strongly in connection with two directions of investigation. The first of these is associated with the theory of the nonlinear interaction of characteristic plasma oscillations [2-7, 11-13] or, in other words, with the theory of a weakly turbulent plasma. The second direction (nonlinear optics) arose as a result of the investigation of certain nonlinear optical problems [14, 15]. These two directions are similar in that they use the nonlinear material equation (1.8) and construct approximate equations for the electromagnetic fields that describe weakly nonlinear phenomena. They differ in that in the theory of a weakly turbulent plasma one generally solves problems of statistical electrodynamics, in which the phase relations between the interacting waves are not fixed. In the problems of nonlinear optics, the phases of the waves are fixed, and in a number of phenomena it is the evolution of the phases that determines the principal effects. For example, many of the problems associated with the study of lasers are of this kind.

The general equations on which we base our treatment can be used to solve both optical problems with fixed phase relations and problems of a weakly turbulent plasma. The present chapter consists of four sections. In the first (§1) we write down and discuss the basic field equations (1.2) and the material equation (1.8). In § 2 we study some general properties of the multi-index tensors of the complex permittivity (2.4), which play in the nonlinear electrodynamics of a homogeneous stationary medium the same important role as the ordinary permittivity tensor in linear theory; we also derive the truncated field equations (2.23). Then, in §3, we go over from the truncated equations, which take into account the phase relations, to the limit of random disordered phases [Eq. (3.22)]. In the concluding § 4, we discuss the results, and we derive the nonlinear dispersion relation (4.12).

By the electric field strength **E** and the magnetic induction **B** we shall understand vectors that define as follows the force

$$\mathbf{F} = e\left\{\mathbf{E} + \frac{1}{c}[\mathbf{vB}]\right\}, \tag{1.1}$$

acting on a point charge e that moves with velocity **v**.

The electric field strength $\mathbf{E}(\mathbf{r}, t)$ and the magnetic induction $\mathbf{B}(\mathbf{r}, t)$ satisfy the equations of the electromagnetic field, which can be written in the form

$$\operatorname{div}\mathbf{E} = 4\pi(\rho + \rho_0), \qquad \operatorname{rot}\mathbf{E} = -\frac{1}{c}\frac{\partial \mathbf{B}}{\partial t},$$
$$\operatorname{rot}\mathbf{B} = \frac{1}{c}\frac{\partial \mathbf{E}}{\partial t} + \frac{4\pi}{c}(\mathbf{j} + \mathbf{j}_0), \qquad \operatorname{div}\mathbf{B} = 0. \tag{1.2}$$

Here ρ and $\mathbf{j}$ are the charge density and current density induced in the medium; their arising when an electromagnetic field is applied is the characteristic feature that distingushes the electrodynamics of a material medium from the electrodynamics of the vacuum; ρ_0 and $\mathbf{j}_0$ are the corresponding densities of the external field sources. Below, we shall be interested in processes that take place in the medium irrespective of the actual form of the external field sources, and in what follows we shall therefore assume that they are completely absent:

$$\rho_0 = 0, \quad \mathbf{j}_0 = 0. \tag{1.3}$$

If we are to be able to use Eq. (1.2) effectively to solve certain electrodynamical problems and, in particular, problems of the nonlinear interaction of waves, we must know the relationship between the charge and current densities induced in the medium and the electromagnetic field. In other words, we must know the material equations. The authors of the different handbooks on electrodynamics write the material equations in very different ways. For the problems in which we are interested, it is not expedient to split the induced current into parts and introduce, for example, the concept of magnetization and the related concept of the magnetic field intensity.* Moreover, we shall not juxtapose an induced current with the displacement current $(1/4\pi)(\partial E/\partial t)$.

Therefore, using the relation ([1], p. 8, (1.8))

$$\mathbf{D}'(\mathbf{r}, t) = \mathbf{E}(\mathbf{r}, t) + 4\pi \int_{-\infty}^{t} dt' \mathbf{j}(\mathbf{r}, t') \tag{1.4}$$

we introduce the vector $\mathbf{D}'$, which enables us to unite the displacement current and the induced current. Namely, on the right-hand side of (1.4) the first term arises from the displacement current, and the second corresponds to the induced current $\mathbf{j}(\mathbf{r}, t)$. Obviously, the displacement current makes a contribution to only the linear part of $\mathbf{D}'$ the electric field strength $\mathbf{E}(\mathbf{r}, t)$. The nonlinear part of $\mathbf{D}'$ is due exclusively to the induced current. The field equations (1.2) in the absence of external charge and current sources (1.3) can be written in the form ([1], p. 9 (1.9)):

$$\operatorname{div}\mathbf{D}' = 0, \qquad \operatorname{rot}\mathbf{E} = -\frac{1}{c}\frac{\partial \mathbf{B}}{\partial t},$$
$$\operatorname{rot}\mathbf{B} = \frac{1}{c}\frac{\partial \mathbf{D}'}{\partial t}, \qquad \operatorname{div}\mathbf{B} = 0. \tag{1.5}$$

This form of expression is possible because of the continuity equation

$$\frac{\partial \rho}{\partial t} + \operatorname{div}\mathbf{j} = 0, \tag{1.6}$$

which relates the induced current and charge.

---

* In this connection, it is quite common in plasma theory to make no distinction between magnetic induction and the magnetic field intensity. In particular, in the present text, this is reflected in the fact that we shall also call **B** the magnetic field.

Thus, with regard to the field equations (1.5), the derivation of a material equation reduces to determining D' as a function of the electromagnetic field. More concretely, we shall speak of a dependence of D' only on the electric but not on the magnetic field, since there is a simple relationship between the magnetic and the electric field [the second equation of the system (1.5)]:

$$-\frac{1}{c}\frac{\partial \mathbf{B}}{\partial t} = \operatorname{rot} \mathbf{E}, \tag{1.7}$$

which enables us to eliminate **B**.

In writing down the material equation we take into account the fact that there are relaxation and transport processes in the medium, which make the induced current at a given time and given point of space dependent on the field at the other points of space and at earlier times. As is well known, this leads to time and spatial dispersion and the relationship between D' and E is integral and nonlocal. On the other hand, as we are interested in the problem of the interaction of electromagnetic waves, we shall, in fact, restrict ourselves to comparatively small amplitudes of the electric field. All this leads to a material equation in the form of an power series in the field:*

$$D'_i(\mathbf{r}, t) = \sum_{n=1}^{\infty} \int_{-\infty}^{t} dt_1 \int d\mathbf{r}_1 \int_{-\infty}^{t_1} dt_2 \int d\mathbf{r}_2 \ldots \int_{-\infty}^{t_{n-1}} dt_n \int d\mathbf{r}_n \times$$

$$\times \varepsilon_{ij(1)\ldots j(n)}(t-t_1, \mathbf{r}-\mathbf{r}_1; \ldots; t_{n-1}-t_n, \mathbf{r}_{n-1}-\mathbf{r}_n; t_n, \mathbf{r}_n) E_{j(1)}(t_1, \mathbf{r}_1) \ldots E_{j(n)}(t_n, \mathbf{r}_n). \tag{1.8}$$

The dependence on the last arguments t_n and $\mathbf{r}_n$ in the kernel $\varepsilon_{ij(1)\cdots j(n)}$ is important only when the medium is nonstationary (there is a dependence on t_n) and spatially inhomogeneous (dependence on $\mathbf{r}_n$).

For if there is no such dependence, the kernels in the integrals (1.8) are functions of only the differences of the space-time coordinates, which corresponds to translational symmetry of space–time. We emphasize that a dependence of the kernels $\varepsilon_{ij(1)\cdots j(n)}$ on the last arguments t_n and $\mathbf{r}_n$ may correspond to an interaction of the electromagnetic waves in the medium as a result of the medium's being nonstationary (parametric coupling of the waves) or because of its being inhomogeneous (transformation of one kind of wave into another on inhomogeneities of the medium) even in the linear approximation in the electric field strength. Such a linear approximation corresponds to retaining on the right-hand side of (1.8) only one term, the first, of the sum:

$$D'_i(\mathbf{r}, t) = \int_{-\infty}^{t} dt_1 \int d\mathbf{r}_1 \varepsilon_{ij}(t-t_1, \mathbf{r}-\mathbf{r}_1; t_1, \mathbf{r}_1) E_j(t_1, \mathbf{r}_1). \tag{1.9}$$

It is therefore natural to distinguish the effects of nonlinear interaction of waves in a medium that correspond to the presence on the right-hand side of (1.8) of at least one term that is nonlinear in the field **E** from the interaction of waves in a nonstationary inhomogeneous medium. This remark can be expressed formally in the fact that in all that follows below we shall omit, without specifically stating so, the last arguments t_n and $\mathbf{r}_n$ in the kernels $\varepsilon_{ij(1)\cdots j(n)}$:

$$\varepsilon_{ij(1)\ldots j(n)}(t-t_1, \mathbf{r}-\mathbf{r}_1; \ldots; t_{n-1}-t_n, \mathbf{r}_{n-1}-\mathbf{r}_n) =$$

$$= \varepsilon_{ij(1)\ldots j(n)}(t-t_1, \mathbf{r}-\mathbf{r}_1; \ldots t_{n-1}-t_n, \mathbf{r}_{n-1}-\mathbf{r}_n; t_n, \mathbf{r}_n). \tag{1.10}$$

*Wherever limits are not indicated, the integration is from $-\infty$ to $+\infty$.

In problems of the nonlinear interaction of waves in a medium one actually uses only some of the first terms of the power series in the material equation (1.8). Below, we shall take into account only the first three terms of such an expansion: the linear, quadratic, and cubic terms in the field, and this, of course, immediately restricts the class of possible nonlinear processes, but, as will be evident from the subsequent exposition (see Chapter II-V), this approximation allows us, in our opinion, to consider a large number of the most important features of the nonlinear interaction of waves in a medium.

The relation (1.8) is very convenient for constructing the theory of a weakly nonlinear interaction of waves; indeed, one can even say that it is specially adapted for the solution of the problems of such a theory. On the other hand, for phenomena in which the field is very strong, the expansion (1.8) may be impossible and the problem of obtaining the material equation must be solved differently. Study of such phenomena is beyond the scope of this monograph.

§ 2. Multi-Index Permittivity Tensors and Truncated Equations of the Electromagnetic Field

The equations of the electromagnetic field (1.5) in conjunction with the material equation (1.8) form a closed system that can be used, in particular, to study the nonlinear interaction of waves in a medium. In accordance with what we have said, we restrict our treatment to the interaction of waves in a homogeneous and stationary medium. In this case it is convenient to expand the field in the Fourier integral

$$\mathbf{E}(\mathbf{r}, t) = \int d\omega d\mathbf{k} e^{-i\omega t + i\mathbf{k}\mathbf{r}} \mathbf{E}(\omega, \mathbf{k}). \tag{2.1}$$

It is obvious [since $\mathbf{E}(\mathbf{r}, t)$ is real] that

$$\mathbf{E}^*(\omega, \mathbf{k}) = \mathbf{E}(-\omega, -\mathbf{k}). \tag{2.2}$$

In a homogeneous and stationary medium, it follows from (1.10) that the dependence of the kernel of the material equation (1.8),

$$\varepsilon_{ij(1)\ldots j(n)}(t - t_1, \mathbf{r} - \mathbf{r}_1; \ldots; t_{n-1} - t_n, \mathbf{r}_{n-1} - \mathbf{r}_n; t_n \mathbf{r}_n), \tag{2.3}$$

on the last space–time argument is not important. For this reason, it is convenient to use multi-index permittivity tensors:

$$\varepsilon_{ij(1)\ldots j(n)}(\omega, \mathbf{k}; \omega_1, \mathbf{k}_1; \ldots; \omega_{n-1}, \mathbf{k}_{n-1}) = \int_0^\infty dt e^{i\omega t} \int d\mathbf{r} e^{-i\mathbf{k}\mathbf{r}} \int_0^\infty dt_1 e^{i\omega_1 t_1} \int d\mathbf{r}_1 e^{-i\mathbf{k}_1\mathbf{r}_1} \ldots \int_0^\infty dt_{n-1} e^{i\omega_{n-1} t_{n-1}} \times$$
$$\times \int d\mathbf{r}_{n-1} e^{-i\mathbf{k}_{n-1}\mathbf{r}_{n-1}} \varepsilon_{ij(1)\ldots j(n)}(t, \mathbf{r}; t_1, \mathbf{r}_1; \ldots; t_{n-1}, \mathbf{r}_{n-1}). \tag{2.4}$$

In particular, the two-index tensor $\varepsilon_{ij}(\omega, \mathbf{k})$ of one pair of arguments $(\omega, \mathbf{k})$ is the ordinary permittivity tensor used in the linear theory of waves; it allows for both the frequency (dependence on ω) and the spatial (dependence on $\mathbf{k}$) dispersion.

By the meaning of the material equation (1.8) and the quantities related to it, the kernels (2.3) are real, and from the definition (2.4) we therefore obtain a symmetry property of the complex tensors ε:

$$\varepsilon^*_{ij(1)\ldots j(n)}(\omega, \mathbf{k}; \omega_1 \mathbf{k}_1; \ldots; \omega_{n-1}, \mathbf{k}_{n-1}) = \varepsilon_{ij(1)\ldots j(n)}(-\omega, -\mathbf{k}; -\omega_1, -\mathbf{k}_1; \ldots; -\omega_{n-1}, -\mathbf{k}_{n-1}), \tag{2.5}$$

which follows from (2.4) in the same way as (2.2) does from (2.1). We shall use this property of the tensors ε below. Using these tensors, we can rewrite the material equation (1.8) in

Fourier components:

$$D'_i(\omega, \mathbf{k}) = \varepsilon_{ij}(\omega, \mathbf{k}) E_j(\omega, \mathbf{k}) + \sum_{n=2}^{\infty} \int d\omega_1 d\mathbf{k}_1 \ldots d\omega_{n-1} d\mathbf{k}_{n-1} \times$$
$$\times \varepsilon_{ij(1)\ldots j(n)}(\omega, \mathbf{k}; \omega_1, \mathbf{k}_1; \ldots; \omega_{n-1}, \mathbf{k}_{n-1}) E_{j(1)}(\omega - \omega_1, \mathbf{k} - \mathbf{k}_1) \ldots$$
$$\ldots E_{j(n-1)}(\omega_{n-2} - \omega_{n-1}, \mathbf{k}_{n-2} - \mathbf{k}_{n-1}) E_{j(n)}(\omega_{n-1}, \mathbf{k}_{n-1}). \quad (2.6)$$

The form of expression of the right-hand side of this relation enables one to deduce readily other symmetry properties of the tensors $\varepsilon_{ij(1)\cdots j(n)}$. In particular,

$$\varepsilon_{ij(1)j(2)}(\omega, \mathbf{k}; \omega_1, \mathbf{k}_1) = \varepsilon_{ij(2)j(1)}(\omega, \mathbf{k}; \omega - \omega_1, \mathbf{k} - \mathbf{k}_1), \quad (2.7)$$

$$\varepsilon_{ij(1)j(2)j(3)}(\omega, \mathbf{k}; \omega_1, \mathbf{k}_1; \omega_2, \mathbf{k}_2) = \varepsilon_{ij(3)j(2)j(1)}(\omega, \mathbf{k}; \omega - \omega_2, \mathbf{k} - \mathbf{k}_2; \omega - \omega_1, \mathbf{k} - \mathbf{k}_1) =$$
$$= \varepsilon_{ij(1)j(3)j(2)}(\omega, \mathbf{k}; \omega_1, \mathbf{k}_1; \omega_1 - \omega_2, \mathbf{k}_1 - \mathbf{k}_2) = \varepsilon_{ij(2)j(1)j(3)}(\omega, \mathbf{k}; \omega + \omega_2 - \omega_1, \mathbf{k} + \mathbf{k}_2 - \mathbf{k}_1; \omega_2, \mathbf{k}_2). \quad (2.8)$$

To illustrate the method that should be used to derive the symmetry properties of the tensor $\varepsilon_{ij(1)\cdots j(n)}$ of arbitrary rank, let us prove, for example, (2.7). Namely, for the second term of the right-hand side of the material equation (2.6) we have the identity

$$\int d\omega_1 d\mathbf{k}_1 \varepsilon_{ij(1)j(2)}(\omega, \mathbf{k}; \omega_1, \mathbf{k}_1) E_{j(1)}(\omega - \omega_1, \mathbf{k} - \mathbf{k}_1) E_{j(2)}(\omega_1, \mathbf{k}_1) =$$
$$= \int d\omega_1 d\mathbf{k}_1 \varepsilon_{ij(2)j(1)}(\omega, \mathbf{k}; \omega - \omega_1, \mathbf{k} - \mathbf{k}_1) E_{j(1)}(\omega - \omega_1, \mathbf{k} - \mathbf{k}_1) E_{j(2)}(\omega_1, \mathbf{k}_1), \quad (2.9)$$

whose right-hand side follows from the left-hand side by successive relabeling of the dummy indices $j_{(1)} \rightleftharpoons j_{(2)}$ and the variables of integration $(\omega_1, \mathbf{k}_1) \rightleftharpoons (\omega - \omega_1, \mathbf{k} - \mathbf{k}_1)$. Thus, the symmetry property (2.7) holds in the sense of the identity (1.9), since the tensor $\varepsilon_{ij(1)j(2)}$ is to be studied in the given theory only as the kernel of the integrals in (2.9). Therefore, when calculating the explicit form of the multi-index permittivity tensors in a specific medium by means of a definite model one must symmetrize them appropriately, since they can arise, in general, in an asymmetry form (for more details see Chapters II and IV).

The relation (2.6) in conjunction with the equations of the magnetic field (1.5) after elimination of the magnetic field $\mathbf{B}$

$$\operatorname{rot}\operatorname{rot}\mathbf{E} + \frac{1}{c^2}\frac{\partial^2 \mathbf{D}'}{\partial t^2} = 0 \quad (2.10)$$

by means of the equation

$$(\operatorname{rot}\operatorname{rot}\mathbf{E})_i = \int d\omega dk e^{-i\omega t + i\mathbf{k}\mathbf{r}} k^2 \left(\delta_{ij} - \frac{k_i k_j}{k^2}\right) E_j(\omega, \mathbf{k}) \quad (2.11)$$

leads to the following nonlinear equation for the Fourier components of the electric field [12]:

$$\left\{\frac{c^2 k^2}{\omega^2}\left(\delta_{ij} - \frac{k_i k_j}{k^2}\right) - \varepsilon_{ij}(\omega, \mathbf{k})\right\} E_j(\omega, \mathbf{k}) =$$
$$= \sum_{n=2}^{\infty} \int d\omega_1 d\mathbf{k}_1 \ldots d\omega_{n-1} d\mathbf{k}_{n-1} \varepsilon_{ij(1)\ldots j(n)}(\omega, \mathbf{k}; \omega_1, \mathbf{k}_1; \ldots; \omega_{n-2}, \mathbf{k}_{n-2}; \omega_{n-1} \mathbf{k}_{n-1}) \times$$
$$\times E_{j(1)}(\omega - \omega_1, \mathbf{k} - \mathbf{k}_1) \ldots E_{j(n-1)}(\omega_{n-2} - \omega_{n-1}, \mathbf{k}_{n-2} - \mathbf{k}_{n-1}) E_{j(n)}(\omega_{n-1}, \mathbf{k}_{n-1}). \quad (2.12)$$

Equation (2.12) can be taken as the basis of nonlinear electrodynamics, in the framework of which one should study the nonlinear interaction of electromagnetic waves in a plasma (see

Chapter II-V). It has been obtained by the expansion (2.1) of the field strengths in plane waves. Such an expansion is effective if plane monochromatic waves are characteristic oscillations of the media. On the other hand, the actual concept of plane monochromatic waves is exact only in linear electrodynamics. In our treatment, we shall assume that this representation is approximately correct, which is, of course, possible under conditions when the nonlinear terms of Eq. (2.12), i.e., its right-hand side, are comparatively small. For this to be the case, one must assume that the wave amplitudes are small. Under these restrictions, the dispersion properties of electromagnetic waves in the medium are determined by the linear approximation of Eq. (2.12):

$$\left\{\varepsilon_{ij}(\omega, \mathbf{k}) - \frac{c^2k^2}{\omega^2}\left(\delta_{ij} - \frac{k_ik_j}{k^2}\right)\right\} E_j(\omega, \mathbf{k}) = 0. \tag{2.13}$$

The condition of solvability of (2.13) leads to the dispersion equation ([1], p. 37 (5.18))

$$\left|\varepsilon_{ij}(\omega, \mathbf{k}) - \frac{c^2k^2}{\omega^2}\left(\delta_{ij} - \frac{k_ik_j}{k^2}\right)\right| = 0, \tag{2.14}$$

which corresponds to vanishing of the determinant of the system of equations (2.12) without nonlinear terms. The linear permittivity tensor $\varepsilon_{ij}(\omega, \mathbf{k})$ in Eq. (2.14) can be represented as a sum of Hermitian and anti-Hermitian parts:

$$\varepsilon_{ij}(\omega, \mathbf{k}) = \varepsilon^H_{ij}(\omega, \mathbf{k}) + \varepsilon^a_{ij}(\omega, \mathbf{k}); \tag{2.15}$$

$$\varepsilon^H_{ij}(\omega, \mathbf{k}) = \varepsilon^{H*}_{ji}(\omega, \mathbf{k}); \quad \varepsilon^a_{ij}(\omega, \mathbf{k}) = -\varepsilon^{a*}_{ji}(\omega, \mathbf{k}). \tag{2.15a}$$

Then solutions of Eq. (2.14) correspond to weakly damped electromagnetic waves in the medium, when the anti-Hermitian part of the permittivity, ε^a_{ij}, is relatively small (compared with the Hermitian part ε^H_{ij}). The presence of damping is manifested by the frequency of the electromagnetic waves having a nonvanishing imaginary part. However, because the imaginary part is small, the frequency $\omega(\mathbf{k})$ can be assumed approximately real and determined from Eq. (2.14), in which only the Hermitian part of the permittivity is retained:

$$\left|\varepsilon^H_{ij}(\omega, \mathbf{k}) - \frac{c^2k^2}{\omega^2}\left(\delta_{ij} - \frac{k_ik_j}{k^2}\right)\right| = 0. \tag{2.16}$$

The restriction that the imaginary correction to the real frequency $\omega(\mathbf{k})$ of the electromagnetic wave be small is natural in the framework of the proposed theory, which allows for only weak nonlinearity, since for the nonlinear interaction of waves one requires the waves to exist for at least the duration of the interaction, which must be shorter than the damping time.

The nonlinear interaction of the electromagnetic waves causes their amplitudes and phases to change in time and space. It is therefore meaningful to speak of (complex) amplitudes of almost monochromatic plane waves:

$$\mathbf{E}(\mathbf{r}, t) = \mathbf{E}(\mathbf{r}, t; \omega, \mathbf{k}) e^{-i\omega t + i\mathbf{k}\mathbf{r}} + \mathbf{E}^*(\mathbf{r}, t; \omega, \mathbf{k}) e^{i\omega t - i\mathbf{k}\mathbf{r}}. \tag{2.17}$$

Here, ω and $\mathbf{k}$ is the real frequency and wave vector of an electromagnetic wave, and the first pair of arguments $\mathbf{r}$ and t on the right-hand side reflect the slow dependence on the coordinates $\mathbf{r}$ and t compared with the rapidly oscillating dependence in the form $\exp\{\mp i\omega t \pm i\mathbf{k}\mathbf{r}\}$.

To obtain equations describing the space–time variation of such amplitudes, we proceed in the same manner as in ordinary linear electrodynamics ([1], §4; [13], §3, pp. 43-48), the more so since the amplitudes of the almost monochromatic waves also vary because of the presence of linear dissipative effects. Assuming that the characteristic time of variation of the amplitudes of the almost monochromatic waves is long compared with the oscillation period

$2\pi/\omega$ in the wave, and also assuming that the characteristic scale of variations of the amplitudes is large compared with the wavelength $2\pi/k$, we write down the following approximate relation (see [1], §4, p. 30):

$$\frac{\partial D'_i(\mathbf{r}, t)}{\partial t} \simeq -i\omega e^{-i\omega t+i\mathbf{kr}}\varepsilon_{ij}(\omega, \mathbf{k}) E_j(\mathbf{r}, t; \omega, \mathbf{k}) + i\omega e^{i\omega t-i\mathbf{kr}}\varepsilon_{ij}(-\omega, -\mathbf{k}) E^*_j(\mathbf{r}, t; \omega, \mathbf{k}) +$$
$$+e^{-i\omega t+i\mathbf{kr}}\frac{\partial}{\partial\omega}[\omega\varepsilon_{ij}(\omega, \mathbf{k})]\frac{\partial}{\partial t}E_j(\mathbf{r}, t; \omega, \mathbf{k}) + e^{i\omega t-i\mathbf{kr}}\frac{\partial}{\partial\omega}[\omega\varepsilon_{ij}(-\omega, -\mathbf{k})]\frac{\partial}{\partial t}E^*_j(\mathbf{r}, t; \omega, \mathbf{k}) -$$
$$-e^{-i\omega t+i\mathbf{kr}}\frac{\partial}{\partial\mathbf{k}}[\omega\varepsilon_{ij}(\omega, \mathbf{k})]\frac{\partial}{\partial\mathbf{r}}E_j(\mathbf{r}, t; \omega, \mathbf{k}) - e^{i\omega t-i\mathbf{kr}}\frac{\partial}{\partial\mathbf{k}}[\omega\varepsilon_{ij}(-\omega, -\mathbf{k})]\frac{\partial}{\partial\mathbf{r}}E^*_j(\mathbf{r}, t; \omega, \mathbf{k}) + \frac{\partial\delta D'_i}{\partial t}. \quad (2.18)$$

The last term on the right-hand side of this relation contains $\delta\mathbf{D}'$, the nonlinear part of $\mathbf{D}'$. We assume that this nonlinear part is comparatively small, and we shall therefore ignore in it the departure of the amplitudes of the monochromatic waves from constants. More precisely, we shall ignore the derivatives with respect to the slow time and slow coordinates of the nonlinear terms. Further, bearing in mind the relative smallness of the space and time derivatives of the amplitudes of the almost monochromatic waves, we shall ignore [see the last four linear terms on the right-hand side of (2.18)] the anti-Hermitian part of the permittivity in the factors in front of them; this anti-Hermitian part is small for weakly damped electromagnetic waves in accordance with our above discussion. Then Eq. (2.18) becomes

$$\frac{\partial D'_i(\mathbf{r}, t)}{\partial t} \simeq -i\omega e^{-i\omega t+i\mathbf{kr}}\varepsilon_{ij}(\omega, \mathbf{k}) E_j(\mathbf{r}, t; \omega, \mathbf{k}) + i\omega e^{i\omega t-i\mathbf{kr}}\varepsilon^*_{ij}(\omega, \mathbf{k}) E^*_j(\mathbf{r}, t; \omega, \mathbf{k}) +$$
$$+e^{-i\omega t+i\mathbf{kr}}\frac{\partial}{\partial\omega}[\omega\varepsilon^H_{ij}(\omega, \mathbf{k})]\frac{\partial}{\partial t}E_j(\mathbf{r}, t; \omega, \mathbf{k}) + e^{i\omega t-i\mathbf{kr}}\frac{\partial}{\partial\omega}[\omega\varepsilon^H_{ji}(\omega, \mathbf{k})]\frac{\partial}{\partial t}E^*_j(\mathbf{r}, t; \omega, \mathbf{k}) -$$
$$-e^{-i\omega t+i\mathbf{kr}}\frac{\partial}{\partial k}[\omega\varepsilon^H_{ij}(\omega, \mathbf{k})]\frac{\partial}{\partial r}E_j(\mathbf{r}, t; \omega, \mathbf{k}) - e^{i\omega t-i\mathbf{kr}}\frac{\partial}{\partial\mathbf{k}}[\omega\varepsilon^H_{ji}(\omega, \mathbf{k})]\frac{\partial}{\partial\mathbf{r}}E^*_j(\mathbf{r}, t; \omega, \mathbf{k}) -$$
$$-i\omega\delta D'_i(\omega, \mathbf{k}) + i\omega\delta D'^*_i(\omega, \mathbf{k}). \quad (2.19)$$

Here we have used the property (2.5), in particular,

$$\varepsilon_{ij}(\omega, \mathbf{k}) = \varepsilon^*_{ij}(-\omega, -\mathbf{k}) \quad (2.5a)$$

and the first of the properties (2.15a)

$$\varepsilon^H_{ij}(-\omega, -\mathbf{k}) = \varepsilon^{H*}_{ij}(\omega, \mathbf{k}) = \varepsilon^H_{ji}(\omega, \mathbf{k}). \quad (2.20)$$

We now substitute Eq. (2.19) into Poynting's theorem:

$$\frac{1}{4\pi}\left\{\mathbf{E}\frac{\partial\mathbf{D}'}{\partial t} + \mathbf{B}\frac{\partial\mathbf{B}}{\partial t}\right\} = -\frac{c}{4\pi}\operatorname{div}[\mathbf{EB}], \quad (2.21)$$

which is an obvious consequence of the field equations (1.5). At the same time, we allow for the material equation (2.6):

$$\delta D'_i(\omega, \mathbf{k}) = \sum_{n=2}^{\infty}\int d\omega_1 d\mathbf{k}_1 \ldots d\omega_{n-1}d\mathbf{k}_{n-1}\varepsilon_{ij(1)\ldots j(n)}(\omega, \mathbf{k}; \omega_1, \mathbf{k}_1; \ldots; \omega_{n-1}, \mathbf{k}_{n-1}) \times$$
$$\times E_{j(1)}(\omega-\omega_1, \mathbf{k}-\mathbf{k}_1)\ldots E_{j(n-1)}(\omega_{n-2}-\omega_{n-1}, \mathbf{k}_{n-2}-\mathbf{k}_{n-1}) E_{j(n)}(\omega_{n-1}, \mathbf{k}_{n-1}). \quad (2.22)$$

Then, after averaging over a time interval that appreciably exceeds the oscillation period of the waves, we obtain

$$\frac{\partial}{\partial t}\left\{\left[\frac{\partial\omega\varepsilon^H_{ij}(\omega, \mathbf{k})}{\partial\omega} + \frac{c^2k^2}{\omega^2}\left(\delta_{ij} - \frac{k_ik_j}{k^2}\right)\right]\frac{1}{4\pi}E_j(\omega, \mathbf{k})E^*_i(\omega, \mathbf{k})\right\} -$$
$$-\frac{\partial}{\partial r_s}\left\{\left[\frac{\partial\omega\varepsilon^H_{ij}(\omega, \mathbf{k})}{\partial k_s} + \frac{c^2}{\omega}(k_i\delta_{sj} + k_j\delta_{is} - 2k_s\delta_{ij})\right]\frac{1}{4\pi}E_j(\omega, \mathbf{k})E^*_i(\omega, \mathbf{k})\right\} =$$

$$=\frac{i\omega}{4\pi}\{\varepsilon_{ij}(\omega, \mathbf{k})E_j(\omega, \mathbf{k})E_i^*(\omega, \mathbf{k})-\varepsilon_{ij}^*(\omega, \mathbf{k})E_j^*(\omega, \mathbf{k})E_i(\omega, \mathbf{k})\}+$$

$$+\frac{i\omega}{4\pi}\sum_{n=2}^{\infty}\int d\omega_1 d\mathbf{k}_1\ldots d\omega_{n-1}d\mathbf{k}_{n-1}\varepsilon_{ij(1)\ldots j(n)}(\omega, \mathbf{k};\ \omega_1, \mathbf{k}_1;\ \ldots;\ \omega_{n-1}, \mathbf{k}_{n-1})\times E_i^*(\omega, \mathbf{k})E_{j(1)}(\omega-\omega_1, \mathbf{k}-\mathbf{k}_1)\ldots$$

$$\ldots E_{j(n-1)}(\omega_{n-2}-\omega_{n-1}, \mathbf{k}_{n-2}-\mathbf{k}_{n-1})\times E_{j(n)}(\omega_{n-1}, \mathbf{k}_{n-1})-$$

$$-\frac{i\omega}{4\pi}\sum_{n=2}^{\infty}\int d\omega_1 d\mathbf{k}_1\ldots d\omega_{n-1}d\mathbf{k}_{n-1}\,\varepsilon^*_{ij(1)\ldots j(n)}(\omega, \mathbf{k};\omega_1, \mathbf{k}_1;\ \ldots;\ \omega_{n-1}, \mathbf{k}_{n-1})E_i(\omega, \mathbf{k})E^*_{j(1)}(\omega-\omega_1, \mathbf{k}-\mathbf{k}_1)\ldots$$

$$\ldots E^*_{j(n-1)}(\omega_{n-2}-\omega_{n-1}, \mathbf{k}_{n-2}-\mathbf{k}_{n-1})E_{j(n)}(\omega_{n-1}, \mathbf{k}_{n-1}). \tag{2.23}$$

Here the magnetic field is eliminated by means of the relation

$$\mathbf{B}(\omega, \mathbf{k})=\frac{c}{\omega}[\mathbf{k}\mathbf{E}(\omega, \mathbf{k})], \tag{2.24}$$

which follows from (1.7) by the Fourier expansion (2.1).

Equation (2.23) expresses the energy conservation. It corresponds to the truncated field equations generally employed in nonlinear optics ([14], p. 90 (2.37); p. 134 (2.176); [15], p. 298 (4.12)). The use of such equations is helpful when one is solving problems of nonlinear electrodynamics in which the phase relations between the different interacting waves are fixed in a definite manner. We shall consider the case when the phases of the waves are random, this being an opposite case in a certain sense. The use of this assumption enables us to take the next step, which is associated with appropriate averaging.

§ 3. Averaging of the Truncated Equations of the Electromagnetic Field with Respect to the Statistical Ensemble

In statistical electrodynamics (or, in other words, in the theory of fluctuations of electromagnetic fields), which is suitable for the case of disordered wave phases, one can advance further by averaging Eq. (2.23) over the statistical ensemble.

This procedure is successful because the nonlinear effects are relatively small and one can restrict oneself to a few terms of the nonlinear power series of the material equation (2.6). In our exposition, we shall restrict ourselves to studying processes for which it is sufficient to retain in Eq. (2.23) terms up to the fourth power of the field [see the remark in § 1 after formula (1.10)]. We shall consider below the equation

$$\frac{\partial}{\partial t}\frac{1}{4\pi}\frac{\partial\omega M^H_{ij}(\omega, \mathbf{k})}{\partial\omega}\langle E_j(\omega, \mathbf{k})E_i^*(\omega, \mathbf{k})\rangle-\frac{\partial}{\partial \mathbf{r}}\frac{1}{4\pi}\frac{\partial\omega M^H_{ij}(\omega, \mathbf{k})}{\partial \mathbf{k}}\langle E_j(\omega, \mathbf{k})E_i^*(\omega, \mathbf{k})\rangle=$$

$$=i\frac{\omega}{4\pi}\{\varepsilon_{ij}(\omega, \mathbf{k})\langle E_j(\omega, \mathbf{k})E_i^*(\omega, \mathbf{k})\rangle-\varepsilon_{ij}^*(\omega, \mathbf{k})\langle E_i(\omega, \mathbf{k})E_j^*(\omega, \mathbf{k})\rangle\}+$$

$$+i\frac{\omega}{4\pi}\int d\omega' d\mathbf{k}'\{\varepsilon_{ijs}(\omega, \mathbf{k};\omega', \mathbf{k}')\langle E_i^*(\omega, \mathbf{k})E_j(\omega-\omega', \mathbf{k}-\mathbf{k}')E_s(\omega', \mathbf{k}')\rangle-$$

$$-\varepsilon^*_{ijs}(\omega, \mathbf{k};\ \omega', \mathbf{k}')\langle E_i(\omega, \mathbf{k})E_j^*(\omega-\omega', \mathbf{k}-\mathbf{k}')E_s^*(\omega', \mathbf{k}')\rangle\}+$$

$$+i\frac{\omega}{4\pi}\int d\omega' d\mathbf{k}' d\omega'' d\mathbf{k}''\{\varepsilon_{ijsr}(\omega, \mathbf{k};\ \omega', \mathbf{k}';\ \omega'', \mathbf{k}'')\langle E_i^*(\omega, \mathbf{k})E_j(\omega-\omega', \mathbf{k}-\mathbf{k}')E_s(\omega'-\omega'', \mathbf{k}'-\mathbf{k}'')E_r(\omega'', \mathbf{k}'')\rangle-$$

$$-\varepsilon^*_{ijsr}(\omega, \mathbf{k};\ \omega', \mathbf{k}';\ \omega'', \mathbf{k}'')E_i(\omega, \mathbf{k})E_j^*(\omega-\omega', \mathbf{k}-\mathbf{k}')E_s^*(\omega'-\omega'', \mathbf{k}'-\mathbf{k}'')E_r^*(\omega'', \mathbf{k}'')\rangle\}, \tag{3.1}$$

in which the angular brackets $\langle\cdots\rangle$ denote statistical averaging and we use the notation for the Maxwell tensor

$$M_{ij}(\omega, \mathbf{k})\equiv\varepsilon_{ij}(\omega, \mathbf{k})-\frac{c^2k^2}{\omega^2}\left(\delta_{ij}-\frac{k_ik_j}{k^2}\right), \tag{3.2}$$

so that

$$M^H_{ij}(\omega, \mathbf{k}) = \varepsilon^H_{ij}(\omega, \mathbf{k}) - \frac{c^2k^2}{\omega^2}\left(\delta_{ij} - \frac{k_ik_j}{k^2}\right). \tag{3.3}$$

For the statistical averaging of the binary product of Fourier components of the electric field strength in the case of stationary and spatially homogeneous states in the theory of fluctuations, one usually writes down the relation ([1], § 9; [16], § 90)

$$\langle E_j(\omega, \mathbf{k}) E^*_i(\omega', \mathbf{k}')\rangle = \delta(\omega - \omega')\,\delta(\mathbf{k} - \mathbf{k}')\,(E_jE_i)_{\omega, \mathbf{k}}, \tag{3.4}$$

where $(E_j E_i)_{\omega,k}$ is the spectral density of the fluctuations of the electric field (or, in other words, the spectral function of the field). We emphasize that the relation (3.4) is derived in the handbooks under the assumption that the mean values of the electric field strengths vanish in the medium, namely,

$$\langle \mathbf{E}(\omega, \mathbf{k})\rangle = 0. \tag{3.5}$$

In all that follows we shall also assume that Eq. (3.5) holds.

In accordance with (2.17) and (3.1), the amplitudes of almost monochromatic waves in Eq. (2.23) vary slowly in space and time. In this sense, the state of the electromagnetic field in the medium is not exactly stationary and homogeneous, as is assumed when one writes down the expression (3.4), but is quasistationary in time and quasihomogeneous in space. Therefore, because the field amplitudes change slowly on account, as we have seen above, of the smallness of the linear damping and the nonlinear effects, we can now assume that Eq. (3.4) is satisfied approximately.

We note further that the operation of statistical averaging presupposes a definite distribution function of the electric field strengths in the medium. The determination of such a distribution function would amount to the complete solution of the problem of statistical electrodynamics. However, for our purposes, such a solution is not necessary since we have set ourselves a more limited task: to obtain equations describing the evolution of the spectral functions $(E_jE_i)_{\omega,k}$ of the field. In the language of probability theory, this corresponds to finding some of the first moments of random variables and not their distribution functions, which, when known, enable one in principle to calculate the moments of any order.

In linear electrodynamics, which corresponds to ignoring all the nonlinear terms on the right-hand side of (2.23), the equation for the spectral functions of the field has the form

$$\frac{\partial \omega M^H_{ij}(\omega, \mathbf{k})}{\partial \omega}\frac{\partial}{\partial t}\frac{(E_jE_i)_{\omega, \mathbf{k}}}{4\pi} - \frac{\partial \omega M^H_{ij}(\omega, \mathbf{k})}{\partial \mathbf{k}}\frac{\partial}{\partial \mathbf{r}}\frac{(E_jE_i)_{\omega, \mathbf{k}}}{4\pi} = i\,(\omega/2\pi)\,\varepsilon^a_{ij}(\omega, \mathbf{k})\,(E_jE_i)_{\omega, \mathbf{k}}. \tag{3.6}$$

The first term on the left-hand side of this equation is the time derivative of the spectral density of the energy of the electromagnetic field:

$$W(\omega, \mathbf{k}) \equiv \frac{\partial \omega M^H_{ij}(\omega, \mathbf{k})}{\partial \omega}\frac{(E_jE_i)_{\omega, \mathbf{k}}}{4\pi}, \tag{3.7}$$

and the second is the divergence of the spectral density of the energy flux (the Poynting vector):

$$S(\omega, \mathbf{k}) \equiv -\frac{\partial \omega M^H_{ij}(\omega, \mathbf{k})}{\partial \mathbf{k}}\frac{(E_jE_i)_{\omega, \mathbf{k}}}{4\pi}. \tag{3.8}$$

The right-hand side of (3.6) arises from the first term of the right-hand side of (2.23) by means of Eqs. (2.15) and (2.15a) and corresponds to the damping of the electromagnetic waves:

$$\gamma(\omega,\ \mathbf{k})\, W(\omega,\ \mathbf{k}) \equiv -i\omega\varepsilon^{a}_{ij}(\omega,\ \mathbf{k})\, \frac{(E_j E_i)_{\omega,\mathbf{k}}}{4\pi}. \tag{3.9}$$

Thus, Eq. (3.6) can be written in a slightly different form:

$$\frac{\partial W(\omega,\ \mathbf{k})}{\partial t} + \operatorname{div} \mathbf{S}(\omega,\ \mathbf{k}) = -2\gamma(\omega,\ \mathbf{k})\, W(\omega,\ \mathbf{k}). \tag{3.10}$$

From the energy conservation equation (3.6) we obtain an obvious property of the spectral functions in the framework of linear electrodynamics. Namely, such functions for different ω and $\mathbf{k}$ are not correlated. Therefore, because of (3.5) and the absence of correlations between waves with different ω and $\mathbf{k}$, we obtain the following relation for the correlation function of three amplitudes of the electric fields of linear electrodynamics:

$$\langle E_i(\omega,\ \mathbf{k})\, E_j(\omega',\ \mathbf{k}')\, E_s(\omega'',\ \mathbf{k}'')\rangle = 0. \tag{3.11}$$

The correlation function of four amplitudes of linear electrodynamics can be written in the form

$$\begin{aligned}
&\langle E_i(\omega,\ \mathbf{k})\, E_s(\omega',\ \mathbf{k}')\, E_j(\omega'',\ \mathbf{k}'')\, E_r(\omega''',\ \mathbf{k}''')\rangle = \langle E_i(\omega,\ \mathbf{k})\, E_s(\omega',\ \mathbf{k}')\rangle \times \\
&\times \langle E_j(\omega'',\ \mathbf{k}'')\, E_r(\omega''',\ \mathbf{k}''')\rangle + \langle E_i(\omega,\ \mathbf{k})\, E_j(\omega'',\ \mathbf{k}'')\rangle \langle E_s(\omega',\ \mathbf{k}')\, E_r(\omega''',\ \mathbf{k}''')\rangle + \\
&+ \langle E_i(\omega,\ \mathbf{k})\, E_r(\omega''',\ \mathbf{k}''')\rangle \langle E_s(\omega',\ \mathbf{k}')\, E_j(\omega'',\ \mathbf{k}'')\rangle + \{\langle E_i(\omega,\ \mathbf{k})\, E_s(\omega',\ \mathbf{k}') \times \\
&\times E_j(\omega'',\ \mathbf{k}'')\, E_r(\omega''',\ \mathbf{k}''')\rangle - \langle E_i(\omega,\ \mathbf{k})\, E_s(\omega',\ \mathbf{k}')\rangle \langle E_j(\omega'',\ \mathbf{k}'')\, E_r(\omega''',\ \mathbf{k}''')\rangle\} \times \\
&\times (\delta_{\omega,\omega''}\delta_{\mathbf{k},\mathbf{k}''})(\delta_{\omega',\omega'''}\delta_{\mathbf{k}',\mathbf{k}'''})(\delta_{\omega,-\omega'}\delta_{\mathbf{k},-\mathbf{k}'}) + \{\langle E_i(\omega,\ \mathbf{k})\, E_s(\omega',\ \mathbf{k}')\, E_j(\omega'',\ \mathbf{k}'') \times \\
&\times E_r(\omega''',\ \mathbf{k}''')\rangle - \langle E_i(\omega,\ \mathbf{k})\, E_j(\omega'',\ \mathbf{k}'')\rangle \langle E_s(\omega',\ \mathbf{k}')\, E_r(\omega''',\ \mathbf{k}''')\rangle\} \times \\
&\times (\delta_{\omega,\omega'''}\delta_{\mathbf{k},\mathbf{k}'''})(\delta_{\omega'',\omega'}\delta_{\mathbf{k}'',\mathbf{k}'})(\delta_{\omega,-\omega''}\delta_{\mathbf{k},-\mathbf{k}''}) + \{\langle E_i(\omega,\ \mathbf{k})\, E_s(\omega',\ \mathbf{k}')\, E_j(\omega'',\ \mathbf{k}'') \times \\
&\times E_r(\omega''',\ \mathbf{k}''')\rangle - \langle E_i(\omega,\ \mathbf{k})\, E_r(\omega''',\ \mathbf{k}''')\rangle \langle E_s(\omega',\ \mathbf{k}')\, E_j(\omega'',\ \mathbf{k}'')\rangle\} \times \\
&\times (\delta_{\omega,\omega'}\delta_{\mathbf{k},\mathbf{k}'})(\delta_{\omega''',\omega''}\delta_{\mathbf{k}''',\mathbf{k}''})(\delta_{\omega,-\omega'''}\delta_{\mathbf{k},-\mathbf{k}'''}),
\end{aligned} \tag{3.12}$$

where $\delta_{a,b}$ is the Kronecker delta,

$$\delta_{a,b} = \begin{cases} 1, & a = b \\ 0, & a \neq b, \end{cases} \tag{3.13}$$

whose appearance in formula (3.12) reflects the fact that the amplitudes of identical electromagnetic waves are correlated, or rather, amplitudes with the same arguments $(\omega,\ \mathbf{k})$ are correlated.

In nonlinear electrodynamics, formulas (3.11) and (3.12) do not, strictly speaking, hold, since it follows from the very meaning of the nonlinear interaction of different waves that their amplitudes are correlated. However, in the case of weak nonlinearity, which is all that we consider here, when only a few terms of the series in powers of the field need be retained on the right-hand side of the relations (2.6), (2.12), and (2.23), such a difference is, first, relatively small and, secondly, bearing in mind the difference between the corresponding formulas of nonlinear electrodynamics and (3.11) and (3.12), they can be replaced by more exact equations. For our subsequent derivation of the equation that describes the evolution of the spectral functions of the field, we must obtain a formula for the correlation function of three amplitudes, this making (3.11) more precise for the case when in the nonlinear field equation

(2.12) one allows for the first nonlinear term, which is quadratic in the amplitudes of the electric field:

$$M_{ij}(\omega, \mathbf{k}) E_j(\omega, \mathbf{k}) = -\int d\omega' d\mathbf{k}' \varepsilon_{ijs}(\omega, \mathbf{k}; \omega', \mathbf{k}') E_j(\omega - \omega', \mathbf{k} - \mathbf{k}') E_s(\omega', \mathbf{k}'). \tag{3.14}$$

The right-hand side of Eq. (3.14) corresponds to effects of the nonlinear interaction of waves. Namely, if the right-hand side of this equation were to vanish, its solutions would correspond to noninteracting waves:

$$E_i(\omega, \mathbf{k}) \equiv E_i^0(\omega, \mathbf{k}), \tag{3.15}$$

where the superscript 0 indicates an amplitude E_i^0 that is a solution of Eq. (2.13) of linear electrodynamics. In this connection, we rewrite (3.14) in a slightly different form:

$$E_i(\omega, \mathbf{k}) = E_i^0(\omega, \mathbf{k}) - A_{ia}(\omega, \mathbf{k}) \int d\omega' d\mathbf{k}' \varepsilon_{ajs}(\omega, \mathbf{k}; \omega', \mathbf{k}') E_j(\omega - \omega', \mathbf{k} - \mathbf{k}') E_s(\omega', \mathbf{k}'), \tag{3.16}$$

where $A_{ij}(\omega, \mathbf{k})$ is the inverse of the Maxwell tensor (3.2):

$$A_{ij}(\omega, \mathbf{k}) \equiv M_{ij}^{-1}(\omega, \mathbf{k}) = \left\{ \varepsilon_{ij}(\omega, \mathbf{k}) - \frac{c^2 k^2}{\omega^2}\left(\delta_{ij} - \frac{k_i k_j}{k^2}\right)\right\}^{-1} \tag{3.17}$$

in the sense of the ordinary definition

$$A_{ij}(\omega, \mathbf{k}) M_{js}(\omega, \mathbf{k}) = \delta_{is}. \tag{3.18}$$

For brevity, we shall simply refer to this tensor as the inverse tensor.

The field equation in the form (3.16) (or, more general form allowing for not only the quadratic but also the higher terms in the powers of the field) is convenient for solving (2.12) in perturbation theory, in which the field $\mathbf{E}^0(\omega, \mathbf{k})$ can be taken as the zeroth approximation. For our purposes, the first approximation is sufficient, this corresponding to the substitution into the second (nonlinear) term of the right-hand side of (3.16) of the fields $\mathbf{E}^0$ of the zeroth approximation. Such a procedure can be used to write down the desired relation that generalizes (3.11) with an accuracy to terms of the fourth power in the amplitudes of the electric field:

$$\begin{aligned}
&\langle E_i(\omega, \mathbf{k}) E_s(\omega', \mathbf{k}') E_j(\omega'', \mathbf{k}'')\rangle = \langle E_i^0(\omega, \mathbf{k}) E_s^0(\omega', \mathbf{k}') E_j^0(\omega'', \mathbf{k}'')\rangle - \\
&- \int d\omega_1 d\mathbf{k}_1 \{A_{ia}(\omega, \mathbf{k}) \varepsilon_{abc}(\omega, \mathbf{k}; \omega_1, \mathbf{k}_1) \langle E_b^0(\omega - \omega_1, \mathbf{k} - \mathbf{k}_1) E_c^0(\omega_1, \mathbf{k}_1) \times \\
&\times E_s^0(\omega', \mathbf{k}') E_j^0(\omega'', \mathbf{k}'')\rangle + A_{sc}(\omega', \mathbf{k}') \varepsilon_{cba}(\omega', \mathbf{k}'; \omega_1, \mathbf{k}_1) \langle E_b^0(\omega' - \omega_1, \mathbf{k}' - \mathbf{k}_1) \times \\
&\times E_a^0(\omega_1, \mathbf{k}_1) E_i^0(\omega, \mathbf{k}) E_j^0(\omega'', \mathbf{k}'')\rangle + A_{jb}(\omega'', \mathbf{k}'') \varepsilon_{bca}(\omega'', \mathbf{k}''; \omega_1, \mathbf{k}_1) \times \\
&\times \langle E_c^0(\omega'' - \omega_1, \mathbf{k}'' - \mathbf{k}_1) E_a^0(\omega_1, \mathbf{k}_1) E_i^0(\omega, \mathbf{k}) E_s^0(\omega', \mathbf{k}')\rangle\}.
\end{aligned} \tag{3.19}$$

In (3.19) the first term on the right vanishes in accordance with (3.11) since the vectors $\mathbf{E}^0$ are the amplitudes of uncorrelated waves. The remaining terms of the right-hand side of (3.19) contain, like (3.12), the product of four amplitudes of the electric fields of the zeroth approximation (3.15). Therefore, in obtaining an equation for the evolution of the spectral functions of the field the terms containing the third and fourth power of the field in Eq. (3.1) [the second and the third term on the right-hand side of (3.1)] lead to effects that are, in general, of the same order of magnitude.

Restricting ourselves in the desired equation to an accuracy to the fourth power of the field, we need not make more precise formula (3.12) for the correlation function of four amplitudes. Moreover, at this accuracy, we need not distinguish between the amplitudes of the electric fields E_i and E_i^0 in the bilinear combinations of the spectral functions of the field. Finally, we restrict our treatment to the study of nonlinear interaction of different waves and we shall not allow for the possible self-interaction of a wave. Formally, this is reflected in our ignoring the last three of the six terms on the right-hand side of (3.12) which contain Kronecker deltas of the frequencies and the wave vectors and are responsible for such an effect. Therefore, without allowance for the self-interaction of waves, we replace formula (3.12) by

$$\langle E_i(\omega, \mathbf{k}) E_s(\omega', \mathbf{k}') E_j(\omega'', \mathbf{k}'') E_r(\omega''', \mathbf{k}''')\rangle = (E_iE_s)_{\omega,\mathbf{k}} (E_jE_r)_{\omega'',\mathbf{k}''}\delta(\omega+\omega') \times \times \delta(\mathbf{k}+\mathbf{k}')\delta(\omega''+\omega''')\delta(\mathbf{k}''+\mathbf{k}''') + (E_iE_j)_{\omega,\mathbf{k}}(E_sE_r)_{\omega',\mathbf{k}'}\delta(\omega+\omega'') \times \times \delta(\mathbf{k}+\mathbf{k}'')\delta(\omega'+\omega''')\delta(\mathbf{k}'+\mathbf{k}''') + (E_iE_r)_{\omega,\mathbf{k}}(E_sE_j)_{\omega',\mathbf{k}'}\delta(\omega+\omega''')\delta(\mathbf{k}+\mathbf{k}''') \times \times \delta(\omega'+\omega'')\delta(\mathbf{k}'+\mathbf{k}''). \tag{3.20}$$

Taking into account formula (3.20) and our above remarks, we rewrite the relation (3.19) for the correlation function of three amplitudes in the form

$$\langle E_i(\omega, \mathbf{k}) E_s(\omega', \mathbf{k}') E_j(\omega'', \mathbf{k}'')\rangle = -\delta(\omega+\omega'+\omega'')\delta(\mathbf{k}+\mathbf{k}'+\mathbf{k}'') \times \times \{A_{ia}(\omega, \mathbf{k})[\varepsilon_{acb}(\omega, \mathbf{k}; -\omega'', -\mathbf{k}'') + \varepsilon_{abc}(\omega, \mathbf{k}; -\omega', -\mathbf{k}')](E_sE_c)_{\omega',\mathbf{k}'} \times \times (E_jE_b)_{\omega'',\mathbf{k}''} + A_{sc}(\omega', \mathbf{k}')[\varepsilon_{cab}(\omega', \mathbf{k}'; -\omega'', -\mathbf{k}'') + \varepsilon_{cba}(\omega', \mathbf{k}'; -\omega, -\mathbf{k})] \times \times (E_iE_a)_{\omega,\mathbf{k}}(E_jE_b)_{\omega'',\mathbf{k}''} + A_{jb}(\omega'', \mathbf{k}'')[\varepsilon_{bac}(\omega'', \mathbf{k}''; -\omega', -\mathbf{k}') + \varepsilon_{bca}(\omega'', \mathbf{k}''; -\omega, -\mathbf{k})](E_iE_a)_{\omega,\mathbf{k}}(E_sE_c)_{\omega',\mathbf{k}'} + [\delta(\omega)\delta(\mathbf{k})A_{ia}(0, 0)(E_sE_j)_{\omega',\mathbf{k}'} + \delta(\omega')\delta(\mathbf{k}') \times \times A_{sa}(0, 0)(E_iE_j)_{\omega,\mathbf{k}} + \delta(\omega'')\delta(\mathbf{k}'')A_{ja}(0, 0)(E_iE_s)_{\omega,\mathbf{k}}] \times \times \int d\omega_1 d\mathbf{k}_1 \varepsilon_{abc}(0, 0; \omega_1, \mathbf{k}_1)(E_cE_b)_{\omega_1,\mathbf{k}_1}\}. \tag{3.21}$$

The last square brackets on the right-hand side of (3.21) contains inverse tensors with zero values of the arguments and for the problems of the nonlinear interaction of electromagnetic waves with which we are concerned they are not important, since ω and $\mathbf{k}$ do not vanish simultaneously for waves. We note in passing that similar expressions, which also do not correspond to nonlinear interaction of waves, also arise when (3.20) is substituted into Eq. (3.1). We shall also omit them. As a result, substituting (3.20) and (3.21) instead of the correlation functions of four and three amplitudes into Eq. (3.1), we obtain [12]

$$\frac{\partial}{\partial t}\frac{1}{\omega}\frac{\partial \omega M_{ij}^H(\omega, \mathbf{k})}{\partial \omega}(E_jE_i)_{\omega,\mathbf{k}} - \frac{\partial}{\partial \mathbf{r}}\frac{1}{\omega}\frac{\partial \omega M_{ij}^H(\omega, \mathbf{k})}{\partial \mathbf{k}}(E_jE_i)_{\omega,\mathbf{k}} = 2i\varepsilon_{ij}^a(\omega, \mathbf{k})(E_jE_i)_{\omega,\mathbf{k}} + + \mathrm{Im}\int d\omega' d\mathbf{k}' \{A_{ia}^*(\omega, \mathbf{k}) S_{ijs}(\omega, \mathbf{k}; \omega', \mathbf{k}') S_{abc}^*(\omega, \mathbf{k}; \omega', \mathbf{k}')(E_sE_c)_{\omega',\mathbf{k}'}(E_jE_b)_{\omega-\omega',\mathbf{k}-\mathbf{k}'} + + 2A_{jb}(\omega-\omega', \mathbf{k}-\mathbf{k}') S_{ijs}(\omega, \mathbf{k}; \omega', \mathbf{k}') S_{bca}(\omega-\omega', \mathbf{k}-\mathbf{k}'; \omega, \mathbf{k}) \times \times (E_sE_c)_{\omega',\mathbf{k}'}(E_aE_i)_{\omega,\mathbf{k}} - 2V_{isac}(\omega, \mathbf{k}; \omega', \mathbf{k}')(E_aE_i)_{\omega,\mathbf{k}}(E_cE_s)_{\omega',\mathbf{k}'}\}. \tag{3.22}$$

Here, the tensors $M_{ij}^H(\omega, \mathbf{k})$, $(E_jE_i)_{\omega,\mathbf{k}}$, $\varepsilon_{ij}^a(\omega, \mathbf{k})$ and $A_{ij}(\omega, \mathbf{k})$ are determined by Eqs. (3.3), (3.4), (2.15), and (3.17), respectively, and we have used the notation

$$V_{isab}(\omega, \mathbf{k}; \omega', \mathbf{k}') \equiv \varepsilon_{isab}(\omega, \mathbf{k}; \omega+\omega', \mathbf{k}+\mathbf{k}'; \omega', \mathbf{k}') + \varepsilon_{isba}(\omega, \mathbf{k}; \omega+\omega', \mathbf{k}+\mathbf{k}'; \omega, \mathbf{k}) = = 2\varepsilon_{isba}(\omega, \mathbf{k}; \omega+\omega', \mathbf{k}+\mathbf{k}'; \omega, \mathbf{k}) = 2\varepsilon_{iasb}(\omega, \mathbf{k}; 0, 0; \omega', \mathbf{k}'); \tag{3.23}$$

$$S_{ijs}(\omega, \mathbf{k}; \omega', \mathbf{k}') \equiv \varepsilon_{ijs}(\omega, \mathbf{k}; \omega', \mathbf{k}') + \varepsilon_{isj}(\omega, \mathbf{k}; \omega-\omega', \mathbf{k}-\mathbf{k}') = 2\varepsilon_{ijs}(\omega, \mathbf{k}; \omega', \mathbf{k}'). \tag{3.24}$$

The last equations in (3.23) and (3.24) are due to the symmetry properties (2.7) and (2.8) of the nonlinear tensors.

Having in mind the procedure for obtaining Eq. (3.22), in accordance with which the equation for the bilinear combinations of the amplitudes (3.23) arises from (2.12) by scalar multiplication by the electric field $E_i^*(\omega, k)$, it is obvious that for different polarizations of such a field and mode of the electromagnetic wave we obtain, in general, different concrete equations. Namely, suppose that an almost monochromatic wave which participates in a nonlinear interaction is a high-frequency transverse electromagnetic oscillation (for example, light) polarized linearly along a unit vector **e**. Then multiplication of (2.12) by

$$\mathbf{E}^*(\omega, \mathbf{k}) = E^*(\omega, \mathbf{k})\mathbf{e} \tag{3.25}$$

leads to Eq. (3.22), which describes the evolution of the energy density of an electromagnetic wave with the spectral function

$$(E_j E_i)_{\omega, \mathbf{k}} = (E_{tr}^2)_{\omega, \mathbf{k}} e_j e_i, \tag{3.26}$$

where the symbol tr on the right-hand side emphasizes the already mentioned transversality property:

$$\mathbf{ke} = 0. \tag{3.27}$$

This means that on the left- and right-hand sides of Eq. (3.22) the spectral function $(E_j E_i)_{\omega,k}$ must be replaced by the right-hand side of Eq. (3.26). The inverse tensor $A_{i\alpha}^*(\omega, k)$ on the right-hand side of (3.22) must, in accordance with formulas (3.14) and (3.16) and the transversality of the polarization (3.27) of the electromagnetic oscillation under consideration, also be transverse with respect to the wave vector **k**, and ω and k in this tensor must be related by the appropriate dispersion law [for light $\omega = ck$; for more detail see Chapter III, § 10, in particular, Eq. (10.7)]. If the light is not polarized but is natural light in which the different polarizations are represented equally, then, summing over such polarizations in (3.26), we find that Eq. (3.22) then contains a spectral function of the form [see Chapter III, §10 (10.2)]

$$(E_j E_i)_{\omega, \mathbf{k}} = \frac{1}{2} (E_{tr}^2)_{\omega, \mathbf{k}} \left(\delta_{ij} - \frac{k_i k_j}{k^2}\right). \tag{3.28}$$

In the same way we describe the evolution of partly polarized light, which in a certain sense is intermediate between (3.26) and (3.28) [see Chapter II, § 6 (6.76)]. The special example we have considered explains the form of the spectral functions that occur in Eq. (3.22) for not only the case of the polarization (3.26), when several different polarizations can correspond to one and the same dispersion law, but also for the "nondegenerate" polarization (3.28), which corresponds to a unique dispersion law and vice versa. As we shall see below, spectral functions of this last type can also be characterized in the general case of a magnetoactive plasma by the tensor part of the reciprocal Maxwell tensor [see, for example, (4.3) or, for more detail, Chapter IV, § 17, in particular, (17.50)].

Returning to the deciphering of Eq. (3.22) for the characteristic plasma oscillations with different polarization, we emphasize once more the key role played by the electric field strength vector $E_i^*(\omega, k)$, with which we multiplied Eq. (2.12) in the derivation of (3.22). Namely, the direction of the vector $E_i^*(\omega, k)$ (which is, in genral, complex) completely determines the type of polarization of the electromagnetic oscillation with spectral function $(E_j E_i)_{\omega, \mathbf{k}}$, which evolves in accordance with Eq. (3.22). In this sense, we shall say that the subscript i in the nonlinear equation (3.26) "recalls" the polarization of the electromagnetic wave that participates in the nonlinear interaction.

Equation (3.22) is the main result of this chapter. It will be used in the kinetic theory of the nonlinear interaction of electromagnetic waves. This equation, which, as we shall see below (Chapters II-V), enables us to obtain definite kinetic equations for the various types of electromagnetic wave, can be naturally called the generalized kinetic field equation in statistical nonlinear electrodynamics.

§4. Discussion of Results. Equation for the Frequencies of Interacting Electromagnetic Waves

We should point out that Eq. (3.22) is not complete in a certain sense. In particular, it is well known that kinetic equations for waves contain an inhomogeneous part that does not depend on the intensity of the oscillations of the electromagnetic field and is due to spontaneous emmission [see, for example, § 36 on p. 565, formula (36.17) in [17], which gives the radiative transfer equation; § 70, p. 396, formula (70.3) of [18]).

The absence in our equation (3.22) of a similar inhomogeneous part renders it, strictly speaking, suitable for the description of only those processes in which the wave energy appreciably exceeds the level of thermal noise. It is problems of just this kind that arise under the conditions of excitation of oscillatory instabilities in a plasma and also in the interaction of strong external waves with artificially excited plasma oscillations. Therefore, the main part of our review (Chapters II-V) will be devoted to the application of (3.22) to the description of a weakly turbulent plasma, or, in other words, the scattering of electromagnetic waves. In fact, on the basis of the generalized kinetic field equation (3.22) one can also obtain results for the interaction of waves with a thermal energy density. However, to do this one must invoke arguments that, although they are trivial, are nevertheless intuitive.

A generalized kinetic equation for waves leading to the allowance of spontaneous radiation was derived for the case of longitudinal plasma oscillations by one of the authors in [19]. In this paper, as also in the later paper of Rostoker and Matsuda [20], a generalized kinetic equation for the waves was obtained by developing Bogolyubov's dynamical method [21, 22], which had earlier been applied to other problems of statistical mechanics. A method similar in spirit to our derivation of Eq. (3.22) has been used by Kadomtsev [4] and Petviashvili [23] to describe weak stationary turbulence due to longitudinal waves in a plasma.

Equation (3.22) can also be augmented from another side. Indeed, from a purely electrodynamical point of view it is, as we have already mentioned, the energy conservation law of interacting magnetic waves. The energy corresponding to an electromagnetic wave of given type may change either because of linear effects determined by the anti-Hermitian part of the permittivity and related to the imaginary corrections to the frequencies $\omega(\mathbf{k})$ of the waves, or as a result of nonlinear effects described by the curly brackets on the right-hand side of (3.22). In this sense, these two qualitatively different effects compete in such a way that the dominant effect is the one that proceeds faster. And Eq. (3.22) enables one to calculate explicitly the characteristic time of the corresponding nonlinear process and to determine its energy evolution. The dispersion properties of an electromagnetic wave that participates in a nonlinear interaction are lost from view in a theory that uses only the generalized kinetic field equation (3.22). However, using the methods developed in § 2 and § 3, this gap in the description of the nonlinear properties of waves can be readily eliminated. We multiply the nonlinear equation for the Fourier components of the electric field (2.12) scalarly by $E_j^*(\omega, \mathbf{k})$ and add the resulting equation term by term to its complex conjugate. We then obtain

$$\left\{2\frac{c^2k^2}{\omega^2}\left(\delta_{ij}-\frac{k_ik_j}{k^2}\right)-[\varepsilon_{ij}(\omega,\ \mathbf{k})+\varepsilon^*_{ji}(\omega,\ \mathbf{k})]\right\}E_j(\omega,\ \mathbf{k})E_i^*(\omega,\ \mathbf{k})=$$
$$=\sum_{n=2}^{\infty}\int d\omega_1 d\mathbf{k}_1\ldots d\omega_{n-1}d\mathbf{k}_{n-1}\varepsilon_{ij(1)\ldots j(n)}(\omega,\ \mathbf{k};\ \omega_1,\ \mathbf{k}_1;\ \ldots;\ \omega_{n-1},\ \mathbf{k}_{n-1})E_i^*(\omega,\ \mathbf{k})\times$$
$$\times E_{j(1)}(\omega-\omega_1,\ \mathbf{k}-\mathbf{k}_1)\ldots E_{j(n-1)}(\omega_{n-2}-\omega_{n-1},\ \mathbf{k}_{n-2}-\mathbf{k}_{n-1})E_{j(n)}(\omega_{n-1},\ \mathbf{k}_{n-1})+$$
$$+\sum_{n=2}^{\infty}\int d\omega_1 d\mathbf{k}_1\ldots d\omega_{n-1}d\mathbf{k}_{n-1}\varepsilon^*_{ij(1)\ldots j(n)}(\omega,\ \mathbf{k};\ \omega_1,\ \mathbf{k}_1;\ \ldots;\ \omega_{n-1},\ \mathbf{k}_{n-1})E_i(\omega,\ \mathbf{k})\times$$
$$\times E^*_{j(1)}(\omega-\omega_1,\mathbf{k}-\mathbf{k}_1)\ldots E^*_{j(n-1)}(\omega_{n-2}-\omega_{n-1},\mathbf{k}_{n-2}-\mathbf{k}_{n-1})E^*_{j(n)}(\omega_{n-1},\mathbf{k}_{n-1}),\tag{4.1}$$

which goes over in the case of vanishing right-hand side, with allowance for (2.15) and (2.15a), into

$$\left\{\varepsilon_{ij}^{H}(\omega,\ \mathbf{k}) - \frac{c^2k^2}{\omega^2}\left(\delta_{ij} - \frac{k_ik_j}{k^2}\right)\right\} E_j(\omega,\ \mathbf{k}) E_i^*(\omega,\ \mathbf{k}) = 0. \tag{4.2}$$

From (4.2) there follows Eq. (2.16), which determines the real part of the frequency $\omega(\mathbf{k})$ of an electromagnetic wave that does not participate in a nonlinear interaction, i.e., in the linear approximation. Such a transition can be followed most simply by noting that the spectral function of the field $(E_jE_i)_{\omega,\mathbf{k}}$, just like $E_j(\omega,\mathbf{k})\,E_i^*(\omega,\mathbf{k})$, is proportional in the case of non-degenerate polarization to the Hermitian part of the inverse tensor (3.17) [compare with (3.28)]

$$(E_jE_i)_{\omega,\mathbf{k}} = \mathrm{const}\, e_{ji}(\omega,\ \mathbf{k}). \tag{4.3}$$

Here, the constant is a scalar proportional to the spectral density of the energy of the electromagnetic wave [cf. (3.7)] with spectral function $(E_jE_i)_{\omega,\mathbf{k}}$

$$W(\mathbf{k}) = (2\pi)^3 \int_0^\infty d\omega \frac{(E_jE_i)_{\omega,\mathbf{k}}}{4\pi} \frac{\partial \omega M_{ij}^H(\omega,\ \mathbf{k})}{\partial\omega}, \tag{4.4}$$

and the matrix e_{ji} is the tensor part of the Hermitian inverse tensor $A_{ji}^H(\omega,\mathbf{k})$

$$e_{ji}(\omega,\ \mathbf{k}) = \lim_{\Delta\to 0} \Delta'(\omega,\ \mathbf{k}) A_{ji}^H(\omega,\ \mathbf{k}), \tag{4.5}$$

where for brevity we have denoted by $\Delta'(\omega,\mathbf{k})$ the left-hand side of Eq. (2.16):

$$\Delta'(\omega,\ \mathbf{k}) \equiv \left| \varepsilon_{ij}^H(\omega,\ \mathbf{k}) - \frac{c^2k^2}{\omega^2}\left(\delta_{ij} - \frac{k_ik_j}{k^2}\right)\right|. \tag{4.6}$$

Substituting (4.3) into Eq. (4.2) averaged over the ensemble, we obtain the dispersion relation (2.16),

$$\Delta'(\omega,\ \mathbf{k}) = 0, \tag{4.7}$$

since by the definition of the inverse tensor (3.18),

$$M_{ij}^H(\omega,\ \mathbf{k})\, e_{ji}(\omega,\ \mathbf{k}) = 3\Delta'(\omega,\ \mathbf{k}). \tag{4.8}$$

As in the derivation of the generalized kinetic equation (3.22), we shall henceforth retain only the first two terms of the series on the right-hand side of (4.1):

$$M_{ij}^H(\omega,\ \mathbf{k}) E_j(\omega,\ \mathbf{k}) E_i^*(\omega,\ \mathbf{k}) = -\mathrm{Re}\int d\omega' d\mathbf{k}' \varepsilon_{ijs}(\omega,\ \mathbf{k};\ \omega',\ \mathbf{k}') E_i^*(\omega,\ \mathbf{k})\, E_j(\omega-\omega',\ \mathbf{k}-\mathbf{k}')\, E_s(\omega',\ \mathbf{k}') -$$
$$-\mathrm{Re}\int d\omega' d\mathbf{k}' d\omega'' d\mathbf{k}'' \varepsilon_{ijsr}(\omega,\mathbf{k};\omega',\mathbf{k}';\omega'',\mathbf{k}'')\, E_i^*(\omega,\ \mathbf{k}) E_j(\omega-\omega',\ \mathbf{k}-\mathbf{k}') E_s(\omega'-\omega'',\ \mathbf{k}'-\mathbf{k}'') E_r(\omega'',\ \mathbf{k}''). \tag{4.9}$$

Then, averaging (4.9) over the statistical ensemble by means of the rules (3.20) and (3.21) discussed in §3, i.e., ignoring the self-interaction of the waves and omitting in (3.21) the last square brackets with simultaneously vanishing frequencies and wave vectors, we obtain the analog of (4.2) in nonlinear statistical electrodynamics:

$$(E_jE_i)_{\omega,\mathbf{k}} M_{ij}^H(\omega,\ \mathbf{k}) + \mathrm{Re}\,\{(E_aE_i)_{\omega,\mathbf{k}} \int d\omega' d\mathbf{k}' (E_sE_c)_{\omega',\mathbf{k}'}\ [V_{icas}(\omega,\ \mathbf{k};\ \omega',\ \mathbf{k}') -$$
$$- A_{jb}(\omega-\omega',\ \mathbf{k}-\mathbf{k}')\, S_{ijs}(\omega,\mathbf{k};\omega',\mathbf{k}')\, S_{bca}(\omega-\omega',\ \mathbf{k}-\mathbf{k}';\ \omega,\ \mathbf{k})]\} = 0. \tag{4.10}$$

At the same time it is expedient to use the identity

$$\mathrm{Re}\, S_{ijs}(\omega,\mathbf{k};\omega',\mathbf{k}')\, S_{abc}^*(\omega,\mathbf{k};\omega',\mathbf{k}')\, A_{ia}^*(\omega,\mathbf{k}) (E_jE_b)_{\omega-\omega',\mathbf{k}-\mathbf{k}'} (E_sE_c)_{\omega',\mathbf{k}'} \equiv 0, \tag{4.11}$$

which is obvious because electromagnetic waves with frequency ω and wave vector k satisfy (4.2) in the linear approximation. All the tensors in (4.10) have been previously defined and do not require further explanation. Replacing in (4.10) the spectral function of the field $(E_j E_i)_{\omega,\mathbf{k}}$ by the right-hand side of (4.3), we finally obtain

$$\Delta'(\omega, \mathbf{k}) + \mathrm{Re}\Big\{e_{ai}(\omega, \mathbf{k}) \int d\omega' d\mathbf{k}' (E_s E_c)_{\omega', \mathbf{k}'} [V_{icas}(\omega, \mathbf{k}; \omega', \mathbf{k}') - A_{jb}(\omega - \omega', \mathbf{k} - \mathbf{k}') S_{ijs}(\omega, \mathbf{k}; \omega', \mathbf{k}') S_{bca}(\omega - \omega', \mathbf{k} - \mathbf{k}'; \omega, \mathbf{k})]\Big\} = 0. \quad (4.12)$$

In the light of our above remarks, Eq. (4.12) may naturally be called the dispersion equation in nonlinear statistical electrodynamics. It enables us to calculate the real corrections to the frequency $\omega(\mathbf{k})$ of the electromagnetic wave that depend on the spectral function of the field.

Summarizing, we emphasize once more the fundamental role of the generalized kinetic equation (3.22) and the nonlinear dispersion relation (4.12) in a theory devoted to weakly nonlinear electromagnetic effects in a medium with the material equation (1.8). Under the conditions formulated in this chapter, these equations are fully adequate to describe nonlinear effects in an arbitrary material medium. However, all the subsequent exposition will be devoted to a study, using Eqs. (3.22) and (4.12), of the nonlinear interaction of characteristic oscillations of a homogeneous high-temperature plasma that is isotropic (Chapters II and III) or magnetoactive (Chapters IV and V), for it is precisely in plasma theory, in our opinion, that the possibilities inherent in these equations are most fully revealed.

CHAPTER II

MULTI-INDEX PERMITTIVITY TENSORS OF A HOMOGENEOUS ISOTROPIC PLASMA

In the present chapter we use Vlasov's self-consistent system of equations to obtain explicit expressions for the multi-index permittivity tensors of an isotropic plasma, i.e., a plasma without external fields; these form the basis of the nonlinear theory. We study in detail the first two nonlinear tensors, S_{ijs} and V_{iajb}, in the generalized kinetic equation (3.22) and the nonlinear dispersion equation (4.12), which we have derived in Chapter I.

The present chapter consists of four sections (§5-8). In §5, we find a solution, which is nonlinear in the field, of the kinetic equation in an isotropic collisionless plasma, this solution giving rise to nonlinearity of the material equation. The solution is sought as a power series in the electric field. The procedure for obtaining such a solution reduces to iteration of the recursion relation for the nonequilibrium corrections to the distribution function in an equilibrium state of the plasma unperturbed by electromagnetic fields. Such a recursion relation arises from the chain of coupled equations to which the kinetic equation is reduced. The possibility of making an expansion in the field is due to the assumption of smallness of the Lorentz force, which acts on a particle and appears in one of the terms of the kinetic equation responsible for the interaction. The smallnesss of this term guarantees that the oscillations of the plasma interact weakly. The main result of §5 is the expression (5.32) for the total distribution function of the plasma particles in terms of the equilibrium distribution fuction f_0 and the electric fields. The generality of the equation is restricted by the assumption that the field is switched on adiabatically in the infinitely distant past.

In §6, we use the general expression obtained in §5 for the distribution function of the plasma particles to find the explicit form of the multi-index tensors $\varepsilon_{ij(1)\cdots j(n)}$. We study in detail the simplest example of such tensors – the permittivity tensor $\varepsilon_{ij}(\omega, \mathbf{k})$ of an isotropic plasma. The majority of the formulas given here have an ancillary nature (see, for example, [1]). Namely, we give here expressions for the spectra and damping constants of electron Langmuir, ion-acoustic, and transverse (nonelectrostatic) oscillations. We discuss in detail the expressions for the spectral functions of the electric fields in an isotropic plasma in the transparency regions and we give the explicit form of the inverse tensor A_{ij}, which occurs in the basic equations (3.22) and (4.12) of the nonlinear theory.

The last two sections (§7 and 8) of this chapter are devoted to a detailed study of the first two nonlinear tensors; $S_{ijs}(\omega, \mathbf{k}; \omega', \mathbf{k}')$ (§7) and $V_{i\alpha jb}(\omega, \mathbf{k}; \omega', \mathbf{k}')$ (§8). The features of the structure of these tensors determine essentially the basic characteristic features of nonlinear processes in the plasma. Generally speaking, this is already evident from the basic equations (3.22) and (4.12). But it should also be emphasized that the role of each of the tensors S and V is different in nonlinear processes of different types. Namely, all the nonlinear interaction processes studied in the present paper in which three plasma oscillations participate can be split into two classes. One of them constitutes decay processes or, in other words, processes of induced combination scattering. All three of the plasma oscillations participating in the decay are weakly damped, i.e., they are characteristic plasma oscillations. Their spectra $\omega(\mathbf{k})$, $\omega'(\mathbf{k}')$, and $\omega''(\mathbf{k}'')$ and wave vectors $\mathbf{k}$, $\mathbf{k}'$, $\mathbf{k}''$ satisfy characteristic conservation laws (decay conditions): $\omega(\mathbf{k}) = \omega'(\mathbf{k}') + \omega''(\mathbf{k}'')$, $\mathbf{k} = \mathbf{k}' + \mathbf{k}''$.

In an isotropic plasma, the total number of decay processes is restricted by selection rules that arise because of the nondecay nature of the spectra of characteristic oscillations: one cannot have decay processes in which three oscillations with one spectrum participate. In this connection, in the three-wave approximation adopted here, one can distinguish five kinds of decay process (if one does not distinguish between decay and coalescence): coalescence of ion sound and an electron Langmuir oscillation into an electron Langmuir oscillation and a transverse oscillation; coalescence of ion sound and a transverse oscillation into a transverse oscillation; coalescence of two electron Langmuir oscillations into a transverse oscillation and coalescence of an electron Langmuir oscillation and a long-wavelength transverse oscillation into a transverse oscillation. To describe decay processes in which three oscillations participate, it is sufficient to know only the first nonlinear tensor S_{ijs}. Then from the generalized kinetic field equation (3.22) one can distinguish groups of terms that are responsible for the evolution of the spectral function of the electric field in the plasma as a result of the decay processes alone:

$$\left\{\frac{1}{\omega}\frac{\partial\omega M^H_{ij}(\omega,\mathbf{k})}{\partial\omega}\frac{\partial}{\partial t}(E_jE_i)_{\omega,\mathbf{k}} - \frac{1}{\omega}\frac{\partial\omega M^H_{ij}(\omega,\mathbf{k})}{\partial\mathbf{k}}\frac{\partial}{\partial\mathbf{r}}(E_jE_i)_{\omega,\mathbf{k}}\right\}^{\text{decay}} =$$
$$= \mathrm{Im}\int d\omega' d\mathbf{k}' d\omega'' d\mathbf{k}''\delta(\omega-\omega'-\omega'')\delta(\mathbf{k}-\mathbf{k}'-\mathbf{k}'')\times$$
$$\times\{S_{ijs}(\omega,\mathbf{k};\omega',\mathbf{k}')S^*_{abc}(\omega,\mathbf{k};\omega',\mathbf{k}')A^{*\text{res}}_{ia}(\omega,\mathbf{k})(E_jE_b)_{\omega'',\mathbf{k}''}\times$$
$$\times(E_sE_c)_{\omega',\mathbf{k}'} + S_{ijs}(\omega,\mathbf{k};\omega',\mathbf{k}')S_{bca}(\omega'',\mathbf{k}'';\omega,\mathbf{k})(E_aE_i)_{\omega,\mathbf{k}}\times$$
$$\times A^{\text{res}}_{jb}(\omega'',\mathbf{k}'')(E_sE_c)_{\omega',\mathbf{k}'} + S_{isj}(\omega,\mathbf{k};\omega'',\mathbf{k}'')S_{cba}(\omega',\mathbf{k}';\omega,\mathbf{k})\times$$
$$\times(E_aE_i)_{\omega,\mathbf{k}}(E_jE_b)_{\omega'',\mathbf{k}''}A^{\text{res}}_{sc}(\omega',\mathbf{k}')\}. \qquad (3.22a)$$

The inverse tensors that occur in this equation consist only of their resonance parts, which correspond to zeros of the dispersion equation, i.e., to transparency regions of the plasma [for more details see §6, formulas (6.82) and (6.83)]. We emphasize that Eq. (3.22a), which is a direct consequence of the generalized kinetic equation (3.22) is valid not only for an isotropic or magnetoactive plasma but also for a larger class of media with corresponding material equation.

Another class of nonlinear processes consists of processes of nonlinear scattering or, in other words, induced scattering of oscillations of a plasma on its particles. In induced scattering, as in decay processes, three plasma oscillations participate, but with the important difference that only two of them are weakly damped (characteristic) plasma oscillations, that is, the scattering and the scattered oscillation. The intermediate third oscillation interacts strongly with the particles of the plasma and in this sense it is strongly damped. We shall call it a virtual oscillation. In the generalized kinetic equation (3.22), weakly damped characteristic plasma oscillations correspond to the spectral functions of the electric fields and the virtual oscillation corresponds to the inverse tensors A_{ij}. Both the nonlinear tensors S and V participate in the description of induced scattering processes. As also for the decay processes, we can distinguish in the generalized kinetic equation (3.22) the terms responsible solely for induced scattering on particles:

$$\left\{\frac{1}{\omega}\frac{\partial\omega M^H_{ij}(\omega,\ \mathbf{k})}{\partial\omega}\frac{\partial}{\partial t}(E_jE_i)_{\omega,\mathbf{k}}-\frac{1}{\omega}\frac{\partial\omega M^H_{ij}(\omega,\ \mathbf{k})}{\partial\mathbf{k}}\frac{\partial}{\partial\mathbf{r}}(E_jE_i)_{\omega,\mathbf{k}}\right\}^{\text{scat}}=$$
$$=2i\varepsilon^a_{ij}(\omega,\ \mathbf{k})(E_jE_i)_{\omega,\mathbf{k}}-2\,\mathrm{Im}\int d\omega' d\mathbf{k}'\,(E_aE_i)_{\omega,\mathbf{k}}(E_sE_c)_{\omega',\mathbf{k}'}\times$$
$$\times\{V_{icas}(\omega,\ \mathbf{k};\ \omega',\ \mathbf{k}')-S_{ijs}(\omega,\ \mathbf{k};\ \omega',\ \mathbf{k}')A_{jb}(\omega-\omega',\ \mathbf{k}-\mathbf{k}')\times$$
$$\times S_{bca}(\omega-\omega',\ \mathbf{k}-\mathbf{k}';\ \omega,\ \mathbf{k})\}. \tag{3.22b}$$

Here, the spectral function $(E_jE_i)_{\omega,k}$ corresponds to the scattered oscillation and $(E_sE_c)_{\omega',k'}$ to the scattering oscillation. The intermediate virtual oscillation has frequency $\omega'' \equiv \omega - \omega'$ and wave vector $\mathbf{k}'' \equiv \mathbf{k} - \mathbf{k}'$. In this sense, the frequencies and wave vectors of all three oscillations that participate in the induced scattering are related by the equations $\omega'(\mathbf{k}') + \omega'' = \omega(\mathbf{k})$, $\mathbf{k}' + \mathbf{k}'' = \mathbf{k}$, which differ from the decay conditions in that the frequency ω'' and wave vector $\mathbf{k}''$ of the virtual oscillation do not satisfy a dispersion law: $\omega'' \neq \omega''(\mathbf{k}'')$.

§ 5. Solution of the Kinetic Equation for an Isotropic Plasma as a Power Series in the Self-Consistent Field

The basis of the theory of an isotropic collisionless plasma is a system of equations with self-consistent electromagnetic field. The distribution function $f(\mathbf{p}, \mathbf{r}, t)$ of charged particles with momentum $\mathbf{p}$ at the point with radius vector $\mathbf{r}$ at the time t satisfies Vlasov's transport equation:

$$\frac{\partial f}{\partial t}+\mathbf{v}\frac{\partial f}{\partial \mathbf{r}}+e\left\{\mathbf{E}+\frac{1}{c}[\mathbf{vB}]\right\}\frac{\partial f}{\partial \mathbf{p}}=0. \tag{5.1}$$

The electric and magnetic fields $\mathbf{E}(\mathbf{r}, t)$ and $\mathbf{B}(\mathbf{r}, t)$ satisfy Maxwell's equations [cf. (1.2) subject to the condition (1.3) or (1.5)]:

$$\begin{aligned}&\operatorname{div}\mathbf{E}=4\pi\rho, && \operatorname{rot}\mathbf{E}=-\frac{1}{c}\frac{\partial\mathbf{B}}{\partial t},\\ &\operatorname{rot}\mathbf{B}=\frac{1}{c}\frac{\partial\mathbf{E}}{\partial t}+\frac{4\pi}{c}\mathbf{j}, && \operatorname{div}\mathbf{B}=0,\end{aligned} \tag{5.2}$$

in which the current density $\mathbf{j}(\mathbf{r}, t)$ and the charge density $\rho(\mathbf{r}, t)$ are in their turn determined by means of the distribution function:

$$\mathbf{j}=e\int d\mathbf{p}\mathbf{v}f(\mathbf{p},\ \mathbf{r},\ t);\qquad \rho=e\int d\mathbf{p}f(\mathbf{p},\ \mathbf{r},\ t). \tag{5.3}$$

Note that on the right-hand sides of Eqs. (5.3) we understand summation over the particle species in the plasma and (5.1) is a system of equations for the distribution functions $f(\mathbf{p}, \mathbf{r}, t)$ of particles of different species.

From the point of view of the arguments of §1, Eqs. (5.1) and (5.3) are the basic relations for finding the explicit form of the material equation. It can be seen from Eqs. (5.3) that this problem resides in expressing the distribution function $f(\mathbf{p}, \mathbf{r}, t)$ in terms of the electric field $\mathbf{E}(\mathbf{r}, t)$. To carry out such a procedure, we represent $f(\mathbf{p}, \mathbf{r}, t)$ as a sum of infinitely many terms,

$$f(\mathbf{p}, \mathbf{r}, t) = f_0(\mathbf{p}, \mathbf{r}, t) + f_1 + f_2 + \dots + f_n + \dots, \tag{5.4}$$

each of which, f_n, is less than the foregoing f_{n-1} ($f_n \ll f_{n-1}$; n = 1, 2, ...) and contains one power of the field E less than the subsequent one. Symbolically, we can write that in order of magnitude

$$f_n = O(E^n), \quad n = 0, 1, 2, \dots. \tag{5.5}$$

The first term $f_0(\mathbf{p}, \mathbf{r}, t)$ on the right-hand side of (5.4) does not depend on the electromagnetic field and therefore determines the unperturbed distribution function. In the theory we are developing we shall assume that $f_0(\mathbf{p}, \mathbf{r}, t)$ is given. The system of equations for the non-equilibrium parts $f_n(\mathbf{p}, \mathbf{r}, t)$ ($n \geq 1$), which are perturbed by the electromagnetic field, arises from the transport equation (5.1) when the term that depends on the Lorentz force (1.1) and is responsible for the interaction is small, the interaction appearing here in the form of coupling of the equations:

$$\frac{\partial f_n}{\partial t} + \mathbf{v}\frac{\partial f_n}{\partial \mathbf{r}} = -e\left\{\mathbf{E} + \frac{1}{c}[\mathbf{v}\mathbf{B}]\right\}\frac{\partial f_{n-1}}{\partial \mathbf{p}}, \quad n = 1, 2, \dots \tag{5.6}$$

The first equation (n = 1) of this chain corresponds to the linear approximation

$$\frac{\partial f_1}{\partial t} + \mathbf{v}\frac{\partial f_1}{\partial \mathbf{r}} = -e\left\{\mathbf{E} + \frac{1}{c}[\mathbf{v}\mathbf{B}]\right\}\frac{\partial f_0}{\partial \mathbf{p}}, \tag{5.7}$$

used in the linear theory of oscillations of an isotropic plasma, and all the subsequent equations ($n \geq 2$) are related to the nonlinear effects.

We obtain a solution of the linearized transport equation (5.7) in the approximation of adiabatic switching on of the field in the infinitely distant past:

$$f_1(\mathbf{p}, \mathbf{r}, t = -\infty) = 0 \tag{5.8}$$

after transition to Fourier components [see (2.1)]

$$f_n(\mathbf{p}, \mathbf{r}, t) = \int d\omega d\mathbf{k} e^{-i\omega t + i\mathbf{k}\mathbf{r}} f_n(\mathbf{p}, \omega, \mathbf{k}), \quad n = 0, 1, 2, \dots \tag{5.9}$$

Let us consider first the case when f_0 does not depend on the time t or the coordinates $\mathbf{r}$. Then (5.7) can be written in the form

$$-i\omega f_1(\mathbf{p}, \omega, \mathbf{k}) + i\mathbf{k}\mathbf{v} f_1(\mathbf{p}, \omega, \mathbf{k}) = -e\left\{\mathbf{E}(\omega, \mathbf{k}) + \frac{1}{c}[\mathbf{v}\mathbf{B}(\omega, \mathbf{k})]\right\}\frac{\partial f_0}{\partial \mathbf{p}}, \tag{5.10}$$

which enables us to obtain immediately the solution

$$f_1(\mathbf{p}, \omega, \mathbf{k}) = -ie\frac{\left\{\mathbf{E} + \frac{1}{c}[\mathbf{v}\mathbf{B}]\right\}}{\omega - \mathbf{k}\mathbf{v}}\frac{\partial f_0}{\partial \mathbf{p}}, \tag{5.11}$$

in which the denominator $(\omega - \mathbf{k}\mathbf{v})^{-1}$ must be understood in the sense of the δ_+ function:

$$\frac{1}{\omega - \mathbf{k}\mathbf{v}} = \frac{P}{\omega - \mathbf{k}\mathbf{v}} - i\pi\delta(\omega - \mathbf{k}\mathbf{v}). \tag{5.12}$$

We emphasize that the appearance of the δ_+ function (5.12) is a direct consequence of the assumption (5.8) of adiabatic switching on of the field. That the δ_+ function arises is due to the field's being switched on in the past ($t = -\infty$) and not in the future ($t = +\infty$). On the other hand, Eq. (5.12) corresponds formally to an infinitesimally small imaginary positive correction to the frequency: $\omega \to \omega + i\Delta,\ \Delta \to +0$. The letter P in (5.12) is the principal value symbol:

$$P\frac{1}{x} = \begin{cases} \frac{1}{x}, & x \neq 0, \\ 0, & x = 0, \end{cases} \tag{5.13}$$

and $\delta(\omega - \mathbf{kv})$ is the Dirac δ function. Using Faraday's law (1.7) to eliminate $\mathbf{B}$,

$$\mathbf{B}(\omega, \mathbf{k}) = \frac{c}{\omega}[\mathbf{kE}], \tag{5.14}$$

We can represent the Fourier component of the Lorentz force (1.1) in (5.11) differently:

$$F_i(\omega, \mathbf{k}) = e\left\{\mathbf{E}(\omega, \mathbf{k}) + \frac{1}{c}[\mathbf{vB}(\omega, \mathbf{k})]\right\}_i = e\alpha_{ij}(\omega, \mathbf{k}, \mathbf{v}) E_j(\omega, \mathbf{k}). \tag{5.15}$$

Here we have introduced the tensor

$$\alpha_{ij}(\omega, \mathbf{k}, \mathbf{v}) \equiv \frac{1}{\omega}\{k_i v_j + \delta_{ij}(\omega - \mathbf{kv})\}. \tag{5.16}$$

In this notation, the solution (5.11) takes the form

$$f_1(\mathbf{p}, \omega, \mathbf{k}) = -ie\frac{\alpha_{ij}(\omega, \mathbf{k}, \mathbf{v})}{\omega - \mathbf{kv}}\frac{\partial f_0}{\partial p_i} E_j(\omega, \mathbf{k}) \tag{5.17}$$

generally used in the linear theory of oscillations [1]. If the unperturbed distribution function f_0 is isotropic, i.e., it depends only on the absolute magnitude p of the particle momentum $\mathbf{p}$, or, which is the same thing, on the particle energy $\mathscr{E}$, the tensor $\alpha_{ij}(\omega, \mathbf{k}, \mathbf{v})$ disappears from the solution (5.17),

$$\alpha_{ij}(\omega, \mathbf{k}, \mathbf{v})\frac{\partial f_0(\mathscr{E})}{\partial p_i} = \frac{\partial f_0(\mathscr{E})}{\partial p_j}, \tag{5.18}$$

so that

$$f_1(\mathbf{p}, \omega, \mathbf{k}) = -ie\frac{\mathbf{E}(\omega, \mathbf{k})}{\omega - \mathbf{kv}}\frac{\partial f_0}{\partial \mathbf{p}}. \tag{5.19}$$

The relations (5.17) and (5.19) completely settle the question of the solution of the linearized transport equation (5.7) in the framework of the rule (5.12) for circumventing the poles [i.e., when the condition (5.8) holds] and a stationary homogeneous unperturbed distribution function f_0. Otherwise, when f_0 depends on the time and the coordinates, the solution of Eq. (5.7) is not the algebraic solution (5.17), but an integral dependence of the first nonequilibrium distribution function $f_1(\mathbf{p}, \omega, \mathbf{k})$ on the electric field $\mathbf{E}(\omega, \mathbf{k})$:

$$f_1(\mathbf{p}, \omega, \mathbf{k}) = -ie\int d\omega' d\mathbf{k}' d\omega'' d\mathbf{k}'' \delta(\omega - \omega' - \omega'')\delta(\mathbf{k} - \mathbf{k}' - \mathbf{k}'')\frac{\alpha_{ij}(\omega', \mathbf{k}', \mathbf{v})}{\omega - \mathbf{kv}}\frac{\partial}{\partial p_i} f_0(\mathbf{p}, \omega'', \mathbf{k}'') E_j(\omega', \mathbf{k}'). \tag{5.20}$$

This formula arises naturally from (5.7) if on the transition to Fourier components we use the relation

$$\begin{gathered} C(\omega, \mathbf{k}) = \int d\omega' d\mathbf{k}' A(\omega', \mathbf{k}') B(\omega - \omega', \mathbf{k} - \mathbf{k}') = \\ = \int d\omega' d\mathbf{k}' d\omega'' d\mathbf{k}'' \, \delta(\omega - \omega' - \omega'')\delta(\mathbf{k} - \mathbf{k}' - \mathbf{k}'') A(\omega', \mathbf{k}') B(\omega'', \mathbf{k}''); \\ C(\mathbf{r}, t) = A(\mathbf{r}, t) B(\mathbf{r}, t). \end{gathered} \tag{5.21}$$

Note that (5.20) goes over into (5.17) if one returns to a stationary and homogeneous distribution f_0:

$$f_0(\mathbf{p}, \omega, \mathbf{k}) = f_0(\mathbf{p})\,\delta(\omega)\,\delta(\mathbf{k}). \tag{5.22}$$

After this brief excursion into linear plasma theory without external fields, the solution of the chain of equations (5.6) for n ≥ 2 is not difficult; for Eqs. (5.6) differ formally from the linearized transport equation (5.7) only by the presence of the subscripts n and n − 1 of the nonequilibrium distribution functions $f_n(\mathbf{p}, \mathbf{r}, t)$ and $f_{n-1}(\mathbf{p}, \mathbf{r}, t)$. Therefore, using (5.20), we can represent the solution of the chain (5.6) as the recursion relation

$$f_n(\mathbf{p}, \omega, \mathbf{k}) = -ie \int d\omega' d\mathbf{k}' d\omega'' d\mathbf{k}'' \frac{a_{ij}(\omega', \mathbf{k}', \mathbf{v})}{\omega - \mathbf{k}\mathbf{v}} \delta(\omega - \omega' - \omega'')\,\delta(\mathbf{k} - \mathbf{k}' - \mathbf{k}'') \frac{\partial}{\partial p_i} f_{n-1}(\mathbf{p}, \omega'', \mathbf{k}'') E_j(\omega', \mathbf{k}'),$$
$$n = 1, 2, \ldots \tag{5.23}$$

The use of adiabatic switching on of the field in the infinitely distant past,

$$f_n(\mathbf{p}, \mathbf{r}, t = -\infty) = 0, \quad n = 1, 2, \ldots, \tag{5.24}$$

i.e., the neglect of the initial values of the nonequilibrium distribution functions f_n, prevents us from constructing a complete statistical theory of the field [24, 25]. However, the solutions (5.23) are suitable for obtaining the nonlinear material equation that, together with Maxwell's equations, describes the higher nonlinear processes.

For n = 2 we obtain from the recursion relation (5.23), replacing f_1 by its expression (5.20), the second nonequilibrium distribution function $f_2(\mathbf{p}, \omega, \mathbf{k})$, which is a bilinear functional of the electric field:

$$f_2(\mathbf{p}, \omega, \mathbf{k}) = (-ie)^2 \int d\omega_1 d\mathbf{k}_1 d\omega_2 d\mathbf{k}_2 \frac{a_{i1j1}(\omega - \omega_1, \mathbf{k} - \mathbf{k}_1, \mathbf{v})}{\omega - \mathbf{k}\mathbf{v}} \times$$
$$\times \frac{\partial}{\partial p_{i1}} \frac{a_{i2j2}(\omega_1 - \omega_2, \mathbf{k}_1 - \mathbf{k}_2, \mathbf{v})}{\omega_1 - \mathbf{k}_1\mathbf{v}} \frac{\partial}{\partial p_{i2}} f_0(\mathbf{p}, \omega_2, \mathbf{k}_2) E_{j1}(\omega - \omega_1, \mathbf{k} - \mathbf{k}_1) E_{j2}(\omega_1 - \omega_2, \mathbf{k}_1 - \mathbf{k}_2). \tag{5.25}$$

The operators of differentiation with respect to the momenta in this equation act on all the functions to the right of them that depend on the momentum. Similarly, setting n = 3 in (5.23) and using (5.25) for f_2, we obtain the following explicit form of the third nonequilibrium distribution function $f_3(\mathbf{p}, \omega, \mathbf{k})$:

$$f_3(\mathbf{p}, \omega, \mathbf{k}) = (-ie)^3 \int d\omega_1 d\mathbf{k}_1 d\omega_2 d\mathbf{k}_2 d\omega_3 d\mathbf{k}_3 \frac{a_{i1j1}(\omega - \omega_1, \mathbf{k} - \mathbf{k}_1, \mathbf{v})}{\omega - \mathbf{k}\mathbf{v}} \times$$
$$\times \frac{\partial}{\partial p_{i1}} \frac{a_{i2j2}(\omega_1 - \omega_2, \mathbf{k}_1 - \mathbf{k}_2, \mathbf{v})}{\omega_1 - \mathbf{k}_1\mathbf{v}} \frac{\partial}{\partial p_{i2}} \frac{a_{i3j3}(\omega_2 - \omega_3, \mathbf{k}_2 - \mathbf{k}_3, \mathbf{v})}{\omega_2 - \mathbf{k}_2\mathbf{v}} \times$$
$$\times \frac{\partial}{\partial p_{i3}} f_0(\mathbf{p}, \omega_3, \mathbf{k}_3) E_{j1}(\omega - \omega_1, \mathbf{k} - \mathbf{k}_1) E_{j2}(\omega_1 - \omega_2, \mathbf{k}_1 - \mathbf{k}_2) E_{j3}(\omega_2 - \omega_3, \mathbf{k}_2 - \mathbf{k}_3). \tag{5.26}$$

From the expressions (5.25) and (5.26) for the second and third distribution functions f_2 and f_3, and retaining the recursion relation (5.23), we can readily express the n-th nonequilibrium distribution function $f_n(\mathbf{p}, \omega, \mathbf{k})$ in terms of the unperturbed equilibrium distribution function $f_0(\mathbf{p}, \omega, \mathbf{k})$ and the vectors $\mathbf{E}(\omega, \mathbf{k})$:

$$f_n(\mathbf{p}, \omega, \mathbf{k}) = (-ie)^n \int d\omega_1 d\mathbf{k}_1 \ldots d\omega_n d\mathbf{k}_n g \Gamma_{j1} g_1 \Gamma_{j2} \ldots g_{n-1} \Gamma_{jn} \times$$
$$\times f_0(\mathbf{p}, \omega_n, \mathbf{k}_n) E_{j1}(\omega - \omega_1, \mathbf{k} - \mathbf{k}_1) E_{j2}(\omega_1 - \omega_2, \mathbf{k}_1 - \mathbf{k}_2) \ldots E_{jn}(\omega_{n-1} - \omega_n, \mathbf{k}_{n-1} - \mathbf{k}_n), \quad n = 1, 2, \ldots \tag{5.27}$$

Here we have used the following notation for the Green's functions g of the particles of an isotropic plasma moving uniformly and rectilinearly with constant velocity **v**:

$$g \equiv \frac{1}{\omega - \mathbf{kv}}, \quad g_n \equiv \frac{1}{\omega_n - \mathbf{k}_n\mathbf{v}}, \quad n \geqslant 1. \tag{5.28}$$

As in the linear theory, the singularities that arise in (5.27) because of (5.28) are understood in the sense of the rule (5.12). The vertex parts Γ_j are operators of differentiation with respect to the particle momenta:

$$\begin{aligned}\Gamma_{jn} \equiv \alpha_{injn}(\omega_{n-1} - \omega_n, \mathbf{k}_{n-1} - \mathbf{k}_n, \mathbf{v})\frac{\partial}{\partial p_{in}} = \frac{1}{\omega_{n-1} - \omega_n}\{((\mathbf{k}_{n-1} - \mathbf{k}_n)_{in} v_{jn} + \\ + \delta_{injn}[\omega_{n-1} - \omega_n - (\mathbf{k}_{n-1} - \mathbf{k}_n, \mathbf{v})]\}\, \partial/\partial p_{in}, \quad n \geqslant 2; \\ \Gamma_{j1} \equiv \alpha_{i1j1}(\omega - \omega_1, \mathbf{k} - \mathbf{k}_1, \mathbf{v})\frac{\partial}{\partial p_{i1}}, \quad n = 1.\end{aligned} \tag{5.29}$$

These operators act on all functions of the momenta and velocities in (5.27) to their right. We note that the general expression (5.27) obviously goes over into the forms (5.20), (5.25), and (5.26) of the distribution functions for n = 1, 2, 3.

In the limit of a nonrelativistic isotropic plasma, when the relationship between the momentum p of a particle with rest mass m and its velocity **v** is especially simple:

$$\mathbf{p} = m\mathbf{v}, \tag{5.30}$$

the right-hand side of the solution (5.27) can be simplified, by expressing it in terms of the velocity **v**, after p has been eliminated by means of (5.30):

$$\begin{aligned}f_n(\mathbf{p}, \omega, \mathbf{k}) = \left(-i\frac{e}{m}\right)^n \int d\omega_1 d\mathbf{k}_1 \ldots d\omega_n d\mathbf{k}_n \frac{\alpha_{i1j1}(\omega - \omega_1, \mathbf{k} - \mathbf{k}_1\mathbf{v})}{\omega - \mathbf{kv}} \times \\ \times \frac{\partial}{\partial v_{i1}} \frac{\alpha_{i2j2}(\omega_1 - \omega_2, \mathbf{k}_1 - \mathbf{k}_2, \mathbf{v})}{\omega_1 - \mathbf{k}_1\mathbf{v}} \frac{\partial}{\partial v_{i2}} \ldots \frac{\alpha_{injn}(\omega_{n-1} - \omega_n, \mathbf{k}_{n-1} - \mathbf{k}_n, \mathbf{v})}{\omega_n - \mathbf{k}_n\mathbf{v}} \times \\ \times \frac{\partial}{\partial v_{in}} f_0(\mathbf{v}, \omega_n, \mathbf{k}_n) E_{j1}(\omega - \omega_1, \mathbf{k} - \mathbf{k}_1) E_{j2}(\omega_1 - \omega_2, \mathbf{k}_1 - \mathbf{k}_2) \ldots E_{jn}(\omega_{n-1} - \omega_n, \mathbf{k}_{n-1} - \mathbf{k}_n).\end{aligned} \tag{5.31}$$

Taking into account the expansion (5.4) and returning in the nonequilibrium terms (5.27) of this expansion from the Fourier variables ω and **k** to the time t and coordinates **r**, we obtain the distribution function f(**p**, **r**, t) of the given species of plasma particles as the solution of the transport equation (5.1) in the approximation of adiabatic switching on of the field in the infinitely distant past [12]:

$$\begin{aligned}f(\mathbf{p}, \mathbf{r}, t) = f_0(\mathbf{p}, \mathbf{r}, t) + \sum_{n=1}^{\infty}(-ie)^n \int d\omega d\mathbf{k} e^{-i\omega t + i\mathbf{kr}} \int d\omega_1 d\mathbf{k}_1 \ldots \\ \ldots d\omega_n d\mathbf{k}_n g\Gamma_{j1} g_1 \Gamma_{j2} \ldots g_{n-1}\Gamma_{jn} f_0(\mathbf{p}, \omega_n, \mathbf{k}_n) E_{j1}(\omega - \omega_1, \mathbf{k} - \mathbf{k}_1) \ldots \\ \ldots E_{jn}(\omega_{n-1} - \omega_n, \mathbf{k}_{n-1} - \mathbf{k}_n).\end{aligned} \tag{5.32}$$

This solution is the basic relation by means of which we shall construct below the theory of the nonlinear interaction of characteristic oscillations of a homogeneous isotropic plasma. We emphasize the simplicity of the derivation of Eq. (5.32): to derive f(**p**, **r**, t) we required only the recursion formula (5.23), which differs from the well-known relation (5.20) of the linear theory of plasma oscillations only by the introduction of the index n, the number of the nonequilibrium term in the infinite sum (5.4). Formula (5.32) can also be represented differently if one does not make a transition to the Fourier components of the fields when solving the chain of equations (5.6):

$$\begin{aligned}f(\mathbf{p}, \mathbf{r}, t) = f_0(\mathbf{p}, \mathbf{r}, t) + \sum_{n=1}^{\infty}(-e)^n \int_{-\infty}^{0} d\tau_0 \ldots \int_{-\infty}^{0} d\tau_{n-1} \times \\ \times \left\{E_{j_1}(\mathbf{r} + \mathbf{v}\tau_0, t + \tau_0) + \frac{1}{c}[\mathbf{vB}(\mathbf{r} + \mathbf{v}\tau_0, t + \tau_0)]_{j_1}\right\} \frac{\partial}{\partial p_{j_1}} \ldots\end{aligned}$$

$$\ldots\Big\{E_{j_n}(\mathbf{r}+\mathbf{v}(\tau_0+\ldots+\tau_{n-1}),\ t+\tau_0+\ldots+\tau_{n-1})+$$
$$+\frac{1}{c}[\mathbf{vB}(\mathbf{r}+\mathbf{v}(\tau_0+\ldots+\tau_{n-1}),\ t+\tau_0+\ldots+\tau_{n-1})]_{j_n}\Big\}\times$$
$$\times\frac{\partial}{\partial p_{j_n}}f_0(\mathbf{p},\ \mathbf{r}+\mathbf{v}(\tau_0+\ldots+\tau_{n-1}),\ t+\tau_0+\ldots+\tau_{n-1}). \tag{5.33}$$

§ 6. Permittivity Tensor of an Isotropic Plasma as the Simplest Example of the Multi-Index Tensor $\varepsilon_{ij1\ldots jn}$ for n = 1. Spectra and Polarization of the Characteristic Oscillations of an Isotropic Plasma

As we have already pointed out in §5, the explicit form of the distribution function $f(\mathbf{p}, \mathbf{r}, t)$ fully determines the material equation and therefore the multi-index tensors of the complex permittivity of an isotropic plasma (see Chapter I, § 2). Indeed, the expansion (5.4) of the distribution function and the explicit form (5.27) of the nonequilibrium terms f_n enable us, on account of (5.3), to represent the induced current density $\mathbf{j}$ and the charge density ρ as a sum of known terms:

$$\begin{aligned}\mathbf{j}(\omega,\ \mathbf{k})&=\mathbf{j}_0(\omega,\ \mathbf{k})+\mathbf{j}_1(\omega,\ \mathbf{k})+\ldots+\mathbf{j}_n(\omega,\ \mathbf{k})+\ldots,\\ \rho(\omega,\ \mathbf{k})&=\rho_0(\omega,\ \mathbf{k})+\rho_1(\omega,\ \mathbf{k})+\ldots+\rho_n(\omega,\ \mathbf{k})+\ldots.\end{aligned} \tag{6.1}$$

Here, the nonequilibrium terms j_n and ρ_n $(n \geq 1)$ are given by the equations

$$\mathbf{j}_n(\omega,\ \mathbf{k})=e(-ie)^n\int d\mathbf{p}\mathbf{v}\int d\omega_1 d\mathbf{k}_1\ldots d\omega_n d\mathbf{k}_n g\Gamma_{j1}g_1\Gamma_{j2}\ldots$$
$$\ldots g_{n-1}\Gamma_{jn}f_0(\mathbf{p},\ \omega_n,\ \mathbf{k}_n)E_{j1}(\omega-\omega_1,\ \mathbf{k}-\mathbf{k}_1)\ldots E_{jn}(\omega_{n-1}-\omega_n,\ \mathbf{k}_{n-1}-\mathbf{k}_n); \tag{6.2}$$

$$\rho_n(\omega,\ \mathbf{k})=e(-ie)^n\int d\mathbf{p}\int d\omega_1 d\mathbf{k}_1\ldots d\omega_n d\mathbf{k}_n g\Gamma_{j1}g_1\Gamma_{j2}\ldots$$
$$\ldots g_{n-1}\Gamma_{jn}f_0(\mathbf{p},\ \omega_n,\ \mathbf{k}_n)E_{j1}(\omega-\omega_1,\ \mathbf{k}-\mathbf{k}_1)\ldots E_{jn}(\omega_{n-1}-\omega_n,\ \mathbf{k}_{n-1}-\mathbf{k}_n),\qquad n\geqslant 1, \tag{6.3}$$

and the equilibrium unperturbed current and charge densities

$$\mathbf{j}_0=e\int d\mathbf{p}\mathbf{v}f_0=0;\qquad \rho_0=e\int d\mathbf{p}f_0=0 \tag{6.4}$$

are assumed to vanish [see formulas (1.3) and the remark on them].

Note that the nonequilibrium nonlinear corrections $\rho_n(\omega, \mathbf{k})$ to the charge density are related to the corresponding nonlinear terms $j_n(\omega, \mathbf{k})$ in the current density by the same relation as in the linear theory:

$$\mathbf{k}\mathbf{j}_n(\omega,\ \mathbf{k})=\omega\rho_n(\omega,\ \mathbf{k}),\ n\geqslant 1. \tag{6.5}$$

Its validity can be proved by integrating by parts with respect to the momenta (6.2) on the left-hand side of (6.5) with allowance for the equations

$$(\mathbf{kv})g+1=\omega g,\ \frac{\partial}{\partial p_i}\alpha_{ij}(\omega,\ \mathbf{k},\ \mathbf{v})=0. \tag{6.6}$$

On the other hand, Eq. (6.5) can be interpreted as an obvious consequence of the continuity equation (1.6),

$$\mathbf{k}\mathbf{j}(\omega,\ \mathbf{k})=\omega\rho(\omega,\ \mathbf{k}), \tag{6.7}$$

which arises from (6.7) when one makes an expansion in a power series in the electric field $\mathbf{E}(\omega, \mathbf{k})$.

Explicit expressions for the multi-index permittivity tensors can be obtained by combining Eq. (1.4),

$$D_i'(\omega, \mathbf{k}) = E_i(\omega, \mathbf{k}) + i\frac{4\pi}{\omega} j_i(\omega, \mathbf{k}) \tag{6.8}$$

with the first of the expansions (6.1). As a result

$$\varepsilon_{ij(1)\ldots j(n)}(\omega, \mathbf{k}; \omega_1, \mathbf{k}_1; \ldots; \omega_{n-1}, \mathbf{k}_{n-1}; \omega_n, \mathbf{k}_n) = \delta_{n1}\delta_{ij(1)}\delta(\omega_1)\delta(\mathbf{k}_1) - \\ - 4\pi(-ie)^{n+1}\int d\mathbf{p}\frac{v_i}{\omega} g\Gamma_{j(1)}g_1\Gamma_{j(2)}\ldots g_{n-1}\Gamma_{j(n)}f_0(\mathbf{p}, \omega_n, \mathbf{k}_n). \tag{6.9}$$

The tensors we have introduced differ by a pair of arguments from those studied phenomenologically in §2. In agreement with the standpoint adopted in Chapter I concerning the stationary and homogeneity of the nonlinear medium, we shall consider instead of (6.9) the tensors

$$\varepsilon_{ij1\ldots jn}(\omega, \mathbf{k}; \omega_1, \mathbf{k}_1; \ldots; \omega_{n-1}, \mathbf{k}_{n-1}) = \delta_{n1}\delta_{ij1} - 4\pi(-ie)^{n+1}\int dp\frac{v_i}{\omega} g\Gamma_{j1}g_1\Gamma_{j2}\ldots g_{n-1}\Gamma_{jn}f_0(\mathbf{p}), \\ n \geqslant 1; \quad \omega_0 \equiv \omega, \quad \mathbf{k}_0 \equiv \mathbf{k}, \quad g_0 \equiv g, \quad n = 1, \tag{6.10}$$

assuming the distribution function f_0 in the form (5.22), i.e., assuming that the unperturbed equilibrium state of the plasma is stationary and homogeneous. An important special case of the multi-index tensors (6.10) is the permittivity tensor ε_{ij1} of an isotropic plasma:

$$\varepsilon_{ij1}(\omega, \mathbf{k}) = \delta_{ij1} + \frac{4\pi e^2}{\omega}\int d\mathbf{p} v_i g\Gamma_{j1}f_0(\mathbf{p}), \tag{6.11}$$

for whose derivation it is sufficient to know the first nonequilibrium distribution function (5.17). Substituting the explicit expressions (5.28) and (5.29) for the Green's function g and the vertex part Γ_{j1}, we can represent (6.11) in the more usual form

$$\varepsilon_{ij}(\omega, \mathbf{k}) = \delta_{ij} + \frac{4\pi e^2}{\omega}\int d\mathbf{p} v_i \frac{\alpha_{sj}(\omega, \mathbf{k}, \mathbf{v})}{\omega - \mathbf{kv}}\frac{\partial f_0}{\partial p_s}. \tag{6.12}$$

The expression (6.12) differs from formula (11.11) of [1] by the replacement of the Kronecker delta δ_{sj} in (11.11) by the tensor $\alpha_{sj}(\omega, \mathbf{k}, \mathbf{v})$. This tensor appears because we allow for the second term $1/c\,[\mathbf{vB}]$ in the Lorentz force (5.15). We give below some well-known [1] special cases of the permittivity tensor (6.12), since we shall use them subsequently to study nonlinear effects in an isotropic plasma. We restrict ourselves to the isotropic distribution function f_0. Then, in accordance with (5.18)

$$\varepsilon_{ij}(\omega, \mathbf{k}) = \delta_{ij} + \frac{4\pi e^2}{\omega}\int d\mathbf{p} v_i \frac{1}{\omega - \mathbf{kv}}\frac{\partial f_0}{\partial p_j}, \tag{6.13}$$

which agrees with (11.11) of [1]. In all that follows below in all concrete applications of these general relations, in both the linear and nonlinear theory, we shall take f_0 to be the nonrelativistic Maxwell distribution:

$$f_0 = \frac{N}{(2\pi)^{3/2} v_T^3} e^{-\frac{v^2}{2v_T^2}} \tag{6.14}$$

with thermal velocity of the particles

$$v_T = (\varkappa T/m)^{1/2} \tag{6.15}$$

and density (the number of particles in 1 cm^3)

$$N=\int d\mathbf{v}f_0. \tag{6.16}$$

Here and in what follows, T is the temperature of the given species of particles and m is the particle mass. Substituting the permittivity tensor (6.13) in the form of the decomposition

$$\varepsilon_{ij}(\omega,\ \mathbf{k})=\varepsilon^l(\omega,\ \mathbf{k})\varkappa_i\varkappa_j+\varepsilon^{tr}(\omega,\ \mathbf{k})(\delta_{ij}-\varkappa_i\varkappa_j) \tag{6.17}$$

with respect to the two elementary tensors

$$(\delta_{ij}-\varkappa_i\varkappa_j),\qquad \varkappa_i\varkappa_j, \tag{6.18}$$

composed of the Kronecker delta δ_{ij} and the unit vector $\varkappa$ along the wave vector $\mathbf{k}=\varkappa|\mathbf{k}|\equiv\varkappa k$, we obtain explicit expressions for the two scalars ε^l and ε^{tr}, which characterize ε_{ij}, in terms of the isotropic distribution function:

$$\begin{aligned}\varepsilon^l(\omega,\ \mathbf{k})&=1+\frac{4\pi e^2}{\omega}\int d\mathbf{p}\frac{(\varkappa\mathbf{v})^2}{\omega-\mathbf{kv}}\frac{\partial f_0}{\partial\mathscr{E}},\\ \varepsilon^{tr}(\omega,\ \mathbf{k})&=1+\frac{2\pi e^2}{\omega}\int d\mathbf{p}\frac{[\varkappa\mathbf{v}]^2}{\omega-\mathbf{kv}}\frac{\partial f_0}{\partial\mathscr{E}}.\end{aligned} \tag{6.19}$$

The elementary tensors (6.18) determine the tensor properties of the permittivity of every isotropic nongyrotropic medium (without external fields) with spatial dispersion and, in particular, an isotropic plasma. The scalars ε^l and ε^{tr} are known as the longitudinal and transverse permittivity of the isotropic plasma, respectively. In the special case of the Maxwellian distribution (6.14), ε^l and ε^{tr} are expressed in terms of the function ([1], p. 90)

$$J_+(\beta)=\beta e^{-\beta^2/2}\int\limits_{i\infty}^{\beta}dte^{t^2/2}=\beta e^{-\beta^2/2}\int\limits_0^{\beta}dte^{t^2/2}-i\sqrt{\frac{\pi}{2}}\beta e^{-\beta^2/2} \tag{6.20}$$

as follows (see (12.6) in [1]):

$$\varepsilon^l(\omega,\ \mathbf{k})=1+\frac{\omega_L^2}{k^2v_T^2}\{1-J_+(\beta)\};\quad \varepsilon^{tr}(\omega,\ \mathbf{k})=1-\frac{\omega_L^2}{\omega^2}J_+(\beta); \tag{6.21}$$

$$\beta\equiv\frac{\omega}{kv_T},\qquad \omega_L\equiv\left(\frac{4\pi Ne^2}{m}\right)^{1/2}. \tag{6.22}$$

In the second terms on the right-hand sides of Eqs. (6.21) summation over the particle species is understood. For example, for an electron–ion plasma consisting of electrons and one species of ion, Eq. (6.1) is equivalent to the expanded equations

$$\varepsilon^l(\omega,\ \mathbf{k})=1+\frac{\omega_{Le}^2}{k^2v_{T_e}^2}\left\{1-J_+\left(\frac{\omega}{kv_{T_e}}\right)\right\}+\frac{\omega_{Li}^2}{k^2v_{T_i}^2}\left\{1-J_+\left(\frac{\omega}{kv_{T_i}}\right)\right\}; \tag{6.23}$$

$$\varepsilon^{tr}(\omega,\ \mathbf{k})=1-\frac{\omega_{Le}^2}{\omega^2}J_+\left(\frac{\omega}{kv_{T_e}}\right)-\frac{\omega_{Li}^2}{\omega^2}J_+\left(\frac{\omega}{kv_{T_i}}\right). \tag{6.24}$$

Here and in what follows the subscripts e and i are attached to quantities that characterize electrons and ions, respectively. For example, $\omega_{Le}=(4\pi N_e e^2/m)^{1/2}$ is the plasma frequency of the electrons with density N_e and $\omega_{Li}=(4\pi N_i e_i^2/M)^{1/2}$ is the plasma frequency of the ions with density N_i, charge e_i, and mass M.

In accordance with the last of Eqs. (6.20), the function $J_+(\beta)$ is complex. This results in the presence of not only real but also imaginary parts of the longitudinal and transverse permittivities (6.21):

$$\operatorname{Im} \varepsilon^{l}(\omega, \mathbf{k}) \equiv \varepsilon^{l\prime\prime}(\omega, \mathbf{k}) = \sqrt{\frac{\pi}{2}} \frac{\omega_L^2}{k^2 v_T^2} \frac{\omega}{k v_T} \exp\left\{-\frac{1}{2}\frac{\omega^2}{k^2 v_T^2}\right\}; \tag{6.25}$$

$$\operatorname{Im} \varepsilon^{tr}(\omega, \mathbf{k}) \equiv \varepsilon^{tr\prime\prime}(\omega, \mathbf{k}) = \sqrt{\frac{\pi}{2}} \frac{\omega_L^2}{\omega k v_T} \exp\left\{-\frac{1}{2}\frac{\omega^2}{k^2 v_T^2}\right\}. \tag{6.26}$$

The function $J_+(\beta)$ is complex and the imaginary parts (6.25)-(6.26) of the permittivities arise directly because of the application of Eq. (5.12), which is used to determine the denominators in formulas (6.19); for the imaginary parts (6.25)-(6.26) are related to the second term,

$$-i\pi\delta(\omega - \mathbf{k}\mathbf{v}), \tag{6.27}$$

on the right-hand side of Eq. (5.12). In this connection, we have a restriction on the applicability of the formulas (6.25)-(6.26) to values of the phase velocities of the oscillations of the isotropic plasma that do not exceed the maximal velocities of the plasma particles, which are less than the velocity of light in vacuum. Formally, this is due to the fact that it is only at such phase velocities ω/k that the δ function (6.27) makes a contribution to the calculated integrals. This obvious assertion follows from the relativistic treatment (cf. §13 in [1]).

The expressions (6.23)-(6.24) in conjunction with the dispersion relation (2.14) enables us to calculate the explicit form of the spectra of characteristic (weakly damped) oscillations of an isotropic plasma. In an isotropic plasma with Maxwellian distribution function (6.14) there exist three characteristic oscillations: two electrostatic (longitudinal: $[\mathbf{k}\mathbf{E}(\omega, \mathbf{k})] = 0$) – the electron Langmuir oscillation and ion sound – and one nonelectrostatic oscillation [transverse: $(\mathbf{k}\mathbf{E}(\omega, \mathbf{k})) = 0$]. The electron Langmuir oscillation has a phase velocity appreciably exceeding the thermal velocity of the electrons and ions:

$$\frac{\omega}{k} \gg v_{T_e}, \quad \frac{\omega}{k} \gg v_{T_i}, \tag{6.28}$$

and exists (is weakly damped) in both an isothermal ($T_e = T_i$) and nonisothermal plasma. Substituting the tensor (6.17) into the dispersion relation (2.14), we obtain for an isotropic plasma

$$\varepsilon^{l}(\omega, \mathbf{k})\left\{\varepsilon^{tr}(\omega, \mathbf{k}) - \frac{c^2 k^2}{\omega^2}\right\}^2 = 0. \tag{6.29}$$

In accordance with the condition that the oscillation is longitudinal,

$$[\mathbf{k}\mathbf{E}] = 0 \tag{6.30}$$

and Maxwell's equations (2.13), longitudinal oscillations correspond to the equation

$$\varepsilon^{l}(\omega, \mathbf{k}) = 0, \tag{6.31a}$$

which makes the left-hand side of the dispersion relation (6.29) vanish because of the first factor. Using the inequality (6.28) for asymptotic expansion of the function $J_+(\beta)$ at large values of the argument ([1], p. 91)

$$J_+(\beta) = 1 + \frac{1}{\beta^2} + \frac{3}{\beta^4} + \frac{15}{\beta^6} + \ldots - i\sqrt{\frac{\pi}{2}}\beta e^{-\beta^2/2}, \quad \beta \gg 1 \tag{6.32}$$

and substituting the explicit expression (6.23) into (6.31a), we obtain for the determination of the spectrum the equation

$$1 - \frac{\omega_{Le}^2}{\omega^2} - 3k^2 v_{T_e}^2 \frac{\omega_{Le}^2}{\omega^4} - \frac{\omega_{Li}^2}{\omega^2} - 3k^2 v_{T_i}^2 \frac{\omega_{Li}^2}{\omega^4} = 0. \tag{6.33}$$

From this, ignoring small terms, we find the actual spectrum of the electron Langmuir oscillation:

$$\omega^2(\mathbf{k}) = \omega_{L_e}^2 + 3k^2 v_{T_e}^2 + \omega_{L_i}^2. \tag{6.34a}$$

Strictly speaking, the expression (6.34a) for the frequency of this oscillation makes only the real part of the longitudinal permittivity vanish:

$$\mathrm{Re}\,\varepsilon^l(\omega, \mathbf{k}) = 0. \tag{6.35}$$

Equations (6.31a) and (6.35) are equivalent only if one ignores the exponentially small imaginary part (6.25), which is important for determining the Landau damping of the electron Langmuir oscillation. Indeed, introducing explicitly the imaginary part of the frequency ω'' ($\omega = \omega' + i\omega''$), which is small compared with the real part $\omega' \equiv \omega(\mathbf{k})$ ($\omega'' \ll \omega'$), the vanishing of the complex longitudinal permittivity (6.31a) can be represented approximately in the form

$$\mathrm{Re}\,\varepsilon^l(\omega', \mathbf{k}) + i\,\mathrm{Im}\,\varepsilon^l(\omega', \mathbf{k}) + i\omega'' \frac{\partial}{\partial\omega'} \mathrm{Re}\,\varepsilon^l(\omega', \mathbf{k}) \cong 0. \tag{6.31b}$$

Hence, with allowance for (6.35) and (6.25), we obtain an expression for the damping $\gamma(\mathbf{k})$ of the electrostatic oscillations of an isotropic plasma:

$$\gamma(\mathbf{k}) \equiv -\omega'' = \mathrm{Im}\,\varepsilon^l(\omega', \mathbf{k}) \left\{ \frac{\partial}{\partial\omega'} \mathrm{Re}\,\varepsilon^l(\omega', \mathbf{k}) \right\}^{-1}, \tag{6.25a}$$

in which, in accordance with (6.35), the real part of the frequency ω' is determined by the corresponding spectrum. In particular, for the damping $\gamma_l(\mathbf{k})$ of the electron Langmuir oscillation, we obtain, ignoring the small contribution of the ion terms, the relation ([1], p. 89)

$$\gamma_l(\mathbf{k}) = \sqrt{\frac{\pi}{8}} \frac{\omega_{L_e}^4}{(kv_{T_e})^3} \exp\left\{ -\frac{3}{2} - \frac{1}{2} \frac{\omega_{L_e}^2}{k^2 v_{T_e}^2} \right\}, \tag{6.34b}$$

which augments (6.34a) naturally. Comparison of the damping (6.34b) with the spectrum (6.34a) shows that the conditions (6.28) on the phase velocity of the electron Langmuir oscillation are indeed necessary and sufficient for its existence in an isotropic plasma.

Ion sound exists only in a nonisothermal plasma with hot electrons and cold ions:

$$T_e \gg T_i \tag{6.36}$$

and has a phase velocity greater than the ion thermal velocity and less than the electron thermal velocity:

$$v_{T_i} \ll \frac{\omega}{k} \ll v_{T_e}. \tag{6.37}$$

Equation (6.35) with allowance for the asymptotic behavior of the function $J_+(\beta)$ in (6.23) in the region of small values of the argument,

$$J_+(\beta) = \beta^2 - \frac{\beta^4}{3} + \frac{\beta^6}{15} - \ldots - i\sqrt{\frac{\pi}{2}}\beta e^{-\beta^2/2}, \quad \beta \ll 1 \tag{6.38}$$

simplifies appreciably to

$$1 + \frac{\omega_{L_e}^2}{k^2 v_{T_e}^2} - \frac{\omega_{L_i}^2}{\omega^2} - 3k^2 v_{T_i}^2 \frac{\omega_{L_i}^2}{\omega^4} = 0 \tag{6.39}$$

and leads to the spectrum

$$\omega^2(\mathbf{k}) = \frac{\omega_{L_i}^2}{1 + \dfrac{\omega_{L_e}^2}{k^2 v_{T_e}^2}} + 3k^2 v_{T_i}^2. \tag{6.40a}$$

Here, the second term on the right-hand side must be taken into account only in the limit of short wavelengths,

$$\omega_{L_e}^2 \ll k^2 v_{T_e}^2, \tag{6.41}$$

which are short compared with the electron Debye radius:

$$r_{D_e} \equiv v_{T_e}/\omega_{L_e}. \tag{6.42}$$

In this case, the spectrum (6.40a) takes the form

$$\omega^2(\mathbf{k}) = \omega_{L_i}^2 + 3k^2 v_{T_i}^2 - \omega_{L_i}^2 (k r_{D_e})^{-2} \tag{6.43}$$

and under the conditions

$$3(k r_{D_e})^2 (k r_{D_i})^2 \gg 1 \tag{6.44}$$

is identical after the replacement of the ion quantities by the electron quantities with (6.34a):

$$\omega^2(\mathbf{k}) = \omega_{L_i}^2 + 3k^2 v_{T_i}^2. \tag{6.45}$$

Electrostatic oscillations with the spectrum (6.45) are sometimes called ion Langmuir oscillations. If the ion temperature is so low (6.36) that the wavelengths of the acoustic oscillations are appreciably greater than the ion Debye radius,

$$(k r_{D_i})^2 \ll 1, \quad (k r_{D_e})^2 \gg 1, \tag{6.46}$$

but remain less than the electron Debye radius (6.41), then the inequality (6.44) can be replaced by the opposite:

$$3(k r_{D_i})^2 (k r_{D_e})^2 \ll 1. \tag{6.47}$$

Under these conditions, the spectrum of short-wavelength ion sound takes the form

$$\omega^2(\mathbf{k}) - \omega_{L_i}^2 \{1 - (k r_{D_e})^{-2}\}. \tag{6.48}$$

In the limit of long wavelengths greater than the electron Debye radius,

$$(k r_{D_e})^2 \ll 1, \tag{6.49}$$

the spectrum of the ion-acoustic oscillations (6.40a) can be written in the form typical of sound:

$$\omega^2(\mathbf{k}) = (k v_s)^2, \tag{6.50a}$$

where $v_s = (\varkappa T_e/M)^{1/2}$ is the velocity of sound (for $e_i = |e|$).

The damping $\gamma_s(\mathbf{k})$ of the ion sound is determined by means of formula (6.25a)

$$\gamma_s(\mathbf{k}) = \omega_{L_i}\left[1 + \frac{\omega_{L_e}^2}{k^2 v_{T_e}^2}\right]^{-2} \left\{ \sqrt{\frac{\pi}{8}} \frac{\omega_{L_e}^2}{k^2 v_{T_e}^2} \frac{\omega_{L_i}}{k v_{T_e}} + \sqrt{\frac{\pi}{8}} \left(\frac{\omega_{L_i}}{k v_{T_i}}\right)^3 \exp\left[-\frac{3}{2} - \frac{1}{2}\frac{\omega_{L_i}^2}{k^2 v_{T_i}^2}\left(1 + \frac{\omega_{L_e}^2}{k^2 v_{T_e}^2}\right)^{-1}\right]\right\} \tag{6.40b}$$

and augments the spectrum (6.40a) naturally. In particular, long-wavelength (6.49) ion sound (6.50a) is damped with the damping constant

$$\gamma_s(\mathbf{k}) = \sqrt{\frac{\pi}{8}\frac{m}{M}}\,(k v_s), \tag{6.50b}$$

which is determined by the first electron term in the curly brackets on the right-hand side of (6.40b). At the same time, the exponentially small damping on ions $(\pi/8)^{1/2}(kv_s)(T_e/T_i)^{3/2} \times \exp\{-\frac{1}{2}(T_e/T_i)\}$ is negligible compared with (6.50b) because of the condition (6.36). Comparison of the spectrum (6.40a) and the damping (6.40b) justifies the introduction of the conditions (6.36)-(6.37), which guarantee the existence of ion-acoustic oscillations of an isotropic plasma.

Nonelectrostatic oscillations, like electron Langmuire oscillations, exist (their damping is equal to zero; see (6.56)) in a plasma with an arbitrary relation between the electron and ion temperatures, and they have a phase velocity that appreciably exceeds the thermal velocity of the electrons and ions (6.28). The transversality condition

$$(\mathbf{kE}) = 0 \tag{6.51}$$

means in accordance with Maxwell's equations (2.13) that the dispersion equation (6.29) holds because of the vanishing of the second factor on the left-hand side:

$$\left\{\varepsilon^{tr}(\omega, \mathbf{k}) - \frac{c^2k^2}{\omega^2}\right\}^2 = 0. \tag{6.52}$$

The square of the curly brackets in Eq. (6.52) corresponds to the presence of two independent transverse oscillations with electric field vectors $\mathbf{E}(\omega, \mathbf{k})$ orthogonal to each other in the plane perpendicular to the wave vector **k**. To the two such independent oscillations there corresponds only a single spectrum,*

$$\omega^2(\mathbf{k}) = \omega_{L_e}^2 + c^2k^2, \tag{6.53}$$

which arises from Eq. (6.52), or, and this is the same thing, from the equation

$$\varepsilon^{tr}(\omega, \mathbf{k}) - \frac{c^2k^2}{\omega^2} = 0, \tag{6.54}$$

which simplifies under the conditions (6.28) to

$$1 - \frac{\omega_{L_e}^2}{\omega^2} - \frac{\omega_{L_i}^2}{\omega^2} - \frac{c^2k^2}{\omega^2} = 0. \tag{6.55}$$

Transverse oscillations with the spectrum (6.53) have a phase velocity ω/k that is greater than the velocity of light in vacuum. Therefore, in agreement with the remark above concerning formulas (6.25)-(6.26) for the imaginary parts $\varepsilon^{l\,''}(\omega, \mathbf{k})$ and $\varepsilon^{tr\,''}(\omega, \mathbf{k})$, the imaginary part of the transverse permittivity for such oscillations vanishes identically:

$$\operatorname{Im} \varepsilon^{tr}(\omega, \mathbf{k}) = 0, \quad \frac{\omega}{k} > c, \tag{6.56}$$

and the dispersion relation (6.54) is equivalent to

$$\operatorname{Re} \varepsilon^{tr}(\omega, \mathbf{k}) - \frac{c^2k^2}{\omega^2} = 0. \tag{6.57}$$

In the limit of short wavelengths

$$\omega_{L_e}^2 \ll c^2k^2 \tag{6.58}$$

the spectrum of nonelectrostatic oscillations (6.53) corresponds to the dispersion law of an electromagnetic wave in vacuum:

$$\omega^2(\mathbf{k}) = c^2k^2. \tag{6.59}$$

---

* Such an effect, when to one and the same spectrum of a strictly transverse electromagnetic wave there correspond two independent polarizations of the wave, i.e., two noncollinear electric field vectors, is naturally called polarization degeneracy. If an external magnetic field is applied to the plasma, the polarization degeneracy is lifted: waves with independent polarization (the ordinary and the extraordinary) are described by different spectra in a magnetoactive plasma.

Long-wavelength oscillations,

$$\omega_{L_e}^2 \gg c^2k^2, \tag{6.60}$$

occur with frequency near the electron plasma frequency:

$$\omega(\mathbf{k}) = \omega_{L_e}\left\{1 + \frac{1}{2}\frac{c^2k^2}{\omega_{L_e}^2}\right\}. \tag{6.61}$$

Apart from the oscillation spectra, we require in a nonlinear theory of the oscillations of an isotropic plasma the spectral functions $(E_j E_i)_{\omega,\mathbf{k}}$ of waves propagating in such a plasma. As in the case of the permittivity tensor (6.17), the spectral function $(E_j E_i)_{\omega,\mathbf{k}}$ of characteristic oscillations of an isotropic plasma, can, as a rule, be represented as a linear combination of the elementary tensors (6.18) with appropriate scalar weights:

$$(E_j E_i)_{\omega,\mathbf{k}} = (E_l^2)_{\omega,\mathbf{k}}\varkappa_i\varkappa_j + \frac{1}{2}(E_{tr}^2)_{\omega,\mathbf{k}}(\delta_{ij} - \varkappa_i\varkappa_j). \tag{6.62}$$

The first term on the right-hand side of (6.62) corresponds to electrostatic oscillations, the second to transverse, nonelectrostatic oscillations. The factor 1/2 in the second term reflects the presence of two transverse oscillations with independent electric field vectors $\mathbf{E}(\omega, \mathbf{k})$ in agreement with the remark on formula (6.52). Turning to the physical meaning of the scalar factors $(E_l^2)_{\omega,\mathbf{k}}$ and $(E_{tr}^2)_{\omega,\mathbf{k}}$ in (6.62), we note that for oscillations with given spectrum $\omega(\mathbf{k})$ they are proportional to the δ function $\delta(\omega^2 - \omega^2(\mathbf{k}))$. Further, using the definition of the spectral energy density $W(\mathbf{k})$ of the corresponding oscillation in an isotropic plasma [cf. (4.4) in § 4]:

$$W(\mathbf{k}) = (2\pi)^3\int_0^\infty d\omega \frac{(E_j E_i)_{\omega,\mathbf{k}}}{4\pi}\frac{\partial}{\partial\omega}\left\{\omega\left[\varepsilon'_{ij}(\omega, \mathbf{k}) - \frac{c^2k^2}{\omega^2}(\delta_{ij} - \varkappa_i\varkappa_j)\right]\right\}, \tag{6.63}$$

where $\varepsilon'_{ij}(\omega, \mathbf{k})$ is the real part of the permittivity tensor (6.17), we can obtain relations that reveal the physical meaning of $(E_l^2)_{\omega,\mathbf{k}}$ and $(E_{tr}^2)_{\omega,\mathbf{k}}$:

$$(E_l^2)_{\omega,\mathbf{k}} = \frac{1}{2\pi^2}\sum W_l(\mathbf{k})\left\{\frac{\partial\omega\varepsilon^{l\prime}}{\partial\omega}\right\}^{-1}\delta(\omega - \omega(\mathbf{k})); \tag{6.64}$$

$$(E_{tr}^2)_{\omega,\mathbf{k}} = \frac{1}{2\pi^2}W_{tr}(\mathbf{k})\sum\left\{\frac{\partial\omega\left(\varepsilon^{tr\prime} - \frac{c^2k^2}{\omega^2}\right)}{\partial\omega}\right\}^{-1}\delta(\omega - \omega(\mathbf{k})). \tag{6.65}$$

Here, $\varepsilon^{l\prime}$ and $\varepsilon^{tr\prime}$ are the real parts of the longitudinal and the transverse permittivity, respectively, and the summation sign corresponds to summation over the possible values of the oscillation spectrum. For example, the spectral function of electrostatic oscillations of an isotropic plasma can be represented as a sum of spectral functions of the electron Langmuir oscillation:

$$(E_j E_i)^l_{\omega,\mathbf{k}} = \frac{1}{(2\pi)^2}W_l(\mathbf{k})\varkappa_i\varkappa_j\left\{\delta\left(\omega - \omega_{L_e} - \frac{3}{2}\omega_{L_e}k^2r_{D_e}^2\right) + \delta\left(\omega + \omega_{L_e} + \frac{3}{2}\omega_{L_e}k^2r_{D_e}^2\right)\right\} \tag{6.66}$$

and ion-sound oscillation:

$$(E_j E_i)^s_{\omega,\mathbf{k}} = \frac{1}{(2\pi)^2}W_s(\mathbf{k})\frac{(kr_{D_e})^2}{1+(kr_{D_e})^2}\varkappa_i\varkappa_j\left\{\delta\left(\omega - \sqrt{\omega_{L_i}^2[1+(kr_{D_e})^{-2}]^{-1} + 3(kv_{T_i})^2}\right) + \right.$$
$$\left. + \delta\left(\omega + \sqrt{\omega_{L_i}^2[1+(kr_{D_e})^{-2}]^{-1} + 3(kv_{T_i})^2}\right)\right\}. \tag{6.67}$$

The subscripts l and tr in (6.64)-(6.65) and l and s in (6.66)-(6.67) of the spectral densities $W(\mathbf{k})$ of the energy indicate the type of oscillation that has the given spectral function. A transverse oscillation can be described by the spectral function

$$(E_jE_i)^{tr}_{\omega,\mathbf{k}} = \frac{1}{(2\pi)^2} W_{tr}(\mathbf{k}) \frac{1}{2} (\delta_{ij} - \varkappa_i\varkappa_j) \{\delta(\omega - \sqrt{\omega_{L_e}^2 + c^2k^2}) + \delta(\omega + \sqrt{\omega_{L_e}^2 + c^2k^2})\}. \quad (6.68)$$

Thus, the expression (6.62) of the spectral function of the oscillations of an isotropic plasma is equivalent to (6.66)-(6.68), i.e.,

$$(E_jE_i)_{\omega,\mathbf{k}} = (E_jE_i)^l_{\omega,\mathbf{k}} + (E_jE_i)^s_{\omega,\mathbf{k}} + (E_jE_i)^{tr}_{\omega,\mathbf{k}}. \quad (6.69)$$

With regard to the spectral function of a transverse oscillation of an isotropic plasma, we make the following comment. Equation (6.68) corresponds to natural polarization of the transverse electromagnetic wave. Only in this case can the polarization tensor be expressed solely in terms of the wave vector and represented in the form $\sim (\delta_{ij} - \varkappa_i\varkappa_j)$. For if the transverse wave is linearly polarized, i.e., the vector of its electric field $\mathbf{E}(\omega, \mathbf{k})$ is along a fixed unit vector $\mathbf{e}$ in the plane perpendicular to the direction of propagation, the polarization tensor is proportional to e_ie_j, and the corresponding spectral function, in contrast to (6.68), has the form

$$(E_jE_i)^{tr}_{\omega,\mathbf{k}} = \frac{1}{(2\pi)^2} W_{tr}(\mathbf{k})\, e_ie_j \{\delta(\omega - \sqrt{\omega_{L_e}^2 + c^2k^2}) + \delta(\omega + \sqrt{\omega_{L_e}^2 + c^2k^2})\}, \quad \boldsymbol{\varkappa}\mathbf{e} = 0. \quad (6.70)$$

The spectral function (6.70) goes over into (6.68) after summation over all possible linear polarizations, corresponding to the transition to natural polarization:

$$\overline{e_ie_j} = \frac{1}{2} (\delta_{ij} - \varkappa_i\varkappa_j). \quad (6.71)$$

Formally, the bar above the tensor e_ie_j on the left-hand side of Eq. (6.71) corresponds to averaging over the angles φ between the vector $\mathbf{e}$ of linear polarization and some coordinate axis in the plane perpendicular to $\mathbf{k}$:

$$\overline{e_ie_j} \equiv \frac{1}{2\pi} \int_0^{2\pi} d\varphi\, e_ie_j. \quad (6.72)$$

In view of what we have said, the region of applicability of the representation (6.62) of the spectral function of the oscillations of an isotropic plasma as a decomposition with respect to the elementary tensors (6.18) is restricted to the case of natural polarization of the transverse wave. This is also obvious from the geometrical point of view, since in the case of linear polarization of nonelectrostatic oscillations, i.e., when there are two vectors $\boldsymbol{\varkappa}$ and $\mathbf{e}$ that characterize the spectral function $(E_jE_i)_{\omega,\mathbf{k}}$, the elementary tensors of second rank are not the same as the tensors (6.18) but are composed by means of the single vector $\boldsymbol{\varkappa}$. In addition, it must be remembered that in the general case the polarization of the transverse electromagnetic wave (for example, of light with frequency $\omega \approx ck$) propagating in an isotropic plasma can be much more complicated than in the two specific cases (6.68) and (6.70) of natural and linear polarization. Indeed, in optics it is well known [26, 27] that partly polarized light is characterized by not one but four parameters (the Stokes parameters). At the same time, using the language we have adopted here, we can express the spectral function $(E_jE_i)^{tr}_{\omega,\mathbf{k}}$ of the transverse electromagnetic waves as a decomposition with respect to the four elementary tensors

$$(\delta_{ij} - \varkappa_i\varkappa_j), \quad e_ie_j - [\boldsymbol{\varkappa}\mathbf{e}]_i[\boldsymbol{\varkappa}\mathbf{e}]_j, \quad e_i[\boldsymbol{\varkappa}\mathbf{e}]_j + e_j[\boldsymbol{\varkappa}\mathbf{e}]_i, \quad e_{isj}\varkappa_s, \quad (6.73)$$

made up of the unit vectors $\boldsymbol{\varkappa}$ and $\mathbf{e}$ ($\boldsymbol{\varkappa}\mathbf{e} = 0$). Namely, (cf. [28], p. 68, formula (6.10)):

$$(E_jE_i)^{tr}_{\omega,\mathbf{k}} = \frac{1}{2} \{W(\omega, \mathbf{k})(\delta_{ij} - \varkappa_i\varkappa_j) + M(\omega, \mathbf{k})(e_je_i - [\boldsymbol{\varkappa}\mathbf{e}]_j[\boldsymbol{\varkappa}\mathbf{e}]_i) +$$
$$+ C(\omega, \mathbf{k})(e_j[\boldsymbol{\varkappa}\mathbf{e}]_i + e_i[\boldsymbol{\varkappa}\mathbf{e}]_j) - iS(\omega, \mathbf{k})\, e_{jsi}\varkappa_s\}. \quad (6.74)$$

This expression enables one to understand most readily the meaning of the Stokes parameters W, M, C, S, and their relation to the previously introduced quantities. Setting M = C = S = 0 in (6.74) and comparing the results with the second term on the right-hand side of (6.62), we find

$$W(\omega, \mathbf{k}) \equiv (E_{tr}^2)_{\omega, \mathbf{k}}. \tag{6.75}$$

Thus, the first term on the right-hand side of (6.74) describes a naturally polarized transverse wave. Going over in (6.74) to the spectral densities $W_{tr}(\mathbf{k})$, $M(\mathbf{k})$, $C(\mathbf{k})$, $S(\mathbf{k})$ of W, M, C, S, we can write (6.74) in the form

$$(E_jE_i)^{tr}_{\omega, \mathbf{k}} = \frac{1}{8\pi^2}\{W_{tr}(\mathbf{k})(\delta_{ij}-\varkappa_i\varkappa_j) + M(\mathbf{k})\,(e_je_i - [\varkappa\mathbf{e}]_i[\varkappa\mathbf{e}]_j) +$$
$$+ C(\mathbf{k})(e_j[\varkappa\mathbf{e}]_i + e_i[\varkappa\mathbf{e}]_j) - iS(\mathbf{k})e_{jsi}\varkappa_s\}\{\delta(\omega-\sqrt{\omega_{Le}^2+c^2k^2}) + \delta(\omega+\sqrt{\omega_{Le}^2+c^2k^2})\}. \tag{6.76}$$

The physical meaning of the scalars $M(\mathbf{k})$, $C(\mathbf{k})$, and $S(\mathbf{k})$ as the coefficients in front of the corresponding elementary tensors in (6.76) is obvious from the geometrical properties of the elementary tensors: (M/W_{tr}) and (C/W_{tr}) determine the degree of linear polarization (in the corresponding planes through $\varkappa$) of transverse electromagnetic waves with the spectral function (6.76), and the ratio (S/W_{tr}) serves as a measure of the circular polarization (the tensor $e_{jsi}\varkappa_s$ describes "rotation" around the direction of propagation $\varkappa$ of the wave).*

Summarizing our discussion of the properties of the spectral function of characteristic oscillations of an isotropic plasma, we emphasize that the general expression for such a function is given by Eq. (6.69), in which the first and the second terms of the right-hand side are defined in (6.66)-(6.67) and the last, third term, is described, in general, in (6.76).

To conclude our review of the properties of an isotropic plasma in the framework of linear oscillation theory, we give an explicit expression for the inverse tensor $A_{ij}(\omega, \mathbf{k})$ (3.17). To do this, we decompose $A_{ij}(\omega, \mathbf{k})$ with respect to the elementary tensors (6.18):

$$A_{ij}(\omega, \mathbf{k}) = A^l(\omega, \mathbf{k})\varkappa_i\varkappa_j + A^{tr}(\omega, \mathbf{k})(\delta_{ij}-\varkappa_i\varkappa_j) \tag{6.77}$$

and we substitute the permittivity tensor (6.17) of an isotropic plasma and the right-hand side of (6.77) into the definition (3.18). The solution of the resulting system of equations for the coefficients A^l and A^{tr}:

$$\begin{gathered} A^l\varepsilon^l - A^{tr}\left(\varepsilon^{tr} - \frac{c^2k^2}{\omega^2}\right) = 0, \\ A^{tr}\left(\varepsilon^{tr} - \frac{c^2k^2}{\omega^2}\right) = 1 \end{gathered} \tag{6.78}$$

allows us to write the desired expression in terms of the longitudinal and transverse permittivity:

$$A_{ij}(\omega, \mathbf{k}) = \frac{\varkappa_i\varkappa_j}{\varepsilon^l(\omega, \mathbf{k})} + \frac{\delta_{ij}-\varkappa_i\varkappa_j}{\varepsilon^{tr}(\omega, \mathbf{k}) - \frac{c^2k^2}{\omega^2}}. \tag{6.79}$$

In particular, for small values of the imaginary parts $\varepsilon^{l''}$ and $\varepsilon^{tr''}$ of the complex permittivities $\varepsilon^l(\omega, \mathbf{k})$ and $\varepsilon^{tr}(\omega, \mathbf{k})$ the denominators in (6.79) can be represented approximately

---

* Comparison with the textbook ([27], p. 163 (50.13)) shows that after the transition in the tensor equation (6.76) to a corresponding coordinate system, the Stokes parameters W_{tr}, M, C, and S are related to the components of the corresponding polarization tensor by the equations $\xi_1 = (C/W_{tr})$, $\xi_2 = -(S/W_{tr})$, $\xi_3 = (M/W_{tr})$.

by the relations*

$$\frac{1}{\varepsilon^{l}(\omega,\ \mathbf{k})} \simeq \frac{P}{\varepsilon^{l\prime}(\omega,\ \mathbf{k})} - i\pi\,\mathrm{sgn}\,\varepsilon^{l\prime\prime}(\omega,\ \mathbf{k})\,\delta\left(\varepsilon^{l\prime}(\omega,\ \mathbf{k})\right); \tag{6.80}$$

$$\frac{1}{\varepsilon^{tr}(\omega,\ \mathbf{k}) - \frac{c^2k^2}{\omega^2}} \simeq \frac{P}{\varepsilon^{tr\prime}(\omega,\mathbf{k}) - \frac{c^2k^2}{\omega^2}} - i\pi\,\mathrm{sgn}\,\varepsilon^{tr\prime\prime}(\omega,\ \mathbf{k})\,\delta\left(\varepsilon^{tr\prime}(\omega,\ \mathbf{k}) - \frac{c^2k^2}{\omega^2}\right). \tag{6.81}$$

In this connection, it is convenient to decompose the inverse tensor $A_{ij}(\omega, \mathbf{k})$ into resonance and nonresonance parts:

$$A_{ij}^{\mathrm{res}}(\omega,\ \mathbf{k}) = -i\pi\left\{\varkappa_i\varkappa_j\,\mathrm{sgn}\,\varepsilon^{l\prime\prime}(\omega,\ \mathbf{k})\,\delta\left(\varepsilon^{l\prime}(\omega,\ \mathbf{k})\right) + (\delta_{ij} - \varkappa_i\varkappa_j)\,\mathrm{sgn}\,\varepsilon^{tr\prime\prime}(\omega,\ \mathbf{k})\,\delta\left(\varepsilon^{tr\prime}(\omega,\ \mathbf{k}) - \frac{c^2k^2}{\omega^2}\right)\right\}; \tag{6.82}$$

$$A_{ij}^{\mathrm{nonres}}(\omega,\ \mathbf{k}) = \varkappa_i\varkappa_j\frac{P}{\varepsilon^{l\prime}(\omega,\ \mathbf{k})} + (\delta_{ij} - \varkappa_i\varkappa_j)\frac{P}{\varepsilon^{tr\prime}(\omega,\ \mathbf{k}) - \frac{c^2k^2}{\omega^2}}. \tag{6.83}$$

As we shall see below, the resonance part (6.82) of the inverse Maxwell tensor will be very important in the study of decay processes in the nonlinear theory of the oscillations of an isotropic plasma described by the generalized kinetic field equation (3.22) by means of the three-index tensor S_{ijs} [see also (3.22a)].

§ 7. Three-Index Tensor $S_{ijs}(\omega,\ \mathbf{k};\ \omega',\ \mathbf{k}')$ in an Isotropic Plasma

The next and very important special case of the complex permittivity tensors (6.10) of an isotropic plasma is the three-index tensor $\varepsilon_{ijs}(\omega, \mathbf{k}; \omega', \mathbf{k}')$. It is the first tensor that arises only in the nonlinear theory of oscillations and is absent in linear theory, i.e., it is the first nonlinear tensor:

$$\varepsilon_{ijs}(\omega,\ \mathbf{k};\ \omega',\ \mathbf{k}') = -i\frac{4\pi e^3}{\omega}\int d\mathbf{p}\,\frac{v_i}{\omega - \mathbf{kv}}\,\alpha''_{mj}\,\frac{\partial}{\partial p_m}\,\frac{\alpha'_{ns}}{\omega' - \mathbf{k}'\mathbf{v}}\,\frac{\partial f_0}{\partial p_n}. \tag{7.1}$$

Here and in what follows we use an abbreviated notation for the tensors α_{ij} [see (5.16)]:

$$\alpha_{ij} \equiv \alpha_{ij}(\omega,\ \mathbf{k},\ \mathbf{v}),\qquad \alpha'_{ij} \equiv \alpha_{ij}(\omega',\ \mathbf{k}',\ \mathbf{v}),\qquad \alpha''_{ij} = \alpha_{ij}(\omega'',\ \mathbf{k}'',\ \mathbf{v}), \tag{7.2}$$

and the frequencies and wave vectors are related by the equations

$$\omega'' = \omega - \omega',\qquad \mathbf{k}'' = \mathbf{k} - \mathbf{k}'. \tag{7.3}$$

In the generalized kinetic equation (3.22) we have the sum S_{ijs} of the two corresponding three-index tensors. The present section is devoted to a systematic study of this sum $S_{ijs}(\omega, \mathbf{k}; \omega', \mathbf{k}')$, and not of the individual terms of the form (7.1). Such a study is greatly simplified by the fact that the tensor $S_{ijs}(\omega, \mathbf{k}; \omega', \mathbf{k}')$, which, in accordance with the definition (3.24) (which arises in the derivation of the generalized kinetic equation), differs from the ex-

---

* The right-hand sides of Eqs. (6.80)-(6.81) are limiting expressions of fractions of the form

$$[\varepsilon(\omega,\ \mathbf{k})]^{-1} = [\varepsilon'(\omega,\ \mathbf{k}) + i\varepsilon''(\omega,\ \mathbf{k})]^{-1} = [\varepsilon'(\omega,\ \mathbf{k})/(\varepsilon'^2 + \varepsilon''^2)] - i\,[\varepsilon''(\omega,\ \mathbf{k})/(\varepsilon'^2 + \varepsilon''^2)]$$

as $\varepsilon'' \to 0$. The relations (6.80)-(6.81) are approximate because of the finite values of the imaginary parts $\varepsilon^{l''}(\omega, \mathbf{k})$ and $\varepsilon^{tr''}(\omega, \mathbf{k})$ of the permittivity. In this sense, such formulas are more accurate, the smaller are the corresponding imaginary parts.

pression symmetrized with respect to the last two indices

$$\varepsilon_{i(js)}(\omega, \mathbf{k}; \omega', \mathbf{k}') = \frac{1}{2!}\{\varepsilon_{ijs}(\omega, \mathbf{k}; \omega', \mathbf{k}') + \varepsilon_{isj}(\omega, \mathbf{k}; \omega'', \mathbf{k}'')\} \tag{7.4}$$

only by the absence of the factor 1/2. The second term of the tensor $S_{ijs}(\omega, \mathbf{k}; \omega', \mathbf{k}')$ is obtained from the first by transposing the last two indices and replacing the second argument:

$$\varepsilon_{isj}(\omega, \mathbf{k}; \omega'', \mathbf{k}'') = -i\frac{4\pi e^3}{\omega}\int d\mathbf{p}\frac{v_i}{\omega - \mathbf{k}\mathbf{v}}\alpha'_{ms}\frac{\partial}{\partial p_m}\frac{\alpha''_{nj}}{\omega'' - \mathbf{k}''\mathbf{v}}\frac{\partial f_0}{\partial p_m}. \tag{7.5}$$

In the nonrelativistic limit we can replace the momenta by the expression (5.30) in terms of the velocity, so that the tensor (7.1) takes the form*

$$\varepsilon_{ijs}(\omega, \mathbf{k}; \omega', \mathbf{k}') = -i\frac{4\pi e^3}{m^2}\int d\mathbf{v}\frac{v_i}{\omega}\frac{\alpha''_{mj}}{\omega - \mathbf{k}\mathbf{v}}\frac{\partial}{\partial v_m}\frac{\alpha'_{ns}}{\omega' - \mathbf{k}'\mathbf{v}}\frac{\partial f_0}{\partial v_n}. \tag{7.6}$$

In this expression we integrate once by parts (with respect to $d\mathbf{v}$) and add the result to the similar tensor $\varepsilon_{ijs}(\omega, \mathbf{k}; \omega', \mathbf{k}')$. Then for the tensor $S_{ijs}(\omega, \mathbf{k}; \omega', \mathbf{k}')$ we obtain the equation

$$S_{ijs}(\omega, \mathbf{k}; \omega', \mathbf{k}') = i\frac{4\pi e^3}{m^2}\frac{1}{\omega\omega'\omega''}\int d\mathbf{v}\beta_{ni}\{\beta'_{ns}\beta''_{mj} + \beta'_{ms}\beta_{nj} - \beta'_{ns}\beta_{mj}\}\frac{\partial f_0}{\partial v_m}, \tag{7.7}$$

if we use the identity

$$\int d\mathbf{v}\frac{\omega' - \mathbf{k}'\mathbf{v}}{\omega - \mathbf{k}\mathbf{v}}\beta_{ni}\{\beta'_{ns}\beta'_{mj} - \beta'_{ms}\beta'_{nj}\}\frac{\partial f_0}{\partial v_m} \equiv 0. \tag{7.8}$$

The tensors β_{ij} are related to the tensors α_{ij} by the simple equations

$$\omega\alpha_{ij} = (\omega - \mathbf{k}\mathbf{v})\beta_{ij}, \quad \omega'\alpha'_{ij} = (\omega' - \mathbf{k}'\mathbf{v})\beta'_{ij}, \quad \omega''\alpha''_{mj} = (\omega'' - \mathbf{k}''\mathbf{v})\beta''_{mj}. \tag{7.9}$$

The advantage of writing the tensor $S_{ijs}(\omega, \mathbf{k}; \omega', \mathbf{k}')$ in the form (6.7) is that the tensor $S_{jsi}(\omega'', \mathbf{k}''; \omega, \mathbf{k})$ too, which also occurs in the generalized kinetic field equation (3.22), can be represented as an integral form on the right-hand side of Eq. (7.7), but with opposite sign. Thus, we have the important symmetry relation

$$S_{ijs}(\omega, \mathbf{k}; \omega', \mathbf{k}') = -S_{jsi}(\omega'', \mathbf{k}''; \omega, \mathbf{k}), \tag{7.10}$$

which appreciably shortens the calculations. Note that there exist at least two other representations of the tensor S_{ijs} that are completely analogous to (7.7) and, therefore, in accordance with (7.10) there are of the tensor S_{jsi} as well:

$$S_{ijs}(\omega, \mathbf{k}; \omega', \mathbf{k}') = i\frac{4\pi e^3}{m^2}\frac{1}{\omega\omega'\omega''}\int d\mathbf{v}\beta'_{ns}\{\beta_{ni}\beta''_{mj} + \beta'_{nj}\beta_{mi} - \beta'_{mj}\beta_{ni}\}\frac{\partial f_0}{\partial v_m}, \tag{7.11}$$

$$S_{ijs}(\omega, \mathbf{k}; \omega', \mathbf{k}') = i\frac{4\pi e^3}{m^2}\frac{1}{\omega\omega'\omega''}\int d\mathbf{v}\beta''_{nj}\{\beta_{ni}\beta'_{ms} + \beta''_{ns}\beta_{mi} - \beta''_{ms}\beta_{ni}\}\frac{\partial f_0}{\partial v_m}. \tag{7.12}$$

It is not difficult to obtain such formulas in the relativistic approximation as well. To do this it is sufficient to allow in the integration by parts (with respect to the momenta) on the

---

* In a relativistic plasma, the equilibrium unperturbed function is normalized in momentum space: $\int d\mathbf{p}f_0 = N$. On the transition to the nonrelativistic limit it is natural to normalize f_0 in the velocity space: $\int d\mathbf{v}f_0 = N$. Such a normalization will be understood in all that follows when the formulas are simplified for velocities of the plasma particles $v \ll c$.

right-hand side of (7.1) for the following differentiation rule:

$$\frac{\partial v_i}{\partial p_j} = \frac{c^2}{\mathcal{E}}\left\{\delta_{ij} - \frac{v_i v_j}{c^2}\right\}. \tag{7.13}$$

Then the relation similar to (7.7) takes the form

$$S_{ijs}(\omega,\ \mathbf{k};\ \omega',\ \mathbf{k}') = i\frac{4\pi e^3}{\omega\omega'\omega''}\int d\mathbf{p}\,\frac{c^2}{\mathcal{E}}\left(\delta_{nl} - \frac{v_n v_l}{c^2}\right)\beta_{ni}\,\{\beta'_{ls}\beta''_{mj} + \beta'_{ms}\beta_{lj} - \beta'_{ls}\beta_{mj}\}\frac{\partial f_0}{\partial p_m} \tag{7.14}$$

and proves that the symmetry property (7.10) also holds in a relativistic plasma, in which the tensor $S_{jsi}(\omega'', \mathbf{k}''; \omega, \mathbf{k})$ is, as before, determined by the right-hand side of (7.14), but with the opposite sign.

Let us return a nonrelativistic plasma. Integrating (7.7) once more by parts with respect to $d\mathbf{v}$ and using the identity

$$(\beta_{nj} - \beta''_{nj})(\beta_{ni}\beta'_{ms} - \beta_{mi}\beta'_{ns})\left(\frac{k'_m}{\omega' - \mathbf{k}'\mathbf{v}} - \frac{k_m}{\omega - \mathbf{k}\mathbf{v}}\right) \equiv 0, \tag{7.15}$$

we obtain the more symmetric representation of the tensor S_{ijs}:

$$S_{ijs}(\omega,\ \mathbf{k};\ \omega',\ \mathbf{k}') = -\frac{i}{\omega\omega'\omega''}\frac{4\pi e^3}{m^2}\int d\mathbf{v} f_0 \beta_{ai}\beta''_{bj}\beta'_{cs}\left\{\frac{k_a}{\omega_0}\delta_{bc} + \frac{k''_b}{\omega_2}\delta_{ac} + \frac{k'_c}{\omega_1}\delta_{ab}\right\}, \tag{7.16}$$

in which for brevity we have introduced the notation

$$\omega_0 \equiv \omega - \mathbf{k}\mathbf{v};\quad \omega_1 \equiv \omega' - \mathbf{k}'\mathbf{v};\quad \omega_2 \equiv \omega'' - \mathbf{k}''\mathbf{v} \tag{7.17}$$

(in this notation, the tensor β_{ij} is written in the form $\beta_{ij} = \delta_{ij} + k_i v_j/\omega_0$). The expression (7.16) is completely symmetric with respect to a cyclical permutation of the primes and indices, provided (7.3) is observed. Therefore, the symmetry property (7.10) of the tensor $S_{ijs}(\omega, \mathbf{k}; \omega', \mathbf{k}')$ follows from the expression (7.16) by the simple relabeling

$$\begin{array}{lll} i \to j, & \omega \to \omega'', & \mathbf{k} \to \mathbf{k}'', \\ j \to s, & \omega'' \to -\omega', & \mathbf{k}'' \to -\mathbf{k}', \\ s \to i, & \omega' \to \omega, & \mathbf{k}' \to \mathbf{k}. \end{array} \tag{7.18}$$

In addition, using the symmetric representation (7.16) of the tensor S, we can strengthen Eq. (7.10) by including in it the further tensor $S_{sji}(\omega', \mathbf{k}'; \omega, \mathbf{k})$:

$$S_{ijs}(\omega,\ \mathbf{k};\ \omega',\ \mathbf{k}') = -S_{jsi}(\omega'',\ \mathbf{k}'';\ \omega,\ \mathbf{k}) = -S_{sji}(\omega',\ \mathbf{k}';\ \omega,\ \mathbf{k}). \tag{7.19}$$

We shall not here list all the symmetry properties that follow from (7.16), since we shall not require them in what follows. We note only that, using (7.16), we can also obtain the symmetry property (2.7) of the tensor S, which, we may mention, also follows directly from the definition (3.24):

$$S_{ijs}(\omega,\ \mathbf{k};\ \omega',\ \mathbf{k}') = S_{isj}(\omega,\ \mathbf{k};\ \omega'',\ \mathbf{k}''). \tag{7.20}$$

In view of the importance of the relation (7.16), we give a further expression of the tensor S (which follows from (7.16)) in terms of the tensor α_{ij}:

$$S_{ijs}(\omega,\ \mathbf{k};\ \omega',\ \mathbf{k}') = -i\frac{4\pi e^3}{m^2}\int d\mathbf{v} f_0\,\frac{\alpha_{ai}\alpha''_{bj}\alpha'_{cs}}{\omega_0\omega_1\omega_2}\left\{\frac{k_a}{\omega_0}\delta_{bc} + \frac{k''_b}{\omega_2}\delta_{ac} + \frac{k'_c}{\omega_1}\delta_{ab}\right\}. \tag{7.21}$$

An equation of the type (7.16) also exists, of course, in a relativistic plasma:

$$S_{ijs}(\omega,\ \mathbf{k};\ \omega',\ \mathbf{k}') = i\frac{4\pi e^3}{\omega\omega'\omega''}\int d\mathbf{p}f_0\left\{\beta_{ai}\beta''_{bj}\beta'_{cs}\frac{c^2}{\mathcal{E}}\left[v_a\left(\delta_{cb}-\frac{v_cv_b}{c^2}\right)+v_b\left(\delta_{ac}-\frac{v_av_c}{c^2}\right)+\right.\right.$$

$$\left.\left.+v_c\left(\delta_{ab}-\frac{v_av_b}{c^2}\right)\right]-\frac{c^4}{\mathcal{E}^2}\left(\delta_{nb}-\frac{v_nv_b}{c^2}\right)\left(\delta_{ma}-\frac{v_mv_a}{c^2}\right)\left[\frac{k_m}{\omega_0}\beta_{ai}\beta''_{bj}\beta'_{ns}+\frac{k''_m}{\omega_2}\beta_{ni}\beta''_{aj}\beta'_{bs}+\frac{k'_m}{\omega_1}\beta_{bi}\beta''_{nj}\beta'_{as}\right]\right\} \tag{7.22}$$

and goes over in the zeroth approximation in the parameter v/c (formally: $c = \infty$) into the nonrelativistic formula (7.16). Hence, in particular, it follows that the symmetry relations (7.19) are also true in a relativistic plasma. In the limit of a cold plasma ($v_T = 0$) we obtain from the relations (7.16) and (7.21) or (7.22) an expression for the tensor S,

$$S_{ijs}(\omega,\ \mathbf{k};\ \omega',\ \mathbf{k}') = -i\frac{e}{m}\frac{\omega_L^2}{\omega\omega'\omega''}\left\{\frac{k_i}{\omega}\delta_{js}+\frac{k'_s}{\omega'}\delta_{ij}+\frac{k''_j}{\omega''}\delta_{is}\right\}, \tag{7.23}$$

which can be obtained by setting formally $\mathbf{v} = 0$ in those relations. Then the tensors β_{ij} and α_{ij} reduce to the Kronecker deltas:

$$\beta_{ij} = \delta_{ij} + \frac{k_iv_j}{\omega_0} = \delta_{ij}, \qquad \alpha_{ij} = \frac{k_iv_j + \delta_{ij}\omega_0}{\omega} = \delta_{ij}, \qquad \mathbf{v} \equiv 0. \tag{7.24}$$

Formula (7.23) emphasizes one of the differences between the nonlinear tensor S_{ijs} and the permittivity tensor of a nonrelativistic isotropic plasma, in which the dependence on the wave number disappears when the thermal motion does:

$$\varepsilon_{ij}(\omega,\ \mathbf{k}) = \delta_{ij}\left(1-\frac{\omega_L^2}{\omega^2}\right), \qquad kv_T = 0. \tag{7.25}$$

The dependence of the tensor S on the wave vectors **k**, **k'**, **k"** in a cold plasma can be interpreted as spatial dispersion that arises in the nonlinear theory of oscillations. The mechanism of particle transport responsible for the nonlocality of the relation between the current **j** and the electric field **E** is in this case the oscillation of the particles in the fields of the characteristic plasma oscillations and not the thermal motion of the particles.

The relativistic expression (7.22) of the three-index tensor S in the long-wavelength limit $\mathbf{k} = \mathbf{k}' = 0$ does not, in contrast to the nonrelativistic expressions (7.16) and (7.21), in general vanish:

$$S_{ijs}(\omega,\ 0;\ \omega',\ 0) = i\frac{4\pi e^3}{\omega\omega'\omega''}\int d\mathbf{p}f_0\frac{c^2}{\mathcal{E}^2}[v_i\delta_{js}+v_j\delta_{is}+v_s\delta_{ij}-(3/c^2)v_iv_jv_s] \tag{7.26}$$

for an appropriate choice of the unperturbed distribution function f_0. In a nonrelativistic isotropic plasma ($c = \infty$) the tensor S vanishes when the wave vectors do:

$$S_{ijs}(\omega,\ 0;\ \omega',\ 0) = 0, \qquad c = \infty. \tag{7.27}$$

When studying the processes of nonlinear interaction of the characteristic oscillations of an isotropic plasma in the generalized kinetic equation, one is dealing essentially not with the tensor S itself, but only its contradictions. In this sense, the above relations for the tensor S will be used to calculate its contractions with the corresponding elementary tensors that determine the structure of the spectral functions of the fields of the interacting oscillations and the inverse tensor. The simplest contraction is the longitudinal one,

$$S(\omega,\ \mathbf{k};\ \omega',\ \mathbf{k}') \equiv S_{ijs}(\omega,\ \mathbf{k};\ \omega',\ \mathbf{k}')\frac{k_ik''_jk'_s}{kk'k''} =$$

$$= -i\frac{4\pi e^3}{m^2}\frac{1}{kk'k''}\int d\mathbf{v}f_0\frac{1}{\omega_0\omega_1\omega_2}\left\{\frac{k^2}{\omega_0}(\mathbf{k}'\mathbf{k}'')+\frac{k''^2}{\omega_2}(\mathbf{k}\mathbf{k}')+\frac{k'^2}{\omega_1}(\mathbf{k}\mathbf{k}'')\right\}, \tag{7.28}$$

which arises, for example, from the representation (7.16) of the tensor S by allowing for the equation

$$\frac{1}{\omega}\beta_{ij}k_j = \frac{k_i}{\omega_0}. \tag{7.29}$$

In the limit of a cold plasma, we obtain from this the expression

$$S(\omega, \mathbf{k}; \omega', \mathbf{k}') = -i\frac{e}{m}\frac{\omega_L^2}{\omega\omega'\omega''}\frac{1}{kk'k''}\left\{\frac{k^2}{\omega}(\mathbf{k}'\mathbf{k}'') + \frac{k''^2}{\omega''^2}(\mathbf{k}\mathbf{k}') + \frac{k'^2}{\omega'^2}(\mathbf{k}\mathbf{k}'')\right\}, \tag{7.30}$$

which can also be obtained by contracting the tensor (7.23) with the corresponding tensor that is longitudinal with respect to the wave vectors. We give another, no less useful, form of the contraction (7.28):

$$S(\omega, \mathbf{k}; \omega', \mathbf{k}') = -i\frac{4\pi e^3}{m^2}\frac{1}{kk'k''}\int d\mathbf{v}f_0\left\{\frac{(\mathbf{k}\mathbf{k}'')(\mathbf{k}\mathbf{k}')}{(\omega_1\omega_2)^2} - \frac{(\mathbf{k}\mathbf{k}')(\mathbf{k}'\mathbf{k}'')}{(\omega_0\omega_2)^2} - \frac{(\mathbf{k}\mathbf{k}'')(\mathbf{k}'\mathbf{k}'')}{(\omega_0\omega_1)^2}\right\}, \tag{7.31}$$

this is obtained from (7.28) on account of the identity

$$\left\{\frac{k^2}{\omega}(\mathbf{k}'\mathbf{k}'') + \frac{k''^2}{\omega''}(\mathbf{k}\mathbf{k}') + \frac{k'^2}{\omega'}(\mathbf{k}''\mathbf{k})\right\}\frac{1}{\omega\omega'\omega''} \equiv \frac{(\mathbf{k}\mathbf{k}'')(\mathbf{k}\mathbf{k}')}{(\omega'\omega'')^2} - \frac{(\mathbf{k}\mathbf{k}')(\mathbf{k}'\mathbf{k}'')}{(\omega\omega'')^2} - \frac{(\mathbf{k}\mathbf{k}'')(\mathbf{k}'\mathbf{k}'')}{(\omega\omega')^2}, \tag{7.32}$$

which holds if the conditions (7.3), which relate the frequencies and wave vectors, are satisifed. Using the equations

$$(\mathbf{k}'\mathbf{k}'') = \frac{1}{2}(k^2 - k'^2 - k''^2), \quad (\mathbf{k}\mathbf{k}'') = \frac{1}{2}(k^2 - k'^2 + k''^2),$$
$$(\mathbf{k}\mathbf{k}') = \frac{1}{2}(k^2 + k'^2 - k''^2) \tag{7.33}$$

we can reduce the contraction (7.28) to the expression

$$S(\omega, \mathbf{k}; \omega', \mathbf{k}') = -i\frac{2\pi e^3}{m^2}\frac{1}{kk'k''}\int d\mathbf{v}f_0\left\{\frac{1}{\omega_0\omega_1\omega_2}\left[\frac{k^4}{\omega_0} + \frac{k'^4}{\omega_1} + \frac{k''^4}{\omega_2}\right] + \left(\frac{kk'}{\omega_0\omega_1}\right)^2 + \left(\frac{kk''}{\omega_0\omega_2}\right)^2 - \left(\frac{k'k''}{\omega_1\omega_2}\right)^2\right\}, \tag{7.34}$$

which is used in [29] to study the Coulomb interaction of the electrostatic oscillations of an isotropic plasma.

If the interaction is one-dimensional, i.e., the wave vectors of the interacting oscillations are collinear with a certain chosen direction, the longitudinal contraction simplifies appreciably:

$$S(\omega, \mathbf{k}; \omega', \mathbf{k}') = -i\frac{4\pi e^3}{m^2}\frac{k_z k_z' k_z''}{kk'k''}\int d\mathbf{v}f_0\left\{\frac{k_z}{(\omega_1\omega_2)^2} - \frac{k_z'}{(\omega_0\omega_2)^2} - \frac{k_z''}{(\omega_0\omega_1)^2}\right\}. \tag{7.35}$$

Here the index z indicates the projections of the wave vectors onto the chosen direction and the combinations (7.17) contain noncomplete scalar products of the wave vectors with the velocity vector $\mathbf{v}$ and the products of the only nonvanishing components k_z, k'_z, and k''_z and v_z:

$$\omega_0 \equiv \omega - k_z v_z, \quad \omega_1 \equiv \omega' - k_z' v_z, \quad \omega_2 \equiv \omega'' - k_z'' v_z. \tag{7.36}$$

The case of one-dimensional nonlinear interaction has frequently been discussed [29, 30], and therefore, we shall study the contraction (7.35) in more detail. Physically, a one-dimensional interaction can arise, for example, under the influence of a strong external magnetic field. A detailed analysis of the resulting phenomena can be found in Chapters IV and V of this review, which are devoted to a magnetoactive plasma. Let us calculate the integrals that arise on the right-hand side of (7.35), taking the Maxwellian function (6.14) as the distribution function f_0. We shall distinguish two cases depending on the relation between the phase velocities of the interacting oscillations. If the phase velocities of two given oscillations are different, then in accordance with the relation that follows from (7.3):

$$\omega' k_z - \omega k'_z = \omega k''_z - \omega'' k_z = \omega' k''_z - \omega'' k'_z \tag{7.37}$$

they cannot be equal for any of the three pairs of interacting oscillations. Conversely, if for even one pair of oscillations the phase velocities are the same, for example,

$$\frac{\omega}{k_z} = \frac{\omega'}{k'_z}, \tag{7.38}$$

then all three oscillations have the same phase velocity v_{ph}, since in this case

$$\omega' k_z - \omega k'_z = \omega k''_z - \omega'' k_z = \omega' k''_z - \omega'' k'_z = 0 \tag{7.39}$$

and therefore (k_z, k'_z, $k''_z \neq 0$),

$$\frac{\omega}{k_z} = \frac{\omega'}{k'_z} = \frac{\omega''}{k''_z} = v_{ph} \tag{7.40}$$

Let us consider the first case of different phase velocities, when the members of the equations (7.37) are nonvanishing. Then the product of the two denominators can be split into their sum by the identity

$$\frac{1}{\omega_0 \omega_1} = \frac{1}{\omega' k_z - \omega k'_z} \left\{ \frac{k_z}{\omega_0} - \frac{k'_z}{\omega_1} \right\}, \tag{7.41}$$

which leads to the integral

$$\frac{1}{N} \int d\mathbf{v} f_0 \frac{1}{\omega_0 \omega_1} = \frac{1}{\omega' k_z - \omega k'_z} \left\{ \frac{k_z}{\omega} J_+\left(\frac{\omega}{|k_z v_T|} \right) - \frac{k'_z}{\omega'} J_+\left(\frac{\omega'}{|k'_z v_T|} \right) \right\}, \tag{7.42}$$

which can be expressed in terms of the function $J_+(\beta)$ (6.20) because of the equation

$$\int_{-\infty}^{+\infty} \frac{dv_z}{\sqrt{2\pi}\, v_T} e^{-v_z^2/2v_T^2} \frac{1}{\omega - k_z v_z} = \frac{1}{\omega} J_+\left(\frac{\omega}{|k_z v_T|} \right). \tag{7.43}$$

The expression for the integral for the product of the squares of the denominators ω_0^{-1} and ω_1^{-1}, for example, for the last term on the right-hand side of Eq. (7.35), is obtained by differentiating with respect to the frequencies in accordance with the rule [see the definition (6.20)]

$$\frac{d}{d\beta} J_+(\beta) = \frac{1}{\beta} J_+(\beta) - \beta J_+(\beta) + \beta. \tag{7.44}$$

Namely,

$$\begin{aligned} \frac{1}{N} \int d\mathbf{v} f_0 \frac{1}{(\omega_0 \omega_1)^2} = \frac{1}{v_T^2} \frac{1}{(\omega' k_z - \omega k'_z)^2} \Bigg\{ J_+\left(\frac{\omega}{|k_z v_T|} \right) + J_+\left(\frac{\omega'}{|k'_z v_T|} \right) + \\ + 2 \frac{k_z k'_z v_T^2}{\omega' k_z - \omega k'_z} \left[\frac{k'_z}{\omega'} J_+\left(\frac{\omega'}{|k'_z v_T|} \right) - \frac{k_z}{\omega} J_+\left(\frac{\omega}{|k_z v_T|} \right) \right] - 2 \Bigg\}. \end{aligned} \tag{7.45}$$

As a result, replacing each of the three terms on the right-hand side of (7.35) by a corresponding integral of the form (7.45), we obtain a very simple expression for the longitudinal contraction of S in the case of one-dimensional interaction of oscillations with different phase velocities:

$$S(\omega, \mathbf{k}; \omega', \mathbf{k}') = i \frac{e}{m} \frac{k_z k'_z k''_z}{k k' k''} \frac{\omega_L^2}{v_T^2} \frac{1}{(\omega' k_z - \omega k'_z)^2} \left\{ k_z J_+\left(\frac{\omega}{|k_z v_T|} \right) - k'_z J_+\left(\frac{\omega'}{|k'_z v_T|} \right) - k''_z J_+\left(\frac{\omega''}{|k''_z v_T|} \right) \right\}. \tag{7.46}$$

Physically, the case of different phase velocities of the oscillations corresponds, for example, to induced scattering, when the phase velocities ω/k_z and ω'/k'_z of the scattered oscillations, which are weakly damped (or not damped at all) and therefore interact weakly directly with the plasma particles, are large compared with the phase velocity ω''/k''_z of the third, virtual, oscillation, which is near or less than the thermal velocity of the particles.

If the phase velocities of the interacting oscillations are the same (7.40) (second of the analyzed possibilities), the expression (7.35) simplifies appreciably even before integration:

$$S(\omega,\ \mathbf{k};\ \omega',\ \mathbf{k}') = -i\frac{12\pi e^3}{m^2}\frac{1}{kk'k''}\int d\mathbf{v}\frac{f_0}{(v_{\text{ph}}-v_z)^4}. \tag{7.47}$$

Applying the differentiation rule (7.44) three times to the integral (7.43), we obtain the contraction of S, and in this case (the distribution f_0 is Maxwellian)

$$S(\omega,\ \mathbf{k};\ \omega',\ \mathbf{k}') = \frac{i}{2}\frac{e}{m}\frac{\omega_L^2}{v_T^4}\frac{1}{kk'k''}\{(3-\beta^2)J_+(\beta)+(\beta^2-2)\}. \tag{7.48}$$

Here

$$v_T\beta = (\omega/|k_z|) = (\omega'/|k'_z|) = (\omega''/|k''_z|). \tag{7.49}$$

Apart from the one-dimensional interaction, there is another case that is of interest on account of its simplicity: the special geometry of wave vectors of interacting oscillations when the wave vectors are perpendicular:

$$\mathbf{k}\mathbf{k}' = 0. \tag{7.50}$$

Then from the expression (7.31) of the longitudinal contraction of the tensor S we have

$$S(\omega,\ \mathbf{k};\ \omega',\ \mathbf{k}') = i\frac{4\pi e^3}{m^2}\frac{(\mathbf{k}\mathbf{k}'')(\mathbf{k}'\mathbf{k}'')}{kk'k''}\int d\mathbf{v} f_0\frac{1}{(\omega_0\omega_1)^2}. \tag{7.51}$$

Performing the integration with respect to the velocities of the particles of a Maxwellian distribution f_0,

$$\frac{1}{N}\int d\mathbf{v} f_0\frac{1}{\omega_0\omega_1} = \frac{1}{\omega}J_+\left(\frac{\omega}{kv_T}\right)\frac{1}{\omega'}J_+\left(\frac{\omega'}{k'v_T}\right), \qquad \mathbf{k}\mathbf{k}'=0, \tag{7.52}$$

and taking into account the differentiation rule (7.44), we obtain

$$S(\omega,\ \mathbf{k};\ \omega',\ \mathbf{k}') = i\frac{e}{m}\frac{\omega_L^2}{v_T^4}\frac{(\mathbf{k}\mathbf{k}'')(\mathbf{k}'\mathbf{k}'')}{(kk')^2}\frac{1}{kk'k''}\left\{1-J_+\left(\frac{\omega}{kv_T}\right)\right\}\left\{1-J_+\left(\frac{\omega'}{k'v_T}\right)\right\}. \tag{7.53}$$

The factor that determines the angles between the wave vectors can be simplified* under the conditions (7.50),

$$(\mathbf{k}\mathbf{k}'')(\mathbf{k}'\mathbf{k}'') \equiv -(kk')^2, \qquad \mathbf{k}\mathbf{k}'=0, \tag{7.54}$$

so that finally

$$S(\omega,\ \mathbf{k};\ \omega',\ \mathbf{k}') = -i\frac{e}{m}\frac{1}{kk'k''}\frac{\omega_L^2}{v_T^4}\left\{1-J_+\left(\frac{\omega}{kv_T}\right)\right\}\left\{1-J_+\left(\frac{\omega'}{k'v_T}\right)\right\}. \tag{7.55}$$

---

* If the angles between the wave vectors are arbitrary, we can generalize the identity:

$$(\mathbf{k}\mathbf{k}')(\mathbf{k}\mathbf{k}'') - (\mathbf{k}'\mathbf{k}'')(\mathbf{k}'\mathbf{k}) - (\mathbf{k}''\mathbf{k})(\mathbf{k}'\mathbf{k}'') \equiv [\mathbf{k}\mathbf{k}']^2. \tag{7.54a}$$

Thus, between the expressions (7.46) and (7.48) for the contraction S, on the one hand, and (7.55) on the other, there is an important difference: in the case of one-dimensional interaction $[\mathbf{kk}'] = 0$, the longitudinal contraction is a linear combination of the functions J_+, which determine the contributions of denominators of the type ω_0^{-1}, ω_1^{-1}, and ω_2^{-1}, which are associated with the interacting oscillations; in the case of the interaction of waves with perpendicular wave vectors, $(\mathbf{kk}') = 0$, the contraction of S contains a bilinear combination of the functions J_+. If instead of (7.50) orthogonality of the other pair of wave vectors holds,

$$(\mathbf{kk}'') = 0, \tag{7.56}$$

then in accordance with (7.31) the contraction of S takes the form

$$S(\omega, \mathbf{k}; \omega', \mathbf{k}') = -i\frac{e}{m}\frac{\omega_L^2}{v_T^4}\frac{1}{kk'k''}\left\{1 - J_+\left(\frac{\omega}{kv_T}\right)\right\}\left\{1 - J_+\frac{\omega''}{k''v_T}\right)\right\}, \tag{7.57}$$

which is obtained from (7.55) by making the substitution $\omega'' \rightleftharpoons \omega'$, $\mathbf{k}'' \rightleftharpoons \mathbf{k}'$. Note also that in the case of a purely one-dimensional interaction, $[\mathbf{kk}'] = 0$, of oscillations with different phase velocities the identity

$$\frac{k_z}{(\omega_1\omega_2)^2} - \frac{k_z'}{(\omega_0\omega_2)^2} - \frac{k_z''}{(\omega_0\omega_1)^2} \equiv \frac{1}{\omega_0\omega_1\omega_2}\left\{\frac{k_z}{\omega_0} + \frac{k_z'}{\omega_1} + \frac{k_z''}{\omega_2}\right\} \equiv \frac{1}{(\omega' k_z - \omega k_z')^2}\left\{\frac{k_z'^3}{\omega_1^2} + \frac{k_z''^3}{\omega_2^2} - \frac{k_z^3}{\omega_0^2}\right\} \tag{7.58}$$

enables us to represent the longitudinal contraction S in a slightly different form:

$$S(\omega, \mathbf{k}; \omega', \mathbf{k}') = -i\frac{e}{m}\omega_L^2\frac{k_z k_z' k_z''}{kk'k''}\frac{1}{(\omega' k_z - \omega k_z')^2}\int d\mathbf{v} f_0\left\{\frac{k_z'^3}{\omega_1^2} + \frac{k_z''^3}{\omega_2^2} - \frac{k_z^3}{\omega_0^2}\right\}. \tag{7.59}$$

Using the fact that for any unperturbed distribution f_0 that satisfies the condition (for example, isotropic f_0)

$$\int d\mathbf{v}\mathbf{k}\frac{\partial f_0}{\partial \mathbf{v}} = 0, \tag{7.60}$$

the longitudinal permittivity (6.19) of an isotropic plasma can be written in the form

$$\varepsilon^l(\omega, \mathbf{k}) = 1 - \frac{4\pi e^2}{m}\int d\mathbf{v} f_0\frac{1}{\omega_0^2}, \tag{7.61}$$

we can express the contraction (7.59) in terms of the partial longitudinal permittivities $\delta\varepsilon^l(\omega, \mathbf{k}) = [\varepsilon^l(\omega, \mathbf{k}) - 1]$ of the charged particles of the given species:

$$S(\omega, \mathbf{k}; \omega', \mathbf{k}') = i\frac{e}{m}\frac{k_z k_z' k_z''}{kk'k''}\frac{1}{(\omega' k_z - \omega k_z')^2}\{k_z'^3\delta\varepsilon^l(\omega', k_z') + k_z''^3\delta\varepsilon^l(\omega'', k_z'') - k_z^3\delta\varepsilon^l(\omega, k_z)\}. \tag{7.62}$$

Note that in this formula f_0 is any function that satisfies the condition (7.60). In particular, if the velocity distribution is the Maxwellian (6.14), then by means of the equation

$$k_z^3\delta\varepsilon^l(\omega, k_z) = \frac{\omega_L^2}{v_T^2}k_z\left\{1 - J_+\left(\frac{\omega}{|k_z v_T|}\right)\right\} \tag{7.63}$$

we transform the right-hand side of (7.62) into (7.46).

The assumption that the nonlinear interaction of the oscillations of an isotropic plasma is one-dimensional enables us to simplify appreciably not only the longitudinal contraction of the three-index tensor S, but also the complete tensor. Indeed, suppose the wave vectors $\mathbf{k}$, $\mathbf{k}'$, and $\mathbf{k}''$ are collinear with some direction along a unit vector $\mathbf{h}$. Then from the symmetric representation (7.16) we can obtain the much simpler expression

$$S_{ijs}(\omega,\ \mathbf{k};\ \omega',\ \mathbf{k}')=-i\frac{4\pi e^3}{m^2}\int d\mathbf{v}f_0\left\{\frac{h_ih_jh_s}{\omega_0\omega_1\omega_2}\left[\frac{k_z}{\omega_0}+\frac{k'_z}{\omega_1}+\frac{k''_z}{\omega_2}\right]+\right.$$
$$\left.+\frac{h_i}{\omega_0^2}(\delta_{js}-h_jh_s)\left[\frac{k''_z}{\omega''}\frac{1}{\omega_1}+\frac{1}{\omega_2}\frac{k'_z}{\omega'}\right]+\frac{h_j}{\omega_2^2}(\delta_{is}-h_ih_s)\left[\frac{k_z}{\omega}\frac{1}{\omega_1}-\frac{1}{\omega_0}\frac{k'_z}{\omega'}\right]+\frac{h_s}{\omega_1^2}(\delta_{ij}-h_ih_j)\left[\frac{k_z}{\omega}\frac{1}{\omega_2}-\frac{1}{\omega_0}\frac{k''_z}{\omega''}\right]\right\};$$
$$k_z\equiv(\mathbf{kh}),\ k'_z\equiv(\mathbf{k'h}),\ k''_z\equiv(\mathbf{k''h}). \tag{7.64}$$

In particular, for a Maxwellian distribution, we then obtain the tensor

$$S_{ijs}(\omega,\ \mathbf{k};\ \omega',\ \mathbf{k}')=-i\frac{e}{m}\frac{\omega_L^2}{v_T^2}\frac{1}{(\omega'k_z-\omega k'_z)^2}\left\{k'_zJ_+\left(\frac{\omega'}{|k'_zv_T|}\right)+\right.$$
$$\left.+k''_zJ_+\left(\frac{\omega''}{|k''_zv_T|}\right)-k_zJ_+\left(\frac{\omega}{|k_zv_T|}\right)\right\}\left\{h_ih_jh_s+v_T^2\frac{k'_zk''_z}{\omega'\omega''}h_i(\delta_{js}-h_jh_s)+\right.$$
$$\left.+v_T^2\frac{k_zk'_z}{\omega\omega'}h_j(\delta_{is}-h_ih_s)+v_T^2\frac{k_zk''_z}{\omega\omega''}h_s(\delta_{ij}-h_ih_j)\right\}-i\frac{e}{m}\frac{\omega_L^2}{v_T^2}\frac{1}{\omega\omega'\omega''}\left\{\frac{\omega}{k_z}h_i(\delta_{js}-h_jh_s)\left[J_+\left(\frac{\omega}{|k_zv_T|}\right)-1\right]+\right.$$
$$\left.+\frac{\omega''}{k''_z}h_j(\delta_{is}-h_ih_s)\left[J_+\left(\frac{\omega''}{|k''_zv_T|}\right)-1\right]+\frac{\omega'}{k'_z}h_s(\delta_{ij}-h_ih_j)\left[J_+\left(\frac{\omega'}{|k'_zv_T|}\right)-1\right]\right\}, \tag{7.65}$$

which is suitable for studying the nonlinear interaction of the oscillations of an isotropic plasma with different phase velocities. The longitudinal contractions (7.35) and (7.46) arise from the tensors (7.64) and (7.65), respectively, by scalar multiplication by the tensor

$$\frac{k_ik''_jk'_s}{kk'k''}. \tag{7.66}$$

Before we conclude our analysis of the different forms of the tensor S_{ijs} and its contradictions, we give for this tensor two further general expressions. Substituting the definition of the tensors β_{ij} into the right-hand side of (7.16), we obtain an expansion of the tensor in "powers" of the velocity **v**:

$$S_{ijs}(\omega,\ \mathbf{k};\ \omega',\ \mathbf{k}')=-i\frac{4\pi e^3}{m^2}\frac{1}{\omega\omega'\omega''}\int d\mathbf{v}f_0\left\{\sigma_{ijs}+\left[\frac{v_i}{\omega_0}k_a\sigma_{ajs}+\frac{v_j}{\omega_2}k''_b\sigma_{ibs}+\frac{v_s}{\omega_1}k'_c\sigma_{ijc}\right]+\right.$$
$$\left.+\left[\frac{v_iv_j}{\omega_0\omega_2}k_ak''_b\sigma_{abs}+\frac{v_iv_s}{\omega_0\omega_1}k_ak'_c\sigma_{ajc}+\frac{v_jv_s}{\omega_1\omega_2}k'_ck''_b\sigma_{ibc}\right]+v_iv_jv_s\frac{k_ak''_bk'_c}{\omega_0\omega_1\omega_2}\sigma_{abc}\right\}. \tag{7.67}$$

Here, to shorten the notation, we use the tensor σ_{ijs}:

$$\sigma_{ijs}\equiv\frac{k_i}{\omega_0}\delta_{js}+\frac{k''_j}{\omega_2}\delta_{is}+\frac{k'_s}{\omega_1}\delta_{ij}, \tag{7.68}$$

and the singularities of the expressions ω_0^{-1}, ω_1^{-1}, ω_2^{-1} are understood in the sense of the general notation (7.17). Note that the tensor σ_{ijs} at zero velocity ($\mathbf{v}$ = 0) is equal to the tensor in the curly brackets on the right-hand side of Eq. (7.23) for the tensor S_{ijs} in a cold isotropic plasma, so that

$$S_{ijs}(\omega,\ \mathbf{k};\ \omega',\ \mathbf{k}')=-i\frac{e}{m}\frac{\omega_L^2}{\omega\omega'\omega''}\sigma_{ijs}(\mathbf{v}=0),\quad v_T=0. \tag{7.69}$$

A second, more concrete expression for the tensor S_{ijs} can be obtained from the relation (7.7) by expanding the denominators ω_0^{-1} and ω_1^{-1} in the tensor β_{ij} in powers of the ratios

$$[(\mathbf{kv})/\omega]\ll 1,\quad [(\mathbf{k'v})/\omega']\ll 1. \tag{7.70}$$

We then obtain

$$
\begin{aligned}
S_{ijs}(\omega, \mathbf{k}; \omega', \mathbf{k}') = & -i\frac{e}{m}\frac{\omega_L^2}{\omega\omega'}\left\{\frac{1}{\omega''}\left[\delta_{ij}\left(\frac{k_s}{\omega}-\frac{k'_s}{\omega'}\right)-\delta_{js}\left(\frac{k_i}{\omega}-\frac{k'_i}{\omega'}\right)\right]+\right.\\
& +\frac{k''_j}{k''^4}\left[\frac{\mathbf{kk}''}{\omega^2}k''_i k_s+\frac{\mathbf{k}'\mathbf{k}''}{\omega'^2}k'_i k''_s+\frac{\mathbf{kk}'}{\omega\omega'}k''_i k''_s\right]-\left[\frac{k_s}{\omega}\left(\delta_{ij}-\frac{k''_i k''_j}{k''^2}\right)+\right.\\
& \left.+\frac{k'_i}{\omega'}\left(\delta_{js}-\frac{k''_j k''_s}{k''^2}\right)\right]\frac{1}{N}\int d\mathbf{v}f_0\frac{1}{\omega''-\mathbf{k}''\mathbf{v}}+\frac{k''_j}{k''^2}\left[\delta_{is}+\frac{\omega''}{k''^2}\left(\frac{k_s k''_i}{\omega}+\right.\right.\\
& \left.\left.+\frac{k'_i k''_s}{\omega'}\right)+\left(\frac{\omega''}{k''^2}\right)^2\left(\frac{k_s k''_i}{\omega^2}\mathbf{kk}''+\frac{k'_i k''_s}{\omega'^2}\mathbf{k}'\mathbf{k}''+\frac{k''_i k''_s}{\omega\omega'}\mathbf{kk}'\right)\right]\times\\
& \times\frac{1}{N}\int d\mathbf{v}\mathbf{k}''\frac{\partial f_0}{\partial\mathbf{v}}\frac{1}{\omega''-\mathbf{k}''\mathbf{v}}-\left\{\frac{k''_j}{k''^2}\left[\frac{k_s}{\omega^2}\left(k_i-\frac{\mathbf{kk}''}{k''^2}k''_i\right)+\frac{k'_i}{\omega'^2}\left(k'_s-\frac{\mathbf{k}'\mathbf{k}''}{k''^2}k''_s\right)+\right.\right.\\
& \left.+\frac{\mathbf{kk}'}{\omega\omega'}\left(\delta_{is}-\frac{k''_i k''_s}{k''^2}\right)\right]+\frac{k_s}{\omega^2}\frac{1}{k''^2}\left[k''_i\left(k_j-\frac{\mathbf{kk}''}{k''^2}k''_j\right)+\mathbf{kk}'\left(\delta_{ij}-\frac{k''_i k''_j}{k''^2}\right)\right]+\\
& +\frac{k'_i}{\omega'^2}\frac{1}{k''^2}\left[k''_s\left(k'_j-\frac{\mathbf{k}'\mathbf{k}''}{k''^2}k''_j\right)+\mathbf{k}'\mathbf{k}''\left(\delta_{js}-\frac{k''_j k''_s}{k''^2}\right)\right]+\frac{\mathbf{kk}'}{\omega\omega'}\frac{1}{k''^2}\left[k''_i\left(\delta_{js}-\right.\right.\\
& \left.\left.\left.\left.-\frac{k''_s k''_j}{k''^2}\right)+k''_s\left(\delta_{ij}-\frac{k''_i k''_j}{k''^2}\right)\right]\right\}\frac{1}{N}\int d\mathbf{v}f_0(\mathbf{k}''\mathbf{v})\frac{1}{\omega''-\mathbf{k}''\mathbf{v}}\right\}.
\end{aligned}
\tag{7.71}
$$

This expression is very useful for studying the nonlinear interaction (induced scattering) of electron Langmuir and transverse oscillations in an isotropic plasma, since their frequencies satisfy the conditions (7.70). The distribution function f_0 in (7.17) is an arbitrary isotropic function. For the Maxwellian distribution (6.14), the integrals from (7.71) can be expressed in terms of the function J_+:

$$
\frac{1}{N}\int d\mathbf{v}f_0\frac{\mathbf{k}''\mathbf{v}}{\omega''-\mathbf{k}''\mathbf{v}} = -v_T^2\frac{1}{N}\int d\mathbf{v}\mathbf{k}''\frac{\partial f_0}{\partial\mathbf{v}}\frac{1}{\omega''-\mathbf{k}''\mathbf{v}} = \left\{J_+\left(\frac{\omega''}{k''v_T}\right)-1\right\}, \tag{7.72}
$$

$$
\frac{1}{N}\int d\mathbf{v}f_0\frac{1}{\omega''-\mathbf{k}''\mathbf{v}} = \frac{1}{\omega''}J_+\left(\frac{\omega''}{k''v_T}\right).
$$

The expression (7.71) is asymmetric under the substitution $\omega \rightleftharpoons \omega'$, $\mathbf{k} \rightleftharpoons \mathbf{k}'$ because the original formula (7.7) is. In the tensor (7.71) the terms with singularities of the form ω_2^{-1} are completely retained (not expanded in powers of $\mathbf{v}$), since it is these terms that, in accordance with the generalized kinetic field equation (3.22), correspond to a virtual oscillation with wave vector $\mathbf{k}''$ and frequency ω'' that interacts strongly with the plasma particles when their velocities are equal to its phase velocity ω''/k''. The conditions (7.70) in a plasma with a Maxwellian particle distribution in the unperturbed state (6.14) are identical to the conditions

$$
\omega \gg kv_T, \quad \omega' \gg k'v_T. \tag{7.73}
$$

For the strong interaction of a virtual oscillation with the particles, i.e., for it to be damped by the inverse Vavilov–Cherenkov effect (Landau damping) it is therefore necessary that the phase velocity ω''/k'' of the virtual oscillation be appreciably less than the phase velocities of the scattered waves:

$$
\frac{\omega}{k}, \quad \frac{\omega'}{k'} \gg \frac{\omega''}{k''}. \tag{7.74}
$$

Essentially, this inequality is a natural formulation of the difference between the physical nature of the scattered oscillations, for which the linear interaction effects* with individual plasma

* In accordance with (6.25)-(6.26) and (6.25a), linear damping of a characteristic plasma oscillation with phase velocity $\omega/k \gg v_T$ is exponentially small: $\exp(-\omega^2/2k^2v_T^2) \ll 1$. The smallness of the damping constants in the linear theory of oscillations means that one must also ignore such exponentially small terms in the nonlinear oscillation theory constructed as perturbation theory. It is for this reason that in the approximate expression (7.71) for the tensor S_{ijs} under the conditions (7.70) we have everywhere omitted the contributions of the δ functions $-i\pi\delta(\omega-\mathbf{kv})$ and $-i\pi\delta(\omega'-\mathbf{k}'\mathbf{v})$ in the expressions ω_0^{-1} and ω_1^{-1} in accordance with Eq. (5.12).

particles is ignored, and the virtual oscillation, which we have named thus precisely because of its interaction with individual plasma particles. If the inequality (7.74) were violated, for example, if the phase velocities of all three oscillations were of the same order,

$$\frac{\omega}{k} \sim \frac{\omega'}{k'} \sim \frac{\omega''}{k''}, \tag{7.75}$$

we could not ignore the linear terms in Eq. (3.22), which describes the nonlinear scattering. In principle, the relation (7.75) can hold only in decay processes, when all three interacting oscillations are on an equal footing in the sense that they are characteristic (weakly damped) oscillations of the plasma. In its turn, the inequality (7.74) in conjunction with the conditions (7.3) on the frequencies and wave vectors entails smallness of the frequency of the virtual oscillation compared with the frequencies of the scattered waves:

$$\omega'' \ll \omega,\ \omega'. \tag{7.76}$$

Thus, induced scattering of oscillations satisfying the conditions of weak spatial dispersion (7.73) can take place only with the small change of the frequency (7.76). That the change in the frequency is small is, of course, also true for decay processes if the phase velocities of the oscillations that participate in the decay satisfy the inequality (7.74).

To conclude this section we point out that the three-index tensor S_{ijs} studied here completely determines the basic features of the evolution of characteristic oscillations of an isotropic plasma in decay processes [see Eq. (3.22a) in the introduction to this chapter]. To study induced scattering we must also consider the four-index tensor $V_{i\alpha jb}(\omega, \mathbf{k};\ \omega', \mathbf{k}')$ [cf. (3.22a) and (3.22b)].

§ 8. The Four-Index Tensor $V_{i\alpha jb}(\omega, \mathbf{k};\ \omega', \mathbf{k}')$ in an Isotropic Plasma

In accordance with the definition (3.23), which we obtained in the derivation of the generalized kinetic equation (3.22), the tensor $V_{i\alpha jb}$ is the sum of two four-index tensors ε, whose concrete expressions are given by the general relation (6.10). Namely, the first tensor ε has the form

$$\varepsilon_{iajb}(\omega, \mathbf{k};\ \omega+\omega',\ \mathbf{k}+\mathbf{k}';\ \omega', \mathbf{k}') =$$
$$= -\frac{4\pi e^4}{\omega} \int d\mathbf{p} v_i \frac{\alpha_{ma}(\omega', \mathbf{k}', \mathbf{v})}{\omega - \mathbf{k}\mathbf{v}} \frac{\partial}{\partial p_m} \frac{\alpha_{nj}(\omega, \mathbf{k}, \mathbf{v})}{\omega+\omega' - (\mathbf{k}+\mathbf{k}', \mathbf{v})} \frac{\partial}{\partial p_n} \frac{\alpha_{lb}(\omega', \mathbf{k}', \mathbf{v})}{\omega' - \mathbf{k}'\mathbf{v}} \frac{\partial f_0}{\partial p_l}, \tag{8.1}$$

and the second term in (3.23) is obtained from this by transposing the indices $j \rightleftharpoons b$ and the arguments $\omega, \mathbf{k} \rightleftharpoons \omega', \mathbf{k}'$ in the tensors $\alpha_{nj}(\omega, \mathbf{k}, \mathbf{v})$ and $\alpha_{lb}(\omega', \mathbf{k}', \mathbf{v})$ with simultaneous replacement of the last Green's function $(\omega' - \mathbf{k}'\mathbf{v})^{-1}$ by $(\omega - \mathbf{k}\mathbf{v})^{-1}$:

$$\varepsilon_{iabj}(\omega, \mathbf{k};\ \omega+\omega',\ \mathbf{k}+\mathbf{k}';\ \omega, \mathbf{k}) =$$
$$= -\frac{4\pi e^4}{\omega} \int d\mathbf{p} v_i \frac{\alpha_{ma}(\omega', \mathbf{k}', \mathbf{v})}{\omega - \mathbf{k}\mathbf{v}} \frac{\partial}{\partial p_m} \frac{\alpha_{nb}(\omega', \mathbf{k}', \mathbf{v})}{\omega+\omega' - (\mathbf{k}+\mathbf{k}', \mathbf{v})} \frac{\partial}{\partial p_n} \frac{\alpha_{lj}(\omega, \mathbf{k}, \mathbf{v})}{\omega - \mathbf{k}\mathbf{v}} \frac{\partial f_0}{\partial p_l}. \tag{8.2}$$

In contrast to S_{ijs}, the tensor $V_{i\alpha jb}$ is not equal to the corresponding symmetrized combination composed of six four-index tensors ε, if for no other reason that it contains only two terms instead of six. Obviously, in the approximation of random phases a symmetrized combination of this kind can arise only when one considers the higher nonlinear processes in which four or more oscillations participate.

In addition, it should be noted that the explicit expressions (8.1) and (8.2) for the tensors ε, which arise from the solution of the transport equation, do not satisfy the symmetry conditions (3.23), which arise from a "phenomenological" treatment of the current density **j**. Of course, this situation is to a large extent formal and merely indicates the necessity of an appropriate symmetrization of the sum of the tensors (8.1) and (8.2) that make up $V_{i\alpha jb}$ and occur in the generalized kinetic equation (3.22).

In the nonrelativistic limit, it is natural to use the velocities **v** instead of the momenta **p**. Then, $\varepsilon_{i\alpha jb}$ takes the form

$$\varepsilon_{iajb}(\omega,\ \mathbf{k};\ \ \omega+\omega',\ \ \mathbf{k}+\mathbf{k}';\ \ \omega',\ \mathbf{k}')=$$

$$=-\frac{4\pi e^4}{m^3}\int d\mathbf{v}\,\frac{v_i}{\omega}\,\frac{a'_{ma}}{\omega-\mathbf{kv}}\,\frac{\partial}{\partial v_m}\,\frac{a_{nj}}{\omega+\omega'-(\mathbf{k}+\mathbf{k}',\ \mathbf{v})}\,\frac{\partial}{\partial v_n}\,\frac{a'_{lb}}{\omega'-\mathbf{k}'\mathbf{v}}\,\frac{\partial f_0}{\partial v_l}\,. \tag{8.3}$$

The tensors ε and therefore $V_{i\alpha jb}$ contain three kinds of singularity (Green's functions): the two already encountered, ω_0^{-1} and ω_1^{-1}, and a new one: $[\omega + \omega' - (\mathbf{k} + \mathbf{k}', \mathbf{v})]^{-1}$. In accordance with the generalized kinetic equation, the first two correspond to scattered waves, and the third to an intermediate one, i.e., to a virtual oscillation through which the other two are scattered on the particles because of its interaction with them. Clearly, for positive frequencies the phase velocity of such a virtual oscillation must be greater than or of the same order as the phase velocities of the scattered oscillations:

$$\frac{\omega+\omega'}{|\mathbf{k}+\mathbf{k}'|}\gtrsim\frac{\omega}{k},\quad \frac{\omega'}{k'} \tag{8.4}$$

and the resulting nonlinear process differs significantly from induced scattering with a small change in frequency. In quantum language, one can say that a nonlinear process of this kind is the simultaneous emission (or absorption) by a particle of two quanta (oscillations) with frequencies ω and ω' and wave vectors **k** and **k'**. Accordingly, the process of induced scattering with small change in the frequency, i.e., scattering through a virtual wave with frequency $\omega'' = \omega - \omega' \ll \omega, \omega'$, can be interpreted as the process of absorption by a particle of a quantum (oscillation) with frequency ω' and wave vector **k'** with the subsequent emission of a quantum (oscillation) with frequency ω and wave vector **k**. That the process is induced is reflected by the presence of the spectral functions $(E_\alpha E_i)_{\omega,\mathbf{k}}$ as a common factor in the two last terms of the curly brackets on the right-hand side of the kinetic equation (3.22) for the waves [see also Eq. (3.22b) in the introduction to this chapter].

Substituting the actual expressions for the spectral functions of the field into (3.22) in accordance with the general relations of the linear theory of oscillations (see §6), and going over, thus, to positive frequencies ω and ω', we see that, besides the tensor V, which contains the Green's function with total frequency $\omega + \omega'$, we also obtain a tensor with the difference frequency $\omega'' = \omega - \omega'$ (for more detail see §21 in Chapter IV). Direct calculation shows that under the conditions (7.70) and (7.74) the terms with the tensor $V(\omega, \omega + \omega', \omega')$ are small compared with the terms that depend on the difference frequency $\omega'' = \omega - \omega'$. Note that a completely analogous situation obtains in the terms that depend on the tensor S. Namely, besides the tensor $S_{ijs}(\omega, \mathbf{k}; \omega', \mathbf{k}')$ calculated in the foregoing section, which depends on the differences of the frequencies $\omega - \omega'$ and the wave vectors $\mathbf{k} - \mathbf{k}'$ of the scattered oscillations, we obtain, on substitution of the actual expressions for the spectral functions of the field into the generalized kinetic equation (3.22), the tensor S_{ijs}, which contains the denominator $[\omega + \omega' - (\mathbf{k} + \mathbf{k}', \mathbf{v})]^{-1}$ with total frequency $\omega + \omega'$ and wave vector $\mathbf{k} + \mathbf{k}'$. The contribution of terms of the last type is small under the conditions in which we are interested.

We therefore turn to the study of the tensor V, which depends on the differences $\omega - \omega' = \omega''$, $\mathbf{k} - \mathbf{k}' = \mathbf{k}''$ [however, see formulas (8.9) and (8.10) below]. The corresponding four-index tensors are obtained from Eqs. (8.1) and (8.2) by the substitution $\omega' \to -\omega'$, $\mathbf{k}' \to -\mathbf{k}'$:

$$\varepsilon_{iajb}(\omega,\ \mathbf{k};\ \omega'',\ \mathbf{k}'';\ \ -\omega',\ -\mathbf{k}')=\frac{4\pi e^4}{\omega}\int d\mathbf{p}v_i\,\frac{a'_{ma}}{\omega_0}\,\frac{\partial}{\partial p_m}\,\frac{a_{nj}}{\omega_2}\,\frac{\partial}{\partial p_n}\,\frac{a'_{lb}}{\omega_1}\,\frac{\partial f_0}{\partial p_l}\,, \tag{8.5}$$

$$\varepsilon_{iabj}(\omega, \mathbf{k};\ \omega'', \mathbf{k}'';\ \omega, \mathbf{k}) = -\frac{4\pi e^4}{\omega}\int d\mathbf{p}\, v_i \frac{\alpha'_{ma}}{\omega_0}\frac{\partial}{\partial p_m}\frac{\alpha'_{nb}}{\omega_2}\frac{\partial}{\partial p_n}\frac{\alpha_{lj}}{\omega_0}\frac{\partial f_0}{\partial p_l}. \tag{8.6}$$

Here we have used the abbreviated notation (7.2) and (7.17) for the tensors α_{ij} and the denominators ω_0, ω_1, ω_2. On the right-hand sides of Eqs. (8.5) and (8.6) we integrate once in each case by parts with respect to the momenta dp and add the results:

$$V_{iajb}(\omega, \mathbf{k};\ -\omega', -\mathbf{k}') = -4\pi e^4 \int d\mathbf{p}\, \frac{c^2}{\mathcal{E}}\left(\delta_{sm} - \frac{v_s v_m}{c^2}\right)\frac{\alpha_{si}\alpha'_{ma}}{\omega_0^2\omega_2}\left\{\alpha_{nj}\frac{\partial}{\partial p_n}\frac{\alpha'_{lb}}{\omega_1}\frac{\partial f_0}{\partial p_l} - \alpha'_{nb}\frac{\partial}{\partial p_n}\frac{\alpha_{lj}}{\omega_0}\frac{\partial f_0}{\partial p_l}\right\}. \tag{8.7}$$

In the nonrelativistic limit, this expression simplifies somewhat:

$$V_{iajb}(\omega, \mathbf{k};\ -\omega', -\mathbf{k}') = -\frac{4\pi e^4}{m^3}\int d\mathbf{v}\, \frac{\alpha_{si}\alpha'_{ma}}{\omega_0^2\omega_2}\left\{\alpha_{nj}\frac{\partial}{\partial v_n}\frac{\alpha'_{mb}}{\omega_1}\frac{\partial f_0}{\partial v_m} - \alpha'_{nb}\frac{\partial}{\partial v_n}\frac{\alpha_{mj}}{\omega_0}\frac{\partial f_0}{\partial v_m}\right\}. \tag{8.8}$$

Note the asymmetry of the right-hand sides of (8.7) and (8.8) under the substitution $(\omega, \mathbf{k}) \rightleftharpoons (\omega', \mathbf{k}')$. In particular, in a cold isotropic plasma for the sum of the two tensors V, which contain the denominators with both the total, $\omega + \omega'$, and the difference frequency, $\omega - \omega'$, and after threefold integration by parts with respect to dv, we obtain the equation*

$$V_{iajb}(\omega, \mathbf{k};\ \omega', \mathbf{k}') + V_{iajb}(\omega, \mathbf{k};\ -\omega', -\mathbf{k}') = 2\frac{e^2}{m^2}\frac{\omega_L^2}{\omega^2}\frac{1}{\omega^2 - \omega'^2}\left\{\delta_{ia}\left[\frac{k'_j k'_b}{\omega'^2} - \frac{k_j k_b}{\omega^2}\right] + \delta_{jb}\left[\frac{k'_i k'_a}{\omega'^2} - \frac{k_i k_a}{\omega^2}\right] - \right.$$
$$\left. - \delta_{ia}\delta_{jb}\left[\left(\frac{\mathbf{k}}{\omega} - \frac{\mathbf{k}'}{\omega'}\right)^2 + 2\frac{k^2 + k'^2}{\omega^2 - \omega'^2} - 2\frac{\mathbf{k}\mathbf{k}'}{\omega\omega'}\frac{\omega^2 + \omega'^2}{\omega^2 - \omega'^2}\right]\right\}. \tag{8.9}$$

Here the right-hand side without the common factor $1/\omega^2$, is symmetric under the above substitution. Similarly, for the longitudinal contraction of the sum of the two tensors V in a hot nonrelativistic isotropic plasma we obtain the relation

$$\{V_{iajb}(\omega, \mathbf{k};\ \omega', \mathbf{k}') + V_{iajb}(\omega, \mathbf{k};\ -\omega', -\mathbf{k}')\}\frac{k_i k'_a k_j k'_b}{(kk')^2} =$$
$$= \frac{8\pi e^4}{m^3}\left(\frac{\mathbf{k}\mathbf{k}'}{kk'}\right)^2 \int d\mathbf{v} f_0 \frac{1}{\omega_0^2}\left\{\frac{1}{2}\left[\left(\frac{\mathbf{k}+\mathbf{k}'}{\omega_0+\omega_1}\right)^2 - \left(\frac{\mathbf{k}-\mathbf{k}'}{\omega_0-\omega_1}\right)^2\right]\frac{1}{\omega_0\omega_1} - \frac{1}{\omega_0^2 - \omega_1^2}\left[2\left(\frac{k^2}{\omega_0^2} - \frac{k'^2}{\omega_1^2}\right) + \left(\frac{\mathbf{k}}{\omega_0} - \frac{\mathbf{k}'}{\omega_1}\right)^2\right]\right\}, \tag{8.10}$$

on whose right-hand side only the expression in the curly brackets is symmetric. However, it must be borne in mind that in the equations of the nonlinear interaction of the characteristic oscillations of an isotropic plasma one does not, in accordance with (3.22) and (3.22b), have the complete tensor V, but only its imaginary part, since the spectral functions of the electric fields in such a plasma are real. This fact greatly simplifies the calculations, since when the conditions (7.70) hold (and it is these conditions in which we are interested) it means that one must allow for only the second term in the δ_+ function associated with the frequency difference $\omega - \omega'$:

$$\frac{1}{\omega_2} = \frac{P}{\omega_2} - i\pi\delta(\omega_2) \tag{8.11}$$

because the contribution of the first term is real. From (8.7) for an isotropic f_0, we obtain

$$\operatorname{Im} V_{iajb}(\omega, \mathbf{k};\ -\omega', -\mathbf{k}') = 4\pi^2 e^4 \omega'' \int d\mathbf{p}\,\delta(\omega_2)\frac{c^2}{\mathcal{E}}\left(\delta_{sm} - \frac{v_s v_m}{c^2}\right)\frac{c^2}{\mathcal{E}}\left(\delta_{nk} - \frac{v_n v_k}{c^2}\right)\frac{\alpha_{si}\alpha'_{ma}\alpha_{nj}\alpha'_{kb}}{\omega_0^4}\frac{\partial f_0}{\partial \mathcal{E}}. \tag{8.12}$$

---

* It follows from Eq. (8.9) in particular, that, in contrast to the tensor S, the four-index tensor $V_{iajb}(\omega, \mathbf{k}; \omega', \mathbf{k}')$ in a cold plasma is a bilinear form in the wave vectors **k** and **k'** of the interacting oscillations. These and other properties of the multi-index permittivity tensors of a cold plasma can be obtained without recourse to a kinetic description by means of the system of equations of hydrodynamics (see the Appendix).

We rewrite the right-hand side somewhat differently, using the notation (7.9) for the tensors β_{ij} and the fact that the integrand in (8.12) is nonvanishing only on the plane $\omega - \mathbf{kv} = \omega' - \mathbf{k'v}$ in the space of the velocities $\mathbf{v}$:

$$\operatorname{Im} V_{iajb}(\omega,\ \mathbf{k};\ -\omega',\ -\mathbf{k}') = 4\pi^2 e^4 \frac{\omega''}{(\omega'\omega)^2} \int d\mathbf{p} \frac{\partial f_0}{\partial \mathscr{E}} \frac{c^4}{\mathscr{E}^2} \left(\delta_{sm} - \frac{v_s v_m}{c^2}\right)\left(\delta_{rn} - \frac{v_r v_n}{c^2}\right) \beta_{si}\beta'_{ma}\beta_{nj}\beta'_{rb}\delta(\omega'' - \mathbf{k}''\mathbf{v}). \quad (8.13)$$

In the nonrelativistic limit, i.e., in the zeroth approximation in the small ratio v/c, we obtain the simpler expression

$$\operatorname{Im} V_{iajb}(\omega,\ \mathbf{k};\ -\omega',\ -\mathbf{k}') = 4\pi^2 \frac{e^4}{m^2} \frac{\omega''}{(\omega\omega')^2} \int d\mathbf{v} \frac{\partial f_0}{\partial \mathscr{E}} \beta_{mi}\ \beta'_{ma}\beta_{nj}\beta'_{nb}\delta(\omega'' - \mathbf{k}''\mathbf{v}). \quad (8.14)$$

Note a very important feature of formulas (8.12), (8.13), and (8.14): in them, in the Green's functions with frequencies ω and ω' and wave vectors $\mathbf{k}$ and $\mathbf{k}'$ of the scattered waves, only the principal parts (the terms in the sense of the principal value) are allowed for, since the terms proportional to $\delta(\omega_0)$ and $\delta(\omega_1)$ are negligibly small on account of the conditions (7.70). Expanding the principal parts of the Green's functions that occur in the tensor β_{ij} in powers of the small ratios (7.70), we obtain from (8.14) the following approximate relation for the imaginary part of the tensor V:

$$\begin{aligned}
\operatorname{Im} V_{iajb}(\omega,\ \mathbf{k};\ -\omega',\ -\mathbf{k}') = {} & 4\pi^2 \frac{e^4}{m^3} \frac{1}{(\omega\omega')^2} \Big\{\Big\{\delta_{ia}\delta_{jb} + \frac{\omega''}{\omega}\frac{1}{k''^2}[\delta_{ia}(k'_j k''_b + k_b k''_j) + \\
& + \delta_{jb}(k'_i k''_a + k_a k''_i)] + \left(\frac{\omega''}{\omega}\right)^2 \frac{1}{k''^4}[\delta_{ia}(k'_j k''_b \mathbf{kk}'' + k_b k''_j \mathbf{kk}'' + k''_j k''_b \mathbf{kk}') + \\
& + \delta_{jb}(k'_i k''_a \mathbf{kk}'' + k_a k''_i \mathbf{kk}'' + k''_i k''_a \mathbf{kk}') + k_a k_b k''_i k''_j + k_a k'_j k''_i k''_b + \\
& + k'_i k_b k''_a k''_j + k'_i k'_j k''_a k''_b]\Big\} \int d\mathbf{v}\mathbf{k}'' \frac{\partial f_0}{\partial \mathbf{v}} \delta(\omega'' - \mathbf{k}''\mathbf{v}) - \frac{\omega''}{\omega^2}\Big\{\delta_{ia}\Big[k'_j\left(k_b - \frac{\mathbf{kk}''}{k''^2}k''_b\right) + \\
& + k_b\left(k_j - \frac{\mathbf{kk}''}{k''^2}k''_j\right) + \mathbf{kk}'\left(\delta_{jb} - \frac{k''_j k''_b}{k''^2}\right)\Big] + \delta_{jb}\Big[k'_i\left(k_a - \frac{\mathbf{kk}''}{k''^2}k''_a\right) + \\
& + k_a\left(k_i - \frac{\mathbf{kk}''}{k''^2}k''_i\right) + \mathbf{kk}'\left(\delta_{ia} - \frac{k''_i k''_a}{k''^2}\right)\Big] + k_a k_b\left(\delta_{ij} - \frac{k''_i k''_j}{k''^2}\right) + \\
& + k_a k'_j\left(\delta_{ib} - \frac{k''_i k''_b}{k''^2}\right) + k'_i k_b\left(\delta_{aj} - \frac{k''_a k''_j}{k''^2}\right) + \\
& + k'_i k'_j\left(\delta_{ab} - \frac{k''_a k''_b}{k''^2}\right)\Big\} \int d\mathbf{v} f_0 \delta(\omega'' - \mathbf{k}''\mathbf{v})\Big\}. \quad (8.15)
\end{aligned}$$

For a Maxwellian distribution, the integrals that occur here have the form

$$\frac{1}{N} \int d\mathbf{v} f_0 \delta(\omega'' - \mathbf{k}''\mathbf{v}) = \frac{1}{\sqrt{2\pi}\, k'' v_T} \exp\left\{-\frac{1}{2}\left(\frac{\omega''}{k'' v_T}\right)^2\right\}; \quad (8.16)$$

$$\frac{1}{N} \int d\mathbf{v}\mathbf{k}'' \frac{\partial f_0}{\partial \mathbf{v}} \delta(\omega'' - \mathbf{k}''\mathbf{v}) = -\frac{\omega''}{\sqrt{2\pi}\, k'' v_T^3} \exp\left\{-\frac{1}{2}\left(\frac{\omega''}{k'' v_T}\right)^2\right\}. \quad (8.17)$$

Turning to the different special cases of the contractions of the tensor V, we note that in a nonrelativistic plasma the incomplete longitudinal contractions $V_{iajb}k_i k'_a$ is proportional to the scalar product $\mathbf{kk}'$ and therefore vanishes in the case of scattering of characteristic oscillations with perpendicular wave vectors. This already follows from the general formula (8.8). The complete longitudinal contraction of the tensor V in the case of a one-dimensional interaction has, in accordance with (8.8), the following form in a nonrelativistic plasma:

$$V_{iajb}(\omega,\ \mathbf{k};\ -\omega',\ -\mathbf{k}') \frac{k_i k'_a k_j k'_b}{(kk')^2} = \frac{4\pi e^4}{m^3}\left(\frac{k_z k'_z}{kk'}\right)^2 \int d\mathbf{v} \frac{\partial f_0}{\partial v_z} \left\{\frac{2k_z}{\omega_0^4 \omega_1} + \frac{k''_z}{\omega_0^3 \omega_1 \omega_2}\right\}, \quad (8.18)$$

where the subscript z indicates the projections of the vectors onto the direction along which the interaction occurs and the denominators ω_0, ω_1, ω_2 are determined by Eqs. (7.36). Allowing for the contribution from only the Green's function ω_2^{-1} with the frequency difference $\omega - \omega'$ and wave vector $\mathbf{k} - \mathbf{k}'$, we obtain for the imaginary part of the longitudinal contraction in the case of a Maxwellian distribution f_0

$$\operatorname{Im} V_{iajb}(\omega, \mathbf{k}; -\omega', -\mathbf{k}') \frac{k_i k'_a k_j k'_b}{(kk')^2} = \sqrt{\frac{\pi}{2}} \frac{e^2}{m^2} \frac{\omega_L^2}{v_T^2} \left(\frac{k_z k'_z}{kk'}\right)^2 \frac{k''^4_z}{(\omega' k_z - \omega k'_z)^4} \frac{\omega''}{|k''_z v_T|} \exp\left\{-\frac{1}{2}\left(\frac{\omega''}{k''_z v_T}\right)^2\right\}. \quad (8.19)$$

Note that the expression $(\omega' k_z - \omega k'_z)^4$, which occurs in the denominator on the right-hand side of (8.19), is nonvanishing, since the contraction we are studying is used exclusively to determine the evolution of the process of induced scattering with small frequency change, when the inequalities (7.74) hold and therefore, $(\omega' k_z - \omega k'_z)^4 = (\omega k''_z - \omega'' k_z)^4 \simeq (\omega k''_z)^4 \neq 0$. Thus, the right-hand side of Eq. (8.19) can be rewritten slightly differently:

$$\sqrt{\frac{\pi}{2}} \frac{e^2}{m^2} \frac{\omega_L^2}{v_T^2} \left(\frac{k_z k'_z}{kk'}\right)^2 \frac{1}{\omega^4} \frac{\omega''}{|k''_z v_T|} \exp\left\{-\frac{1}{2}\left(\frac{\omega''}{k''_z v_T}\right)^2\right\}. \quad (8.20)$$

In this simple example, we see a very important property of the imaginary part of the longitudinal contraction of Im V, which also holds for the imaginary parts $\operatorname{Im} V_{iajb}$ of the tensor, namely, antisymmetry under the substitution $\omega \rightleftharpoons \omega'$, $\mathbf{k} \rightleftharpoons \mathbf{k}'$, which corresponds to the principle of detailed balance: the probability of scattering of an oscillation with frequency ω' and wave vector $\mathbf{k}'$ into an oscillation with frequency ω and wave vector $\mathbf{k}$ is equal to the probability of the inverse scattering process: $(\omega, \mathbf{k})$ into $(\omega', \mathbf{k}')$. Of course, this is true only if one does not take into account the contributions from the poles that determine the Green's functions with frequencies ω and ω' of the scattered oscillations, which introduce a strong asymmetry into the tensor $\operatorname{Im} V_{iajb}$. In contrast to the tensor $\operatorname{Im} V_{iajb}(\omega, \mathbf{k}; -\omega', -\mathbf{k})$, which we have considered in detail, the tensor $\operatorname{Im} V_{iajb}(\omega, \mathbf{k}; \omega', \mathbf{k}')$ is symmetric under this substitution. For example, its longitudinal contraction in the case of a one-dimensional interaction has the form

$$\operatorname{Im} V_{iajb}(\omega, \mathbf{k}; \omega', \mathbf{k}') \frac{k_i k'_a k_j k'_b}{(kk')^2} = \sqrt{\frac{\pi}{2}} \frac{e^2}{m^2} \frac{\omega_L^2}{v_T^2} \left(\frac{k_z k'_z}{kk'}\right)^2 \frac{(k_z + k'_z)^4}{(\omega' k_z - \omega k'_z)^4} \frac{\omega + \omega'}{|k_z + k'_z| v_T} \exp\left\{-\frac{1}{2}\left(\frac{\omega + \omega'}{(k_z + k'_z) v_T}\right)^2\right\}, \quad (8.21)$$

from which we obtain the symmetry.

To conclude this section, we give two very useful approximate expressions. The first is the contraction of the product of the two tensors S that occur in the second term of the curly brackets of the generalized kinetic equation (3.22) with a tensor that is transverse with respect to the vector $\mathbf{k}''$ of the virtual wave:

$$S_{ijs}(\omega, \mathbf{k}; \omega', \mathbf{k}') S_{boa}(\omega'', \mathbf{k}''; \omega, \mathbf{k}) \left(\delta_{bj} - \frac{k''_b k''_j}{k''^2}\right) = \frac{1}{(\omega\omega'\omega'')^2} \left(\frac{e}{m}\right)^2 \frac{\omega_L^4}{\omega^2} \left\{\delta_{ij} k''_s - \delta_{js} k''_i - (k_s \delta_{ij} + k'_i \delta_{js}) \frac{\omega''}{N} \times \right.$$

$$\left. \times \int d\mathbf{v} \frac{f_0}{\omega'' - \mathbf{k}''\mathbf{v}}\right\} \left\{\delta_{ab} k''_o - \delta_{bo} k''_a - (k_o \delta_{ba} + k'_a \delta_{bo}) \frac{\omega''}{N} \int d\mathbf{v} \frac{f_0}{\omega'' - \mathbf{k}''\mathbf{v}}\right\} \left(\delta_{bj} - \frac{k''_b k''_j}{k''^2}\right). \quad (8.22)$$

This equation follows directly from (7.71) and is very convenient for studying induced scattering through a transverse virtual oscillation. We shall use the second relation to investigate induced scattering through a longitudinal virtual oscillation in an electron plasma:

$$\operatorname{Im}\left\{V_{ioas}(\omega, \mathbf{k}; -\omega', -\mathbf{k}') - S_{ijs}(\omega, \mathbf{k}; \omega', \mathbf{k}') S_{boa}(\omega'', \mathbf{k}''; \omega, \mathbf{k}) \frac{k''_j k''_b}{k''^2} [\varepsilon^l(\omega'', \mathbf{k}'')]^{-1}\right\} =$$

$$= \frac{4\pi^2 e^4}{m^3} \frac{\omega''}{(\omega\omega')^3} \left\{k_o k_s \left(\delta_{ia} - \frac{k''_i k''_a}{k''^2}\right) + k_o k'_a \left(\delta_{is} - \frac{k''_i k''_s}{k''^2}\right) + k'_i k_s \left(\delta_{ao} - \frac{k''_a k''_o}{k''^2}\right) + k'_i k'_a \left(\delta_{os} - \frac{k''_o k''_s}{k''^2}\right)\right\} \int d\mathbf{v} f_0 \delta(\omega'' - \mathbf{k}''\mathbf{v}). \quad (8.23)$$

It is a direct consequence of formulas (7.71) and (8.15) for the tensors S and V under the conditions (7.70).

CHAPTER III

EXAMPLES OF NONLINEAR PROCESSES IN AN ISOTROPIC PLASMA

The generalized kinetic equation (3.22) for waves in conjunction with the actual expressions for the nonlinear tensors S and V in an isotropic plasma studied in Chapter II enable us to consider the most varied special cases of nonlinear interaction of oscillations of an isotropic plasma. This chapter, in which we carry out such an investigation, consists of six sections (§ 9-14).

In § 9, on the basis of [12, 29, 31], we discuss aspects of the nonlinear interaction of electrostatic oscillations of an isotropic plasma. This section consists of four parts. First, we obtain the general equation (9.1), which describes the evolution of only longitudinal waves in an isotropic plasma because of their interaction. Using this equation, we study Coulomb induced scattering of Langmuir oscillations on plasma particles (i.e., scattering through an electrostatic virtual oscillation). The procedure of this treatment reduces to calculating the kernel of the integrodifferential equation for the spectral density of the energy of the waves and a subsequent analysis of the process (elucidation of the way in which energy is transferred within the spectrum, the conservation laws, the possibility of nonlinear damping on particles). We show that in the first nonvanishing approximation in the ratio of the electron Debye radius to the wavelength the total energy of the Langmuir oscillations is not dissipated but conserved, being redistributed within the spectrum. The direction of such redistribution (from short to long wavelengths) is established for the example of the interaction of two infinitesimally narrow oscillation packets. In the second part of § 9, we analyze the effect of the fields of the virtual oscillation being nonelectrostatic on the induced scattering of the Langmuir waves. By comparing the characteristic times of the processes, we show that the scattering of Langmuir waves with allowance for retardation can take place faster than Coulomb scattering if the wavelength of the oscillations is greater than the ratio of the velocity of light in vacuum to the electron plasma frequency. As in the case of Coulomb scattering, the total energy is conserved. Then (in § 9) we consider the interaction of longitudinal waves with the spectrum (9.70) in a system of two interpenetrating electron plasmas (9.68) moving with velocities $\mathbf{u}_1$ and $\mathbf{u}_2$ [see (9.69)]. A distinctive feature of this nonlinear interaction is the possibility of nonlinear excitation of longitudinal waves in the case when the temperature T_1 of the plasma at rest ($\mathbf{u}_1 = 0$) is higher than the temperature T_2 of the beam ($\mathbf{u}_2 \neq 0$) and the energy is transferred from the lower to the higher frequencies. In the final and fourth part of § 9, we discuss the induced scattering of waves of short-wavelength ion sound (6.48) with frequency near the ion plasma frequency.

In contrast to the induced scattering of Langmuir waves, the transfer of the energy of short-wavelength ion-acoustic oscillations is by no means the only result of their nonlinear interaction. Namely, from the integrodifferential equation (9.77) obtained for the spectral density of the energy $W_s(\mathbf{k})$ of sound it follows that under the conditions (9.82), for example, when the wavelengths of the ion-acoustic oscillations are greater than the cubic root of the product of the square of the electron Debye radius and the ion Debye radius, the waves are damped, i.e., there is a transfer of energy from the waves to individual particles.

The next section (§ 10) opens with a general description of the generalized kinetic equations for longitudinal and transverse waves in an isotropic plasma, these following from (3.22) as special cases. The first pair of equations (10.1)-(10.2) of the evolution of interacting oscillations takes into account only the natural polarization of the transverse electromagnetic waves that participate in the interaction with longitudinal waves. Equations (10.5) and (10.7) are suitable for the study of the nonlinear interaction of longitudinal and linearly polarized transverse waves and, after averaging over all possible linear polarizations, go over into Eqs. (10.1) and (10.2), respectively.

As one of the specific examples of equations (10.1)-(10.2) we study the induced scattering of Langmuir waves into long-wavelength naturally polarized transverse oscillations with frequency near the electron plasma frequency and the corresponding "inverse" scattering of transverse oscillations into Langmuir oscillations. In contrast to the nonlinear interaction of Langmuir waves with one another, this scattering can take place through a virtual oscillation with a phase velocity that is either low or high compared with the electron thermal velocity. In the case of induced scattering of Langmuir and transverse waves through a fast virtual oscillations (its phase velocity still remains very low compared with the phase velocities of the waves) it is important to allow for this oscillation's being nonelectrostatic. Conversely, the most rapid (with shortest characteristic time) scattering through a slow virtual photon is as a rule Coulomb scattering. The general laws that the equations for induced scattering of Langmuir and long-wavelength transverse waves satisfy are characterized by the conservation of the total energy of the interacting oscillations (in the first approximation) and transfer of energy density from higher to lower frequencies. At the end of § 10, we consider the coalescence of a Langmuir oscillation and a longwavelength transverse oscillation into a transverse wave.

Sections 10 and 11 are augmented and completed by § 11; here, we consider the nonlinear interaction of naturally polarized transverse electromagnetic waves in an isotropic plasma on the basis of the generalized kinetic equation (11.1), which follows from Eq. (3.22). It is shown that in the adopted approximation the total energy of the transverse waves is conserved, being transferred from oscillations with high frequency to oscillations with low frequency. Coulomb induced scattering on particles, in agreement with [32], is discussed for any wavelength of the transverse oscillations together with the limit of short and long wavelengths. Unlike scattering of Langmuir oscillations into transverse oscillations, which we consider in § 10, scattering with allowance for retardation of transverse waves into transverse waves through a fast virtual oscillation takes place much slower than Coulomb scattering. In its turn, the Coulomb scattering of long-wavelength transverse oscillations on electrons is less important (it occurs with a longer characteristic time) than scattering with allowance for retardation in the case of a slow virtual oscillation. In all the analyzed special cases (§ 9-11), scattering on ions, if it plays a role, takes place much faster than scattering on electrons. At the end of § 11, we derive equation (11.22), which describes the nonlinear interaction of linearly polarized transverse waves; this equation, like the basic equation (11.1) of this section, follows from the general equation (3.22). In addition, from Eq. (11.22) we also obtain the two relations (11.23) and (11.24), which determine, respectively, the evolution of naturally polarized transverse electromagnetic waves when they interact nonlinearly with linearly polarized waves and the change in the energy density of the linearly polarized oscillations during the process of nonlinear interaction with the naturally polarized oscillations.

In the second half of the chapter, two sections (§ 12 and 14) are devoted to the question of the cross sections of the wave scattering processes in an isotropic plasma. In § 12, in accordance with the method developed in [31, 33] on the basis of the classical limit of the quantum kinetic equation, the scattering cross sections of waves in a plasma are obtained as a natural consequence of the results of the theory of nonlinear interaction for a number of the processes considered in § 9-11. It is shown that to within factors such cross sections are equal to the corresponding kernels of the integrodifferential equations of the nonlinear processes. For the transverse waves that participate in the scattering we allow for not only natural but also linear polarization.

In contrast, the relations of § 14 do not follow methodologically in a direct manner from the theory of the nonlinear interaction of waves in a plasma set forth in Chapter I, though they do lead as a whole to the same final expressions for the scattering cross sections. This paragraph is based on [34], in which it is shown, by a method adopted in scattering theory [8, 9], that there is complete equivalence between the results of the theory of scattering of waves in a plasma and certain of the results of the theory of the nonlinear interaction formulated in § 12 in the form of a list of formulas for the cross sections. The expression (14.32) obtained in § 14 for the cross section of Coulomb scattering of oscillations of an isotropic plasma with high phase velocities is, in our opinion, of considerable interest from the point of view of elucidating the physical picture of the fluctuation processes in a plasma that make the greatest contribution to the cross section. For the specific example of Coulomb scattering on electrons of naturally polarized high-frequency transverse waves it is shown, in particular, that in a number of cases [see the conditions (14.36)-(14.38)] scattering on solenoidal fluctuations of the electron velocity dominate over scattering on fluctuations of their density.

Finally, the content of § 13 is related to the content of the second part of § 4 of Chapter I. In it we discuss the nonlinear dispersion properties of the oscillations of an isotropic plasma, although, of course, the general relations without particularization of the form of the nonlinear tensors can also be used for a larger class of material media. The nonlinear dispersion relations take into account natural, linear, and (the most general case) partly polarized transverse electromagnetic waves, which affect the spectrum of plasma oscillations in the process of nonlinear interaction. In general form, the nonlinear corrections to the spectra are a functional that is linear in the energy density of the oscillation. We emphasize that although such additional terms in the spectra of the waves are small correction, they can lead to qualitatively new effects in the processes of nonlinear interaction of these waves, as has been shown in [35] for the example of the Coulomb scattering of Langmuir oscillations on ions.

§ 9. Nonlinear Interaction of Longitudinal Oscillations of an Isotropic Plasma

The fact that in the linear theory of oscillations of an isotropic plasma the characteristic oscillations can be divided neatly into longitudinal and transverse (electrostatic and nonelectrostatic) makes it possible to separate the terms in the generalized kinetic equation (3.22) in accordance with the polarization of the virtual and the interacting oscillations, for which we must bear in mind the remark made at the end of § 3 that the index i recalls the polarization.

In this section we shall study in detail the nonlinear interaction of characteristic longitudinal oscillations with one another. Substituting the expressions (6.62)-(6.79) for the spectral functions of the fields in the plasma and the inverse tensor into the generalized kinetic equation (3.22) and retaining in the spectral functions only the terms are longitudinal with respect to the wave vectors, we obtain [12]

$$\frac{\partial \varepsilon^{l\prime}(\omega, \mathbf{k})}{\partial \omega} \frac{\partial}{\partial t}(E_l^2)_{\omega, \mathbf{k}} - \frac{\partial \varepsilon^{l\prime}(\omega, \mathbf{k})}{\partial \mathbf{k}} \frac{\partial}{\partial \mathbf{r}}(E_l^2)_{\omega, \mathbf{k}} = -2\varepsilon^{l\prime\prime}(\omega, \mathbf{k})(E_l^2)_{\omega, \mathbf{k}} +$$
$$+ 2(E_l^2)_{\omega, \mathbf{k}} \int d\omega' d\mathbf{k}' (E_l^2)_{\omega', \mathbf{k}'} \frac{k_i k_a k'_c k'_s}{(kk')^2} \operatorname{Im}\left\{\int d\omega'' d\mathbf{k}'' \delta(\omega - \omega' - \omega'') \times\right.$$
$$\left. \times \delta(\mathbf{k} - \mathbf{k}' - \mathbf{k}'') \frac{1}{\varepsilon^l(\omega'', \mathbf{k}'')} S_{ijs}(\omega, \mathbf{k}; \omega', \mathbf{k}') S_{bca}(\omega'', \mathbf{k}''; \omega, \mathbf{k}) \frac{k''_j k''_b}{k''^2} - V_{icas}(\omega, \mathbf{k}; \omega', \mathbf{k}')\right\} +$$
$$+ \pi \operatorname{sign} \varepsilon^{l\prime\prime}(\omega, \mathbf{k})\, \delta[\varepsilon^{l\prime}(\omega, \mathbf{k})] \int d\omega' d\mathbf{k}' d\omega'' d\mathbf{k}'' \delta(\omega - \omega' - \omega'') \delta(\mathbf{k} - \mathbf{k}' - \mathbf{k}'') \times$$
$$\times \left| S_{ijs}(\omega, \mathbf{k}; \omega', \mathbf{k}') \frac{k_i k''_j k'_s}{kk'k''} \right|^2 (E_l^2)_{\omega', \mathbf{k}'} (E_l^2)_{\omega'', \mathbf{k}''} + 2 \int d\omega' d\mathbf{k}' d\omega'' d\mathbf{k}'' \delta(\omega - \omega' - \omega'') \delta(\mathbf{k} - \mathbf{k}' - \mathbf{k}'') (E_l^2)_{\omega, \mathbf{k}} \times$$
$$\times (E_l^2)_{\omega', \mathbf{k}'} \operatorname{Im}\left\{ \frac{\delta_{bj} - k''_j k''_b / k''^2}{\varepsilon^{tr}(\omega'', \mathbf{k}'') - \frac{c^2 k''^2}{\omega''^2}} \frac{k_i k_a k'_s k'_c}{(kk')^2} S_{ijs}(\omega, \mathbf{k}; \omega', \mathbf{k}') S_{bca}(\omega'', \mathbf{k}''; \omega, \mathbf{k})\right\}. \tag{9.1}$$

On the right-hand side of this equation there are four terms. The first term corresponds to linear damping of electrostatic oscillations on account of the inverse Cherenkov effect (Landaudamping). The second and the fourth describe processes of induced scattering of longitudinal waves through an intermediate (virtual) oscillation: longitudinal and transverse, respectively. The third term describes the coalescence of two longitudinal oscillations into a third.

Note that decay processes of only longitudinal waves in an isotropic plasma can be studied in accordance with (3.22) by means of the equation (see § 21 for more details)

$$\left\{\frac{\partial \varepsilon^{l\prime}(\omega, \mathbf{k})}{\partial \omega} \frac{\partial}{\partial t}(E_l^2)_{\omega, \mathbf{k}} - \frac{\partial \varepsilon^{l\prime}(\omega, \mathbf{k})}{\partial \mathbf{k}} \frac{\partial}{\partial \mathbf{r}}(E_l^2)_{\omega, \mathbf{k}}\right\}^{\text{decay}} =$$
$$= -\pi \operatorname{Re} \int d\omega' d\mathbf{k}' d\omega'' d\mathbf{k}'' \delta(\omega - \omega' - \omega'') \delta(\mathbf{k} - \mathbf{k}' - \mathbf{k}'') \times$$
$$\times \{-S_{ijs}(\omega, \mathbf{k}; \omega', \mathbf{k}') S^*_{abc}(\omega, \mathbf{k}; \omega', \mathbf{k}') (E_l^2)_{\omega', \mathbf{k}'} \times$$
$$\times (E_l^2)_{\omega'', \mathbf{k}''} \delta[\varepsilon^{l\prime}(\omega, \mathbf{k})] \operatorname{sign} \varepsilon^{l''}(\omega, \mathbf{k}) + S_{ijs}(\omega, \mathbf{k}; \omega', \mathbf{k}') \times$$
$$\times S_{bca}(\omega'', \mathbf{k}''; \omega, \mathbf{k}) (E_l^2)_{\omega, \mathbf{k}} (E_l^2)_{\omega', \mathbf{k}'} \delta[\varepsilon^{l\prime}(\omega'', \mathbf{k}'')] \times$$
$$\times \operatorname{sign} \varepsilon^{l''}(\omega'', \mathbf{k}'') + S_{isj}(\omega, \mathbf{k}; \omega'', \mathbf{k}'') S_{cba}(\omega', \mathbf{k}'; \omega, \mathbf{k}) \times$$
$$\times (E_l^2)_{\omega, \mathbf{k}} (E_l^2)_{\omega'', \mathbf{k}''} \delta[\varepsilon^{l\prime}(\omega', \mathbf{k}')] \operatorname{sign} \varepsilon^{l''}(\omega', \mathbf{k}')\} \frac{k_i k_a k''_j k''_b k'_s k'_c}{(k k' k'')^2}. \qquad (9.2)$$

Equation (9.1) was derived in [12]. Without the last term on the right-hand side it corresponds to Eq. (10) of [19]. There is a further difference between these equations. As we have already mentioned in § 4, the right-hand side of (9.1) does not contain a term describing the emission of longitudinal waves by particles, which is contained in Eq. (10) of [19]. This term has the form

$$2\pi\delta[\varepsilon^{l\prime}(\omega, \mathbf{k})] \frac{4\pi e^2}{k^2} \int d\mathbf{p} f_0(\mathbf{p}) \delta(\omega - \mathbf{k}\mathbf{v}). \qquad (9.3)$$

Without the last term on the right and for a time-independent spectral function, the equation (9.1) of nonlinear interaction goes over into the equation used in [29].

We shall use the expressions "Coulomb scattering" and "scattering with allowance for retardation" for the nonlinear interaction through a longitudinal and a transverse virtual oscillation, respectively, when one can distinguish parts of the inverse tensor $A_{jb}(\omega'', \mathbf{k}'')$ that are longitudinal and transverse with respect to the wave vector $\mathbf{k}'' = \mathbf{k} - \mathbf{k}'$. In this sense, the second term on the right-hand side of (9.1) is responsible for Coulomb scattering and the last, fourth term, for scattering with allowance for retardation.

We begin our treatment of the various nonlinear processes by studying the nonlinear interaction of electron Langmuir waves; this is described by means of the second term on the right-hand side of the generalized kinetic equation (9.1) for the longitudinal waves. Going over to the energy spectral density (6.64) and restricting ourselves to the largest terms in powers of the Debye radius [cf. (8.23)],

$$(k r_{D_e})^2 \ll 1, \quad (k' r_{D_e})^2 \ll 1 \qquad (9.4)$$

in the expansion of the tensors S and V with the Maxwell distribution function (6.14), we obtain an equation for the change of the spectral density of the energy of electron Langmuir waves due to Coulomb scattering on particles [32]:

$$\left\{\frac{dW_l(\mathbf{k})}{dt}\right\}^{\text{Coul}} = -\frac{1}{16\pi^3} \omega_{L_e} \frac{W_l(\mathbf{k})}{N_e \varkappa T_e} \int d\mathbf{k}' W_l(\mathbf{k}') \left(\frac{\mathbf{k}\mathbf{k}'}{k k'}\right)^2 \frac{(k'' r_{D_e})^2}{|\varepsilon^l(\omega'', \mathbf{k}'')|^2} \times$$
$$\times \left\{\delta\varepsilon_e^{l''}\left[4[\mathbf{k}\mathbf{k}']^2 \frac{r_{D_e}^2}{k''^2}(2\delta\varepsilon_e^{l\prime} + 2\delta\varepsilon_e^{l\prime}\delta\varepsilon_i^{l\prime} + |\delta\varepsilon_e^l|^2) + |1 + \delta\varepsilon_i^l|^2\right] + \delta\varepsilon_i^{l''}|\delta\varepsilon_e^l|^2\right\}. \qquad (9.5)$$

Here the kernel of the nonlinear integrodifferential equation on the right-hand side of (9.5) is expressed in terms of the imaginary, $\delta\varepsilon^{l''}$, and real, $\delta\varepsilon^{l'}$, parts of the partial polarizabilities of the electron, $\delta\varepsilon_e^l$, and ion, $\delta\varepsilon_i^l$, components of the plasma that occur in the longitudinal permittivity $\varepsilon^l(\omega'', \mathbf{k}'')$ at the frequency ω'' of the virtual oscillation:

$$\varepsilon^l(\omega'', \mathbf{k}'') = 1 + \delta\varepsilon_e^l(\omega'', \mathbf{k}'') + \delta\varepsilon_i^l(\omega'', \mathbf{k}'') \equiv 1 + \delta\varepsilon_e^l + \delta\varepsilon_i^l. \tag{9.6}$$

Explicit expressions for these quantities can be obtained by means of formulas (6.23) and (6.20) of the foregoing chapter. Note that the frequency ω'' of the virtual oscillation has, on account of the expression (6.34a) for the spectra of the interacting Langmuir waves, the following form in (9.5)

$$\omega'' = \frac{3}{2}(k^2 - k'^2)\, r_{D_e} v_{T_e}. \tag{9.7}$$

On the left-hand side of Eq. (9.5), the spectral density $W_l(\mathbf{k})$ of the energy of the Langmuir waves has a total derivative with respect to the time applied to it. Such a total (substantial) derivative with respect to the time will also be used below to give an abbreviated expression of the change of $W_l(\mathbf{k})$ in time and space. In the given case, the total derivative is defined by

$$\frac{d}{dt} W_l(\mathbf{k}) \equiv \frac{\partial}{\partial t} W_l(\mathbf{k}) - \frac{\partial \varepsilon^{l'}(\omega, \mathbf{k})}{\partial \mathbf{k}} \left[\frac{\partial \varepsilon^{l'}(\omega, \mathbf{k})}{\partial \omega}\right]^{-1} \frac{\partial}{\partial \mathbf{r}} W_l(\mathbf{k}), \tag{9.8}$$

which, using an equation that follows from the dispersion relation (6.35):

$$\frac{\partial \varepsilon^{l'}(\omega, \mathbf{k})}{\partial \mathbf{k}} = -\frac{d\omega(k)}{d\mathbf{k}} \frac{\partial \varepsilon^{l'}(\omega, \mathbf{k})}{\partial \omega} \tag{9.9}$$

can be written, if we use the spectrum (6.34a), in the form

$$\frac{d}{dt} W_l(\mathbf{k}) = \frac{\partial}{\partial t} W_l(\mathbf{k}) + 3 v_{T_e} (\mathbf{k} r_{D_e}) \frac{\partial}{\partial \mathbf{r}} W_l(\mathbf{k}). \tag{9.10}$$

The superscript "Coul" on the left-hand side of Eq. (9.5) indicates explicitly that from the generalized kinetic equation (9.1) for longitudinal waves we have retained only the second term on the right-hand side, which describes the Coulomb scattering.

The antisymmetry of the kernel $Q(\mathbf{k}, \mathbf{k}')$ of Eq. (9.5):

$$Q(\mathbf{k}, \mathbf{k}') \equiv -\frac{1}{16\pi^3} \frac{\omega_{L_e}}{N_e \varkappa T_e} \left(\frac{\mathbf{k}\mathbf{k}'}{kk'}\right)^2 \frac{(k'' r_{D_e})^2}{|\varepsilon^l(\omega'', \mathbf{k}'')|^2} \Big\{\delta\varepsilon_e^{l''}\Big[4[\mathbf{k}\mathbf{k}']^2 \frac{r_{D_e}^2}{k''^2}(2\delta\varepsilon_e^{l'} + 2\delta\varepsilon_e^{l'}\delta\varepsilon_i^{l'} + |\delta\varepsilon_e^l|^2) + $$
$$+ |1 + \delta\varepsilon_i^l|^2\Big] + \delta\varepsilon_i^{l''} |\delta\varepsilon_e^l|^2\Big\} \tag{9.11}$$

under the replacement $\mathbf{k}$ by $\mathbf{k}'$ and $\mathbf{k}'$ by $\mathbf{k}$,

$$Q(\mathbf{k}, \mathbf{k}') = -Q(\mathbf{k}', \mathbf{k}) \tag{9.12}$$

entails conservation of the quantity

$$W_0 = \int d\mathbf{k} W_l(\mathbf{k}) = \text{const}. \tag{9.13}$$

In other words, in this approximation, the induced scattering of Langmuir waves on particles does not change the total energy of these waves. Therefore, a decrease in the spectral energy density in one region of wavelengths must be accompanied by an increase in the energy density of the oscillations with other wavelengths. Such a process may naturally be called a transfer of energy within the spectrum.

The direction in which this tensor takes place can be established by considering, for example, the nonlinear interaction of two infinitesimally narrow wave packets. For this, we represent $W_l(\mathbf{k})$ as a sum of two terms:

$$W_l(\mathbf{k}, \mathbf{r}, t) = W_1(\mathbf{r}, t)\delta(\mathbf{k}-\mathbf{k}_1) + W_2(\mathbf{r}, t)\delta(\mathbf{k}-\mathbf{k}_2) \tag{9.14}$$

and we substitute (9.14) into (9.5). Then the integrodifferential equation (9.5) reduces to a system of two first-order partial differential equations:

$$\begin{aligned}\frac{\partial W_1}{\partial t} + 3v_{T_e}(\mathbf{k}_1 r_{D_e})\frac{\partial W_1}{\partial \mathbf{r}} &= +Q(\mathbf{k}_1, \mathbf{k}_2)W_1W_2,\\ \frac{\partial W_2}{\partial t} + 3v_{T_e}(\mathbf{k}_2 r_{D_e})\frac{\partial W_2}{\partial \mathbf{r}} &= -Q(\mathbf{k}_1, \mathbf{k}_2)W_1W_2.\end{aligned} \tag{9.15}$$

For simplicity, we here ignore the spatial variation of the energy densities W_1 and W_2:

$$\begin{aligned}\frac{\partial W_1}{\partial t} &= Q(\mathbf{k}_1, \mathbf{k}_2)W_1W_2,\\ \frac{\partial W_2}{\partial t} &= -Q(\mathbf{k}_1, \mathbf{k}_2)W_1W_2.\end{aligned} \tag{9.16}$$

This neglect corresponds to a representation that is even simpler than (9.14) [29]:

$$W_l(\mathbf{k}, \mathbf{r}, t) = W_1(t)\delta(\mathbf{k}-\mathbf{k}_1) + W_2(t)\delta(\mathbf{k}-\mathbf{k}_2). \tag{9.17}$$

The solution of the system (9.16) can be written down by means of the equations

$$\begin{aligned}&W_1(t) + W_2(t) = W_1(t_0) + W_2(t_0) = W_0,\\ &\frac{W_1(t)}{W_2(t)} = \frac{W_1(t_0)}{W_2(t_0)}\exp\{Q(\mathbf{k}_1, \mathbf{k}_2)W_0(t-t_0)\},\\ &\frac{W_1(t)}{W_1(t_0)} = W_0\{W_1(t_0) + W_2(t_0)\exp[-Q(\mathbf{k}_1, \mathbf{k}_2)W_0(t-t_0)]\}^{-1}.\end{aligned} \tag{9.18}$$

Since $Q(\mathbf{k}_1, \mathbf{k}_2)$ is negative for $k_1 > k_2$, we can assert in accordance with (9.18) that for the Coulomb scattering of two Langmuir oscillations energy is transferred from short to long wavelengths. This energy redistribution differs appreciably from the damping of waves in linear theory due to the inverse Cherenkov effect or particle collisions, when there is a decrease in the amplitude of the oscillations at any wavelength on account of energy being transferred to individual particles.

Using the language of turbulence theory, the transfer of energy of the Langmuir oscillations from the shorter to longer wavelengths can be interpreted as transfer of energy from small-scale pulsations to large-scale pulsations. Applied to the evolution in time of a wave packet of finite width, we must not, in accordance with what we have said, speak of damping of a wave packet but of a change of its form due to the change of its spectrum. The simplest characteristic of such a change of form is the time of transfer of energy density within the spectrum:

$$\tau = -[Q(\mathbf{k}_1, \mathbf{k}_2)W_0]^{-1}, \quad k_1 > k_2, \tag{9.19}$$

for the case of two Langmuir oscillations with slightly different wave vectors $\mathbf{k}_1 \simeq \mathbf{k}_2$, $k_1 > k_2$.

The expression (9.11) for the kernel $Q(\mathbf{k}, \mathbf{k}')$ of Eq. (9.5) can be appreciably simplified in a number of limiting cases. The phase velocity ω''/k'' of the virtual oscillation [see (9.5) for ω''] is small compared with the electron thermal velocity, which leads to the inequalities

$$\delta\varepsilon_e^{l'} \gg \delta\varepsilon_e^{l''}, \quad \delta\varepsilon_e^{l'} \simeq (k'' r_{D_e})^{-2} \gg 1. \tag{9.20}$$

Conversely, for ions, because of their large mass ($M \gg m$) compared with the electron mass, the ratio of the phase velocity of the virtual oscillation to the thermal velocity v_{Ti} may be arbitrary and, in particular, greater than unity. In this sense,

$$\frac{\omega''}{k'' v_{T_i}} \tag{9.21}$$

can be used to express a quantitative criterion of the role of ions in the process of Coulomb scattering.* We shall say that Coulomb scattering occurs on electrons it the eletron imaginary part of the longitudinal permittivity, $\delta\varepsilon_e^{l''}(\omega'', \mathbf{k}'')$, exceeds the ion imaginary part:

$$\delta\varepsilon_e^{l''}(\omega'', \mathbf{k}'') \gg \delta\varepsilon_i^{l''}(\omega'', \mathbf{k}''). \tag{9.22}$$

This inequality holds if the ratio (9.21) is not too small:

$$\omega''^2 \gg k''^2 v_{T_i}^2 \ln\left\{\left(\frac{T_e}{T_i}\right)^3 \frac{M}{m}\frac{e_i^2}{e^2}\right\} \tag{9.23}$$

or, with allowance for the explicit expression (9.7) for ω'':

$$\frac{9}{4}\frac{(k^2 - k'^2)^2}{(\mathbf{k} - \mathbf{k}')^2} r_{D_e}^2 \gg \frac{m}{M}\frac{T_i}{T_e} \ln\left\{\left(\frac{T_e}{T_i}\right)^3 \frac{M}{m}\frac{e_i^2}{e^2}\right\}. \tag{9.24}$$

Conversely, Coulomb scattering occurs on ions if the inequality (9.22) is violated:

$$\delta\varepsilon_e^{l''} \lesssim \delta\varepsilon_i^{l''}. \tag{9.25}$$

Let us consider first a case when the ratio (9.21) is very samll:

$$\frac{\omega''}{k'' v_{T_i}} \ll 1. \tag{9.26}$$

Then in the expression (9.11) for the kernel Q the last term, which is proportional to $\delta\varepsilon_i^{l''}$, is the largest in the curly brackets, and Eq. (9.5) takes the form

$$\frac{dW_l(\mathbf{k})}{dt} = -\frac{3}{8(2\pi)^{5/2}}\omega_{L_e}\frac{W_l(\mathbf{k})}{N_e \varkappa T_e} r_{D_e}\frac{v_{T_e}}{v_{T_i}}\left[\frac{r_{D_i}/r_{D_e}}{1 + (r_{D_i}/r_{D_e})^2}\right]^2 \int d\mathbf{k}' W_l(\mathbf{k}')\left(\frac{\mathbf{k}\mathbf{k}'}{kk'}\right)^2 \frac{k^2 - k'^2}{|\mathbf{k} - \mathbf{k}'|}. \tag{9.27}$$

Obviously, the condition (9.25) is then satisfied and the scattering occurs on ions. As a generalization of (9.27) we can take the equation (which also describes scattering on ions)

$$\frac{dW_l(\mathbf{k})}{dt} = -\frac{3}{8(2\pi)^{5/2}}\omega_{L_e}\frac{W_l(\mathbf{k})}{N_e \varkappa T_e} r_{D_e}\frac{v_{T_e}}{v_{T_i}} \int d\mathbf{k}' W_l(\mathbf{k}')\left(\frac{\mathbf{k}\mathbf{k}'}{kk'}\right)^2 \frac{k^2 - k'^2}{|\mathbf{k} - \mathbf{k}'|}\left[\frac{r_{D_i}/r_{D_e}}{F + (r_{D_i}/r_{D_e})^2}\right]^2 \exp\left\{-\frac{9}{8}\frac{v_{T_e}^2}{v_{T_i}^2} r_{D_e}^2 \frac{(k^2 - k'^2)^2}{(\mathbf{k} - \mathbf{k}')^2}\right\}, \tag{9.28}$$

* Since the nonlinear tensors S and V are inversely proportional to the square and the cube of the particle mass, the small ion terms ($M \gg m$) drop out on the summation over the particle species on the right-hand sides of the expressions that determine these tensors if the main and most important part of these tensors depends weakly on the particle species. Just such a situation obtains in a plasma with moderately different temperatures of the electrons and ions in the case of the nonlinear interaction of oscillations with high phase velocities ω/k, $\omega'/k' \gg v_T$. The kernel Q given here is determined by the contraction of only the electron parts of the nonlinear tensors, so that the ions enter Q only through $\varepsilon^l(\omega'', \mathbf{k}'')$.

in which the function F is determined by the equations

$$F=\begin{cases}0, & 1\gg\left[v_{T_i}^2(\mathbf{k}-\mathbf{k}')^2/(\omega-\omega')^2\right]\gg\left[\ln\left(\frac{T_e^3}{T_i^3}\frac{M}{m}\frac{e_i^2}{e^2}\right)\right]^{-1},\\ 1, & v_{T_i}^2(\mathbf{k}-\mathbf{k}')^2\gg(\omega-\omega')^2.\end{cases} \tag{9.29}$$

If, in contrast to (9.26), the phase velocity of the virtual oscillation is high compared with the ion thermal velocity:

$$\frac{\omega''}{k''v_{T_i}}\to\infty, \tag{9.30}$$

then the term proportional to $\delta\varepsilon_i^{l''}$ in the kernel of (9.11) can be ignored compared with the term $\sim\delta\varepsilon_i^{l''}$. In this limiting case, the kernel (9.11) takes the form [32]

$$Q(\mathbf{k},\mathbf{k}')=-\frac{3}{8(2\pi)^{5/2}}\frac{\omega_{L_e}}{N_e\varkappa T_e}r_{D_e}\frac{k^2-k'^2}{|\mathbf{k}-\mathbf{k}'|}\left(\frac{\mathbf{k}\mathbf{k}'}{kk'}\right)^2\left[1+(k''r_{D_e})^2\frac{\omega_{L_i}^2}{\omega''^2}\right]^{-2}\times$$
$$\times\left\{4\frac{r_{D_e}^2}{k''^2}[\mathbf{k}\mathbf{k}']^2\left[1+2(k''r_{D_e})^2\frac{\omega_{L_i}^2}{\omega''^2}\right]+(k''r_{D_e})^4\left(1+\frac{\omega_{L_i}^2}{\omega''^2}\right)^2\right\}. \tag{9.31}$$

For the same values of the wave vectors $\mathbf{k}$ and $\mathbf{k}'$ of interacting Langmuir oscillations for which the frequency ω'' of the virtual oscillation is high compared with the ion plasma frequency ω_{L_i}, and the angle between $\mathbf{k}$ and $\mathbf{k}'$ is not too small,

$$|\omega''|\gg\omega_{L_i},\qquad [\mathbf{k}\mathbf{k}']^2\gg(k'')^6r_{D_e}^2, \tag{9.32}$$

the kernel (9.31) simplifies even further and the equation (9.5) of the nonlinear interaction can be represented in the form [12, 29]

$$\frac{dW_l(\mathbf{k})}{dt}=-\frac{3}{2(2\pi)^{5/2}}\omega_{L_e}\frac{W_l(\mathbf{k})}{N_e\varkappa T_e}r_{D_e}^3\int d\mathbf{k}'W_l(\mathbf{k}')\frac{k^2-k'^2}{|\mathbf{k}-\mathbf{k}'|^3}[\mathbf{k}\mathbf{k}']^2\left(\frac{\mathbf{k}\mathbf{k}'}{kk'}\right)^2. \tag{9.33}$$

We emphasize that in accordance with (9.33) Coulomb scattering of Langmuir oscillations is determined solely by the electron terms. In this connection, the equation (9.33) of nonlinear interaction can be obtained in an electron plasma by means of the contraction (8.23). Clearly, the inequality (9.23) is satisfied in this case on account of (9.30), and the scattering occurs on electrons. Note that scattering on electrons can also occur in accordance with a different law. Namely, under the conditions

$$(k''r_{D_e})\ll(\omega''/\omega_{L_i})\ll1,\qquad \omega''^2\ll\omega_{L_i}^2r_{D_e}(k''^3/|[\mathbf{k}\mathbf{k}']|) \tag{9.34}$$

the kernel (9.31) is determined by the equation

$$Q(\mathbf{k},\mathbf{k}')=-\frac{1}{6(2\pi)^{5/2}}\frac{\omega_{L_e}}{N_e\varkappa T_e}\frac{\omega_{L_i}^2}{\omega_{L_e}^2}r_{D_e}\frac{|\mathbf{k}-\mathbf{k}'|^3}{(k^2-k'^2)}\left(\frac{\mathbf{k}\mathbf{k}'}{kk'}\right)^2 \tag{9.35}$$

and corresponds to an equation of nonlinear interaction that is very different from (9.33).

Let us consider in more detail the Coulomb scattering of electron Langmuir oscillations, proceeding on the basis of Eq. (9.33). We note first that in an electron plasma allowance for the first term on the right-hand side of (9.1), which describes Landau damping on electrons, and the term (9.3), which is responsible for the spontaneous emission of waves by electrons, leads to the equation [31]

$$\frac{\partial W_l(\mathbf{k})}{\partial t}+3v_{T_e}(\mathbf{k}r_{D_e})\frac{\partial W_l(\mathbf{k})}{\partial \mathbf{r}}=\sqrt{\frac{\pi}{2}}\frac{\omega_{L_e}}{(kr_{D_e})^3}\exp\left\{-\frac{3}{2}-\frac{1}{2(kr_{D_e})^2}\right\}\{\varkappa T_e-W_l(\mathbf{k})\}-$$
$$-\frac{3}{2(2\pi)^{5/2}}\omega_{L_e}\frac{W_l(\mathbf{k})}{N_e\varkappa T_e}r^3_{D_e}\int d\mathbf{k}'W_l(\mathbf{k}')\frac{k^2-k'^2}{|\mathbf{k}-\mathbf{k}'|^3}[\mathbf{k}\mathbf{k}']^2\left(\frac{\mathbf{k}\mathbf{k}'}{kk'}\right)^2, \quad (9.36)$$

which differs from (9.33) by the first term on the right. From this form of expression it is clear that the spontaneous emission of Langmuir waves in an electron plasma with the Maxwellian distribution (6.14) can be ignored* if the spectral density of the energy of these waves is large compared with the electron temperature:

$$W_l(\mathbf{k})\gg\varkappa T_e. \quad (9.37)$$

In its turn, the linear dissipation of Langmuir waves due to the inverse Cherenkov effect is very small in the limit of long waves [see the explicit expression for the damping constant (6.34b)]. Therefore, it can be ignored in (9.36) if it is assumed that the wavelengths λ of the Langmuir oscillations are greater than λ_{cr} ([31], p. 74)

$$\lambda>\lambda_{cr}=r_{D_e}\sqrt{\ln\{N_e r^3_{D_e}\varkappa T_e/W_l(\mathbf{k})\}}. \quad (9.38)$$

Under the conditions (9.37) and (9.38), Eqs. (9.33) and (9.36) are completely equivalent.

In deriving the tensor quantities in the generalized kinetic equation for the waves (3.22) and therefore (9.1) we have restricted ourselves to a collisionless plasma. Let us see on the basis of the example of the Coulomb scattering of Langmuir waves in an electron plasma when one can ignore the damping of these waves due to electron–ion collisions. The effective frequency of collisions, which characterizes the reciprocal damping time of the waves, is determined by [see (1), p. 114, (16.13)]

$$\nu_{eff}=\frac{L}{3(2\pi)^{3/2}}\frac{\omega_{L_e}}{N_e r^3_{D_e}}\frac{\sum_i N_i e_i^2}{N_e e^2}, \quad (9.39)$$

where the sum over i denotes summation over all ion species and L is the Coulomb logarithm:

$$L=\ln(4\pi N_e r^3_{D_e}). \quad (9.40)$$

Comparison of this expression for the effective collision frequency with the right-hand side of Eq. (9.33) shows that the nonlinear interaction is more important then the linear effect of collisions of the plasma particles with one another at a sufficiently high energy density of the waves:

$$W_l(\mathbf{k})\gg L\varkappa T_e. \quad (9.41)$$

In the whole of this exposition we shall assume that this inequality holds. Therefore, in particular, we can ignore the spontaneous terms in the generalized kinetic equation for the waves [see (9.37) and (9.41) for $L \gg 1$].

A feature of the Coulomb scattering of Langmuir oscillations on electrons described by Eq. (9.33) is the vanishing of the nonlinear interaction between waves with parallel and perpendicular wave vectors. This feature and also the conservation law (9.13) for the total energy of the oscillations are characteristic of the first nonvanishing approximation in the expansion

* Compared with the term that describes the linear dissipation.

of the nonlinear tensors in powers of the square of the ratio of the Debye radius to the wavelength. Allowance for the next term in kr_D leads to the appearance of an additional term on the right-hand side of (9.33):

$$-\frac{3}{2(2\pi)^{5/2}}\omega_{L_e}\frac{W_l(\mathbf{k})}{N_e\varkappa T_e}r_{D_e}^3\int d\mathbf{k}'W_l(\mathbf{k}')\frac{k^2-k'^2}{|\mathbf{k}-\mathbf{k}'|}\left\{\frac{3}{4}r_{D_e}^2(k^2-k'^2)\left(\frac{\mathbf{kk}'}{kk'}\right)^2\frac{[\mathbf{kk}']^2}{|\mathbf{k}-\mathbf{k}'|^2}+\frac{r_{D_e}^2}{4}\frac{[\mathbf{kk}']^4+9(\mathbf{kk}')^4}{(kk')^2}\right\}. \quad (9.42)$$

Because of this term, the Langmuir waves interact with parallel (and perpendicular) wave vectors, and the wave energy is no longer conserved. At the same time, the part of the kernel in (9.42) that is symmetric with respect to k and k' determines the change in the energy of the waves, whereas the interaction of waves with parallel (and perpendicular) wave vectors is determined by the antisymmetric part. Equation (9.33) with the additional term (9.42) on the right-hand side again has an integral. However, the conserved quantity in this case is not the total energy of the waves. It is not difficult to show that with the accuracy we have chosen in the expansion in powers of the square of the ratio of the Debye radius to the wavelength we can say that the following quantity is conserved:

$$\int d\mathbf{k}\,\frac{W_l(\mathbf{k})}{\omega(\mathbf{k})}=\text{const}. \quad (9.43)$$

In quantum language, this conservation law means that the total number $d\mathbf{k}N_\mathbf{k}$ of quanta* is unchanged, $W_l(\mathbf{k}) = \hbar\omega(\mathbf{k})N_\mathbf{k}$, and in classical language it corresponds to conservation of an adiabatic invariant.

For the spectral density of the energy, which depends only on the absolute magnitude of the wave vector, Eq. (9.33) in conjunction with (9.42) takes the form

$$\frac{dW_l(k)}{dt}=-\frac{2}{35(2\pi)^{5/2}}\omega_{L_e}\frac{W_l(k)}{N_e\varkappa T_e}\frac{r_{D_e}^2}{k^3}\left\{\int_0^k dk'k'^4W_l(k')(7k^4+5k^2k'^2-12k'^4)-\right.$$
$$\left.-\int_k^{k_{\max}}\frac{dk'}{k'}k^5(7k'^4+5k^2k'^2-12k^4)W_l(k')\right\}\left[1+\frac{3}{4}(k^2-k'^2)r_{D_e}^2\right]. \quad (9.44)$$

The factor $[1 + (3/4)(k^2 - k'^2)r_{D_e}^2]$ in the square bracket on the right-hand side of (9.44) arises because of the term (9.42) and is responsible for the energy not being conserved or, in other words, for nonlinear dissipation. And, for example, in the problem of the interaction of two infinitesimally narrow wave packets one again obtains a solution of the form (9.18), which corresponds to spectral transformation with almost the same characteristic time (9.19) that determines the energy transfer in the case of Coulomb scattering of Langmuir waves on electrons without allowance for (9.42)†

$$\tau\simeq\frac{L}{\nu_{\text{eff}}}\frac{\varkappa T_e}{(W_0/k^3)}\frac{1}{r_{D_e}^6k^5\Delta k}, \quad (9.45)$$

where k is of order of k_1 and k_2, and $\Delta k = |k_1 - k_2|$. Therefore, with increasing time the spectral density of the energy of the short waves again tends to zero while the energy density of the

* As we shall see below (see §21) conservation of the number of oscillations participating in scattering on particles is an exact law in the framework of the generalized kinetic equation for waves (3.22) [see (3.22b)].

† For example, for the case when $(kr_D)^2 \sim 0.1$, $L \sim 10$, $\Delta k \sim k$, we find from (9.45) that $\tau \sim 10^4\,\nu_{\text{eff}}^{-1}(\varkappa T_e k^3/W_0)$. Therefore, for $(W_0/k^3) \sim 10^5\varkappa T_e$, which is evidently a fully realistic value, the time of spectral redistribution is shorter by an order of magnitude than the time of damping of the Langmuir oscillations due to the Coulomb collisions of the plasma particles.

long waves tends to a finite limit. However, because in accordance with (9.43) the wave number dkN_k is conserved, the energy of the waves decreases with increasing wavelength. In the problem of the interaction of two infinitesimally narrow wave packets, the energy lost by the Langmuir oscillations tends asymptotically to the value ([31], p. 79, (12, 18))

$$\Delta W = \hbar(\omega_1 - \omega_2) N_{k_1}(t_0) \simeq \frac{3}{2}(k_1^2 - k_2^2) r_{D_e}^2 W_1(t_0). \tag{9.46}$$

This makes it obvious that the nonlinear damping of Langmuir waves is weak, since the fraction of energy given up by the oscillations to the particles for the case considered here, that is, for wavelengths greater than the Debye radius, is small.

If it is not assumed that the spectral density of the energy of the plasma oscillations is isotropic, $W_l(\mathbf{k}) = W_l(k)$, the spectral transfer (9.33) due to Coulomb scattering on electrons can be characterized by the time

$$\tau \simeq \frac{(2\pi)^{5/2}}{3} \frac{1}{\omega_{Le}} \frac{N_e \varkappa T_e}{W_0} \frac{1}{(kr_{D_e})^3} \frac{1}{\sin^2\theta \left|\cos\theta + \frac{1}{2}\frac{\Delta k}{k}\right|}, \tag{9.47}$$

in which, in accordance with (9.19), $k = \mathbf{k}_1$, $\Delta\mathbf{k} = \mathbf{k}_2 - \mathbf{k}_1$, $\Delta k = |\Delta\mathbf{k}| \ll k_1, k_2$, and θ is the angle between $\mathbf{k}$ and $\Delta\mathbf{k}$. The characteristic time of Coulomb scattering on ions described by Eq. (9.27) can be represented approximately in the form

$$\tau \sim \frac{L}{\nu_{\text{eff}}} \frac{\varkappa T_e}{W_l} \frac{1}{r_{D_e}^4 k^3 \Delta k} \frac{v_{T_i}}{v_{T_e}}. \tag{9.48}$$

Comparison of formulas (9.48) on the one hand, and (9.45) and (9.47) on the other shows that in the region of long waves, in which scattering on ions is important, the spectral redistribution of the energy density takes place more rapidly than at shorter wavelengths, where scattering on electrons is predominant.

With these remarks we conclude our study of the Coulomb scattering of Langmuir waves in an isotropic collisionless plasma, and we turn to the problem of allowing for transverse, nonelectrostatic fields on the nonlinear interaction of such waves. As we have already pointed out, this is described by the last term on the right-hand side of Eq. (9.1).

The first effect that is due to this last term is the process of coalescence of two longitudinal waves into one transverse one. The corresponding contribution arises because of the vanishing of the denominator [cf. (6.81)]

$$\varepsilon^{tr}(\omega'', \mathbf{k}'') - \frac{c^2k''^2}{\omega''^2} = 0. \tag{9.49}$$

The coalescence of two Langmuir waves with frequencies of order ω_{Le} leads to the appearance of a transverse electromagnetic wave with twice the frequency, $2\omega_{Le}$, and, as follows from the dispersion law (6.53), with wave number equal to $\sqrt{3}\omega_{Le}/c$. All the phase velocities of the waves that participate in such a decay process are large compared with the electron thermal velocity. This enables us to use the expression (7.23) for the tensor S_{ijs} in a cold ($v_T = 0$) isotropic plasma. Substituting (7.23) into the last term on the right-hand side of (9.1) and replacing the denominator (9.49) by the δ function:

$$\frac{1}{\varepsilon^{tr}(\omega'', k'') - \frac{c^2k''^2}{\omega''^2}} \simeq -i\pi\delta\left[\varepsilon^{tr'}(\omega'', \mathbf{k}'') - \frac{c^2k''^2}{\omega''^2}\right] \operatorname{sgn} \varepsilon^{tr''}(\omega''\mathbf{k}''), \tag{9.50}$$

we obtain an equation that describes the rate of decrease of the energy density of the Langmuir waves due to their coalescing with Langmuir waves and their transformation into transverse electromagnetic waves:*

$$\left\{\frac{dW_l(\mathbf{k})}{dt}\right\}^{c} = -\frac{1}{12(2\pi)^2}\frac{\omega_{L_e}}{N_e r_{D_e}^3}\frac{W_l(\mathbf{k})}{\varkappa T_e}\frac{c^2}{v_{T_e}^2} r_{D_e}^7 \int d\mathbf{k}'\delta\left[3-\frac{c^2}{\omega_{L_e}^2}(\mathbf{k}+\mathbf{k}')^2\right] W_l(\mathbf{k}')\left(\frac{[\mathbf{k}\mathbf{k}']}{kk'}\right)^2(\mathbf{k}^2-\mathbf{k}'^2)^2. \quad (9.51)$$

The superscript c on the left-hand side of (9.51) emphasizes that $W_l(\mathbf{k})$ changes here because of the coalescene. If the energy density of the Langmuir waves with wavelengths of order $\lambda_0 = (c/\omega_{Le})$ depends only on the wave number and not on the direction of the wave vector, the characteristic time of decrease of $W_l(\mathbf{k})$ due to the formation fo transverse oscillations can be represented in accordance with (9.19) in the form (cf. [36]; [8], pp. 133-135, § 18)

$$\tau \simeq 10^3 \frac{N_e r_{D_e}^3}{\omega_{L_e}} \frac{\varkappa T_e}{W_l(\mathbf{k})}\left(\frac{c}{v_{T_e}}\right)^5. \quad (9.52)$$

The same time characterizes the coalescene of Langmuir oscillations in wave packets with large wave numbers but width $\sim\omega_{Le}/c$.

Under conditions when Coulomb scattering with the participation of ions is dominant, one must compare (9.52) with the time of spectral transfer (9.48) of the energy density due to scattering on ions. This time (9.48) for wavelengths $\sim\lambda_0 = (c/\omega_{Le})$ is of the order

$$\tau \sim 10^2 \frac{N_e r_{D_e}^3}{\omega_{L_e}} \frac{\varkappa T_e}{W_l(\mathbf{k})}\left(\frac{c}{v_{T_e}}\right)^4 \frac{v_{T_i}}{v_{T_e}} \quad (9.53)$$

and is shorter than the time of coalescence (9.52). Thus, in this case when there is nonlinear relaxation of the Langmuir oscillations only a fraction of their energy goes over into transverse waves as a result of coalescence, while the major part of the energy is redistributed among the longer wavelength Langmuir oscillations. The small fraction of this energy is characterized by the ratio of the time of spectral transfer due to scattering on ions to the time of coalescence, this ratio being equal in accordance with (9.52)-(9.53) to the ratio v_{Ti}/c of the thermal velocity of the ions to the velocity of light. However, since the energy density of the Langmuir waves can appreciably exceed the energy density of the equilibrium thermal fluctuations, the electromagnetic radition formed by the coalescence of such electrostatic oscillations may be very great.

For wavelengths of order $\lambda_0 = c/\omega_{Le}$, the inequality (9.24), whose fulfilment leads to Coulomb scattering on electrons being predominant over scattering on ions, can be represented in the form

$$T_e \frac{T_e}{T_i} \gg \frac{m}{M} mc^2 \ln\left[\frac{T_e^3}{T_i^3}\frac{M}{m}\frac{e_i^2}{e^2}\right]. \quad (9.54)$$

In particular, for a hydrogen plasma, scattering on ions is unimportant if $(T_e/T_i)T_e \gtrsim 2 \cdot 10^{6}$°K and for mercury $(T_e/T_i)T_e \gtrsim 10^{4}$°K. Then the spectral transfer is determined by Coulomb scattering on electrons and, as follows from (945), occurs for wavelengths $\lambda_0 = c/\omega_{Le}$ in a time of order

$$\tau \sim 10^2 \frac{N_e r_{D_e}^3}{\omega_{L_e}} \frac{\varkappa T_e}{W_l}\left(\frac{c}{v_{T_e}}\right)^6. \quad (9.55)$$

* This decay process can be studied in more detail by means of Eq. (3.22a), which enables one to follow the evolution of the spectral density of the energy of the two other waves, that is, the Langmuir and the transverse wave.

Comparing (9.55) and (9.52), we can see that under such conditions transfer of the energy density of the Langmuir oscillations within the spectrum is unimportant. Note, however, that in this case it is difficult to realize conditions under which the coalescence time (9.52) is appreciably shorter than the time of free flight of an electron in the plasma.

Apart from this coalescence effect, the last term on the right-hand side of the generalized kinetic equation (9.1) for longitudinal waves is responsible for a further form of nonlinear interaction, which we have called in the beginning of this section scattering of waves on particles with allowance for retardation. This interaction comes from the regions of ω'' and $\mathbf{k}''$ values for which the expression on the left-hand side of (9.49) is nonvanishing, i.e., Eq. (9.50) is not satisfied. We are here concerned with the contribution to the induced scattering of electrostatic oscillations of the nonelectrostatic interaction of the plasma particles. One can characterize such scattering as scattering through a transverse virtual (intermediate) oscillation.

In accordance with Eq. (9.1), the contribution to the rate of change of the energy density of the Langmuir waves due to this kind of scattering has the form

$$\left\{\frac{dW_l(\mathbf{k})}{dt}\right\}^{\text{ret}} = -\frac{1}{6(2\pi)^{5/2}}\frac{\omega_{Le}}{N_e r_{D_e}^3}\frac{W_l(\mathbf{k})}{\varkappa T_e} r_{D_e}^2 \int d\mathbf{k}' W_l(\mathbf{k}') \frac{[\mathbf{k}\mathbf{k}']^2}{(kk')^2}\left[1 + 4\frac{c^2}{\omega_{Le}^2}(\mathbf{k}\mathbf{k}')\right] \times$$
$$\times \frac{|\mathbf{k}-\mathbf{k}'|}{k^2-k'^2}\left|\frac{4}{9}\frac{c^4(\mathbf{k}-\mathbf{k}')^4}{v_{T_e}^4(k^2-k'^2)^2} + 2\frac{c^2}{v_{T_e}^2} + \frac{\pi}{2}\frac{1}{r_{D_e}^2(\mathbf{k}-\mathbf{k}')^2}\right|^{-1}. \tag{9.56}$$

Here the superscript "ret" on the left-hand side emphasizes that Eq. (9.56) describes scattering with allowance for retardation through a transverse virtual oscillation. In deriving (9.56) we have ignored the contribution of the ions, which is possible for a plasma that is not strongly nonisothermal:

$$\frac{T_e}{T_i}\frac{m}{M} \ll 1. \tag{9.57}$$

This can be readily proved by noting that, as in the case of (9.25), scattering on ions through the virtual transverse oscillation can be significant only if

$$\delta\varepsilon_i^{tr''}(\omega'', \mathbf{k}'') \gg \delta\varepsilon_e^{tr''}(\omega'', \mathbf{k}''), \tag{9.58}$$

which, in accordance with (6.26),

$$\omega''^2 < k''^2 v_{T_i}^2 \ln\left\{\frac{T_e}{T_i}\frac{m}{M}\frac{e_i^2}{e^2}\right\} \tag{9.59}$$

requires the plasma to be much more strongly nonisothermal, so that the electron temperature T_e exceeds the ion temperature T_i by at least the amount by which the ion mass is greater than the electron mass.* We shall assume that the inequality (9.57) is satisfied.

For wavelengths short compared with $\lambda_0 = c/\omega_{Le}$, Eq. (9.56) simplifies somewhat:

$$\frac{dW_l(\mathbf{k})}{dt} = -\frac{3}{2}\frac{1}{(2\pi)^{5/2}}\frac{\omega_{Le}}{N_e r_{D_e}^3}\frac{W_l(\mathbf{k})}{\varkappa T_e}\left(\frac{v_{T_e}}{c}\right)^2 r_{D_e}^4 \int d\mathbf{k}' W_l(\mathbf{k}')(\mathbf{k}\mathbf{k}')\frac{[\mathbf{k}\mathbf{k}']^2}{(kk')^2}\frac{k^2-k'^2}{|\mathbf{k}-\mathbf{k}'|^3}. \tag{9.60}$$

---

*A direct calculation shows that in the case of scattering through a transverse virtual wave the terms corresponding to the ion corrections for scattering on electrons are smaller by a factor M/m, and the terms corresponding to scattering on ions are smaller by a factor $(MT_i/mT_e)^{1/2}$, than the expression considered in (9.56).

The time of spectral transfer of the energy density of Langmuir waves that corresponds to this equation is always much longer than the time of spectral redistribution due to scattering on electrons with allowance for only the Coulomb interaction between the plasma particles. From this we conclude that for such wavelengths allowance for the transverse field in the interaction of particles leads to only small corrections.

The situation is different for wavelengths greater than $\lambda_0 = c/\omega_{Le}$. Let us first assume that the difference $\Delta k = |k - k'|$ of the wave numbers of the interacting Langmuir oscillations is not too small:

$$\omega_{L_e} v_{T_e} \Delta k \gg c^2 k^2. \tag{9.61}$$

Such an inequality for $\Delta k \sim k$ is satisifed for wave numbers $k \ll (\omega_{Le}/c)(v_{Te}/c)$. The right-hand side of Eq. (9.56) then takes the form

$$-\frac{2}{3(2\pi)^{7/2}} \frac{\omega_{L_e}}{N_e r_{D_e}^3} \frac{W_l(\mathbf{k})}{\varkappa T_e} r_{D_e}^4 \int d\mathbf{k}' W_l(\mathbf{k}') \frac{[\mathbf{k}\mathbf{k}']^2}{(kk')^2} \frac{|\mathbf{k}-\mathbf{k}'|^3}{k^2 - k'^2}. \tag{9.62}$$

This expression, as in the case of Coulomb scattering, corresponds to the fact that the energy of the Langmuir oscillations is not transferred to the particles, but is redistributed within the spectrum, passing from the shorter to the longer waves. The corresponding time of spectral redistribution,

$$\tau \sim 10^3 \frac{N_e r_{D_e}^3}{\omega_{L_e}} \frac{\varkappa T_e}{W_l(\mathbf{k})} \frac{\Delta k}{k^5 r_{D_e}^4}, \tag{9.63}$$

is in a number of cases much shorter than the time (9.45), which is characteristic of the Coulomb scattering of Langmuir waves on electrons. In its turn, the latter is short compared with the time (9.48) of scattering on ions. This follows, for example, from the inequality (9.24) for such wavelengths subject to the condition $k \sim k' \sim \Delta k$ and $v_{Ti} c^2 \ll v_{Te}^3$. In this case, the time (9.63) is shorter than the time of Coulomb scattering on electrons for wave numbers $k \gg r_{De}^{-1}(v_{Ti}/v_{Te})$ and the scattering of Langmuir waves through a transverse virtual oscillation is the main effect.

On the other hand, if the time of spectral transfer due to scattering on ions is shorter than the time of Coulomb scattering on electrons, then (9.63) is shorter than the time (9.48) of scattering on ions if $\Delta k < k(v_{Ti}/v_{Te})^{1/2}$.

Let us now suppose that the difference $\Delta k = |k - k'|$ of the wave numbers is small and that [cf. (9.61)]

$$\omega_{L_e} v_{T_e} \Delta k \ll c^2 k^2. \tag{9.64}$$

Then Eq. (9.56) takes the form

$$\frac{dW_l(\mathbf{k})}{dt} = -\frac{3}{8(2\pi)^{5/2}} \frac{\omega_{L_e}}{N_e r_{D_e}^3} \frac{W_l(\mathbf{k})}{\varkappa T_e} r_{D_e}^2 \left(\frac{v_{T_e}}{c}\right)^4 \int d\mathbf{k}' W_l(\mathbf{k}') \frac{[\mathbf{k}\mathbf{k}']^2}{(kk')^2} \frac{k^2 - k'^2}{|\mathbf{k}-\mathbf{k}'|^3} \tag{9.65}$$

and it describes the transfer of spectral energy density of the Langmuir waves from shorter to longer waves with the characteristic time

$$\tau \sim 10^2 \frac{N_e r_{D_e}^3}{\omega_{L_e}} \frac{\varkappa T_e}{W_l(\mathbf{k})} \left(\frac{c}{v_{T_e}}\right)^4 \frac{1}{r_{D_e}^2 k \Delta k}. \tag{9.66}$$

This time is appreciably shorter than the time of induced scattering of waves on electrons with allowance for only the Coulomb interaction of the particles. The time (9.66) characterizes the spectral redistribution of the energy density of the Langmuir waves in the case when the time of scattering on ions is long compared with the time of Coulomb scattering on electrons (at the same time $v_{Ti}c^2 \ll v_{Te}^3$). However, the time (9.66) may also play a decisive role in the spectral redistribution in the opposite case of a short time of scattering on ions (9.48) compared with (9.45) if

$$c^2k^2 \ll 0.1\omega_{L_e}^2 (v_{T_i} v_{T_e}/c^2). \tag{9.67}$$

In this case, the time (9.66) is shorter than (9.48) and scattering with allowance for retardation again plays the principal role.

Concluding our study of the induced scattering of Langmuir oscillations on particles of an isotropic plasma through a transverse virtual oscillation, let us make a remark on the relativistic corrections. It is not entirely clear from general arguments that the relativistic corrections to the Coulomb scattering are not of the same order as the effects associated with the transverse intermediate wave. However, direct calculations by means of the relativistic expressions (7.22) and (8.13) for the tensors S and V show* that for nondistinguished directions of the interaction $(\mathbf{kk}') \neq 0$, $[\mathbf{kk}'] \neq 0$) the relativistic corrections ($v_T^2 \ll c^2$) to the Coulomb scattering always remain corrections, whereas the scattering associated with allowance for the fields being nonelectrostatic in the case of interaction of particles may become the principal effect (and this is confirmed by the examples considered above) determining the spectral redistribution of the energy density of the Langmuir waves.

Using, as before, the generalized kinetic equation for the electrostatic oscillations (9.1), which is our point of departure in this section, let us now consider the nonlinear interaction of longitudinal waves in a system of two interpenetrating electron plasmas [29] with equilibrium distribution function of the electrons in the form

$$f_0 = \frac{N_1}{(2\pi)^{3/2} v_{T_1}^3} e^{-\frac{1}{2}\frac{(\mathbf{v}-\mathbf{u}_1)^2}{v_{T_1}^2}} + \frac{N_2}{(2\pi)^{3/2} v_{T_2}^3} e^{-\frac{1}{2}\frac{(\mathbf{v}-\mathbf{u}_2)^2}{v_{T_2}^2}}. \tag{9.68}$$

Assuming that the directed velocities $\mathbf{u}_1$ and $\mathbf{u}_2$ of the electrons are small compared with the thermal velocities $v_{T_1} = (\varkappa T_1/m)^{1/2}$ and $v_{T_2} = (\varkappa T_2/m)^{1/2}$, and that the phase velocities of the oscillations are correspondingly large,

$$\frac{\omega}{k}, \quad \frac{\omega'}{k'} \gg v_{T_1}, \quad v_{T_2} \gg u_1, \quad u_2, \tag{9.69}$$

we obtain, using the longitudinal part of the permittivity (6.12), which generalizes $\varepsilon^l(\omega, \mathbf{k})$ from (6.21) to the case of the distribution function (9.68), which differs from (6.14),† the spectrum of longitudinal waves

$$\omega(\mathbf{k}) = \omega_L + \mathbf{ku} + \frac{3}{2}k^2\frac{\varkappa T}{m\omega_L}. \tag{9.70}$$

Here

$$N_1 + N_2 = N, \quad \mathbf{u} = \frac{N_1}{N}\mathbf{u}_1 + \frac{N_2}{N}\mathbf{u}_2,$$
$$T = \frac{N_1}{N}T_1 + \frac{N_2}{N}T_2, \quad \omega_L^2 = \frac{4\pi N e^2}{m}. \tag{9.71}$$

* Such calculations can be found in Appendix VII of [31].

† The distribution function (9.68) is an exception to the rule adopted in this expositon: to use in all specific special cases the Maxwell distribution (6.14) without directed drift of the particles.

As in the case of Langmuir waves, the decay conditions

$$\omega(\mathbf{k}) = \omega'(\mathbf{k}') + \omega''(\mathbf{k}'') \quad (9.72)$$

for three waves with the dispersion law (9.70) cannot be satisfied and the third term on the right-hand side of (9.1) must be ignored. In addition, as we wish to study the interaction of waves with wavelengths less than $\lambda_0 = c/\omega_L$, we shall not take into account scattering through a transverse virtual oscillation, and on the right-hand side of (9.1) we shall consider only the first nonlinear term (the second term).* As a result, taking into account (9.68) and (9.70) in the expressions for the tensors S and V, we obtain from (9.1) an equation that describes the nonlinear interaction of longitudinal waves with the spectrum (9.70):

$$\frac{dW_l(\mathbf{k})}{dt} = -\frac{1}{(2\pi)^{5/2}} \frac{W_l(\mathbf{k})}{N^3 m\omega_L} \int d\mathbf{k}' W_l(\mathbf{k}') \frac{[\mathbf{k}\mathbf{k}']^2}{|\mathbf{k}-\mathbf{k}'|^3} \left(\frac{\mathbf{k}\mathbf{k}'}{kk'}\right)^2 \times$$
$$\times \left\{ N_1 N_2 \left(\frac{1}{v_{T_1}} - \frac{1}{v_{T_2}}\right)(\mathbf{k}-\mathbf{k}', \mathbf{u}_2 - \mathbf{u}_1) + \frac{3}{2} \frac{k^2 - k'^2}{\omega_L} \frac{N\varkappa T}{m} \left(\frac{N_1}{v_{T_1}} + \frac{N_2}{v_{T_2}}\right) \right\}. \quad (9.73)$$

It is similar to Eq. (9.33) (on account of the conditions $|\omega - \omega'| \gg |\mathbf{k} - \mathbf{k}'| v_{T_i}$ and $|\omega - \omega'| \gg \omega_{L_i}$ the ions make a negligible contribution to the scattering). In (9.73), we have retained only the largest terms in powers of the ratio of the directed velocities of the electrons to the thermal velocities and the ratio of the thermal velocities to the phase velocities. In this approximation, the wave energy (9.13) is conserved.

The most important difference between Eq. (9.73) and (9.33) is manifested under conditions when the directed velocity of the electrons appreciably exceeds the ratio of the square of the particle thermal velocity to the first power of the phase velocity of the waves. One can then ignore the last term in the curly brackets on the right-hand side of Eq. (9.73). Then, for example, in the problem of the nonlinear interaction of two infinitesimally narrow wave packets (9.17) one can use the formula (9.18) with the kernel

$$Q(\mathbf{k}_1, \mathbf{k}_2) = -\frac{1}{(2\pi)^{5/2}} \frac{\omega_L}{N r_D^3} \frac{N_1 N_2}{N^2} \frac{1}{\varkappa T} \left(\frac{1}{v_{T_1}} - \frac{1}{v_{T_2}}\right)(\mathbf{k}_1 - \mathbf{k}_2, \mathbf{u}_2 - \mathbf{u}_1) r_D^5 \frac{[\mathbf{k}_1\mathbf{k}_2]^2}{|\mathbf{k}_1 - \mathbf{k}_2|^3} \left(\frac{\mathbf{k}_1\mathbf{k}_2}{k_1 k_2}\right)^2. \quad (9.74)$$

Here, it is important that there is a dependence on the ratio of the temperatures of the electron beams. Assuming, for example, $\mathbf{u}_1 = 0$ and that the temperatures of the plasma at rest is higher than the temperature of the beam ($T_1 > T_2$), we readily find that there is transfer of energy density of waves with low values of the projection of the wave vector onto the direction of the velocity $\mathbf{u}_2$ of the beam to waves with higher values or, and this is the same thing, from lower to higher frequencies. For in this case $Q(\mathbf{k}_1, \mathbf{k}_2)$ is negative if $\mathbf{k}_1\mathbf{u}_2 < \mathbf{k}_2\mathbf{u}_2$. Conversely, if the temperature of the plasma at rest is lower than the temperature of the beam, energy is transferred from higher values of the projection of the vector onto the direction of the beam velocity to smaller values. In other words, in this case the energy is redistributed from higher to lower frequencies.

In fact, these two such variants differ even more if one allows for nonconservation of the energy. For this we go over to a more accurate equation than (9.73). Namely, assuming that the directed velocity of the electrons is much less than the thermal velocity, but at the same time much greater than the ratio of the square of the electron thermal velocity to the phase velocity of the oscillation, we can make Eq. (9.73) more precise by obtaining a correction that leads to violation of the conservation law (9.13) for the oscillation energy:

---

* The first term on the right-hand side of (9.1) is small on account of the conditions (9.69), which guarantee weak damping of the longitudinal waves (9.70) in an electron plasma [cf. (6.28), (6.34a), and (6.34b)].

$$\frac{dW_l(\mathbf{k})}{dt} = -\frac{1}{(2\pi)^{5/2}} \frac{W_l(\mathbf{k})}{Nm\omega_L} \frac{N_1N_2}{N^2}\left(\frac{1}{v_{T_1}} - \frac{1}{v_{T_2}}\right) \int d\mathbf{k}' W_l(\mathbf{k}') \frac{[\mathbf{k}\mathbf{k}']^2}{|\mathbf{k}-\mathbf{k}'|^3}\left(\frac{\mathbf{k}\mathbf{k}'}{kk'}\right)^2 (\mathbf{k}-\mathbf{k}', \mathbf{u}_2-\mathbf{u}_1)\left[1+\frac{1}{2\omega_L}(\mathbf{k}-\mathbf{k}', \mathbf{u})\right]. \tag{9.75}$$

With allowance for such a correction [second term in the square brackets on the right-hand side of (9.75)] the wave energy is no longer conserved but the integral (9.43), which determines the total number of longitudinal waves, is. Then, as in the case of (9.46), we can obtain the expression

$$\Delta W = (\omega_1 - \omega_2)\frac{1}{\omega_1} W_1(t_0) \simeq W_1(t_0)\frac{(\mathbf{k}_1 - \mathbf{k}_2, \mathbf{u})}{\omega_L} \tag{9.76}$$

for the difference of the values of the energies of two waves at the initial instant of time and in the infinitely distant future, when the oscillation with wave vector $\mathbf{k}_2$ is damped and there remains only the oscillation with wave vector $\mathbf{k}_1$ ($k_1 > k_2$).

We now turn to the question of the change of the wave energy (9.76) in the above examples of a plasma at rest and a beam ($\mathbf{u}_1 = 0$, $\mathbf{u}_2 \neq 0$). In the case when the temperature T_1 of the plasma at rest is lower than the beam temperature T_2 and the energy is transferred from higher to lower frequencies, the energy of the longitudinal waves, which have the spectrum (9.70), decreases in accordance with (9.76). Conversely, if the temperature T_1 of the plasma at rest is higher than the beam temperature and energy is transferred from lower to higher frequencies, the energy of the longitudinal waves increases. Thus, in the latter case we have not only a redistribution of the energy within the spectrum but also an increase in the energy of the waves, i.e., we have a nonlinear excitation of electrostatic oscillations of the plasma.

The example considered here of nonlinear excitation of waves by a slow beam corresponds to the fundamental possibility of such excitation when there is induced scattering of waves on particles, as was noted in [37]. This, in particular, reveals an important difference between nonlinear and linear excitation of waves. For whereas in the linear theory of oscillations one does not obtain any restrictions on the amplitude (or energy) of the increasing waves, in this variant of nonlinear excitation the wave energy increases by a finite amount.

To conclude this section, let us consider the nonlinear interaction of short-wavelength ion sound with the spectrum (6.48), proceeding from Eq. (9.1). The dispersion law (6.48) does not allow the decay interaction of three ion-acoustic waves.* It is also obvious that the scattering of ion-acoustic oscillations through an intermediate virtual oscillation is unimportant compared with Coulomb scattering, since the phase velocities of the oscillations are much less than the thermal velocity of the electrons (6.37), and in all the variants of scattering studied above with retardation such scattering was shown to be of fundamental importance at wave phase velocities greater than the velocity of light in vacuum (wavelengths greater than $\lambda_0 = c/\omega_{Le}$). In this connection, in the generalized kinetic equation (9.1) for longitudinal waves, we omit the last two terms on the right-hand side, neglecting, at the same time, the small change in the energy density of the ion-acoustic oscillations (6.48) due to their coalescing with Langmuir oscillations (6.34a) to form long-wavelength transverse electromagnetic waves (6.53). This effect is described by the last, fourth, term in (9.1) by means of (9.50). As a result, taking into account in (9.1) the spontaneous emission (9.3) of ion-acoustic waves by electrons and ions and also their damping on particles, we obtain the following approximate equation, which describes the

* Therefore, the third term on the right-hand side of (9.1), which is responsible for decay processes with the participation of three longitudinal waves, and the decay equation (9.2) can describe only the coalescence of a Langmuir oscillation (6.34a) and ion sound (6.40) [or, in particular, (6.48)] into a Langmuir oscillation.

Coulomb scattering of short-wavelength ion sound (6.48) [29]:

$$\frac{dW_s(\mathbf{k})}{dt}+\sqrt{\frac{\pi}{2}}\frac{\omega_{L_i}^2}{k^3}\left[\frac{W_s(\mathbf{k})-\varkappa T_e}{r_{D_e}^2 v_{T_e}}+\frac{W_s(\mathbf{k})-\varkappa T_i}{r_{D_i}^2 v_{T_i}}\exp\left(-\frac{1}{2}\frac{\omega_{L_i}^2}{k^2 v_{T_i}^2}\right)\right]=$$

$$=-\frac{1}{(2\pi)^{3/2}}\frac{e_i^2}{M^2}\frac{W_s(\mathbf{k})}{\omega_{L_i}^2 r_{D_e}^2 v_{T_i}}\int d\mathbf{k}' W_s(\mathbf{k}')\frac{|\mathbf{k}\mathbf{k}'|^2}{(kk')^2}\left(\frac{\mathbf{k}\mathbf{k}'}{kk'}\right)^2\frac{k^2-k'^2}{|\mathbf{k}-\mathbf{k}'|^2}\exp\left[-\frac{(k^2-k'^2)^2}{8r_{D_e}^4 r_{D_i}^2 k^4 k'^4(\mathbf{k}-\mathbf{k}')^2}\right]-\frac{4}{9(2\pi)^{3/2}}\frac{e_i^2}{M^2}\frac{W_s(\mathbf{k})}{\omega_{L_i}^2 r_{D_e}^2 v_{T_e}}\times$$

$$\times\int d\mathbf{k}' W_s(\mathbf{k}')\frac{k^2k'^2}{|\mathbf{k}+\mathbf{k}'|^5}\left[2+\frac{3}{2}\mathbf{k}\mathbf{k}'\left(\frac{1}{k^2}+\frac{1}{k'^2}\right)+\left(\frac{\mathbf{k}\mathbf{k}'}{kk'}\right)^2\right]^2\left\{1+\frac{r_{D_e}^2}{r_{D_i}^2}\frac{v_{T_e}}{v_{T_i}}\exp\left[-\frac{2}{r_{D_i}^2(\mathbf{k}+\mathbf{k}')^2}\right]\right\}. \qquad (9.77)$$

Note that in the derivation of this equation we have used in (9.1) the nonlinear tensors S and V, which depend not only on the differences $\omega-\omega'$ and $\mathbf{k}-\mathbf{k}'$ (as was the case in our study of the nonlinear interaction of Langmuir waves) but also on the sums $\omega+\omega'$ and $\mathbf{k}+\mathbf{k}'$ of the frequencies and wave vectors of the scattered ion-acoustic oscillations (see §21). Being interested in the nonlinear interaction of acoustic waves of large amplitude ($W_s \gg \varkappa T_e$), we can ignore the terms on the left-hand side of Eq. (9.77), which describe linear damping and spontaneous emission, compared with the nonlinear terms on the right-hand side [cf. (9.37), (9.38), (9.41)]. Then, for example, for the problem of the nonlinear interaction of two infinitesimally narrow wave packets (9.17)

$$W_s(\mathbf{k})=W_1(t)\delta(\mathbf{k}-\mathbf{k}_1)+W_2(t)\delta(\mathbf{k}-\mathbf{k}_2) \qquad (9.78)$$

we obtain a system of two ordinary differential equations

$$\frac{dW_1}{dt}=-W_1W_2[(Q_1(\mathbf{k}_1,\ \mathbf{k}_2)+Q_2(\mathbf{k}_1,\ \mathbf{k}_2)],$$
$$\frac{dW_2}{dt}=-W_1W_2[-Q_1(\mathbf{k}_1,\ \mathbf{k}_2)+Q_2(\mathbf{k}_1,\ \mathbf{k}_2)], \qquad (9.79)$$

in which the quantities Q_1 and Q_2 are determined by equations that arise from the kernels on the right-hand side of (9.77) when one substitutes into it the sum (9.78):

$$Q_1(\mathbf{k}_1,\ \mathbf{k}_2)\equiv\frac{1}{(2\pi)^{3/2}}\frac{e_i^2}{M^2}\frac{1}{\omega_{L_i}^2 r_{D_e}^2 v_{T_i}}\left(\frac{\mathbf{k}_1\mathbf{k}_2}{k_1k_2}\right)^2\frac{|\mathbf{k}_1\mathbf{k}_2|^2}{(k_1k_2)^2}\frac{\mathbf{k}_1^2-\mathbf{k}_2^2}{|\mathbf{k}_1-\mathbf{k}_2|^3}\exp\left\{-\frac{1}{8r_{D_i}^2 r_{D_e}^4 k_1^4 k_2^4}\frac{(k_1^2-k_2^2)^2}{(\mathbf{k}_1-\mathbf{k}_2)^2}\right\}; \qquad (9.80)$$

$$Q_2(\mathbf{k}_1,\ \mathbf{k}_2)\equiv\frac{4}{9(2\pi)^{3/2}}\frac{e_i^2}{M^2}\frac{1}{\omega_{L_i}^2 r_{D_e}^2 v_{T_e}}\frac{k_1^2k_2^2}{|\mathbf{k}_1+\mathbf{k}_2|^5}\left[2+\frac{3}{2}\mathbf{k}_1\mathbf{k}_2\left(\frac{1}{k_1^2}+\frac{1}{k_2^2}\right)+\left(\frac{\mathbf{k}_1\mathbf{k}_2}{k_1k_2}\right)^2\right]^2\left\{1+\frac{r_{D_e}^2}{r_{D_i}^2}\frac{v_{T_e}}{v_{T_i}}\exp\left[-\frac{2}{r_{D_i}^2(\mathbf{k}_1+\mathbf{k}_2)^2}\right]\right\}. \qquad (9.81)$$

Under conditions when

$$(k_1^2-k_2^2)^2\gg(\mathbf{k}_1-\mathbf{k}_2)^2(r_{D_i}r_{D_e}^2k_1^2k_2^2)^2, \qquad (9.82)$$

we can ignore the coefficient Q_1 in Eqs. (9.79):

$$Q_1(\mathbf{k}_1,\ \mathbf{k}_2)\ll Q_2(\mathbf{k}_1,\ \mathbf{k}_2). \qquad (9.83)$$

Then the amplitudes of both ion-acoustic oscillations decrease: the wave energy decreases on account of its being transferred to the particles, and the solution of the system of equations (9.79) can be written in the form [29]

$$W_1(t)-W_1(t_0)=W_2(t)-W_2(t_0); \qquad (9.84)$$

$$\frac{W_1(t)}{W_2(t)}=\frac{W_1(t_0)}{W_2(t_0)}\exp\{(t-t_0)Q_2(\mathbf{k}_1,\ \mathbf{k}_2)[W_1(t_0)-W_2(t_0)]\},\quad W_1(t_0)\neq W_2(t_0). \qquad (9.85)$$

Thus, although the energy density of both waves decreases, the rates of decrease are different, since the ratio W_1/W_2 of the energies varies exponentially with the time [for $W_1(t_0)\neq W_2(t_0)$].

If at the initial time t_0 the wave energies are equal:

$$W_1(t_0) = W_2(t_0), \tag{9.86}$$

then in accordance with (9.84) they also remain identical at subsequent times, varying in accordance with the law [which replaces (9.85)]

$$\frac{W_1(t)}{W_1(t_0)} = \frac{W_2(t)}{W_2(t_0)} = \{1 + Q_2(\mathbf{k}_1, \mathbf{k}_2)\, W_1(t_0)(t - t_0)\}^{-1}. \tag{9.87}$$

If the initial energies are not equal, the law of variation of the wave energy has, in accordance with (9.84) and (9.85), the form [29]

$$\frac{W_1(t)}{W_1(t_0)} = [W_1(t_0) - W_2(t_0)]\,\{W_1(t_0) - W_2(t_0)\exp[-(t - t_0)\,Q_2(\mathbf{k}_1, \mathbf{k}_2)(W_1(t_0) - W_2(t_0))]\}^{-1}. \tag{9.88}$$

It follows from this that if at the initial time the energy density W_1 exceeds W_2 then with increasing time the amplitude of the wave with lower energy tends to zero, while the energy of the second oscillation tends to a value equal to the difference $W_1 - W_2$ of the initial values. The characteristic time of this nonlinear relaxation of the ion-acoustic waves is approximately [29]

$$\tau \sim 30 \frac{M}{m} \frac{L}{\nu_{\text{eff}}} \frac{1}{(kr_{De})^3} \frac{k^3 \varkappa T_e}{[W_1(t_0) - W_2(t_0)]} \tag{9.89}$$

and in a plasma with Coulomb logarithm $L \simeq 10$ and fairly well developed fluctuations $|W_1(t_0) - W_2(t_0)| \sim 10^4 k^2 \varkappa T_e$ of the short-wavelength ion sound, $(kr_{De})^2 \sim 10$, is shorter by approximately an order of magnitude than the time of free flight of the electrons until they undergo Coulomb collisions.

If the condition (9.83) is violated, for example, the strong inequality of the opposite sense is satisfied:

$$Q_1(\mathbf{k}_1, \mathbf{k}_2) \gg Q_2(\mathbf{k}_1, \mathbf{k}_2), \tag{9.90}$$

then Eqs. (9.79) reduce to (9.16), in which $[-Q(\mathbf{k}_1, \mathbf{k}_2)]$ is determined by the right-hand side of (9.80). In this approximation, one can only have redistribution of energy within the spectrum from short to long wavelengths, as in the case of Coulomb scattering of Langmuir oscillations with conserved total energy. The order of magnitude of the corresponding time of spectral redistribution can be found from the formula [29]

$$\tau \sim \frac{L}{\nu_{\text{eff}}} \sqrt{\frac{MT_i}{mT_e}} \frac{\varkappa T_e}{[W_0/k^3]}, \tag{2.91}$$

in which [cf. (9.13)]

$$W_0 = \int d\mathbf{k} W_s(\mathbf{k}) = \text{const}. \tag{9.92}$$

As in the estimates of the relaxation times in the cases considered above, one can readily see that there is a wide range of parameters of a collisionless isotropic plasma for which such a time is fairly short.

To conclude our analysis, we emphasize that for short ion-acoustic waves the energy redistribution within the spectrum is by no means always decisive. Namely, when $k_1 \sim k_2 \sim \Delta k = |\mathbf{k}_1 - \mathbf{k}_2|$ such oscillations are damped in the process of Coulomb scattering for wavelengths greater than $(r_{De}^2 r_{Di})^{1/3}$.

§ 10. Nonlinear Interaction of Longitudinal and Transverse Oscillations in an Isotropic Plasma

This section is devoted to a detailed analysis of the nonlinear interaction of the electrostatic oscillations of a plasma with long-wavelength transverse electromagnetic waves.

After the spectral function (6.62) has been substituted into the generalized kinetic equation (3.22) for the waves, we obtain two equations: the equation for the evolution of the longitudinal oscillations due to the nonlinear interaction with the transverse waves [12]:

$$
\begin{aligned}
&\frac{\partial}{\partial t}(E_l^2)_{\omega,\mathbf{k}}\frac{\partial\varepsilon^{l'}(\omega,\mathbf{k})}{\partial\omega}-\frac{\partial}{\partial\mathbf{r}}(E_l^2)_{\omega,\mathbf{k}}\frac{\partial\varepsilon^{l'}(\omega,\mathbf{k})}{\partial\mathbf{k}}=-2\varepsilon^{l''}(\omega,\mathbf{k})(E_l^2)_{\omega,\mathbf{k}}-\\
&-(E_l^2)_{\omega,\mathbf{k}}\int d\omega' d\mathbf{k}'(E_{tr}^2)_{\omega',\mathbf{k}'}\,\mathrm{Im}\,\frac{k_ik_j}{k^2}\left(\delta_{ab}-\frac{k'_ak'_b}{k'^2}\right)V_{iajb}(\omega,\mathbf{k};\omega',\mathbf{k}')+\\
&+(E_l^2)_{\omega,\mathbf{k}}\,\mathrm{Im}\int d\omega' d\mathbf{k}' d\omega'' d\mathbf{k}''\delta(\omega-\omega'-\omega'')\delta(\mathbf{k}-\mathbf{k}'-\mathbf{k}'')\times\\
&\times\frac{k_ik_a}{k^2}\left(\delta_{cs}-\frac{k'_ck'_s}{k'^2}\right)(E_{tr}^2)_{\omega',\mathbf{k}'}S_{ijs}(\omega,\mathbf{k};\omega',\mathbf{k}')S_{bca}(\omega'',\mathbf{k}'';\omega,\mathbf{k})\times\\
&\times\left\{\frac{k''_bk''_j}{k''^2}\frac{1}{\varepsilon^l(\omega'',\mathbf{k}'')}+\frac{\delta_{bj}-k''_bk''_j/k''^2}{\varepsilon^{tr}(\omega'',\mathbf{k}'')-\dfrac{c^2k''^2}{\omega''^2}}\right\}+\frac{\pi}{2}\,\mathrm{sign}\,\varepsilon^{l''}(\omega,\mathbf{k})\delta[\varepsilon^{l'}(\omega,\mathbf{k})]\times\\
&\times\int d\omega' d\mathbf{k}' d\omega'' d\mathbf{k}''\delta(\omega-\omega'-\omega'')\delta(\mathbf{k}-\mathbf{k}'-\mathbf{k}'')\times\\
&\times\left\{\frac{k_ik_ak'_sk'_c}{(kk')^2}\left(\delta_{bj}-\frac{k''_bk''_j}{k''^2}\right)(E_l^2)_{\omega',\mathbf{k}'}(E_{tr}^2)_{\omega'',\mathbf{k}''}+\frac{k_ik_ak''_jk''_b}{(kk'')^2}\left(\delta_{sc}-\frac{k'_sk'_c}{k'^2}\right)\times\right.\\
&\left.\times(E_l^2)_{\omega'',\mathbf{k}''}(E_{tr}^2)_{\omega',\mathbf{k}'}+\frac{1}{2}\frac{k_ik_a}{k^2}\left(\delta_{bj}-\frac{k''_bk''_j}{k''^2}\right)\left(\delta_{sc}-\frac{k'_sk'_c}{k'^2}\right)(E_{tr}^2)_{\omega',\mathbf{k}'}(E_{tr}^2)_{\omega'',\mathbf{k}''}\right\}\times\\
&\times S_{ijs}(\omega,\mathbf{k};\omega',\mathbf{k}')S^*_{abc}(\omega,\mathbf{k};\omega',\mathbf{k}')
\end{aligned}
\tag{10.1}
$$

and the equation of the evolution of the transverse oscillations due to the nonlinear interaction with the longitudinal waves [12]:

$$
\begin{aligned}
&\frac{\partial}{\partial t}(E_{tr}^2)_{\omega,\mathbf{k}}\frac{1}{\omega}\frac{\partial}{\partial\omega}\left\{\omega\left[\varepsilon^{tr'}(\omega,\mathbf{k})-\frac{c^2k^2}{\omega^2}\right]\right\}-\frac{\partial}{\partial\mathbf{r}}(E_{tr}^2)_{\omega,\mathbf{k}}\frac{1}{\omega}\frac{\partial}{\partial\mathbf{k}}\left\{\omega\left[\varepsilon^{tr'}(\omega,\mathbf{k})-\frac{c^2k^2}{\omega^2}\right]\right\}=\\
&=-2\varepsilon^{tr''}(\omega,\mathbf{k})(E_{tr}^2)_{\omega,\mathbf{k}}-(E_{tr}^2)_{\omega,\mathbf{k}}\,\mathrm{Im}\int d\omega' d\mathbf{k}' V_{iajb}(\omega,\mathbf{k};\omega',\mathbf{k}')\left(\delta_{ij}-\frac{k_ik_j}{k^2}\right)\times\\
&\times\frac{k'_ak'_b}{k'^2}(E_l^2)_{\omega',\mathbf{k}'}+(E_{tr}^2)_{\omega,\mathbf{k}}\,\mathrm{Im}\int d\omega' d\mathbf{k}' d\omega'' d\mathbf{k}''\delta(\omega-\omega'-\omega'')\delta(\mathbf{k}-\mathbf{k}'-\mathbf{k}'')\times\\
&\times\left\{\frac{k''_bk''_j}{k''^2}\frac{1}{\varepsilon^l(\omega'',\mathbf{k}'')}+\frac{\delta_{bj}-k''_bk''_j/k''^2}{\varepsilon^{tr}(\omega'',\mathbf{k}'')-\dfrac{c^2k''^2}{\omega''^2}}\right\}\left(\delta_{ia}-\frac{k_ik_a}{k^2}\right)S_{ijs}(\omega,\mathbf{k};\omega',\mathbf{k}')\times\\
&\times S_{bca}(\omega'',\mathbf{k}'';\omega,\mathbf{k})\frac{k'_sk'_c}{k'^2}(E_l^2)_{\omega',\mathbf{k}'}+\pi\,\mathrm{sign}\,\varepsilon^{tr''}(\omega,\mathbf{k})\delta\left[\varepsilon^{tr'}(\omega,\mathbf{k})-\frac{c^2k^2}{\omega^2}\right]\times\\
&\times\int d\omega' d\mathbf{k}' d\omega'' d\mathbf{k}''\delta(\omega-\omega'-\omega'')\delta(\mathbf{k}-\mathbf{k}'-\mathbf{k}'')\left(\delta_{ia}-\frac{k_ik_a}{k^2}\right)S_{ijs}(\omega,\mathbf{k};\omega',\mathbf{k}')\times\\
&\times S^*_{abc}(\omega,\mathbf{k};\omega',\mathbf{k}')\left\{\frac{k'_sk'_ck''_bk''_j}{(k'k'')^2}(E_l^2)_{\omega',\mathbf{k}'}(E_l^2)_{\omega'',\mathbf{k}''}+\frac{1}{2}\frac{k'_sk'_c}{k'^2}\left(\delta_{bj}-\frac{k''_bk''_j}{k''^2}\right)(E_l^2)_{\omega',\mathbf{k}'}\times\right.\\
&\left.\times(E_{tr}^2)_{\omega'',\mathbf{k}''}+\frac{1}{2}\frac{k''_jk''_b}{k''^2}\left(\delta_{sc}-\frac{k'_ck'_s}{k'^2}\right)(E_l^2)_{\omega'',\mathbf{k}''}(E_{tr}^2)_{\omega',\mathbf{k}'}\right\}.
\end{aligned}
\tag{10.2}
$$

Let us consider the physical meaning of the terms in this equation. As in (9.1), the terms on the right-hand side of (10.1) can be divided into three groups corresponding to linear damping, induced scattering on particles, and decay processes (combination scattering). The first term is identical to the first term of (9.1) and describes the damping of longitudinal waves. The second, third, and fourth terms describe scattering on particles through a longitudinal virtual oscillation if allowance is made for the third term and through a transverse virtual oscillation if allowance is made for the fourth term. These terms determine the evolution of the longitudinal oscillations due to the nonlinear interaction (scattering) of the longitudinal waves with the transverse ones. It is already evident from the very structure of the equation that contributions to the general effect of induced scattering on particles from Coulomb scattering and from scattering with allowance for retardation can compete, so that the resulting process is the fastest process, as in the case (9.1), of the interaction of only longitudinal waves with one another. Vanishing of the denominators of the third or the fourth term correspond to a decay process as a result of which there is formed a longitudinal and a transverse oscillation with the difference frequency $\omega'' = \omega - \omega'$ and wave vector $\mathbf{k}'' = \mathbf{k} - \mathbf{k}'$. The last group of terms consists of three terms that give the change of the spectral density of the energy of the longitudinal and transverse oscillations and two transverse oscillations. Of course, under conditions when it is necessary to allow for the change of the spectral density of the energy of the longitudinal waves due to their nonlinear interaction with longitudinal waves, Eq. (10.1) must be augmented by appropriate terms from the right-hand side of (9.1). As before, the spontaneous emission of longitudinal waves can be taken into account by adding the expression (9.3) to the right-hand side of (10.1). In this section we shall be interested in the nonlinear interaction of only longitudinal with transverse waves, and we shall therefore restrict ourselves to Eq. (10.1).

The use of spectral field functions in a plasma in the form (6.62) entails, in accordance with the arguments given in §6, a restriction to natural polarization of the transverse electromagnetic waves that interact nonlinearly with the longitudinal waves. Therefore, for example, for linearly polarized transverse electromagnetic waves Eq. (10.1) must be replaced by a different equation, which also arises from (3.22) if the spectral function of the field is represented in the form

$$(E_j E_i)_{\omega, \mathbf{k}} = (E_l^2)_{\omega, \mathbf{k}} \varkappa_i \varkappa_j + (E_{tr}^2)_{\omega, \mathbf{k}} e_i e_j, \tag{10.3}$$

where, in accordance with (6.70)

$$(E_{tr}^2)_{\omega, \mathbf{k}} = \frac{1}{(2\pi)^2} W_{tr}(\mathbf{k}) \left\{ \delta\left(\omega - \sqrt{\omega_{L_e}^2 + c^2k^2}\right) + \delta\left(\omega + \sqrt{\omega_{L_e}^2 + c^2k^2}\right) \right\} \tag{10.4}$$

and $\mathbf{e}$ is a unit polarization vector ($\mathbf{ke} = 0$). Namely, the equation for the evolution of the spectral energy density of the electrostatic oscillations due to their nonlinear interaction with transverse, linearly polarized electromagnetic waves is determined by the equation

$$\begin{aligned}
&\frac{\partial}{\partial t}(E_l^2)_{\omega, \mathbf{k}} \frac{\partial \varepsilon^{l\prime}(\omega, \mathbf{k})}{\partial \omega} - \frac{\partial}{\partial \mathbf{r}}(E_l^2)_{\omega, \mathbf{k}} \frac{\partial \varepsilon^{l\prime}(\omega, \mathbf{k})}{\partial \mathbf{k}} = -2\varepsilon^{l\prime\prime}(\omega, \mathbf{k})(E_l^2)_{\omega, \mathbf{k}} - (E_l^2)_{\omega, \mathbf{k}} \times \\
&\times \int d\omega' d\mathbf{k}' (E_{tr}^2)_{\omega', \mathbf{k}'} 2\,\mathrm{Im}\, \varkappa_i \varkappa_j e_a' e_b' V_{iajb}(\omega, \mathbf{k}; \omega', \mathbf{k}') + \\
&+ (E_l^2)_{\omega, \mathbf{k}} 2\,\mathrm{Im} \int d\omega' d\mathbf{k}' d\omega'' d\mathbf{k}'' \delta(\omega - \omega' - \omega'') \delta(\mathbf{k} - \mathbf{k}' - \mathbf{k}'') \varkappa_i \varkappa_a e_s' e_c' (E_{tr}^2)_{\omega', \mathbf{k}'} \times \\
&\times S_{ijs}(\omega, \mathbf{k}; \omega', \mathbf{k}') S_{bca}(\omega'', \mathbf{k}''; \omega, \mathbf{k}) \left\{ \frac{\varkappa_b'' \varkappa_j''}{\varepsilon^l(\omega'', \mathbf{k}'')} + \frac{\delta_{bj} - \varkappa_b'' \varkappa_j''}{\varepsilon^{tr}(\omega'', \mathbf{k}'') - \frac{c^2 k''^2}{\omega''^2}} \right\} + \\
&+ \pi\, \mathrm{sign}\, \varepsilon^{l\prime\prime}(\omega, \mathbf{k}) \delta[\varepsilon^{l\prime}(\omega, \mathbf{k})] \int d\omega' d\mathbf{k}' d\omega'' d\mathbf{k}'' \delta(\omega - \omega' - \omega'') \delta(\mathbf{k} - \mathbf{k}' - \mathbf{k}'') \times \\
&\times \{ \varkappa_i \varkappa_a \varkappa_s' \varkappa_c' e_b'' e_j'' (E_l^2)_{\omega', \mathbf{k}'} (E_{tr}^2)_{\omega'', \mathbf{k}''} + \varkappa_i \varkappa_a \varkappa_j'' \varkappa_b'' e_s' e_c' (E_l^2)_{\omega'', \mathbf{k}''} (E_{tr}^2)_{\omega', \mathbf{k}'} + \\
&+ \varkappa_i \varkappa_a e_b'' e_j'' e_s' e_c' (E_{tr}^2)_{\omega', \mathbf{k}'} (E_{tr}^2)_{\omega'', \mathbf{k}''} \} S_{ijs}(\omega, \mathbf{k}; \omega', \mathbf{k}') S_{abc}^*(\omega, \mathbf{k}; \omega', \mathbf{k}').
\end{aligned} \tag{10.5}$$

Here, $\varkappa$, $\varkappa'$, and $\varkappa''$ are unit vectors along the wave vectors $\mathbf{k}$, $\mathbf{k}'$, and $\mathbf{k}''$, respectively,

$$\varkappa = \frac{\mathbf{k}}{k}, \quad \varkappa' = \frac{\mathbf{k}'}{k'}, \quad \varkappa'' = \frac{\mathbf{k}''}{k''}, \tag{10.6}$$

and $\mathbf{e}'$ and $\mathbf{e}''$ are real unit vectors along the electric field vectors $\mathbf{E}(\omega', \mathbf{k}')$ and $\mathbf{E}(\omega'', \mathbf{k}'')$. After summation over all possible linear polarizations (6.72), Eq. (10.5) goes over in accordance with (6.71) into (10.1) with natural polarization of the transverse electromagnetic waves.

In this section in our discussion of various special cases of nonlinear interaction of longitudinal and transverse waves, we shall use Eq. (10.1). Equation (10.5) [see also (10.7)] will enable us in §13 to obtain the cross sections of induced scattering of oscillations with allowance for the linear polarization of electromagnetic waves.

Equation (10.2) augments Eq. (10.1) in the sense that it determines the evolution of naturally polarized transverse electromagnetic waves that interact with longitudinal waves and also, in accordance with (10.1). changes the spectral energy density of the longitudinal waves. The first term on the right-hand side of (10.2) corresponds to linear damping of a transverse wave. Damping due to the inverse Cherenkov effect is in this case equal to zero, since the phase velocity of the transverse waves in the isotropic homogeneous and unbounded plasma that we are considering here is greater than c, the velocity of light, i.e., greater than the velocity of any of the plasma particles [see (6.56) and the remark on formulas (6.25)-(6.26) in §6]. The second, third, and fourth terms describe the change in the spectral energy density $(E^2_{tr})_{\omega,\mathbf{k}}$ of transverse oscillations due to their nonlinear interaction (induced scattering on particles) with longitudinal waves with, respectively, the participation of longitudinal and transverse virtual oscillations. The last group of terms – the decay terms – determines the change of $(E^2_{tr})_{\omega,\mathbf{k}}$ in combination scattering, when a transverse oscillation is formed as a result of the interaction of two longitudinal oscillations [this includes the well-known process of the coalescence of two electron Langmuir oscillations into a transverse oscillation, which we have already discussed in part in the foregoing section [(9.51) and (9.52)] and in the interaction of a longitudinal and a transverse wave.

We note a property that is common to all the decay terms in the equations (9.1), (10.1), and (10.2) of nonlinear interaction. They are all proportional to the sign function of the imaginary part of the permittivity which describes the oscillation whose evolution is being studied. In a homogeneous isotropic plasma with isotropic Maxwell distribution of the particles in the equilibrium state, the imaginary parts $\varepsilon^{l\,\prime\prime}$ and $\varepsilon^{tr\,\prime\prime}$ for positive frequencies are positive, which corresponds to sign functions equal to unity. In other words, in such a plasma the characteristic oscillations are stable. In the opposite case of unstable oscillations (we have in mind only the kinetic instability, since one of the basic assumptions under which the equations are derived is that the frequencies of the interacting oscillations are nonvanishing), which can be achieved by an appropriate choice of the distribution function of the ground state, $\varepsilon^{l\,\prime\prime}$ and $\varepsilon^{tr\,\prime\prime}$ are negative, and the sign functions take values equal to −1. One can then have a new nonlinear instability, which may be much more important, i.e., at a sufficiently large amplitudes of the oscillations it can develop much more rapidly than the linear instability.

Concluding our formulation of the fundamentals relating to the nonlinear interaction of longitudinal waves with transverse waves in an isotropic plasma, we also give an equation for the evolution of the spectral energy density of a linearly polarized transverse electromagnetic waves; this augments (10.2) in the same way as (10.5) does (10.1):

$$\frac{\partial}{\partial t}(E^2_{tr})_{\omega,\mathbf{k}}\frac{1}{\omega}\frac{\partial}{\partial\omega}\left\{\omega\left[\varepsilon^{tr}(\omega,\mathbf{k})-\frac{c^2k^2}{\omega^2}\right]\right\}-\frac{\partial}{\partial\mathbf{r}}(E^2_{tr})_{\omega,\mathbf{k}}\frac{1}{\omega}\frac{\partial}{\partial\mathbf{k}}\left\{\omega\left[\varepsilon^{tr}(\omega,\mathbf{k})-\frac{c^2k^2}{\omega^2}\right]\right\}=$$
$$=-2\varepsilon^{tr\prime\prime}(\omega,\mathbf{k})(E^2_{tr})_{\omega,\mathbf{k}}-2(E^2_{tr})_{\omega,\mathbf{k}}\,\mathrm{Im}\int d\omega' d\mathbf{k}' V_{iajb}(\omega,\mathbf{k};\omega',\mathbf{k}')\,e_i e_j \varkappa'_a\varkappa'_b\times$$
$$\times(E^2_l)_{\omega',\mathbf{k}'}+(E^2_{tr})_{\omega,\mathbf{k}}\,2\,\mathrm{Im}\int d\omega' d\mathbf{k}' d\omega'' d\mathbf{k}''\,\delta(\omega-\omega'-\omega'')\,\delta(\mathbf{k}-\mathbf{k}'-\mathbf{k}'')\times$$
$$\times\left\{\frac{\varkappa''_b\varkappa''_j}{\varepsilon^l(\omega'',\mathbf{k}'')}+\frac{\delta_{bj}-\varkappa''_b\varkappa''_j}{\varepsilon^{tr}(\omega'',\mathbf{k}'')-\dfrac{c^2k''^2}{\omega''^2}}\right\}e_i e_a S_{ijs}(\omega,\mathbf{k};\omega',\mathbf{k}')\,S_{bca}(\omega'',\mathbf{k}'';\omega,\mathbf{k})\,\varkappa'_s\varkappa'_c\times$$
$$\times(E^2_l)_{\omega',\mathbf{k}'}+2\pi\,\mathrm{sign}\,\varepsilon^{tr\prime\prime}(\omega,\mathbf{k})\,\delta\left[\varepsilon^{tr\prime}(\omega,\mathbf{k})-\frac{c^2k^2}{\omega^2}\right]\int d\omega' d\mathbf{k}' d\omega'' d\mathbf{k}''\times$$
$$\times\delta(\omega-\omega'-\omega'')\,\delta(\mathbf{k}-\mathbf{k}'-\mathbf{k}'')\,e_i e_a S_{ijs}(\omega,\mathbf{k};\omega',\mathbf{k}')\,S^*_{abc}(\omega,\mathbf{k};\omega',\mathbf{k}')\times$$
$$\times\{\varkappa'_s\varkappa'_c\varkappa''_b\varkappa''_j(E^2_l)_{\omega',\mathbf{k}'}(E^2_l)_{\omega'',\mathbf{k}''}+\varkappa'_s\varkappa'_c e''_b e''_j(E^2_l)_{\omega',\mathbf{k}'}(E^2_{tr})_{\omega'',\mathbf{k}''}+$$
$$+\varkappa''_j\varkappa''_b e'_s e'_c(E^2_l)_{\omega'',\mathbf{k}''}(E^2_{tr})_{\omega',\mathbf{k}'}\}. \qquad (10.7)$$

After summation over all possible linear polarizations of the transverse waves, which corresponds to a transition to natural polarization, this equation is identical with (10.2).

We now turn to the study of specific processes of nonlinear interaction of longitudinal and transverse oscillations of an isotropic plasma. Let us consider first the case of induced scattering on particles, using the same terminology as in §9. We emphasize that, as in the foregoing section, the phase velocities of the oscillations that interact in this case are large compared with the thermal velocities of the electrons and ions (in this respect the scattering of ion-acoustic oscillations in §9 is an exception). We shall also allow for the fact that the process of induced scattering on particles is of greatest importance for waves with low frequencies. Applied to the interaction of Langmuir waves with transverse waves, which is what we are interested in, this means that we consider transverse waves with wavelengths that are long compared with $\lambda_0 = c/\omega_{L_e}$, i.e., we are concerned with the spectrum (6.61).

Taking into account only Coulomb scattering, in the first approximation in the small ratio of the frequency difference of the interacting waves to their frequency, we obtain from Eqs. (10.1) and (10.2), using formula (8.23) for the tensors S and V with the Maxwell distribution function (6.14),

$$\begin{aligned} \left\{\frac{\partial W_l(\mathbf{k})}{\partial t} + 3v_{T_e}(kr_{D_e})\frac{\partial W_l(\mathbf{k})}{\partial \mathbf{r}}\right\}^{\text{Coul}} &= \int d\mathbf{k}' Q(\mathbf{k}, \mathbf{k}') W_l(\mathbf{k}) W_{tr}(\mathbf{k}'), \\ \left\{\frac{\partial W_{tr}(\mathbf{k})}{\partial t} + c(k\lambda_0)\frac{\partial W_{tr}(\mathbf{k})}{\partial \mathbf{r}}\right\}^{\text{Coul}} &= -\int d\mathbf{k}' Q(\mathbf{k}', \mathbf{k}) W_{tr}(\mathbf{k}) W_l(\mathbf{k}'). \end{aligned} \tag{10.8}$$

Here, the kernel Q(**k**, **k'**) as a function of the wave vectors **k** and **k'** is defined by the equation [32]

$$\begin{aligned} Q(\mathbf{k}, \mathbf{k}') = &-\frac{1}{4(2\pi)^3}\frac{\omega_{L_e}}{N_e\varkappa T_e}\frac{(k'' r_{D_e})^2}{|1+\delta\varepsilon_e^l+\delta\varepsilon_i^l|^2}\left\{\delta\varepsilon_e^{l''}\left\{\frac{r_{D_e}^2}{k''^2}\left[\frac{[\mathbf{k}\mathbf{k}']^2}{(kk')^2}(k^2k'^2 - 4(\mathbf{k}\mathbf{k}')^2 + 2k'^2(\mathbf{k}\mathbf{k}')) + \right.\right.\right. \\ &\left.\left.\left. + 2k''^2\frac{(\mathbf{k}\mathbf{k}')^2}{k^2}\right][2\delta\varepsilon_e^{l'}(1+\delta\varepsilon_i^{l'}) + |\delta\varepsilon_e^l|^2] + \frac{[\mathbf{k}\mathbf{k}']^2}{(kk')^2}|1+\delta\varepsilon_i^l|^2\right\} + \delta\varepsilon_i^{l''}|\delta\varepsilon_e^l|^2\frac{[\mathbf{k}\mathbf{k}']^2}{(kk')^2}\right\} \end{aligned} \tag{10.9}$$

and is expressed in terms of the real and the imaginary part of the partial polarizabilities of the plasma [cf. (9.5)] at the frequency of the virtual oscillation [cf. (9.7)]

$$\omega'' = \omega - \omega' = \frac{1}{2}\{3(kr_{D_e})^2 - \lambda_0^2 k'^2\}\omega_{L_e}, \quad \lambda_0 \equiv \frac{c}{\omega_{L_e}}, \tag{10.10}$$

with wave vector

$$\mathbf{k}'' = \mathbf{k} - \mathbf{k}', \quad k'' = |\mathbf{k} - \mathbf{k}'|. \tag{10.11}$$

To shorten the equations, we shall use, as in §9, the total (substantial) derivative with respect to the time:

$$\frac{dW_{tr}}{dt} \equiv \frac{\partial W_{tr}}{\partial t} + c(k\lambda_0)\frac{\partial W_{tr}}{\partial \mathbf{r}}, \quad \frac{dW_l}{dt} = \frac{\partial W_l}{\partial t} + 3v_{T_e}(kr_{D_e})\frac{\partial W_l}{\partial \mathbf{r}}. \tag{10.12}$$

It follows from Eqs. (10.8) that in this approximation the total energy of the longitudinal and transverse waves is conserved. This can be seen by adding these equations term by term:

$$\frac{d}{dt}(W_l + W_{tr}) = \int d\mathbf{k}' \{Q(\mathbf{k}, \mathbf{k}') W_l(\mathbf{k}) W_{tr}(\mathbf{k}') - Q(\mathbf{k}', \mathbf{k}) W_{tr}(\mathbf{k}) W_l(\mathbf{k}')\} \tag{10.13}$$

and integrating with respect to d**k** [cf. (9.13)]

$$\frac{d}{dt}\int d\mathbf{k}\{W_l(\mathbf{k}) + W_{tr}(\mathbf{k})\} = 0. \tag{10.14}$$

Thus, the process of nonlinear interaction is not dissipative in this case: scattering on particles is elastic in the sense that energy is not given up to the particles, but merely passes from one oscillation to another. The direction of the transition or, in other words, redistribution of the energy can also be established by means of Eqs. (10.8). Indeed, since it follows from (10.9) that the kernel is negative:

$$Q(\mathbf{k}, \mathbf{k}') < 0, \quad \omega > \omega' > 0, \tag{10.15}$$

we find that the energy density $W_l(\mathbf{k})$ of the Langmuir wave decreases in accordance with the formal solution

$$W_l(\mathbf{k}) = W_l(\mathbf{k}, t_0) \exp\left\{\int_{t_0}^{t} dt' \int d\mathbf{k}' Q(\mathbf{k}, \mathbf{k}') W_{tr}(\mathbf{k}')\right\} \tag{10.16}$$

if its frequency

$$\omega_{L_e} + \frac{3}{2}\omega_{L_e}(kr_{D_e})^2 \tag{10.17}$$

is greater than the frequency of the long-wavelength transverse oscillation:

$$\omega_{L_e} + \frac{1}{2}\omega_{L_e}(\lambda_0^2 k'^2). \tag{10.18}$$

Then the wavelengths of the interacting oscillations satisfy the inequality

$$k^2 > \frac{1}{3}\frac{c^2}{v_{T_e}^2} k'^2. \tag{10.19}$$

One can arrive at the same conclusion by means of the second equation (10.8) for $W_{tr}(\mathbf{k})$. Hence, in the case of nonlinear interaction of Langmuir oscillations with long-wavelength transverse oscillations through a virtual longitudinal oscillation in an isotropic plasma, energy is transferred from the high- to the low-frequency oscillations. This can be seen in the example of the interaction of infinitesimally narrow packets of longitudinal and transverse waves (as is demonstrated in §9 for Langmuir and ion-acoustic oscillations).

The expression for the kernels (10.9) can be simplified in a number of limiting cases. For example, when

$$\omega'' \ll k'' v_{T_i} \ll k'' v_{T_e}, \tag{10.20}$$

when the phase velocity ω''/k'' of the virtual oscillation is small compared with the thermal velocity of the ions, the last term in $Q(\mathbf{k}, \mathbf{k}')$, which is proportional to the imaginary part of the longitudinal ion permittivity $\delta\varepsilon_e^{l''}(\omega'', \mathbf{k}'')$, is the largest. In this case, the nonlinear interaction is associated with the scattering of waves on ions and is described by the equations

$$\frac{dW_l(\mathbf{k})}{dt} = -\frac{1}{16(2\pi)^{5/2}}\omega_{L_e}\frac{W_l(\mathbf{k})}{N_e \varkappa T_e}\frac{v_{T_e}}{v_{T_i}}\frac{1}{r_{D_e}}\int d\mathbf{k}' W_{tr}(\mathbf{k}')\frac{3k^2 r_{D_e}^2 - \lambda_0^2 k'^2}{|\mathbf{k}-\mathbf{k}'|}\frac{[\mathbf{k}\mathbf{k}']^2}{(kk')^2}\left\{\frac{r_{D_i}/r_{D_e}}{1+(r_{D_i}/r_{D_e})^2}\right\}^2; \tag{10.21}$$

$$\frac{dW_{tr}(\mathbf{k})}{dt} = -\frac{1}{16(2\pi)^{5/2}}\omega_{L_e}\frac{W_{tr}(\mathbf{k})}{N_e \varkappa T_e}\frac{v_{T_e}}{v_{T_i}}\frac{1}{r_{D_e}}\int d\mathbf{k}' W_l(\mathbf{k}')\frac{\lambda_0^2 k^2 - 3k'^2 r_{D_e}^2}{|\mathbf{k}-\mathbf{k}'|}\frac{[\mathbf{k}\mathbf{k}']^2}{(kk')^2}\left\{\frac{r_{D_i}/r_{D_e}}{1+(r_{D_i}/r_{D_e})^2}\right\}^2. \tag{10.22}$$

Using the function (9.29), one can represent scattering on ions by somewhat more general equations. For example, (10.21) will have the form

$$\frac{dW_l(\mathbf{k})}{dt} = -\frac{1}{16(2\pi)^{5/2}}\frac{\omega_{L_e}}{N_e r_{D_e}^3}\frac{W_l(\mathbf{k})}{\varkappa T_e}\frac{v_{T_e}}{v_{T_i}} r_{D_e}^2 \int d\mathbf{k}' W_{tr}(\mathbf{k}')\frac{3k^2 r_{D_e}^2 - \lambda_0^2 k'^2}{|\mathbf{k}-\mathbf{k}'|}\frac{[\mathbf{k}\mathbf{k}']^2}{(kk')^2} \times$$

$$\times \left\{\frac{r_{D_i}/r_{D_e}}{F + (r_{D_i}^2/r_{D_e}^2)}\right\}^2 \exp\left\{-\frac{1}{8}\frac{v_{T_e}^2}{v_{T_i}^2}\frac{(3k^2 r_{D_e}^2 - \lambda_0^2 k'^2)^2}{r_{D_e}^2(\mathbf{k}-\mathbf{k}')^2}\right\}. \tag{10.23}$$

In the case of scattering on electrons one can also have conditions under which the terms with $\delta\varepsilon_i$ can be ignored in the expression (10.9) for the kernel Q(**k**, **k**'). One can then, in practice, speak of scattering in an electron plasma ($M \to \infty$)

$$\frac{dW_l(\mathbf{k})}{dt} = -\frac{1}{16(2\pi)^{5/2}} \frac{\omega_{L_e}}{N_e r_{D_e}^3} \frac{W_l(\mathbf{k})}{\varkappa T_e} r_{D_e}^4 \int d\mathbf{k}' W_{tr}(\mathbf{k}') \frac{3k^2 r_{D_e}^2 - \lambda_0^2 k'^2}{|\mathbf{k}-\mathbf{k}'|^3} \times$$
$$\times \left\{ \frac{[\mathbf{k}\mathbf{k}']^2}{(kk')^2} [k^2 k'^2 - 4(\mathbf{k}\mathbf{k}')^2 + 2k'^2(\mathbf{k}\mathbf{k}')] + 2\frac{(\mathbf{k}\mathbf{k}')^2}{k^2}(\mathbf{k}-\mathbf{k}')^2 \right\}; \qquad (10.24)$$

$$\frac{dW_{tr}(\mathbf{k})}{dt} = -\frac{1}{16(2\pi)^{5/2}} \frac{\omega_{L_e}}{N_e r_{D_e}^3} \frac{W_{tr}(\mathbf{k})}{\varkappa T_e} r_{D_e}^4 \int d\mathbf{k}' W_l(\mathbf{k}') \frac{\lambda_0^2 k^2 - 3k'^2 r_{D_e}^2}{|\mathbf{k}-\mathbf{k}'|^3} \times$$
$$\times \left\{ \frac{[\mathbf{k}\mathbf{k}']^2}{(kk')^2} [k^2 k'^2 - 4(\mathbf{k}\mathbf{k}')^2 + 2k^2(\mathbf{k}\mathbf{k}')] + 2\frac{(\mathbf{k}\mathbf{k}')^2}{k'^2}(\mathbf{k}-\mathbf{k}')^2 \right\}. \qquad (10.25)$$

To compare the processes of Coulomb scattering on ions, (10.21)-(10.22), and on electrons, (10.24)-(10.25), we note that the characteristic time of spectral transfer corresponding to Eqs. (10.21)-(10.22) is approximately equal to

$$\tau \sim 10^3 \left\{ \frac{\omega_{L_e}}{N r_{D_e}^3} \frac{W_l}{\varkappa T_e} \frac{v_{T_e}}{v_{T_i}} (k r_{D_e})^2 (3k^2 r_{D_e}^2 - \lambda_0^2 k'^2) \right\}^{-1}, \quad W_l \sim W_{tr}. \qquad (10.26)$$

The ratio of this time to the time of spectral redistribution described by the system of equations (10.24)-(10.25) is approximately equal to $(kr_{D_e})^2(v_{T_i}/v_{T_e})$, which under the condition $(kr_{D_e}) \gg \lambda_0 k'$ is less than $(v_{T_i}/v_{T_e})^3 \ln\{(e_i^2 M T_e^3)/(e^2 m T_i^3)\}$. Therefore, for an isothermal hydrogen plasma the characteristic time of scattering on ions for wavelengths at which the given scattering is decisive is shorter by four order of magnitude than the characteristic time of scattering on electrons. For a mercury plasma, τ is smaller by approximately eight orders. Thus, scattering on ions, if it occurs [i.e., (9.25) holds], is a much faster process than scattering on electrons.

Hitherto, we have considered the scattering of electron Langmuir oscillations and long-wavelength transverse oscillations only through a longitudinal virtual oscillation, i.e., in the equations (10.1) and (10.2) of the nonlinear interaction, we have taken into account only the first term, which is longitudinal with respect to the vector **k**", in the inverse tensor $A_{jb}(\omega'', \mathbf{k}'')$, which describes the virtual oscillation. We now obtain an equation for scattering with retardation with allowance for the second term in the inverse tensor; in its denominator this contains the expression

$$\varepsilon^{tr}(\omega'', \mathbf{k}'') - \frac{c^2 k''^2}{\omega''^2} \qquad (10.27)$$

and it determines scattering through a transverse oscillation. The scattering on ions, as in the case of the interaction of only Langmuir waves, is small in a plasma that is not too non-isothermal. The contribution of scattering of Langmuir waves on electrons can be found from Eq. (10.1) by means of (8.22):

$$\left\{\frac{dW_l(\mathbf{k})}{dt}\right\}^{\text{ret}} = -\frac{1}{16(2\pi)^{5/2}} \frac{\omega_{L_e}}{N_e r_{D_e}^3} \frac{W_l(\mathbf{k})}{\varkappa T_e} r_{D_e}^4 \int d\mathbf{k}' W_{tr}(\mathbf{k}') \times$$
$$\times \frac{3k^2 r_{D_e}^2 - \lambda_0^2 k'^2}{(kk')^2 |\mathbf{k}-\mathbf{k}'|} [k'^2(\mathbf{k}, \mathbf{k}-\mathbf{k}')^2 + (\mathbf{k}-\mathbf{k}')^2(\mathbf{k}\mathbf{k}')^2] \left\{ \lambda_0^4 (\mathbf{k}-\mathbf{k}')^4 + \right.$$
$$\left. + \frac{1}{2}\frac{c^2}{v_{T_e}^2}\left[1 + \frac{\pi}{4}\lambda_0^2(\mathbf{k}-\mathbf{k}')^2\right][3k^2 r_{D_e}^2 - \lambda_0^2 k'^2]^2 \right\}^{-1}. \qquad (10.28)$$

Similarly, from (10.2) we obtain an equation for the evolution of transverse electromagnetic waves in such a process:

$$\left\{\frac{dW_{tr}(\mathbf{k})}{dt}\right\}^{\text{ret}} = -\frac{1}{16(2\pi)^{5/2}}\frac{\omega_{L_e}}{Nr_{D_e}^3}\frac{W_{tr}(\mathbf{k})}{\varkappa T_e} r_{D_e}^4 \int d\mathbf{k}' W_l(\mathbf{k}') \times$$
$$\times \frac{\lambda_0^2 k^2 - 3(k' r_{D_e})^2}{(kk')^2 |\mathbf{k}-\mathbf{k}'|} [k^2(\mathbf{k}', \mathbf{k}-\mathbf{k}')^2 + (\mathbf{k}-\mathbf{k}')^2(\mathbf{k}\mathbf{k}')^2] \Big\{\lambda_0^4(\mathbf{k}-\mathbf{k}')^4 +$$
$$+ \frac{1}{2}\frac{c^2}{v_{T_e}^2}\left[1 + \frac{\pi}{4}\lambda_0^2(\mathbf{k}-\mathbf{k}')^2\right][3k'^2 r_{D_e}^2 - \lambda_0^2 k^2]^2\Big\}^{-1}, \tag{10.29}$$

and this leads in conjunction with (10.28) to the conservation law (10.14). It must be emphasized that scattering on electrons through a transverse virtual oscillation is small compared with scattering on ions in the region in which the inequality (9.25) is satisfied. Therefore, it is only in the region in which the opposite inequality (9.22) is satisfied, when the scattering on electron proceeds through a longitudinal oscillation, that the interaction described by these equations can be important. And, depending on the relationship between the wavelengths of the longitudinal and the transverse oscillation, either Coulomb scattering (i.e., scattering through a longitudinal oscillation) determined by the system of equations (10.24)-(10.25) or scattering with allowance for retardation (i.e., scattering through a transverse oscillation), which corresponds to Eqs. (10.28) and (10.29), will be more important. For example, when $(v_{Te}/c^2)\omega_{Le} \gg k \gg k'$ and $(kr_{De})^2 \gg (v_{Ti}/v_{Te})^2 \ln\{(e_i^2 MT_e^3)/(e^2 mT_i^3)\}$, the spectral transfer is determined by scattering through a transverse oscillation and takes place with a characteristic time of order

$$\tau \sim 10^3 \frac{Nr_{D_e}^3}{\omega_{L_e}}\frac{\varkappa T_e}{W_l}\frac{k^3}{(kr_{D_e})^4 k'^3}, \quad W_l \sim W_{tr}. \tag{10.30}$$

In all the foregoing relations we have assumed that the virtual oscillation through which the scattering occurs interacts very strongly with the electrons of the plasma, having, as it does, a phase velocity ω''/k'' that is appreciably less than the electron thermal velocity v_{Te} [see, for example, (10.20)]:

$$\omega'' \ll k'' v_{T_e}. \tag{10.31}$$

Then there is strong absorption of the virtual oscillation by the plasma electrons. One can also consider the opposite case of a weaker interaction with the electrons when the phase velocity of the virtual oscillation is appreciably greater than the electron thermal velocity:

$$\omega'' \gg k'' v_{T_e}. \tag{10.32}$$

At the same time, of course, we still have the condition that the frequency ω'' of the virtual oscillation is small compared with the frequencies of the interacting oscillations: $\omega \sim \omega' \gg \omega''$, i.e., as before we have scattering with a small change in the frequency, and the denominators $\varepsilon^l(\omega'', \mathbf{k}'')$ and (10.27) in the inverse tensor $A_{jb}(\omega'', \mathbf{k}'')$ are nonvanishing. In this case the ions always play a negligible role since the ion terms contain the exponentially small factor $\exp\{-(\omega''/2k'' v_{Ti})^2\}$ with exponent that is greater by a factor $(v_{Te}/v_{Ti})^2$ than in the exponent in the electron terms.

Coulomb scattering on electrons differs from the scattering we have analyzed earlier by the presence in the integral with respect to $d\mathbf{k}'$ of the exponential on the right-hand side of the equations [cf. (10.24)-(10.25)]:

$$\frac{dW_l(\mathbf{k})}{dt} = -\frac{1}{16(2\pi)^{5/2}}\frac{\omega_{L_e}}{N_e r_{D_e}^3}\frac{W_l(\mathbf{k})}{\varkappa T_e} r_{D_e}^4 \int d\mathbf{k}' W_{tr}(\mathbf{k}') \frac{3k^2 r_{D_e}^2 - \lambda_0^2 k'^2}{(kk')^2|\mathbf{k}-\mathbf{k}'|^3} \exp\left\{-\frac{r_{D_e}^2}{8|\mathbf{k}-\mathbf{k}'|^2}\left(\frac{k'^2 c^2}{v_{T_e}^2} - 3k^2\right)^2\right\} \times$$
$$\times \{[\mathbf{k}\mathbf{k}']^2[k^2 k'^2 - 4(\mathbf{k}\mathbf{k}')^2 + 2k'^2(\mathbf{k}\mathbf{k}')] + 2(\mathbf{k}\mathbf{k}')^2 k'^2(\mathbf{k}-\mathbf{k}')^2\}; \tag{10.33}$$

$$\frac{dW_{tr}(\mathbf{k})}{dt} = -\frac{1}{16(2\pi)^{5/2}} \frac{\omega_{Le}}{N_e r_{De}^3} \frac{W_{tr}(\mathbf{k})}{\varkappa T_e} r_{De}^4 \int d\mathbf{k}' W_l(\mathbf{k}') \frac{\lambda_0^2 k^2 - 3k'^2 r_{De}^2}{(kk')^2 |\mathbf{k}-\mathbf{k}'|^3} \exp\left\{-\frac{r_{De}^2}{8|\mathbf{k}-\mathbf{k}'|^2}\left(\frac{k^2c^2}{v_{T_e}^2} - 3k'^2\right)^2\right\} \times$$
$$\times \{[\mathbf{kk}']^2 [k^2k'^2 - 4(\mathbf{kk}')^2 + 2k^2(\mathbf{kk}')] + 2(\mathbf{kk}')^2 k^2 (\mathbf{k}-\mathbf{k}')^2\}. \qquad (10.34)$$

As before, the kernels of these equations are such that the conservation law (10.14) for the total energy of the interacting oscillations is satisfied.

The scattering of a transverse oscillation into an electron Langmuir oscillation through a transverse virtual oscillation is described by different equations depending on the ratio between the wavelength of the longitudinal oscillation and $\lambda_0 = c/\omega_{Le}$. If the wavelength of the Langmuir oscillation is greater than λ_0, the dynamics of the nonlinear interaction is determined by the equations

$$\frac{dW_l(\mathbf{k})}{dt} = -\frac{1}{16(2\pi)^{5/2}} \frac{\omega_{Le}}{N_e r_{De}^3} \frac{W_l(\mathbf{k})}{\varkappa T_e} r_{De}^4 \int d\mathbf{k}' W_{tr}(\mathbf{k}') \frac{3k^2 r_{De}^2 - \lambda_0^2 k'^2}{(kk')^2 |\mathbf{k}-\mathbf{k}'|^3} \times$$
$$\times \exp\left\{-\frac{r_{De}^2}{8|\mathbf{k}-\mathbf{k}'|^2}\left(\frac{c^2k'^2}{v_{T_e}^2} - 3k^2\right)^2\right\} \{k'^2 (\mathbf{k}-\mathbf{k}')^2 (\mathbf{k}, \mathbf{k}-\mathbf{k}')^2 +$$
$$+ (\mathbf{k}-\mathbf{k}')^4 (\mathbf{kk}')^2 - k'^2 (\mathbf{k}-\mathbf{k}')^2 (\mathbf{kk}')^2 - [[\mathbf{kk}']^2 + (\mathbf{kk}')(\mathbf{k}', \mathbf{k}-\mathbf{k}')]^2\}; \qquad (10.35)$$

$$\frac{dW_{tr}(\mathbf{k})}{dt} = -\frac{1}{16(2\pi)^{5/2}} \frac{\omega_{Le}}{N_e r_{De}^3} \frac{W_{tr}(\mathbf{k})}{\varkappa T_e} r_{De}^4 \int d\mathbf{k}' W_l(\mathbf{k}') \frac{\lambda_0^2 k^2 - 3k'^2 r_{De}^2}{(kk')^2 |\mathbf{k}-\mathbf{k}'|^3} \times$$
$$\times \exp\left\{-\frac{r_{De}^2}{8|\mathbf{k}-\mathbf{k}'|^2}\left(3k'^2 - \frac{k^2c^2}{v_{T_e}^2}\right)^2\right\} \{k^2 (\mathbf{k}-\mathbf{k}')^2 (\mathbf{k}', \mathbf{k}-\mathbf{k}')^2 +$$
$$+ (\mathbf{k}-\mathbf{k}')^4 (\mathbf{kk}')^2 - k^2 (\mathbf{k}-\mathbf{k}')^2 (\mathbf{kk}')^2 - [[\mathbf{kk}']^2 + (\mathbf{kk}')(\mathbf{k}, \mathbf{k}'-\mathbf{k})]^2\}. \qquad (10.36)$$

In the opposite case, when the wavelength of the electron Langmuir oscillation is less than λ_0, the equation for the change in the energy density $W_l(\mathbf{k})$ is

$$\frac{dW_l(\mathbf{k})}{dt} = -\frac{1}{32(2\pi)^{5/2}} \frac{\omega_{Le}}{N_e r_{De}^3} \frac{W_l(\mathbf{k})}{\varkappa T_e} \left(\frac{v_{T_e}}{c}\right)^2 r_{De}^2 \int d\mathbf{k}' W_{tr}(\mathbf{k}') \frac{3k^2 r_{De}^2 - \lambda_0^2 k'^2}{(kk')^2 |\mathbf{k}-\mathbf{k}'|^3} \exp\left\{-\frac{r_{De}^2}{8|\mathbf{k}-\mathbf{k}'|^2}\left(\frac{c^2k'^2}{v_{T_e}^2} - 3k^2\right)^2\right\} \times$$
$$\times \left\{k^4 \frac{(\mathbf{k}', \mathbf{k}-\mathbf{k}')^2}{(\mathbf{k}-\mathbf{k}')^2} + k^2k'^2(\mathbf{kk}') + \mathbf{k}^2(\mathbf{kk}')(\mathbf{k}', \mathbf{k}-\mathbf{k}')\right\} \qquad (10.37)$$

and, similarly, for $W_{tr}(\mathbf{k})$:

$$\frac{dW_{tr}(\mathbf{k})}{dt} = -\frac{1}{32(2\pi)^{5/2}} \frac{\omega_{Le}}{N_e r_{De}^3} \frac{W_{tr}(\mathbf{k})}{\varkappa T_e} \left(\frac{v_{T_e}}{c}\right)^2 r_{De}^2 \int d\mathbf{k}' W_l(\mathbf{k}') \frac{\lambda_0^2 k^2 - 3k'^2 r_{De}^2}{(kk')^2 |\mathbf{k}-\mathbf{k}'|^3} \exp\left\{-\frac{r_{De}^2}{8|\mathbf{k}-\mathbf{k}'|^2}\left(3k'^2 - \frac{c^2k^2}{v_{T_e}^2}\right)^2\right\} \times$$
$$\times \left\{k'^4 \frac{(\mathbf{k}, \mathbf{k}-\mathbf{k}')^2}{(\mathbf{k}-\mathbf{k}')^2} + (kk')^2(\mathbf{kk}') + k'^2(\mathbf{kk}')(\mathbf{k}, \mathbf{k}'-\mathbf{k})\right\}. \qquad (10.38)$$

These three pairs of equations, which correspond to nonlinear interaction through a virtual oscillation with the large phase velocity (10.32), augment quantitatively the equations obtained previously for the change in the energy density of the oscillations in the opposite case (10.31) of a low phase velocity of the virtual oscillation through which the scattering takes place. We emphasize that the replacement of the inequality (10.31) by (10.32) does not affect the qualitative nature of the process of energy redistribution from higher to lower frequencies.

Comparison of the characteristic times of variation of the energy of the longitudinal oscillations due to the interaction with the transverse oscillations under the conditions (10.32) shows that the fundamental role is always played by the interaction through the transverse virtual oscillation except for the case $k(c/v_{T_e}) \gg k' \gg \lambda_0^{-1}$, when the Coulomb interaction may be of the same order.

To conclude this section, let us consider one further possible process of nonlinear interaction of longitudinal and transverse oscillations. Namely, we have analyzed above processes of scattering on particles. Let us now consider the decay process. A long-wave transverse oscillation and an electron Langmuir oscillation may coalesce, giving rise to a transverse oscillation with twice the electron plasma frequency $2\omega_{Le}$ and wave number $3^{1/2}(\omega_{Le}/c)$. Note that a similar transverse electromagnetic wave is formed as a result of the coalescing of two Langmuir oscillations, which we have considered in § 9 [see Eq. (9.51)]. This case corresponds in the general equations of the nonlinear interaction of longitudinal and transverse oscillations to the vanishing of the denominator of (10.27) in the second term of the inverse tensor $A_{jb}(\omega'', \mathbf{k}'')$ [see (9.49) and (9.50)]. The contribution of such a process to the rate of decrease of the energy $W_l(\mathbf{k})$ of the longitudinal coalesced oscillation is determined from (10.1) by means of the expression (7.23) for the tensor S_{ijs} in a cold isotropic plasma and has the form

$$\left\{\frac{dW_l(\mathbf{k})}{dt}\right\}^{\mathrm{c}} = -\frac{1}{8(2\pi)^2}\frac{\omega_{Le}}{N_e r_{De}^3}\frac{W_l(\mathbf{k})}{\varkappa T_e}\,\delta\left(3-\frac{c^2k^2}{\omega_{Le}^2}\right) r_{De}^5 \int d\mathbf{k}' W_{tr}(\mathbf{k}')\left(k^2+\frac{(\mathbf{k}\mathbf{k}')^2}{k'^2}\right). \tag{10.39}$$

Similarly, for the rate of decrease of the energy of a transverse coalesced oscillation we obtain from Eq. (10.2)

$$\left\{\frac{dW_{tr}(\mathbf{k})}{dt}\right\}^{\mathrm{c}} = -\frac{1}{8(2\pi)^2}\frac{\omega_{Le}}{N_e r_{De}^3}\frac{W_{tr}(\mathbf{k})}{\varkappa T_e} r_{De}^5 \int d\mathbf{k}' W_l(\mathbf{k}')\,\delta\left(3-\frac{c^2k'^2}{\omega_{Le}^2}\right)\left(k'^2+\frac{(\mathbf{k}\mathbf{k}')^2}{k^2}\right). \tag{10.40}$$

Since the wavelength of the transverse electromagnetic wave that participates in the coalescence is appreciably greater than λ_0, we can say, to corresponding accuracy, that the wavelength of the Langmuir oscillation in these two equations is equal to $\lambda_0/3^{1/2}$.

§ 11. Nonlinear Interaction of Transverse Characteristic Oscillations of an Isotropic Plasma

The nonlinear interaction of transverse, naturally polarized electromagnetic waves is described in accordance with the generalized transport equation (3.22) by the relation [12]:

$$\begin{aligned}
&\frac{\partial}{\partial t}(E_{tr}^2)_{\omega,\mathbf{k}}\frac{1}{\omega}\frac{\partial}{\partial\omega}\left\{\omega\left[\varepsilon^{tr'}(\omega,\mathbf{k})-\frac{c^2k^2}{\omega^2}\right]\right\}-\frac{\partial}{\partial\mathbf{r}}(E_{tr}^2)_{\omega,\mathbf{k}}\frac{1}{\omega}\frac{\partial}{\partial\mathbf{k}}\left\{\omega\left[\varepsilon^{tr'}(\omega,\mathbf{k})-\frac{c^2k^2}{\omega^2}\right]\right\}=\\
&=-\frac{1}{2}(E_{tr}^2)_{\omega,\mathbf{k}}\int d\omega' d\mathbf{k}'(E_{tr}^2)_{\omega',\mathbf{k}'}\,\mathrm{Im}\,V_{iajb}(\omega,\mathbf{k};\omega',\mathbf{k}')\left(\delta_{ij}-\frac{k_ik_j}{k^2}\right)\times\\
&\times\left(\delta_{ab}-\frac{k'_ak'_b}{k'^2}\right)+\frac{1}{2}(E_{tr}^2)_{\omega,\mathbf{k}}\,\mathrm{Im}\int d\omega' d\mathbf{k}' d\omega'' d\mathbf{k}''\delta(\omega-\omega'-\omega'')\,\delta(\mathbf{k}-\mathbf{k}'-\mathbf{k}'')\times\\
&\times\left\{\frac{k''_jk''_b}{k''^2}\frac{1}{\varepsilon^l(\omega'',\mathbf{k}'')}+\frac{\delta_{jb}-k''_jk''_b/k''^2}{\varepsilon^{tr}(\omega'',\mathbf{k}'')-\frac{c^2k''^2}{\omega''^2}}\right\}(E_{tr}^2)_{\omega',\mathbf{k}'}S_{ijs}(\omega,\mathbf{k},\omega',\mathbf{k}')\times\\
&\times S_{bca}(\omega'',\mathbf{k}'';\omega,\mathbf{k})\left(\delta_{ia}-\frac{k_ik_a}{k^2}\right)\left(\delta_{sc}-\frac{k'_sk'_c}{k'^2}\right)+\frac{\pi}{4}\,\mathrm{sign}\,\varepsilon^{tr''}(\omega,\mathbf{k})\times\\
&\times\delta\left[\varepsilon^{tr'}(\omega,\mathbf{k})-\frac{c^2k^2}{\omega^2}\right]\int d\omega' d\mathbf{k}' d\omega'' d\mathbf{k}''\,\delta(\omega-\omega'-\omega'')\,\delta(\mathbf{k}-\mathbf{k}'-\mathbf{k}'')\times\\
&\times(E_{tr}^2)_{\omega',\mathbf{k}'}(E_{tr}^2)_{\omega'',\mathbf{k}''}S_{ijs}(\omega,\mathbf{k};\omega',\mathbf{k}')\,S^*_{abc}(\omega,\mathbf{k};\omega',\mathbf{k}')\left(\delta_{ia}-\frac{k_ik_a}{k^2}\right)\left(\delta_{sc}-\frac{k'_sk'_c}{k'^2}\right)\left(\delta_{jb}-\frac{k''_jk''_b}{k''^2}\right).
\end{aligned} \tag{11.1}$$

Such an equation arises when one substitutes into (3.22) the expressions (6.62) and (6.79) for the spectral functions of the fields and the inverse tensor and separates out from the spectral functions only the terms that are transverse with respect to the wave vectors. The physical meaning of the terms on the right-hand side of this equation can be elucidated in the same way as was used for the equations of the nonlinear interaction of Langmuir oscillations with one another (§ 9) and Langmuir oscillations with transverse oscillations (§ 10). The first three terms are responsible for the induced scattering of naturally polarized transverse waves on particles through a longitudinal oscillation (the first and the second terms) and a transverse virtual oscillation if the denominators $\varepsilon^{l}(\omega'', \mathbf{k}'')$ and (10.27) of the inverse tensor do not vanish. Such a possibility can be realized in decay processes to which there also corresponds the last (fourth) term on the right-hand side of (11.1). However, it follows from the dispersion law of the transverse oscillations of an isotropic plasma (6.53) that for them one cannot have processes of coalescence (decay) with the participation of three transverse waves, since the decay condition (9.72) for the spectrum (6.53) is not satisfied. In this connection, the last term of (11.1) and the decay part of the third term [see (9.50)] for waves with the spectrum (6.53) do not contribute to the nonlinear interaction. To summarize, decay processes in Eq. (11.1) are described by the decay part of the second term with

$$\frac{1}{\varepsilon^{l}(\omega'', \mathbf{k}'')} = -i\pi\delta\left[\varepsilon^{l\prime}(\omega'', \mathbf{k}'')\right] \operatorname{sgn} \varepsilon^{l\prime\prime}(\omega'', \mathbf{k}'') \tag{11.2}$$

and corresponds to a loss of energy density of the transverse electromagnetic waves (6.53) with frequency ω and wave vector $\mathbf{k}$ due to their decay into a transverse oscillation and ion sound and into a transverse and a Langmuir oscillation (see the introductory remarks to Chapter II).

In this section we shall consider processes of induced scattering of transverse electromagnetic waves on particles. We begin with the first two terms in the generalized kinetic equation (11.2) for transverse waves, in which we take into account only the Coulomb interaction of the particles. We shall be interested in the case when the frequency difference of the interacting oscillations, i.e., the frequency of the longitudinal virtual oscillation, is small compared with the frequencies themselves:

$$\omega'' = \omega - \omega' = \sqrt{\omega_{Le}^2 + c^2k^2} - \sqrt{\omega_{Le}^2 + c^2k'^2}, \quad \omega'' \ll \omega, \ \omega'. \tag{11.3}$$

Then for the Maxwell distribution function (6.14) we obtain from (11.1), using the relation (8.23), an equation for the evolution of the energy density $W_{tr}(\mathbf{k})$ of the transverse waves [see (10.12)] [32]:

$$\left\{\frac{dW_{tr}(\mathbf{k})}{dt}\right\}^{\text{Coul}} = \int d\mathbf{k}' Q(\mathbf{k}, \mathbf{k}') W_{tr}(\mathbf{k}) W_{tr}(\mathbf{k}'), \tag{11.4}$$

in which the kernel $Q(\mathbf{k}, \mathbf{k}')$

$$\begin{aligned} Q(\mathbf{k}, \mathbf{k}') = &-\frac{1}{8(2\pi)^3} \frac{\omega_{Le}}{N_e \varkappa T_e} \frac{\omega_{Le}^3}{\omega^3} \frac{(k'' r_{De})^2}{|1 + \delta\varepsilon_e^l + \delta\varepsilon_i^l|^2} \left\{ \delta\varepsilon_e^{l\prime\prime} \left[2\frac{v_{Te}^2}{\omega^2} \frac{[\mathbf{k}\mathbf{k}']^2}{(kk')^2} (k'^2 + k^2 - \mathbf{k}\mathbf{k}')(2\delta\varepsilon_e^{l\prime} + 2\delta\varepsilon_e^{l\prime}\delta\varepsilon_i^{l\prime} + |\delta\varepsilon_e^l|^2) + \right.\right. \\ &\left.\left. + \left(1 + \left(\frac{\mathbf{k}\mathbf{k}'}{kk'}\right)^2\right) |1 + \delta\varepsilon_i^l|^2 \right] + \left(1 + \left(\frac{\mathbf{k}\mathbf{k}'}{kk'}\right)^2\right) \delta\varepsilon_i^{l\prime\prime} |\delta\varepsilon_e^l|^2 \right\}, \end{aligned} \tag{11.5}$$

is expressed in terms of the imaginary, $\delta\varepsilon^{l\prime\prime}$, and the real, $\delta\varepsilon^{l\prime}$, parts of the partial polarizabilities of the electron and the ion components of the plasma at the frequency $\omega'' = \omega - \omega'$ of the virtual oscillation with wave vector $\mathbf{k}'' = \mathbf{k} - \mathbf{k}'$ [cf. (9.5) and (10.9)].

The antisymmetry (9.12) of the kernel (11.5) under the transposition of the vectors $\mathbf{k}$ and $\mathbf{k}'$ enables us to conclude that the total energy of the transverse electromagnetic waves is conserved:

$$\int d\mathbf{k} W_{tr}(\mathbf{k}) = \text{const.} \tag{11.6}$$

Thus, in this approximation the evolution of these waves reduces to a redistribution of the energy density from the short to the long wavelengths, or from the high frequencies (6.53) to low frequencies.

For the interaction of the long-wavelength (6.60) transverse oscillations (6.61) (with wavelength greater than $\lambda_0 = c/\omega_{Le}$) with frequency near the electron plasma frequency under the conditions (10.20) of a low phase velocity ω''/k'' of the virtual oscillation:

$$\frac{1}{2}\frac{c^2}{\omega_{L_e}v_{T_e}}\frac{k^2-k'^2}{|\mathbf{k}-\mathbf{k}'|}\ll 1, \tag{11.7}$$

the largest term in the kernel (11.5) is the last one in the curly brackets. Then the equation that arises for $W_{tr}(\mathbf{k})$ describes scattering on ions:

$$\frac{dW_{tr}(\mathbf{k})}{dt}=-\frac{1}{32(2\pi)^{5/2}}\frac{\omega_{L_e}}{N_e r_{D_e}^3}\frac{W_{tr}(\mathbf{k})}{\varkappa T_e}\frac{c^2}{v_{T_e}v_{T_i}}r_{D_e}^4\int d\mathbf{k}'W_{tr}(\mathbf{k}')\frac{k^2-k'^2}{|\mathbf{k}-\mathbf{k}'|}\left[1+\left(\frac{\mathbf{k}\mathbf{k}'}{kk'}\right)^2\right]\left\{\frac{r_{D_i}/r_{D_e}}{1+(r_{D_i}/r_{D_e})^2}\right\}^2 \tag{11.8}$$

with characteristic time of spectral redistribution

$$\tau\sim 10^3\frac{N_e r_{D_e}^3}{\omega_{L_e}}\frac{v_{T_e}v_{T_i}}{c^2}\frac{\varkappa T_e}{W_{tr}}\frac{1}{r_{D_e}^4 k^3\Delta k}. \tag{11.9}$$

Here Δk is the characteristic value of the change in the wave number on scattering:

$$\Delta k=\frac{1}{2}\frac{k^2-k'^2}{|\mathbf{k}-\mathbf{k}'|}. \tag{11.10}$$

A slightly more general equation of scattering on ions can be represented in the form

$$\frac{dW_{tr}(\mathbf{k})}{dt}=-\frac{1}{32(2\pi)^{5/2}}\frac{\omega_{L_e}}{N_e r_{D_e}^3}\frac{W_{tr}(\mathbf{k})}{\varkappa T_e}\frac{c^2}{v_{T_e}v_{T_i}}r_{D_e}^4\int d\mathbf{k}'W_{tr}(\mathbf{k}')\frac{k^2-k'^2}{|\mathbf{k}-\mathbf{k}'|}\times$$
$$\times\left[1+\left(\frac{(\mathbf{k}\mathbf{k}')}{kk'}\right)^2\right]\left[\frac{r_{D_i}/r_{D_e}}{F+(r_{D_i}/r_{D_e})^2}\right]^2\exp\left\{-\frac{c^4(k^2-k'^2)^2}{8v_{T_i}^2\omega_{L_e}^2(\mathbf{k}-\mathbf{k}')^2}\right\}, \tag{11.11}$$

where the function F is given by Eqs. (9.29), in which

$$\omega''=\frac{1}{2}\omega_{L_e}(k^2-k'^2)\lambda_0^2. \tag{11.12}$$

The equation for the Coulomb scattering of long-wavelength transverse oscillations on electrons in, for example, an electron plasma can be obtained as a limiting form of (11.4)-(11.5) if one completely ignores the contribution of the ions ($\omega'' \ll k''v_{Te}$):

$$\frac{dW_{tr}(\mathbf{k})}{dt}=-\frac{1}{16(2\pi)^{5/2}}\frac{\omega_{L_e}}{N_e r_{D_e}^3}\frac{W_{tr}(\mathbf{k})}{\varkappa T_e}\left(\frac{c}{v_{T_e}}\right)^2 r_{D_e}^6\int d\mathbf{k}'W_{tr}(\mathbf{k}')\frac{[\mathbf{k}\mathbf{k}']^2}{(kk')^2}\frac{k^2-k'^2}{|\mathbf{k}-\mathbf{k}'|}(k^2+k'^2-\mathbf{k}\mathbf{k}'). \tag{11.13}$$

In the opposite limit of Coulomb scattering of short-wavelength ($\lambda \ll \lambda_0 = c/\omega_{Le}$) transverse oscillations (6.59) special cases of the kernel (11.5) differ considerably from the cases we have considered above. As an illustration, we give the expression for Q in the case when the most important (i.e., fastest process) is scattering of waves on ions [32]:

$$Q(\mathbf{k},\mathbf{k}')=-\frac{1}{16(2\pi)^{5/2}}\frac{ck}{N_e r_{D_e}^4\varkappa T_e}\left(\frac{v_{T_e}}{c}\right)^3\frac{k-k'}{|\mathbf{k}-\mathbf{k}'|}\frac{1}{(kk')^2}\left[1+\left(\frac{\mathbf{k}\mathbf{k}'}{kk'}\right)^2\right]\left\{\frac{r_{D_i}/r_{D_e}}{1+(r_{D_i}/r_{D_e})^2}\right\}^2. \tag{11.14}$$

In obtaining this relation we have assumed that the electron Debye radius is small compared with the wavelength of the virtual oscillation: $(k''r_{De})^2 \ll 1$.

The induced scattering of transverse electromagnetic waves with allowance for retardation, i.e., through a transverse virtual oscillation, is determined by the third term on the right-hand side of Eq. (11.1). The fact that the slow ($\omega''/k'' \ll v_{Te}$) virtual oscillation is nonelectrostatic is clearly manifested in the example of the interaction of transverse waves with wavelengths much greater than $\lambda_0 = c/\omega_{Le}$:

$$\frac{dW_{tr}(\mathbf{k})}{dt} = -\frac{1}{4(2\pi)^{5/2}} \frac{\omega_{Le}}{N_e r_{D_e}^3} \frac{W_{tr}(\mathbf{k})}{\varkappa T_e} r_{D_e}^4 \left(\frac{\omega_{Le}}{c}\right)^2 \int d\mathbf{k}' W_{tr}(\mathbf{k}') \frac{[\mathbf{k}\mathbf{k}']^2}{(kk')^2} \times$$
$$\times \frac{k^2 + k'^2 - \mathbf{k}\mathbf{k}'}{|\mathbf{k}-\mathbf{k}'|(k^2 - k'^2)} \left\{ 4\frac{(\mathbf{k}-\mathbf{k}')^4}{(k^2-k'^2)^2} + \frac{\pi}{2}\frac{1}{r_{D_e}^2(\mathbf{k}-\mathbf{k}')^2}\right\}^{-1}. \tag{11.15}$$

Under conditions when the characteristic change Δk of the wave number on scattering is small:

$$\frac{\Delta k}{k} \ll kr_{D_e} \ll 1, \tag{11.16}$$

the time of spectral transfer is, in accordance with (11.15),

$$\tau \sim 10^3 \left(\frac{c}{v_{T_e}}\right)^2 \frac{N_e r_{D_e}^3}{\omega_{Le}} \frac{\varkappa T_e}{W_{tr}} \frac{1}{k\Delta k r_{D_e}^2}, \tag{11.17}$$

and in the opposite case

$$\tau \sim 10^3 \left(\frac{c}{v_{T_e}}\right)^2 \frac{N_e r_{D_e}^3}{\omega_{Le}} \frac{\varkappa T_e}{W_{tr}} \frac{\Delta k}{k^5 r_{D_e}^4}, \quad \frac{\Delta k}{k} \gg kr_{D_e}. \tag{11.18}$$

As in the processes of the nonlinear interaction of Langmuir and transverse waves (see § 9 and § 10), the effect of the ions on the scattering with retardation is unimportant and scattering on electrons is decisive. And the scattering on electrons through a transverse virtual oscillation proceeds much faster than Coulomb scattering on electrons with the characteristic time

$$\tau \sim 10^3 \frac{N_e r_{D_e}^3}{\omega_{Le}} \frac{\varkappa T_e}{W_{tr}} \left(\frac{v_{T_e}}{c}\right)^2 \frac{1}{r_{D_e}^6 k^5 \Delta k}, \tag{11.19}$$

which corresponds to Eq. (11.13). This can be seen by a direct comparison of the times (11.17) and (11.18) on the one hand and (11.19) on the other.

A quite different picture is obtained when one considers the effect of the fast ($\omega''/k'' \gg v_{T_e}$) virtual oscillations being nonelectrostatic on the scattering of transverse waves. As before, the scattering of waves on ions is negligibly small, and the scattering on electrons is described by the equation [12]

$$\frac{dW_{tr}(\mathbf{k})}{dt} = -\frac{1}{16(2\pi)^{5/2}} \frac{\omega_{Le}}{N_e r_{D_e}^3} \frac{W_{tr}(\mathbf{k})}{\varkappa T_e} \left(\frac{c}{v_{T_e}}\right)^2 r_{D_e}^6 \int d\mathbf{k}' W_{tr}(\mathbf{k}') \frac{[\mathbf{k}\mathbf{k}']^2}{(kk')^2} \frac{k^2 - k'^2}{|\mathbf{k}-\mathbf{k}'|} \times$$
$$\times (k^2 + k'^2 - \mathbf{k}\mathbf{k}') \left[1 + 8\left(\frac{v_{T_e}}{c}\right)^4 \frac{(\mathbf{k}-\mathbf{k}')^2}{r_{D_e}^2 (k^2-k'^2)^2}\right] \exp\left\{-\frac{1}{8}\left(\frac{c}{v_{T_e}}\right)^4 \frac{(k^2-k'^2)^2 r_{D_e}^2}{(\mathbf{k}-\mathbf{k}')^2}\right\}. \tag{11.20}$$

The second term in the square brackets on the right-hand side of (11.20) arises precisely because we allow for scattering through a transverse virtual oscillation and is always negligibly small compared with unity. Therefore, in this case of a fast ($\omega''/k'' \gg v_{T_e}$) transverse virtual oscillation, the main role is played by scattering on electrons with allowance for only the Coulomb interaction between the particles [which corresponds to the first term – unity – in the square brackets of (11.20)]. The characteristic time of spectral transfer of the energy density of the transverse electromagnetic waves that proceeds in accordance with Eq. (11.20) is of order

$$\tau \sim 10^3 \frac{N_e r_{D_e}^3}{\omega_{Le}} \frac{\varkappa T_e}{W_{tr}} \left(\frac{v_{T_e}}{c}\right)^2 \frac{1}{r_{D_e}^6 k^5 \Delta k} \exp\left\{\frac{1}{8}\left(\frac{c}{v_{T_e}}\right)^4 r_{D_e}^2 (\Delta k)^2\right\} \tag{11.21}$$

and is exponentially large compared with (11.19).

In conclusion, we allow in Eq. (11.1), which was the starting point to the exposition in this section, for the polarization of the transverse waves that participate in the interaction. Namely, if the electromagnetic waves are linearly polarized [see (10.6) and also (10.3) and (10.4) and the explanations to them in § 10], the generalized kinetic equation for the spectral density of their energy is determined in contrast to (11.1), by the equation

$$\frac{\partial}{\partial t}(E_{tr}^2)_{\omega,\mathbf{k}}\frac{1}{\omega}\frac{\partial}{\partial\omega}\left\{\omega\left[\varepsilon^{tr'}(\omega,\mathbf{k})-\frac{c^2k^2}{\omega^2}\right]\right\}-\frac{\partial}{\partial\mathbf{r}}(E_{tr}^2)_{\omega,\mathbf{k}}\frac{1}{\omega}\frac{\partial}{\partial\mathbf{k}}\left\{\omega\left[\varepsilon^{tr'}(\omega,\mathbf{k})-\frac{c^2k^2}{\omega^2}\right]\right\}=$$
$$=-2(E_{tr}^2)_{\omega,\mathbf{k}}\int d\omega'd\mathbf{k}'(E_{tr}^2)_{\omega',\mathbf{k}'}\operatorname{Im}V_{iajb}(\omega,\mathbf{k};\omega',\mathbf{k}')e_ie_je'_ae'_b+$$
$$+2(E_{tr}^2)_{\omega,\mathbf{k}}\operatorname{Im}\int d\omega'd\mathbf{k}'d\omega''d\mathbf{k}''\delta(\omega-\omega'-\omega'')\delta(\mathbf{k}-\mathbf{k}'-\mathbf{k}'')\times$$
$$\times\left\{\frac{\varkappa''_j\varkappa''_b}{\varepsilon^l(\omega'',\mathbf{k}'')}+\frac{\delta_{jb}-\varkappa''_j\varkappa''_b}{\varepsilon^{tr}(\omega'',\mathbf{k}'')-\frac{c^2k''^2}{\omega''^2}}\right\}(E_{tr}^2)_{\omega',\mathbf{k}'}S_{ijs}(\omega,\mathbf{k};\omega',\mathbf{k}')\times$$
$$\times S_{bca}(\omega'',\mathbf{k}'';\omega,\mathbf{k})e_ie_ae'_se'_c+\pi\operatorname{sign}\varepsilon^{tr''}(\omega,\mathbf{k})\delta\left[\varepsilon^{tr'}(\omega,\mathbf{k})-\frac{c^2k^2}{\omega^2}\right]\times$$
$$\times\int d\omega'd\mathbf{k}'d\omega''d\mathbf{k}''\delta(\omega-\omega'-\omega'')\delta(\mathbf{k}-\mathbf{k}'-\mathbf{k}'')(E_{tr}^2)_{\omega',\mathbf{k}'}(E_{tr}^2)_{\omega'',\mathbf{k}''}\times$$
$$\times S_{ijs}(\omega,\mathbf{k};\omega',\mathbf{k}')S^*_{abc}(\omega,\mathbf{k};\omega',\mathbf{k}')(\delta_{ia}-\varkappa_i\varkappa_a)e'_se'_ce''_je''_b. \qquad (11.22)$$

After summation over all possible linear polarizations (6.71) of the two transverse electromagnetic waves that participate in the nonlinear interaction, Eq. (11.22) goes over in accordance with (6.72) into the equation (11.1) used above. From this equation we can also obtain relations that describe the nonlinear interaction of naturally and linearly polarized transverse oscillations if we carry out selective summation over the polarizations. Namely, the equation of evolution of naturally polarized electromagnetic waves due to their nonlinear interaction with linearly polarized waves with frequency ω', wave vector $\mathbf{k}'$, and polarization vector $\mathbf{e}'$ takes the following form after the integration of (6.71)-(6.72) in (11.22):

$$\frac{\partial}{\partial t}(E_{tr}^2)_{\omega,\mathbf{k}}\frac{1}{\omega}\frac{\partial}{\partial\omega}\left\{\omega\left[\varepsilon^{tr'}(\omega,\mathbf{k})-\frac{c^2k^2}{\omega^2}\right]\right\}-\frac{\partial}{\partial\mathbf{r}}(E_{tr}^2)_{\omega,\mathbf{k}}\frac{1}{\omega}\frac{\partial}{\partial\mathbf{k}}\left\{\omega\left[\varepsilon^{tr'}(\omega,\mathbf{k})-\frac{c^2k^2}{\omega^2}\right]\right\}=$$
$$=-(E_{tr}^2)_{\omega,\mathbf{k}}\int d\omega'd\mathbf{k}'(E_{tr}^2)_{\omega',\mathbf{k}'}\operatorname{Im}V_{iajb}(\omega,\mathbf{k};\omega',\mathbf{k}')\times$$
$$\times(\delta_{ij}-\varkappa_i\varkappa_j)e'_ae'_b+(E_{tr}^2)_{\omega,\mathbf{k}}\operatorname{Im}\int d\omega'd\mathbf{k}'d\omega''d\mathbf{k}''\delta(\omega-\omega'-\omega'')\delta(\mathbf{k}-\mathbf{k}'-\mathbf{k}'')\times$$
$$\times\left\{\frac{\varkappa''_j\varkappa''_b}{\varepsilon^l(\omega'',\mathbf{k}'')}+\frac{\delta_{jb}-\varkappa''_j\varkappa''_b}{\varepsilon^{tr}(\omega'',\mathbf{k}'')-\frac{c^2k''^2}{\omega''^2}}\right\}(E_{tr}^2)_{\omega',\mathbf{k}'}S_{ijs}(\omega,\mathbf{k};\omega',\mathbf{k}')S_{bca}(\omega'',\mathbf{k}'';\omega,\mathbf{k})\times$$
$$\times(\delta_{ia}-\varkappa_i\varkappa_a)e'_se'_c+\pi\operatorname{sign}\varepsilon^{tr''}(\omega,\mathbf{k})\delta\left[\varepsilon^{tr'}(\omega,\mathbf{k})-\frac{c^2k^2}{\omega^2}\right]\times$$
$$\times\int d\omega'd\mathbf{k}'d\omega''d\mathbf{k}''\delta(\omega-\omega'-\omega'')\delta(\mathbf{k}-\mathbf{k}'-\mathbf{k}'')(E_{tr}^2)_{\omega',\mathbf{k}'}(E_{tr}^2)_{\omega'',\mathbf{k}''}\times$$
$$\times S_{ijs}(\omega,\mathbf{k};\omega',\mathbf{k}')S^*_{abc}(\omega,\mathbf{k};\omega',\mathbf{k}')(\delta_{ia}-\varkappa_i\varkappa_a)e'_se'_ce''_je''_b. \qquad (11.23)$$

The additional equation for a linearly polarized wave interacting with a naturally polarized wave arises from (11.22) similarly:

$$\frac{\partial}{\partial t}(E_{tr}^2)_{\omega,\mathbf{k}}\frac{1}{\omega}\frac{\partial}{\partial\omega}\left\{\omega\left[\varepsilon^{tr'}(\omega,\mathbf{k})-\frac{c^2k^2}{\omega^2}\right]\right\}-\frac{\partial}{\partial\mathbf{r}}(E_{tr}^2)_{\omega,\mathbf{k}}\frac{1}{\omega}\frac{\partial}{\partial\mathbf{k}}\left\{\omega\left[\varepsilon^{tr'}(\omega,\mathbf{k})-\frac{c^2k^2}{\omega^2}\right]\right\}=$$
$$=-(E_{tr}^2)_{\omega,\mathbf{k}}\int d\omega'd\mathbf{k}'(E_{tr}^2)_{\omega',\mathbf{k}'}\operatorname{Im}V_{iajb}(\omega,\mathbf{k};\omega',\mathbf{k}')e_ie_j(\delta_{ab}-\varkappa'_a\varkappa'_b)+$$
$$+(E_{tr}^2)_{\omega,\mathbf{k}}\operatorname{Im}\int d\omega'd\mathbf{k}'d\omega''d\mathbf{k}''\delta(\omega-\omega'-\omega'')\delta(\mathbf{k}-\mathbf{k}'-\mathbf{k}'')\times$$
$$\times\left\{\frac{\varkappa''_j\varkappa''_b}{\varepsilon^l(\omega'',\mathbf{k}'')}+\frac{\delta_{jb}-\varkappa''_j\varkappa''_b}{\varepsilon^{tr}(\omega'',\mathbf{k}'')-\frac{c^2k''^2}{\omega''^2}}\right\}(E_{tr}^2)_{\omega',\mathbf{k}'}S_{ijs}(\omega,\mathbf{k};\omega'\,\mathbf{k}')S_{bca}(\omega'',\mathbf{k}'';\omega,\mathbf{k})\times$$

$$\times e_i e_a (\delta_{sc} - \varkappa_s' \varkappa_c') + \frac{\pi}{2} \operatorname{sign} \varepsilon^{tr''}(\omega, \mathbf{k}) \, \delta\left[\varepsilon^{tr'}(\omega, \mathbf{k}) - \frac{c^2 k^2}{\omega^2}\right] \int d\omega' d\mathbf{k}' d\omega'' d\mathbf{k}'' \times$$
$$\times \delta(\omega - \omega' - \omega'') \, \delta(\mathbf{k} - \mathbf{k}' - \mathbf{k}'') (E_{tr}^2)_{\omega', \mathbf{k}'} (E_{tr}^2)_{\omega'', \mathbf{k}''} S_{ijs}(\omega, \mathbf{k}; \omega', \mathbf{k}') \times$$
$$\times S_{abe}^*(\omega, \mathbf{k}; \omega', \mathbf{k}') (\delta_{ia} - \varkappa_i \varkappa_a)(\delta_{cs} - \varkappa_c' \varkappa_s') e_j'' e_b''. \qquad (11.24)$$

The generalized kinetic equations (11.22)-(11.24) for polarized waves will be used in the next section to calculate the cross sections for the scattering of waves on particles.

§ 12. Effective Cross Sections of Various Scattering Processes in an Isotropic Plasma

The equations of the nonlinear interaction of the characteristic oscillations of a plasma are essentially kinetic equations for the numbers N_k of oscillations associated with the spectral energy density by the definition

$$N_{\mathbf{k}} = \frac{W(\mathbf{k})}{\hbar\omega(\mathbf{k})}. \qquad (12.1)$$

Since the probabilities of the corresponding processes described by the kinetic equations are uniquely determined by their cross sections, the kernels of the integrodifferential equations obtained in the foregoing sections can be used not only to estimate the characteristic times but also to obtain the scattering cross sections [33].

By scattering cross sections we understand the ratio of the energy of a scattered oscillations (wave vector **k** and frequency ω) to the density N_e of the electrons and the density of the energy flux in the incident oscillation (wave vector **k'** and frequency ω'). In accordance with the foregoing, scattering is understood as a process in which the incident and the scattered oscillation may be of the same type (for example, scattering of transverse oscillations into transverse oscillations), as well as a process in which these oscillations are different. In the latter case we sometimes speak of transformation of oscillations.

The equation for the change in the number N_k of oscillations (number of quanta) with wave vector **k** due to scattering can be represented in the form [31]

$$\frac{dN_{\mathbf{k}}}{dt} = 2\pi\hbar \sum_{\mathbf{k}', \mathbf{p}} |V(\mathbf{p}, \mathbf{k}, \mathbf{k}')|^2 \{[(N_{\mathbf{k}} + 1) N_{\mathbf{k}'} f(\mathbf{p} + \hbar\mathbf{q})(1 - f(\mathbf{p})) -$$
$$- N_{\mathbf{k}}(N_{\mathbf{k}'} + 1) f(\mathbf{p})(1 - f(\mathbf{p} + \hbar\mathbf{q}))] \, \delta[\varepsilon(\mathbf{p} + \hbar\mathbf{q}) - \varepsilon(\mathbf{p}) - \hbar\omega(\mathbf{k}) + \hbar\omega'(\mathbf{k}')] +$$
$$+ [(N_{\mathbf{k}} + 1)(N_{\mathbf{k}'} + 1) f(\mathbf{p} + \hbar\mathbf{q}')(1 - f(\mathbf{p})) - N_{\mathbf{k}} N_{\mathbf{k}'} f(\mathbf{p})(1 - f(\mathbf{p} + \hbar\mathbf{q}'))] \times$$
$$\times \delta[\varepsilon(\mathbf{p} + \hbar\mathbf{q}') - \varepsilon(\mathbf{p}) - \hbar\omega(\mathbf{k}) - \hbar\omega'(\mathbf{k}')]\}. \qquad (12.2)$$

Here $\varepsilon(\mathbf{p})$ and $\hbar\omega(\mathbf{k})$ are, respectively, the energy of a particle and the quantum, V (**p**, **k**, **k'**) is the matrix element corresponding to absorption of the quantum $\hbar\omega(\mathbf{k})$ by a particle with momentum **p** and the emission of a quantum $\hbar\omega'(\mathbf{k}')$; the sum and the difference of wave vectors are denoted differently in this section from all the remainder of the exposition, namely:

$$\mathbf{q} \equiv \mathbf{k} - \mathbf{k}', \qquad \mathbf{q}' \equiv \mathbf{k} + \mathbf{k}'. \qquad (12.3)$$

In the classical limit ($\hbar = 0$, $f \ll 1$), Eq. (12.2) goes over into a relation for the spectral energy density [31]:

$$\frac{dW(\mathbf{k})}{dt} = 2\pi \int d\mathbf{k}' d\mathbf{p} \, |V(\mathbf{p}, \mathbf{k}, \mathbf{k}')|^2 \left\{\left[\frac{1}{\omega'} W(\mathbf{k}) W(\mathbf{k}') \mathbf{q} \frac{\partial f}{\partial \mathbf{p}} + \left(W(\mathbf{k}') \frac{\omega}{\omega'} - W(\mathbf{k})\right) f\right] \delta(\mathbf{q}\mathbf{v} - \omega + \omega') +\right.$$
$$\left. + \left[W(\mathbf{k}) W(\mathbf{k}') \frac{1}{\omega'} \mathbf{q}' \frac{\partial f}{\partial \mathbf{p}} + \left(W(\mathbf{k}) + \frac{\omega}{\omega'} W(\mathbf{k}')\right) f\right] \delta(\mathbf{q}'\mathbf{v} - \omega' - \omega)\right\}, \qquad (12.4)$$

in which the frequencies ω and ω' stand for the spectra $\omega(\mathbf{k}) \equiv \omega$ and $\omega'(\mathbf{k}') \equiv \omega'$ in order to shorten the equation. In this equation (on its right-hand side) the terms that are nonlinear in the energy density describe the induced scattering of waves on particles, which we have considered above by means of the generalized kinetic equation (3.22) for waves in a number of particular cases. However, the cross section can also be determined by means of another term in Eq. (12.4):

$$\delta\left\{\frac{dW(\mathbf{k})}{dt}\right\} = \int d\mathbf{k}' W(\mathbf{k}') R(\mathbf{k}, \mathbf{k}'); \tag{12.5}$$

$$R(\mathbf{k}, \mathbf{k}') = 2\pi \frac{\omega}{\omega'} \int d\mathbf{p} \,|V(\mathbf{p}, \mathbf{k}, \mathbf{k}')|^2 f \{\delta(\mathbf{qv} - \omega + \omega') + \delta(\mathbf{q}'\mathbf{v} - \omega - \omega')\}. \tag{12.6}$$

Equation (12.5) describes the change in the energy density of waves with wave vector **k** due to the transfer of energy from waves with wave vector **k'**. If there is a single monochromatic wave with wave vector **k'**,

$$W(\mathbf{k}') = W' \delta(\mathbf{k}' - \mathbf{k}_1), \tag{12.7}$$

then in Eq. (12.5) the integration is eliminated:

$$\frac{dW(\mathbf{k})}{dt} d\mathbf{k} = W' d\mathbf{k} R(\mathbf{k}, \mathbf{k}'). \tag{12.8}$$

Here we have introduced a volume element dk in the wave-vector space in order to consider scattering with the formation of oscillations of the plasma in a small range of angles and frequencies. Using the quantities we have introduced, we can represent the scattering cross section by the equation

$$d\sigma(\mathbf{k}', \mathbf{k}) = \frac{R(\mathbf{k}, \mathbf{k}')}{N_e\,|\partial\omega'/\partial\mathbf{k}'|} d\mathbf{k}. \tag{12.9}$$

However, the various particular forms of the generalized kinetic equation for waves obtained above (§ 9-11) correspond, as we have already noted, to the nonlinear term in Eq. (12.4):

$$\delta\left\{\frac{dW(\mathbf{k})}{dt}\right\} = \int d\mathbf{k}' W(\mathbf{k}) W(\mathbf{k}') Q(\mathbf{k}, \mathbf{k}'); \tag{12.10}$$

$$Q(\mathbf{k}, \mathbf{k}') = \frac{2\pi}{\omega'} \int d\mathbf{p}\,|V(\mathbf{p}, \mathbf{k}, \mathbf{k}')|^2 \left\{\mathbf{q}\frac{\partial f}{\partial \mathbf{p}} \delta(\mathbf{qv} - \omega + \omega') + \mathbf{q}'\frac{\partial f}{\partial \mathbf{p}} \delta(\mathbf{q}'\mathbf{v} - \omega - \omega')\right\}. \tag{12.11}$$

Therefore, from the nonlinear equations we can find the kernel Q(k, k') and, then, replacing $\mathbf{q}(\partial f/\partial \mathbf{p})$ and $\mathbf{q}'(\partial f/\partial \mathbf{p})$ by ωf, we can find R(k, k'), and thus, the cross section in accordance with the definition (12.9).

In the case of a Maxwell distribution of the plasma particles (6.14), the expression for the cross section can be written down immediately in terms of the kernels of the nonlinear integrodifferential equations for the energy density of the waves:

$$d\sigma(\mathbf{k}', \mathbf{k}) = \frac{d\mathbf{k}}{N_e\,|\partial\omega'/\partial\mathbf{k}'|} T\omega \left\{\frac{Q_1(\mathbf{k}, \mathbf{k}')}{\omega' - \omega} - \frac{Q_2(\mathbf{k}, \mathbf{k}')}{\omega + \omega'}\right\}, \tag{12.12}$$

where Q_1 and Q_2 are, respectively, the first and the second term in (12.11). Using this formula, we derive expressions for the cross sections for a number of processes considered in the earlier sections of this chapter.

Let us consider the scattering of a Langmuir oscillation into a Langmuir oscillation when scattering on ions is predominant. The actual expression for the cross sections is obtained by means of Eq. (9.28) for the evolution of the energy density $W_l(\mathbf{k})$ of the Langmuir oscillations when they undergo Coulomb scattering [33]:

$$d\sigma^{ll}(\mathbf{k}'\ \mathbf{k}) = \left(\frac{e^2}{mv_{T_e}^2}\right)^2 \frac{d\mathbf{k}}{3\sqrt{2\pi}} \frac{T_i}{T_e} \frac{v_{T_e}}{v_{T_i}} \frac{r_{D_e}}{k'\,|\mathbf{k}-\mathbf{k}'|} \left(\frac{\mathbf{kk}'}{kk'}\right)^2 \left\{\frac{r_{D_i}/r_{D_e}}{F+(r_{D_i}/r_{D_e})^2}\right\}^2 \exp\left\{-\frac{9}{8}\frac{v_{T_e}^2}{v_{T_i}^2} r_{D_e}^2 \frac{k^2-k'^2}{(\mathbf{k}-\mathbf{k}')^2}\right\}. \qquad (12.13)$$

The cross section for the scattering of Langmuir waves (with wavelength $\lambda < \lambda_0 = c/\omega_{Le}$) on electrons can be obtained from Eq. (9.33):

$$d\sigma^{ll}(\mathbf{k}', \mathbf{k}) = \left(\frac{e^2}{mv_{T_e}^2}\right)^2 d\mathbf{k} \frac{4}{3\sqrt{2\pi}} r_{D_e}^3 \left(\frac{\mathbf{kk}'}{kk'}\right)^2 \frac{[\mathbf{kk}']^2}{k'\,|\mathbf{k}-\mathbf{k}'|^3}. \qquad (12.14)$$

It follows from this formula that the cross section vanishes for scattering through a right angle and along the direction of the incident oscillation. However, it is nonvanishing in the following approximation in the ratio of the wavelength to the electron Debye radius, as can be readily seen by means of the term (9.42), which augments Eq. (9.33) in this case. Therefore, we can speak only of an abrupt decrease of the scattering cross section in these directions.

The cross sections (12.13)-(12.14) describe the scattering of Langmuir waves on longitudinal (electrostatic) fluctuations of the plasma. For waves with wavelength greater than λ_0 it becomes important to allow for a transverse field in the interaction of the particles, i.e., scattering on transverse, nonelectrostatic fluctuations of the plasma becomes important.* The corresponding cross section arises from Eq. (9.56), which describes the change in the energy density of the Langmuir oscillations during the process of induced scattering through a transverse virtual oscillation:

$$d\sigma^{ll}(\mathbf{k}', \mathbf{k}) = \left(\frac{e^2}{mv_{T_e}^2}\right)^2 d\mathbf{k} \frac{4}{27\sqrt{2\pi}} \frac{[\mathbf{kk}']^2}{(kk')^2} \frac{1}{k' r_{D_e}} \left\{\frac{4}{9}\frac{c^4(\mathbf{k}-\mathbf{k}')^4}{v_{T_e}^4 (k^2-k'^2)^2} + 2\frac{c^2}{v_{T_e}^2} + \frac{\pi}{2}\frac{1}{(\mathbf{k}-\mathbf{k}')^2 r_{D_e}^2}\right\}^{-1}. \qquad (12.15)$$

Thus, depending on the angle between the incident and the scattered wave and also the parameters that characterize the plasma, the cross section for the scattering of Langmuir oscillations into Langmuir oscillations with wavelength greater than λ_0 is determined by either the expression (12.15) or (12.13). For example, for waves with $k \sim k' \sim \Delta k = |k'-k|$ and at not too high temperatures of the ions, $v_{T_i} c^2 \ll v_{T_e}^3$, the cross section of scattering on transverse fluctuations of the plasma is largest if the wave numbers of the Langmuir oscillations are greater than $r_{De}^{-1}(v_{Ti}/v_{Te})$.

To conclude our illustrations of the scattering of Langmuir oscillations by means of Eq. (9.51) for the process of the coalescence of two Langmuir waves into one transverse wave, we give the cross section of combination scattering [33]:

$$d\sigma^{ll}(\mathbf{k}', \mathbf{k}) = \frac{2}{3}\left(\frac{e^2}{mv_{T_e}^2}\right)^2 \delta\left(3 - \frac{c^2k^2}{\omega_{L_e}^2}\right) \frac{r_{D_e}^4}{k'} d\mathbf{k} \frac{W_l(\mathbf{k}-\mathbf{k}')}{\varkappa T_e} \frac{[\mathbf{kk}']^2}{(kk')^2} \frac{(k^2-2\mathbf{kk}')^2}{(\mathbf{k}-\mathbf{k}')^2}. \qquad (12.16)$$

Here $W_l(\mathbf{k}-\mathbf{k}')$ is the energy density of electrostatic fluctuations of the plasma in its transparency region, i.e., the spectral density of the energy of the Langmuir oscillations. Essentially, combination scattering (in the literature it is sometimes called coherent scattering to distinguish it from scattering on particles, which is incoherent) can be interpreted qualitatively as scattering not on individual particles but on a group of particles in the plasma that execute a collective motion with frequency and wave number of the corresponding characteristic oscillation of the plasma. If the scattering occurs on thermal, equilibrium fluctuations, then

* The influence of transverse fluctuations of a plasma on the scattering of transverse waves in a plasma with a beam has been studied in [38] (see also [39] and the last section of this chapter.

$W_l(\mathbf{k}-\mathbf{k}') = \varkappa T_e$, and the cross section (12.16) can be found by means of the results of [40]. Then from (12.16) we can also find the total cross section by integrating over the angles [31]:

$$\sigma^{ll}(k') = \frac{2\pi}{9}\left(\frac{e^2}{mv_{T_e}^2}\right)^2 r_{D_e}^4 \frac{k}{k'} k^2 \left\{\frac{4}{3}(k^2 - k'^2) + \frac{k'^2}{2k^2}(k^2 + k'^2) - \frac{(k^2 - k'^2)^2}{4k^3} k' \ln\left|\frac{k+k'}{k-k'}\right|\right\}, \quad k \equiv \sqrt{3}\frac{\omega_{L_e}}{c}. \qquad (12.17)$$

If the wave number k' of the scattered oscillation is appreciably greater than the wave number k of the Langmuir oscillation that arises in the scattering process, formula (12.17) simplifies appreciably

$$\sigma^{ll}(k') = \frac{32\pi}{25}\frac{k}{k'} r_0^2, \quad k' \gg k, \qquad (12.18)$$

where $r_0 = (e^2/mc^2)$ is the classical electron radius. In the opposite limiting case ($k \gg k'$), the cross section (12.17),

$$\sigma^{ll}(k') = \frac{8\pi}{3}\frac{k}{k'} r_0^2, \quad k \gg k' \qquad (12.19)$$

increases with increasing wavelength of the incident (scattered) Langmuir oscillation ($k' \to 0$).

Let us now consider the scattering of Langmuir oscillations with the formation of transverse linearly polarized electromagnetic waves with wavelength greater than λ_0 (in agreement with the special cases of the nonlinear interaction of longitudinal and transverse oscillations that we studied in § 10). Instead of the generalized kinetic equation (10.1) we shall use Eq. (10.5), which takes into account explicitly the linear polarization of the transverse waves.

Suppose that an electron Langmuir oscillation is scattered on longitudinal fluctuations (on ions) of the plasma into a long-wavelength ($\lambda \gg \lambda_0$) linearly polarized transverse oscillation with polarization vector **e**. Then the cross section of this process can be represented in the form

$$d\sigma^{lt}(\mathbf{k}', \mathbf{k}) = \frac{1}{3\sqrt{2\pi}}\left(\frac{e^2}{mv_{T_e}^2}\right)^2 d\mathbf{k}\, \frac{v_{T_e}}{v_{T_i}}\frac{T_i}{T_e}\frac{r_{D_e}}{k'\,|\mathbf{k}-\mathbf{k}'|}\, \frac{(\mathbf{k}'\mathbf{e})^2}{k'^2}\left[\frac{r_{D_i}/r_{D_e}}{F + (r_{D_i}/r_{D_e})^2}\right]^2 \exp\left\{-\frac{v_{T_e}^2 (3k'^2 r_{D_e}^2 - \lambda_0^2 k^2)^2}{8 v_{T_i}^2 r_{D_e}^2 (\mathbf{k}-\mathbf{k}')^2}\right\}, \qquad (12.20)$$

where the function F is, as usual, determined by Eqs. (9.29). Summing over all possible linear polarizations of the transverse electromagnetic waves, we obtain the cross section of scattering on ions of a Langmuir oscillation with the formation of a naturally polarized transverse oscillation:

$$d\sigma^{lt}(\mathbf{k}', \mathbf{k}) = \frac{1}{3\sqrt{2\pi}}\left(\frac{e^2}{mv_{T_e}^2}\right)^2 d\mathbf{k}\, \frac{v_{T_e}}{v_{T_i}}\frac{T_i}{T_e} r_{D_e} \frac{[\mathbf{k}\mathbf{k}']^2}{(kk')^2}\frac{1}{k'\,|\mathbf{k}-\mathbf{k}'|} \exp\left\{-\frac{v_{T_e}^2}{8v_{T_i}^2}\frac{(3k'^2 r_{De}^2 - \lambda_0^2 k^2)^2}{r_{D_e}^2 (\mathbf{k}-\mathbf{k}')^2}\right\}\left[\frac{r_{D_i}/r_{D_e}}{F + (r_{D_i}/r_{D_e})^2}\right]^2, \qquad (12.21)$$

which is identical to within a factor and the transposition $\mathbf{k} \rightleftharpoons \mathbf{k}'$ with the kernel of Eq. (10.23), which determines the Coulomb scattering of these oscillations on ions. The scattering of Langmuir oscillations on longitudinal fluctuations of the electron plasma is characterized by the cross section

$$d\sigma^{lt}(\mathbf{k}', \mathbf{k}) = \left(\frac{e^2}{mv_{T_e}^2}\right)^2 \frac{1}{3\sqrt{2\pi}} d\mathbf{k}\, \frac{r_{D_e}^3}{k'^3\,|\mathbf{k}-\mathbf{k}'|^3}\{(\mathbf{k}'\mathbf{e})^2[\mathbf{k}\mathbf{k}']^2 + 2(\mathbf{k}\mathbf{k}')(\mathbf{k}, \mathbf{k}-\mathbf{k}')(\mathbf{k}'\mathbf{e})^2 + (\mathbf{k}\mathbf{k}')^2[\mathbf{k}-\mathbf{k}', \mathbf{e}]^2\}, \qquad (12.22)$$

which goes over into the cross section of scattering into a naturally polarized transverse oscillation,

$$d\sigma^{lt}(\mathbf{k}', \mathbf{k}) = \frac{1}{3\sqrt{2\pi}}\left(\frac{e^2}{mv_{T_e}^2}\right)^2 d\mathbf{k}\, \frac{r_{D_e}^3}{k'\,|\mathbf{k}-\mathbf{k}'|^3}\left\{\frac{[\mathbf{k}\mathbf{k}']^2}{(kk')^2}(k^2k'^2 - 4(\mathbf{k}\mathbf{k}')^2 + 2k^2(\mathbf{k}\mathbf{k}')) + 2(\mathbf{k}\mathbf{k}')^2\frac{1}{k'^2}(\mathbf{k}-\mathbf{k}')^2\right\}, \qquad (12.23)$$

after summation over all possible linear polarizations of the scattered wave. The expression in the curly brackets on the right-hand side of (12.23) is identical with the corresponding part of the kernel of the system of the two integrodifferential equations (10.24) and (10.25) for the variation of the energy density of Langmuir and transverse oscillations in the case of Coulomb scattering on electrons.

Allowance for scattering on solenoidal, nonelectrostatic fluctuations of the plasma (i.e., on fluctuations formed by the interaction between particles with allowance for retardation) leads for this process to the cross section

$$d\sigma^{lt}(\mathbf{k}', \mathbf{k}) = \frac{1}{3\sqrt{2\pi}}\left(\frac{e}{mv_{T_e}^2}\right)^2 d\mathbf{k}\, \frac{r_{D_e}^3}{k'^3|\mathbf{k}-\mathbf{k}'|^3}\left\{\lambda_0^4(\mathbf{k}-\mathbf{k}')^4 + \frac{1}{2}\left(\frac{c}{v_{T_e}}\right)^2\left[1+\frac{\pi}{4}\frac{1}{\lambda_0^2(\mathbf{k}-\mathbf{k}')^2}\right][3k'^2r_{D_e}^2 - \lambda_0^2k^2]^2\right\}^{-1} \times$$
$$\times\{[\mathbf{k}\mathbf{k}']^2(\mathbf{k}'\mathbf{e})^2 + 2(\mathbf{k}', \mathbf{k}-\mathbf{k}')(\mathbf{k}, \mathbf{k}-\mathbf{k}')(\mathbf{k}'\mathbf{e})^2 + (\mathbf{k}, \mathbf{k}-\mathbf{k}')^2[\mathbf{k}-\mathbf{k}', \mathbf{e}]^2\}. \tag{12.24}$$

From this we readily obtain the cross section of scattering with the formation of an unpolarized oscillation [cf. (10.28)-(10.29)]:

$$d\sigma^{lt}(\mathbf{k}', \mathbf{k}) = \left(\frac{e^2}{mv_{T_e}^2}\right)^2 \frac{1}{3\sqrt{2\pi}}\, d\mathbf{k}\, \frac{r_{D_e}^3}{k'|\mathbf{k}-\mathbf{k}'|}\left[\frac{(\mathbf{k}, \mathbf{k}-\mathbf{k}')^2}{k^2} +\right.$$
$$\left. + (\mathbf{k}-\mathbf{k}')^2\right]\left\{\lambda_0^4(\mathbf{k}-\mathbf{k}')^4 + \frac{1}{2}\frac{c^2}{v_{T_e}^2}(3k'^2r_{D_e}^2 - \lambda_0^2k^2)^2\left[1+\frac{\pi}{4}\frac{1}{\lambda_0^2(\mathbf{k}-\mathbf{k}')^2}\right]\right\}^{-1}. \tag{12.25}$$

The cross sections of the inverse processes (scattering of a long-wavelength transverse oscillation with the formation of an electron Langmuir oscillation) can also be obtained by means of the kernels of the generalized kinetic equations for the waves. On account of the symmetry of the process, they can be obtained from the cross sections given above by transposing **k** and **k'** (in accordance with the principle of detailed balance).

Finally, we turn to the scattering of linearly polarized transverse electromagnetic waves. Namely, suppose that a transverse oscillation with unit polarization vector **e'** is scattered into a similar oscillation polarized along **e**. Then, using the generalized kinetic equation (11.22), which takes into account explicitly the linear polarization of the interacting transverse waves and also the definition (12.12) in the case of the scattering of long-wavelength ($\lambda \gg \lambda_0$) oscillations on longitudinal fluctuations of the plasma, we find the cross section

$$d\sigma^{tt}(\mathbf{k}', \mathbf{k}) = \left(\frac{e^2}{mc^2}\right)^2 d\mathbf{k}\, \frac{r_{D_e}}{\sqrt{2\pi}}\frac{c^2}{v_{T_e}v_{T_i}}\frac{T_i}{T_e}(\mathbf{e}\mathbf{e}')^2\frac{1}{k'|\mathbf{k}-\mathbf{k}'|} \times \exp\left\{-\frac{c^4}{8v_{T_i}^2\omega_{Le}^2}\frac{(k^2-k'^2)^2}{(\mathbf{k}-\mathbf{k}'^2)}\right\}\left[\frac{r_{D_i}/r_{D_e}}{F+(r_{D_i}/r_{D_e})^2}\right]^2, \tag{12.26}$$

which, after summation over all possible linear polarizations of the scattered wave, takes the form

$$d\sigma^{tt}(\mathbf{k}', \mathbf{k}) = \left(\frac{e^2}{mc^2}\right)^2 d\mathbf{k}\, \frac{r_{D_e}}{\sqrt{2\pi}}\frac{c^2}{v_{T_i}v_{T_e}}\frac{T_i}{T_e}\left[1-\left(\frac{\mathbf{k}\mathbf{e}'}{k}\right)^2\right]\frac{1}{k'|\mathbf{k}-\mathbf{k}'|}\left[\frac{r_{D_i}/r_{D_e}}{F+(r_{D_i}/r_{D_e})^2}\right]^2\exp\left\{-\frac{c^4}{8v_{T_i}^2\omega_{L_e}^2}\frac{(k^2-k'^2)^2}{(\mathbf{k}-\mathbf{k}')^2}\right\}. \tag{12.27}$$

But if not only the scattered but also the incident wave is naturally polarized, the cross section is given by

$$d\sigma^{tt}(\mathbf{k}', \mathbf{k}) = \left(\frac{e^2}{mc^2}\right)^2 d\mathbf{k}\, \frac{r_{D_e}}{2\sqrt{2\pi}}\frac{c^2}{v_{T_e}v_{T_i}}\frac{T_i}{T_e}\left[1+\left(\frac{\mathbf{k}\mathbf{k}'}{kk'}\right)^2\right]\frac{1}{k'|\mathbf{k}-\mathbf{k}'|}\left[\frac{r_{D_i}/r_{D_e}}{F+(r_{D_i}/r_{D_e})^2}\right]^2\exp\left\{-\frac{c^4}{8v_{T_i}^2\omega_{L_e}^2}\frac{(k^2-k'^2)^2}{(\mathbf{k}-\mathbf{k}')^2}\right\}, \tag{12.28}$$

which can be obtained by means of the kernel of the integrodifferential equation (11.11), which characterizes the evolution of the spectral density $W_{tr}(\mathbf{k})$ of naturally polarized electromagnetic waves in the case of induced scattering on ions through an electrostatic virtual oscillation.

We emphasize that the conditions of the calculation (and for the validity) of formulas (12.26)-(12.28) for the cross sections are the same as for Eq. (11.11). Note also that the cross section (12.37) can be obtained from the generalized kinetic equations (11.23)-(11.24), which describes the nonlinear interaction of linearly and naturally polarized transverse oscillations.

In an electron plasma, the cross section for the scattering of linearly polarized transverse waves is determined by

$$d\sigma^{tt}(\mathbf{k}', \mathbf{k}) = \left(\frac{e^2}{mc^2}\right)^2 d\mathbf{k} \frac{r_{De}^3}{\sqrt{2\pi}} \left(\frac{c}{v_{T_e}}\right)^3 \left\{1 + 4\frac{v_{T_e}^5}{c^5} \frac{1}{r_{D_e}^4 (k^2 - k'^2)^2} \left[4\frac{(\mathbf{k}-\mathbf{k}')^4}{(k^2-k'^2)^2} + \right.\right.$$
$$\left.\left. + \frac{\pi}{2} \frac{c^2}{v_{T_e}^2} \frac{1}{\lambda_0^2 (\mathbf{k}-\mathbf{k}')^2}\right]^{-1}\right\} \frac{1}{k' \,|\mathbf{k}-\mathbf{k}'|^3} \{(\mathbf{k}\mathbf{e}')^2 [\mathbf{k}-\mathbf{k}', \mathbf{e}]^2 +$$
$$+ 2(\mathbf{k}\mathbf{e}')(\mathbf{k}'\mathbf{e})[(\mathbf{k}-\mathbf{k}')^2(\mathbf{e}\mathbf{e}') + (\mathbf{k}'\mathbf{e})(\mathbf{k}\mathbf{e}')] + (\mathbf{k}'\mathbf{e})^2[\mathbf{k}-\mathbf{k}', \mathbf{e}']^2\}. \qquad (12.29)$$

Here, the second term in the first curly brackets corresponds to scattering on transverse fluctuations and at certain wavelengths can be greater than the unity, which represents the contribution of scattering on longitudinal fluctuations.

Averaging over the initial polarization and summing over the final polarization of the oscillations, we obtain from (12.29) the cross section for the scattering of naturally polarized long-wavelength transverse oscillations [cf. (11.13) and (11.15)]:

$$d\sigma^{tt}(\mathbf{k}', \mathbf{k}) = \left(\frac{e^2}{mc^2}\right)^2 d\mathbf{k} \frac{r_{D_e}^3}{\sqrt{2\pi}} \frac{c^3}{v_{T_e}^3} \frac{[\mathbf{k}\mathbf{k}']^2}{(kk')^2} \frac{k^2 + k'^2 - \mathbf{k}\mathbf{k}'}{k' \,|\mathbf{k}-\mathbf{k}'|} \times$$
$$\times \left\{1 + 4\frac{v_{T_e}^5}{c^5} \frac{1}{r_{D_e}^4 (k^2-k'^2)^2} \left[4\frac{(\mathbf{k}-\mathbf{k}')^4}{(k^2-k'^2)^2} + \frac{\pi}{2}\frac{c^2}{v_{T_e}^2} \frac{1}{\lambda_0^2(\mathbf{k}-\mathbf{k}')^2}\right]^{-1}\right\}, \qquad (12.30)$$

which corresponds to induced scattering of transverse waves on electrons. The relative magnitude of the cross sections (12.26) and (12.29) is determined in the same way as the values of the kernels in the corresponding equations of the nonlinear interaction.

To conclude this section, we give the cross section of combination scattering corresponding to the coalescence of a Langmuir oscillation with a long-wavelength transverse oscillation to form a linearly polarized transverse oscillation:

$$d\sigma^{lt}(\mathbf{k}', \mathbf{k}) = \frac{1}{3}\left(\frac{e^2}{mv_{T_e}^2}\right)^2 d\mathbf{k} \frac{r_{D_e}^4}{k'} \delta\left(3 - \frac{c^2k^2}{\omega_{L_e}^2}\right) \frac{W_{tr}(\mathbf{k}-\mathbf{k}')}{\varkappa T_e} k'^2 \frac{[\mathbf{k}\mathbf{e}]^2}{k^2}. \qquad (12.31)$$

Here, $W_{tr}(\mathbf{k}-\mathbf{k}')$ is the energy density of the transverse oscillation on which the scattering occurs and $\mathbf{e}$ is its polarization vector. Note that the cross section of the combination scattering of a longitudinal oscillation does not depend on the polarization of the resulting transverse oscillation but is determined by the polarization of the oscillation on which the scattering occurs. Averaging with respect to this polarization gives the cross section

$$d\sigma^{lt}(\mathbf{k}', \mathbf{k}) = \frac{1}{3}\left(\frac{e^2}{mv_{T_e}^2}\right)^2 d\mathbf{k} \frac{r_{D_e}^4}{k'} \delta\left(3 - \frac{c^2k^2}{\omega_{L_e}^2}\right) \frac{W_{tr}(\mathbf{k}-\mathbf{k}')}{\varkappa T_e} k'^2 \left[1 + \frac{(\mathbf{k}, \mathbf{k}-\mathbf{k}')^2}{k^2(\mathbf{k}-\mathbf{k}')^2}\right], \qquad (12.32)$$

which goes over in the case of scattering on thermal transverse fluctuations after integration over the angles into the expression

$$\sigma^{lt}(k') = \frac{\pi}{18}\left(\frac{e^2}{mv_{T_e}^2}\right)^2 r_{D_e}^4 kk' \left\{7k^2 - k'^2 + \frac{(k^2-k'^2)^2}{2kk'} \ln\left|\frac{k+k'}{k-k'}\right|\right\},$$
$$k = \sqrt{3}\frac{\omega_{L_e}}{c}. \qquad (12.33)$$

In the limiting case of long wavelengths of the scattered Langmuir oscillation ($k' \ll k$), the total cross section (12.33) becomes

$$\sigma^{lt}(k') = 4\pi r_0^2 \frac{k'}{k}, \quad k' \ll k, \quad r_0 = \frac{e^2}{mc^2}, \tag{12.34}$$

and in the opposite limiting case of long wavelengths of the scattered transverse oscillation $(k \ll k' \ll r_{D_e}^{-1})$ it reduces to the expression

$$\sigma^{lt}(k') = \frac{8\pi}{3} r_0^2 \frac{k'}{k}, \quad k \ll k' \ll r_{D_e}^{-1}, \quad r_0 = \frac{e^2}{mc^2}. \tag{12.35}$$

Note that formulas (12.16) and (12.31) augment each other naturally and are the individual terms of the complete expression for the cross section of combination scattering of a Langmuir oscillation on longitudinal and transverse fluctuations of the plasma:

$$d\sigma(\mathbf{k}', \mathbf{k}) = \frac{2}{3}\left(\frac{e^2}{mv_{T_e}^2}\right)^2 d\mathbf{k} \frac{r_{D_e}^4}{k'} \delta\left(3 - \frac{c^2k^2}{\omega_{L_e}^2}\right)\left\{\frac{W_l(\mathbf{k}-\mathbf{k}')}{\varkappa T_e} \frac{[\mathbf{k}\mathbf{k}']^2}{(kk')^2} \frac{(k^2 - 2\mathbf{k}\mathbf{k}')^2}{(\mathbf{k}-\mathbf{k}')^2} + \right.$$
$$\left. + \frac{1}{2} \frac{W_{tr}(\mathbf{k}-\mathbf{k}')}{\varkappa T_e} k'^2 \left[1 + \frac{(\mathbf{k}, \mathbf{k}-\mathbf{k}')^2}{k^2(\mathbf{k}-\mathbf{k}')^2}\right]\right\}. \tag{12.36}$$

§ 13. Corrections to the Spectra of the Characteristic Oscillations of an Isotropic Plasma due to Nonlinear Interaction

It is well known that interaction leads to a shift of the energy levels of the states of quantum-mechanical systems. For example, interaction between phonons in a solid results in their frequencies changing. The same situation obtains in the theory of the nonlinear interaction of characteristic oscillations of a plasma. In linear plasma theory, i.e., in a theory with infinestimally small amplitudes of the characteristic oscillations, the wave spectra are functions of variables (the temperature of the plasma, its density, the masses of the particles, an external field, for example, the magnetic field) that do not depend on the wave energy. In a nonlinear theory it follows from the nonlinear dispersion equation (4.12) obtained in § 4 that there are corrections to the wave spectra that depend on their energy. Formally, these nonlinear dispersion corrections are linear functionals of the spectral density of the energy of the oscillations. We emphasize that although the terms in the formulas for the wave spectrum that arise as a result of the nonlinear interaction and depend on the oscillation energy are small corrections to the linear expressions, they are nevertheless of interest because they have a completely different qualitative nature from the large linear term and therefore, as we shall show below, can be manifested in qualitatively different effects.

In the framework of our derivation of the nonlinear dispersion equation (4.12), we give some more specific forms of the equations for the determination of the nonlinear dispersion corrections. Setting

$$(E_jE_s)_{\omega,\mathbf{k}} = (E_l^2)_{\omega,\mathbf{k}}\varkappa_s\varkappa_j, \quad (E_aE_i)_{\omega,\mathbf{k}} = (E_l^2)_{\omega,\mathbf{k}}\varkappa_a\varkappa_i, \tag{13.1}$$

in Eq. (4.10), we obtain a nonlinear dispersion equation for longitudinal oscillations of an isotropic plasma that interact with longitudinal and naturally polarized transverse waves:

$$\varepsilon^{l'}(\omega, \mathbf{k}) + \varkappa_i\varkappa_a \operatorname{Re}\int d\omega' d\mathbf{k}' \left\{(E_l^2)_{\omega',\mathbf{k}'}\varkappa_s'\varkappa_c' + (E_{tr}^2)_{\omega',\mathbf{k}'} \frac{1}{2}(\delta_{sc} - \varkappa_s'\varkappa_c')\right\} \times$$
$$\times \{V_{icas}(\omega, \mathbf{k}; \omega', \mathbf{k}') - A_{jb}(\omega-\omega', \mathbf{k}-\mathbf{k}') S_{ijs}(\omega, \mathbf{k}; \omega', \mathbf{k}') S_{bca}(\omega-\omega', \mathbf{k}-\mathbf{k}'; \omega, \mathbf{k})\} = 0. \tag{13.2}$$

In particular, in the nonlinear interaction of only longitudinal waves in an isotropic plasma,

$$\varepsilon^{l'}(\omega, \mathbf{k}) + \varkappa_i \varkappa_a \operatorname{Re} \int d\omega' d\mathbf{k}' (E_l^2)_{\omega', \mathbf{k}'} \varkappa_s' \varkappa_c' \{V_{icas}(\omega, \mathbf{k}; \omega', \mathbf{k}') - A_{jb}(\omega - \omega', \mathbf{k} - \mathbf{k}') S_{ijs}(\omega, \mathbf{k}; \omega', \mathbf{k}') S_{bca}(\omega - \omega', \mathbf{k} - \mathbf{k}'; \omega, \mathbf{k})\} = 0. \quad (13.3)$$

Note that the nonlinear dispersion correction to the spectrum of electrostatic oscillations of an isotropic plasma that interact with transverse electromagnetic waves depends essentially on the polarization of the latter; for if the transverse wave is linearly polarized along the unit vector $\mathbf{e}'$, then, using Eq. (4.10), we can obtain an equation that differs from (13.2) by the replacement of the tensor $(1/2)(\delta_{sc} - \varkappa_s' \varkappa_c')$ by $e_s' e_c'$:

$$\varepsilon^{l'}(\omega, \mathbf{k}) + \varkappa_i \varkappa_a \operatorname{Re} \int d\omega' d\mathbf{k}' \{(E_l^2)_{\omega', \mathbf{k}'} \varkappa_s' \varkappa_c' + (E_{tr}^2)_{\omega', \mathbf{k}'} e_s' e_c'\} \{V_{icas}(\omega, \mathbf{k}; \omega', \mathbf{k}') - A_{jb}(\omega - \omega', \mathbf{k} - \mathbf{k}') \times S_{ijs}(\omega, \mathbf{k}; \omega', \mathbf{k}') S_{bca}(\omega - \omega', \mathbf{k} - \mathbf{k}'; \omega, \mathbf{k})\} = 0 \quad (13.4)$$

and goes over after summation over all possible linear polarizations $\mathbf{e}'$ into (13.2).

In the general case of partly polarized transverse waves (6.74), their nonlinear interaction with longitudinal oscillations of an isotropic plasma makes more accurate the linear dispersion equation (6.35) and means that the frequencies of the electrostatic oscillations begin to depend on the Stokes parameters:

$$\varepsilon^{l'}(\omega, \mathbf{k}) + \frac{1}{2} \varkappa_i \varkappa_a \operatorname{Re} \int d\omega' d\mathbf{k}' \{W(\omega', \mathbf{k}')(\delta_{sc} - \varkappa_s' \varkappa_c') + M(\omega', \mathbf{k}')(e_s' e_c' - [\boldsymbol{\varkappa}' \mathbf{e}']_s [\boldsymbol{\varkappa}' \mathbf{e}']_c) + C(\omega', \mathbf{k}')(e_s' [\boldsymbol{\varkappa}' \mathbf{e}']_c + e_c' [\boldsymbol{\varkappa}' \mathbf{e}']_s) - iS(\omega', \mathbf{k}') e_{snc} \varkappa_n'\} \times \{V_{icas}(\omega, \mathbf{k}; \omega', \mathbf{k}') - A_{jb}(\omega - \omega', \mathbf{k} - \mathbf{k}') S_{ijs}(\omega, \mathbf{k}; \omega', \mathbf{k}') S_{bca}(\omega - \omega', \mathbf{k} - \mathbf{k}'; \omega, \mathbf{k})\} = 0. \quad (13.5)$$

In its turn, the spectrum of transverse waves also changes because of the interaction with longitudinal waves [see (4.10)]:

$$\left\{\varepsilon^{tr'}(\omega, \mathbf{k}) - \frac{c^2 k^2}{\omega^2}\right\} + \frac{1}{2}(\delta_{ai} - \varkappa_a \varkappa_i) \operatorname{Re} \int d\omega' d\mathbf{k}' (E_l^2)_{\omega', \mathbf{k}'} \varkappa_s' \varkappa_c' \{V_{icas}(\omega, \mathbf{k}; \omega', \mathbf{k}') - A_{jb}(\omega - \omega', \mathbf{k} - \mathbf{k}') \times S_{ijs}(\omega, \mathbf{k}; \omega', \mathbf{k}') S_{bca}(\omega - \omega', \mathbf{k} - \mathbf{k}'; \omega, \mathbf{k})\} = 0. \quad (13.6)$$

This dispersion relation is suitable for finding the correction to the spectrum of naturally polarized electromagnetic waves. If the transverse waves are polarized linearly along $\mathbf{e}$ ($\varkappa \mathbf{e} = 0$), it must be replaced by a different equation:

$$\left\{\varepsilon^{tr'}(\omega, \mathbf{k}) - \frac{c^2 k^2}{\omega^2}\right\} + e_a e_i \operatorname{Re} \int d\omega' d\mathbf{k}' (E_l^2)_{\omega', \mathbf{k}'} \varkappa_s' \varkappa_s' \{V_{icas}(\omega, \mathbf{k}; \omega', \mathbf{k}') - A_{jb}(\omega - \omega', \mathbf{k} - \mathbf{k}') \times S_{ijs}(\omega, \mathbf{k}; \omega', \mathbf{k}') S_{bca}(\omega - \omega', \mathbf{k} - \mathbf{k}'; \omega, \mathbf{k})\} = 0. \quad (13.7)$$

Comparison of (13.6) and (13.7) shows that in the nonlinear theory the spectrum of the transverse wave begins to depend on its type of polarization. There is no such effect in the linear approximation. When allowance is made for the nonlinear interaction of transverse waves with transverse waves, Eqs. (13.6) and (13.7) must be augmented by terms that are proportional to the spectral energy density of the transverse waves. For example, in the case of the nonlinear interaction of naturally polarized transverse waves the nonlinear dispersion equation for the spectrum can be written in the form [see (4.10)]

$$\left\{\varepsilon^{tr'}(\omega, \mathbf{k}) - \frac{c^2 k^2}{\omega^2}\right\} + \frac{1}{4}(\delta_{ai} - \varkappa_a \varkappa_i) \operatorname{Re} \int d\omega' d\mathbf{k}' (E_{tr}^2)_{\omega', \mathbf{k}'} (\delta_{sc} - \varkappa_s' \varkappa_c') \{V_{icas}(\omega, \mathbf{k}; \omega', \mathbf{k}') - A_{jb}(\omega - \omega', \mathbf{k} - \mathbf{k}') S_{ijs}(\omega, \mathbf{k}; \omega', \mathbf{k}') S_{bca}(\omega - \omega', \mathbf{k} - \mathbf{k}'; \omega, \mathbf{k})\} = 0. \quad (13.8)$$

The complete dispersion equation for naturally polarized transverse electromagnetic waves is given by a relation that combines (13.6) and (13.8):

$$\left\{\varepsilon^{tr'}(\omega, \mathbf{k}) - \frac{c^2k^2}{\omega^2}\right\} + \frac{1}{2}(\delta_{ai} - \varkappa_a\varkappa_i) \operatorname{Re} \int d\omega' d\mathbf{k}' \left\{(E_l^2)_{\omega', \mathbf{k}'}\varkappa'_s\varkappa'_c + \frac{1}{2}(E_{tr}^2)_{\omega', \mathbf{k}'}(\delta_{sc} - \varkappa'_s\varkappa'_c)\right\}\times$$

$$\times\{V_{icas}(\omega, \mathbf{k}; \omega', \mathbf{k}') - A_{jb}(\omega - \omega', \mathbf{k} - \mathbf{k}')S_{ijs}(\omega, \mathbf{k}; \omega', \mathbf{k}')S_{bca}(\omega - \omega', \mathbf{k} - \mathbf{k}'; \omega, \mathbf{k})\} = 0. \quad (13.9)$$

The nonlinear interaction of linearly polarized transverse oscillations of an isotropic plasma leads to the dispersion equation

$$\left\{\varepsilon^{tr'}(\omega, \mathbf{k}) - \frac{c^2k^2}{\omega^2}\right\} + e_a e_i e'_s e'_c \operatorname{Re} \int d\omega' d\mathbf{k}' (E_{tr}^2)_{\omega', \mathbf{k}'}\times$$

$$\times\{V_{icas}(\omega, \mathbf{k}; \omega', \mathbf{k}') - A_{jb}(\omega - \omega', \mathbf{k} - \mathbf{k}')S_{ijs}(\omega, \mathbf{k}; \omega', \mathbf{k}')S_{bca}(\omega - \omega', \mathbf{k} - \mathbf{k}'; \omega, \mathbf{k}\} = 0. \quad (13.10)$$

Hence, by summing (6.71) over all possible linear polarizations of both interacting waves, we obtain, in particular, Eq. (13.8) for naturally polarized electromagnetic waves. Partial summation of (6.71) over the polarizations of one of the waves gives the equations

$$\left\{\varepsilon^{tr'}(\omega, \mathbf{k}) - \frac{c^2k^2}{\omega^2}\right\} + \frac{1}{2}e_a e_i \operatorname{Re} \int d\omega' d\mathbf{k}' (E_{tr}^2)_{\omega', \mathbf{k}'}(\delta_{sc} - \varkappa'_s\varkappa'_c)\times$$

$$\times\{V_{icas}(\omega, \mathbf{k}; \omega', \mathbf{k}') - A_{jb}(\omega - \omega', \mathbf{k} - \mathbf{k}')S_{ijs}(\omega, \mathbf{k}; \omega', \mathbf{k}')S_{bca}(\omega - \omega', \mathbf{k} - \mathbf{k}'; \omega, \mathbf{k})\} = 0; \quad (13.11)$$

$$\left\{\varepsilon^{tr'}(\omega, \mathbf{k}) - \frac{c^2k^2}{\omega^2}\right\} + \frac{1}{2}(\delta_{ai} - \varkappa_a\varkappa_i) \operatorname{Re} \int d\omega' d\mathbf{k}' e'_s e'_c (E_{tr}^2)_{\omega', \mathbf{k}'}\times$$

$$\times\{V_{icas}(\omega, \mathbf{k}; \omega', \mathbf{k}') - A_{jb}(\omega - \omega', \mathbf{k} - \mathbf{k}')S_{ijs}(\omega, \mathbf{k}; \omega', \mathbf{k}')S_{bca}(\omega - \omega', \mathbf{k} - \mathbf{k}'; \omega, \mathbf{k})\} = 0, \quad (13.12)$$

of which the first determines the correction to the spectrum of the linearly polarized transverse oscillation due to its interaction with the naturally polarized oscillation and the second enables one to calculate the correction to the spectrum of the naturally polarized electromagnetic waves through the spectral energy density of the linearly polarized transverse oscillation. Finally, using (4.10) we write down the dispersion equation that arises because of the nonlinear interaction of linearly and partly polarized transverse waves:

$$\left\{\varepsilon^{tr'}(\omega, \mathbf{k}) - \frac{c^2k^2}{\omega^2}\right\} + \frac{1}{2}e_a e_i \operatorname{Re} \int d\omega' d\mathbf{k}' \{W(\omega, \mathbf{k})(\delta_{sc} - \varkappa'_s\varkappa'_c) +$$

$$+ M(\omega', \mathbf{k}')(e'_s e'_c - [\boldsymbol{\varkappa}'\mathbf{e}']_s[\boldsymbol{\varkappa}'\mathbf{e}']_c) + C(\omega', \mathbf{k}')(e'_s[\boldsymbol{\varkappa}'\mathbf{e}']_c +$$

$$+ e'_c[\boldsymbol{\varkappa}'\mathbf{e}']_s) - iS(\omega', \mathbf{k}')e_{snc}\varkappa'_n\}\{V_{icas}(\omega, \mathbf{k}; \omega', \mathbf{k}') -$$

$$- A_{jb}(\omega - \omega', \mathbf{k} - \mathbf{k}')S_{ijs}(\omega, \mathbf{k}; \omega', \mathbf{k}')S_{bca}(\omega - \omega', \mathbf{k} - \mathbf{k}'; \omega, \mathbf{k})\} = 0. \quad (13.13)$$

After the replacement of the tensor $e_a e_i$ by the tensor $(1/2)(\delta_{ai} - \varkappa_a\varkappa_i)$, this equation can be used to calculate the correction to the spectrum of a naturally polarized transverse oscillation in terms of the Stokes parameters of the partly polarized wave with which it interacts.

Note that the above splitting of the nonlinear dispersion equations for longitudinal and transverse waves separately is made possible by the rigorous division of the characteristic oscillations of an isotropic plasma into electrostatic and nonelectrostatic. In a magnetoactive plasma, for example, longitudinal and transverse waves are, in general, already related in the linear approximation. In this case, it is more convenient for the calculation of the dispersion corrections to use Eq. (4.12).* In addition, using this equation we can calculate the corrections

* Of course, an exception to the above are transverse electromagnetic waves of very high frequency ($\omega \gg \Omega_e$) for which the tensor form of the spectral function of the fields in the case of degenerate polarization cannot be represented by means of Eq. (4.3), which was essential in the derivation of (4.12).

to the spectra of longitudinal and transverse waves of an isotropic plasma if the transverse waves are naturally polarized. In this sense, Eqs. (13.2), (13.3), (13.6), (13.8), and (13.9) give the same results as are obtained by means of the nonlinear dispersion equation (4.12). In contrast, Eqs. (13.4), (13.5), (13.7), (13.10), (13.12), and (13.13) augment (4.12) in the case of a polarization of the transverse electromagnetic waves that is not natural.

We now illustrate the general arguments put forward above by analyzing the dispersion law of the Langmuir oscillations of an isotropic plasma. In doing so, we shall, following [35], restrict ourselves to the nonlinear interaction of Langmuir waves with Langmuir waves through an intermediate electrostatic oscillation, which corresponds to retaining in the tensor $A_{jb}(\omega - \omega', \mathbf{k} - \mathbf{k}')$ of Eq. (13.3) the term.* That is longitudinal with respect to the vector $\boldsymbol{\varkappa}'' = (\mathbf{k} - \mathbf{k}') \times (|\mathbf{k} - \mathbf{k}'|)^{-1}$:

$$\varepsilon^{l'}(\omega, \mathbf{k}) + \varkappa_i \varkappa_a \operatorname{Re} \int d\omega' d\mathbf{k}' (E_l^2)_{\omega', \mathbf{k}'} \varkappa'_s \varkappa'_c \Big\{ V_{icas}(\omega, \mathbf{k}; \omega', \mathbf{k}') - \\ - \frac{\varkappa''_j \varkappa''_b}{\varepsilon^l(\omega - \omega', \mathbf{k} - \mathbf{k}')} S_{ijs}(\omega, \mathbf{k}; \omega', \mathbf{k}') S_{bca}(\omega - \omega', \mathbf{k} - \mathbf{k}'; \omega, \mathbf{k}) \Big\} = 0. \quad (13.14)$$

In our study of the Coulomb scattering of Langmuir waves in § 9, we have seen that the kernel of the corresponding integrodifferential equation (9.5), which describes this process, is proportional to the frequency difference of the interacting waves. Since the frequencies (6.34a) of such waves differ little from the frequency $(\omega_{Le}^2 + \omega_{Li}^2)^{1/2}$ (which is nearly equal to the electron plasma frequency of account of $\omega_{Le} \gg \omega_{Li}$, $m \ll M$), it is essential to take into account the small corrections to the frequency in order to describe this process.

Indeed, the kernel (9.11) of Eq. (9.5) vanishes when the frequency difference (9.7) of the interacting waves tends to zero. At the same time, under certain conditions the corrections that determine the frequency difference (9.7) (we shall call them linear) may be smaller than the nonlinear corrections to the frequencies of the interacting Langmuir waves determined by the second term on the left-hand side of Eq. (13.14). In this case, the nature of the nonlinear process will be determined by nonlinear corrections to this kind to the spectrum. In [35], it is shown that allowance for the "nonlinear" correction leads to transfer of energy from waves with lower energy to waves with higher energy [and not from short to longer waves, as is the case when allowance is made for only the linear correction, i.e., for the frequency difference (9.7)].

Substituting into Eq. (13.14) the explicit expressions (7.28) and (8.8) for the contractions of the nonlinear tensors S and V with the Maxwell distribution function (6.14) with allowance for (6.23) and the conditions of existence (weak damping) of Langmuir waves in the plasma (6.28), we obtain in the lowest approximation in the ratio of the electron Debye radius to the wavelength ($k^2 r_{De}^2 \ll 1$) the following expression for the nonlinear correction $\Delta\omega$ to the spectrum (6.34a) [[35], formula (12)]:

$$\Delta\omega = -\frac{1}{8\pi^2} \frac{e^2}{m^2} \frac{1}{\omega_{Le}^3} \int d\mathbf{k}' W_l(\mathbf{k}') \left(\frac{\mathbf{k}\mathbf{k}'}{kk'}\right)^2 \left\{ \frac{1}{r_{De}^2} \left[1 - \frac{(k'' r_{De})^{-2}}{1 + (k'' r_{De})^{-2} + \delta\varepsilon_i^{l'}(\omega'', \mathbf{k}'')} \right] + 6k'^2 - k''^2 - \frac{4}{3} \frac{[\mathbf{k}\mathbf{k}']^4}{k''^2 (\mathbf{k}\mathbf{k}')^2} \right\},$$

$$\mathbf{k}'' = \mathbf{k} - \mathbf{k}'; \quad k'' = |\mathbf{k} - \mathbf{k}'|, \quad \omega'' \equiv \omega - \omega'. \quad (13.15)$$

For a sufficiently short wavelength of the interacting oscillations:

$$1 \gg k r_{De} \gg (m/M)^{1/2} \quad (13.16)$$

---

*Equation (13.4) was obtained in [41] and is contained in the review [4] (p. 232 of the Russian original). In [35], which we follow here, it was used to calculate the nonlinear correction to the spectrum of Langmuir waves.

the partial polarizability of the ions in (13.15) can be ignored. Then the nonlinear correction to the frequency (13.15) can be represented in the form

$$\Delta\omega = -\frac{1}{8\pi^2}\frac{e^2}{m^2}\frac{1}{\omega_{L_e}^3}\int d\mathbf{k}' W_l(\mathbf{k})\left(\frac{\mathbf{kk}'}{kk'}\right)^2\left\{6k'^2 - \frac{4}{3}[\mathbf{kk}']^4(\mathbf{kk}')^{-2}(k'')^{-2}\right\} \tag{13.17}$$

This expression, which is proportional to the ratio of the mean square of the displacement of an electron in the field of the oscillations to the square of the wavelength of the oscillations (cf. [42-44]), does not depend on the ion variables, and can be obtained if one completely ignores the ion motion, i.e., in an electron plasma. Such a nonlinear correction is greater than the linear correction only if the Debye radius of the electrons is less than their mean displacement in the field of the oscillations, when the plasma is unstable [45]. Therefore, it is much more interesting to consider the case when the ion polarizability is important:

$$1 \gg (m/M)^{1/2} \gg (kr_{D_e}). \tag{13.18}$$

Then the resulting correction (13.15),

$$\Delta\omega = -\frac{\omega_{L_e}}{32\pi^3}\frac{1}{N_e \varkappa T_e}\frac{r_{D_e}^2}{r_{D_e}^2 + r_{D_i}^2}\int d\mathbf{k}'\left(\frac{\mathbf{kk}'}{kk'}\right)^2 W_l(\mathbf{k}'), \tag{13.19}$$

is greater than the purely electron correction (13.17) by a factor $(kr_{De})^{-2}$ if the electron and ion temperatures are of the same order and may exceed the linear correction $(3/2)(kr_{De})^2\omega_{Le}$ in the spectrum (6.34a). Note that in the case of an isotropic energy density $W_l(k)$ of the Langmuir oscillations the nonlinear correction (13.19) does not depend on the wave vector **k** and cannot therefore change the group velocity of the oscillations.

Under conditions when (13.19) holds, i.e., the spectrum of the Langmuir oscillations can be represented by the equation

$$\omega(\mathbf{k}) = \omega_{L_e}\left\{1 + \frac{1}{2}\frac{\omega_{L_i}^2}{\omega_{L_e}^2} + \frac{3}{2}(kr_{D_e})^2 - \frac{1}{32\pi^3}\frac{r_{D_e}^2}{r_{D_e}^2 + r_{D_i}^2}\int d\mathbf{k}'(\varkappa\varkappa')^2\frac{W_l(\mathbf{k}')}{N_e \varkappa T_e}\right\}, \tag{13.20}$$

induced scattering on ions is the most important factor in the nonlinear interaction of Langmuir waves. In accordance with (9.5), the equation that describes this process has the form [cf. with Eq. (9.27), in which the spectrum (6.34a) is used] [35]

$$\frac{\partial W_l(\mathbf{k})}{\partial t} + 3v_{T_e}(\mathbf{k}r_{D_e})\frac{\partial W_l(\mathbf{k})}{\partial \mathbf{r}} - \frac{1}{16\pi^3}\frac{r_{D_e}^2}{r_{D_e}^2 + r_{D_i}^2}\int d\mathbf{k}'(\varkappa\varkappa')[\varkappa[\varkappa'\varkappa]]\frac{W_l(\mathbf{k}')}{N_e \varkappa T_e}\frac{\omega_{L_e}}{k}\frac{\partial W_l(\mathbf{k})}{\partial \mathbf{r}} =$$

$$= -\frac{1}{4(2\pi)^{5/2}}\frac{1}{N_e r_{D_e}^3}\frac{W_l(\mathbf{k})}{\varkappa T_e}\frac{v_{T_e}}{v_{T_i}}\frac{r_{D_i}^2 r_{D_e}^4}{(r_{D_e}^2 + r_{D_i}^2)^2}\int d\mathbf{k}'(\varkappa\varkappa')^2 W_l(\mathbf{k}')\frac{\omega(\mathbf{k}) - \omega(\mathbf{k}')}{|\mathbf{k} - \mathbf{k}'|}. \tag{13.21}$$

Here, the third term* on the left-hand side is a direct consequence of the contribution of the nonlinear correction (13.19) to the group velocity $(\partial\omega/\partial\mathbf{k})$ of the Langmuir waves. Replacing

* In accordance with [13, 20], the group velocity $d\omega/d\mathbf{k}$ of the Langmuir oscillations is determined by the linear correction to the spectrum if their total energy W_0 is not too large: $1 \gg (kr_{De})^2 \gg (1/48\pi^3)(W_0/N_e\varkappa T_e)$. In the opposite case $1 \gg (1/48\pi^3)(W_0/N_e\varkappa T_e)[r_{De}^2/(r_{De}^2 + r_{Di}^2)] \gtrsim (kr_{De})^2$, the nonlinear correction has an important effect on the group velocity of the Langmuir waves [cf. (13.30)]. Note also that the angular dependence of the part of the group velocity vector that is determined by the nonlinear correction to the spectrum depends essentially on the form of $W_l(\mathbf{k}')$ as a function of $\mathbf{k}'$ and may have a quite different direction from $(3\mathbf{k}v_{Te}^2/\omega_{Le})$. In particular, if $W_l(\mathbf{k}')$ is isotropic $[W_l(\mathbf{k}') \equiv W_l(k')]$, then, as we have already pointed out, the nonlinear correction does not contribute to the group velocity.

$\omega(\mathbf{k})$ and $\omega(\mathbf{k}')$ by the spectrum (13.20) on the right-hand side of (13.21) we obtain

$$\frac{dW_l(\mathbf{k})}{dt} = -\frac{3}{8(2\pi)^{5/2}}\frac{\omega_{L_e}}{N_e r_{D_e}^3}\frac{W_l(\mathbf{k})}{\varkappa T_e}\frac{r_{D_i}^2 r_{D_e}^6}{(r_{D_i}^2 + r_{D_e}^2)^2}\frac{v_{T_e}}{v_{T_i}} \times$$

$$\times \int d\mathbf{k}' W_l(\mathbf{k}')\left(\frac{\mathbf{k}\mathbf{k}'}{kk'}\right)^2 \frac{1}{|\mathbf{k}-\mathbf{k}'|}\left\{(k^2 - k'^2) - \frac{1}{48\pi^3}\frac{1}{N_e \varkappa T_e}\frac{1}{r_{D_e}^2 + r_{D_i}^2}\int d\mathbf{k}'' W_l(\mathbf{k}'')\left[\left(\frac{\mathbf{k}\mathbf{k}''}{kk''}\right)^2 - \left(\frac{\mathbf{k}'\mathbf{k}''}{k'k''}\right)^2\right]\right\}. \quad (13.22)$$

Equation (13.22) describes a process of nonlinear interaction in which the total energy W_0 of the Langmuir waves (9.13) and the total number of oscillations (9.43) are conserved. To bring out the effects to which allowance for the nonlinear dispersion correction (13.19) leads, let us consider the interaction of two wave packets of Langmuir oscillations with widths that can be ignored (9.17). (We recall that we then also ignore the spatial variation of W_l.) Then Eq. (13.22) reduces to a system of two ordinary differential equations:

$$\begin{aligned} \frac{dW_1}{dt} &= -W_1 W_2\{Q_1(\mathbf{k}_1, \mathbf{k}_2) - Q_2(\mathbf{k}_1, \mathbf{k}_2)[W_1 - W_2]\}; \\ \frac{dW_2}{dt} &= W_1 W_2\{Q_1(\mathbf{k}_1, \mathbf{k}_2) - Q_2(\mathbf{k}_1, \mathbf{k}_2)[W_1 - W_2]\}, \end{aligned} \quad (13.23)$$

in which the functions Q_1 and Q_2 arise from the kernel of the integrodifferential equation (13.22):

$$Q_1(\mathbf{k}_1, \mathbf{k}_2) = \frac{3}{8(2\pi)^{5/2}}\frac{\omega_{L_e}}{N_e r_{D_e}^3}\frac{1}{\varkappa T_e}\frac{r_{D_i}^2 r_{D_e}^6}{(r_{D_e}^2 + r_{D_i}^2)^2}\frac{v_{T_e}}{v_{T_i}}\left(\frac{\mathbf{k}_1\mathbf{k}_2}{k_1 k_2}\right)^2\frac{k_1^2 - k_2^2}{|\mathbf{k}_1 - \mathbf{k}_2|}; \quad (13.24)$$

$$Q_2(\mathbf{k}_1, \mathbf{k}_2) = \frac{1}{128\pi^3(2\pi)^{5/2}}\frac{\omega_{L_e}}{N_e r_{D_e}^3}\frac{r_{D_i}^2 r_{D_e}^6}{(r_{D_e}^2 + r_{D_i}^2)^3}\frac{1}{N_e(\varkappa T_e)^2}\frac{v_{T_e}}{v_{T_i}}\frac{[\mathbf{k}_1\mathbf{k}_2]^2}{(k_1 k_2)^2}\left(\frac{\mathbf{k}_1\mathbf{k}_2}{k_1 k_2}\right)^2\frac{1}{|\mathbf{k}_1 - \mathbf{k}_2|}.$$

The solution of Eqs. (13.23) for $(Q_2 W_0/Q_1) \neq 1$ has the form

$$\exp(-Q_1 W_0 t) = \left\{\frac{W_1(t)}{W_1(0)}\right\}^{\frac{1}{1+\alpha}}\left\{\frac{W_0 - W_1(t)}{W_0 - W_1(0)}\right\}^{\frac{1}{\alpha-1}}\left|\frac{Q_1 - 2Q_2 W_1(t) + Q_2 W_0}{Q_1 - 2Q_2 W_1(0) + Q_2 W_0}\right|^{\frac{2\alpha}{1-\alpha^2}}; \quad (13.25)$$

$$\alpha \equiv \frac{Q_2}{Q_1} W_0 \neq 1, \quad W_0 \equiv W_1(t) + W_2(t) = W_1(0) + W_2(0). \quad (13.26)$$

If $\alpha \ll 1$, we obtain from this ($t_0 \equiv 0$) the solution (9.18), which determines the transfer of the energy of the Langmuir oscillations from the short- to the long-wavelength part of the spectrum (a stationary state $t \to \infty$ arising when all the energy is concentrated in the oscillations with the smallest wave number)

If the nonlinear correction to the dispersion law (13.20) is larger than the linear correction, then $\alpha > 1$ and the last factor on the right-hand side of Eq. (13.25) can become infinite if

$$W_1 = W_{1\text{cr}} \equiv \frac{1}{2} W_0\left(1 + \frac{1}{\alpha}\right) < W_0. \quad (13.27)$$

However, the exponential on the left-hand side of (13.25) is bounded. Therefore, $W_1(t)$ can only vary in such a way that $W_1(t) \neq W_{1\text{cr}}$. It follows that when $W_1(0) > W_{1\text{cr}}$ a stationary state is achieved for $W_1(\infty) = W_0$ and $W_2(\infty) = 0$. Conversely, if $W_1(0) < W_{1\text{cr}}$, a stationary state is attained when $W_1(\infty) = 0$ and $W_2(\infty) = W_0$. Thus, if the nonlinear corrections in the dispersion law (13.20) of the Langmuir waves are large, the direction in which their energy is redistributed depends essentially on the initial distribution: energy is transferred from a wave packet with lower energy to a wave packet with higher energy. The time taken for this redistribution can be found by writing the solution (13.25) in the approximation $\alpha \gg 1$:

$$W_1(t) = \frac{W_0}{2} - \frac{W_0 - 2W_1(0)}{2}\left\{\frac{W_0^2 - 4W_1(0)[W_0 - W_1(0)]\exp(-Q_2 W_0^2 t)}{[W_0 - 2W_1(0)]^2}\right\}^{1/2}. \quad (13.28)$$

The characteristic time for a stationary distribution to be established is approximately [cf. (9.19)]

$$\tau \simeq \frac{1}{Q_2 W_0^2} \sim 10^5 \left(\frac{N_e \varkappa T_e}{W_0}\right)^2 \frac{\Delta k r_{D_e}}{\omega_{L_e}} \left(\frac{m}{M}\right)^{1/2}, \qquad \Delta k \equiv |k_1 - k_2|. \tag{13.29}$$

If $(\Delta k r_{De}) \sim 10^{-2}$, $\omega_{Le} \sim 10^{12}$ sec^{-1}, and $(N_e \varkappa T_e / W_0) \sim 10$, then for a hydrogen plasma $\tau \sim 10^{-9}$ sec. We emphasize that the expression (13.29) is true if there is sufficient energy concentrated in the Langmuir oscillations,

$$\frac{W_0}{N_e \varkappa T_e} > (k r_{D_e})^2 \tag{13.30}$$

and for sufficiently long waves (13.18). Note also that in accordance with (13.29) this process of induced scattering of Langmuir waves with allowance for the nonlinear correction to the spectrum (13.19) can take place fairly fast, for example, compared with the process of coalescence of two Langmuir oscillations into a transverse oscillation described by Eq. (9.51). Namely, comparison of the times (13.29) and (9.52),

$$\tau \simeq 10^3 \frac{N_e \varkappa T_e}{W_0} \frac{1}{\omega_{L_e}} \left(\frac{c}{v_{T_e}}\right)^5 (k r_{D_e})^3, \tag{13.31}$$

shows that the coalescence takes place slower if the Langmuir oscillations have sufficiently high energy [35]:

$$\frac{W_0}{N_e \varkappa T_e} \gg 10^2 \left(\frac{v_{T_e}}{c}\right)^5 \left(\frac{m}{M}\right)^{1/2} (k r_{D_e})^{-2}. \tag{13.32}$$

For a hydrogen plasma with thermal velocities of the electrons, $v_{Te} \sim 10^{-2} c$ and Debye radius $(k r_{D_e})^{-2} \sim 10^4$, this condition can be written in the form

$$\frac{W_0}{N_e \varkappa T_e} \gg 10^{-6}. \tag{13.33}$$

Concluding our discussion of the nonlinear dispersion laws of characteristic plasma oscillations obtained in the theory expounded here of the nonlinear interaction of waves, we note that allowance for the first nonlinear corrections to the spectra leads to equations that contain the product of three functions W, which determine the energy density of the oscillations. The study of such equations is intimately related to the next nonlinear interaction with the participation of four oscillations, but this goes beyond the scope of the present work, which is devoted to the study of three-wave interaction.

§ 14. Nonlinear Interaction of Characteristic Oscillations of a Plasma and the Theory of the Scattering of Electromagnetic Waves on Fluctuations of the Plasma

To conclude this chapter, which is devoted to the nonlinear interaction of the characteristic oscillations of an isotropic plasma, we give some results of the theory of scattering of waves in a plasma, following [34]. In the spirit of the exposition, the method employed and the system of notation, this section is an exception to the general arrangement of the review. Nevertheless, some of the particular physical results obtained in [34] agree with some of those considered above in the framework of the nonlinear theory of the interaction of waves.

T... fact clarifies, in our opinion, the relationship between the theory of the scattering of electromagnetic waves on plasma fluctuations and the theory of the nonlinear interaction of characteristic oscillations of a plasma; in view of the considerable number of papers on these two theories, this is of considerable importance.

As is well known [8, 9, 46], the cross section for the scattering of an electromagnetic wave vector **k** and frequency $\omega = \omega(\mathbf{k})$ that gives rise to the formation of a wave with wave vector **k'** can be expressed in terms of the correlation function of the current $\mathbf{j}(\omega', \mathbf{k}')$ producing the scattered wave:

$$d\sigma(\mathbf{k}, \mathbf{k}') = -\frac{V}{(2\pi)^3} \frac{d\mathbf{k}'}{W(\mathbf{k}) \left| \frac{\partial \omega}{\partial \mathbf{k}} \right|} \operatorname{Im} \int_{-\infty}^{+\infty} \frac{d\omega'}{\omega'} A_{ji}(\omega', \mathbf{k}') \langle j_i(\omega' \, \mathbf{k}') \, j_j^*(\omega', \mathbf{k}') \rangle, \tag{14.1}$$

where V is the volume of the scattering system,* W(k) is the spectral density of the energy in the incident wave, equal in accordance with [46] [cf. with (4.4) from § 4 in Chapter I]

$$W(\mathbf{k}) = \frac{\partial \omega^2 \varepsilon_{ij}^H(\omega, \mathbf{k})}{\omega \partial \omega} \frac{E_{0i}^* E_{0j}}{16\pi}, \tag{14.2}$$

E_0 is the amplitude of the incident wave, A_{ji} and ε_{ij}^H are the inverse temperature and the Hermitian part of the permittivity tensor, respectively. The averaging $\langle \cdots \rangle$ in (14.1) is over the fluctuations of the plasma.

The current that produces the scattered radiation can be determined in the case of an isotropic plasma† from the transport equation that describes the perturbation of the distribution function due to the interaction of the incident wave with the fluctuations [8, 39]:

$$\frac{\partial f_a^{\text{scat}}}{\partial t} + \mathbf{v}_a \frac{\partial f_a^{\text{scat}}}{\partial \mathbf{r}} + \frac{e_a}{m_a} \left\{ \mathbf{E} + \frac{1}{c} [\mathbf{v}_a \mathbf{B}] \right\} \frac{\partial f_{a0}}{\partial \mathbf{v}_a} = -\frac{e_a}{m_a} \left\{ \left(\mathbf{E}_0 + \frac{1}{c} [\mathbf{v}_a \mathbf{B}_0] \right) \frac{\partial \delta N_a}{\partial \mathbf{v}_a} + \left(\delta \mathbf{E} + \frac{1}{c} [\mathbf{v}_a \delta \mathbf{B}] \right) \frac{\partial f_a^{\text{inc}}}{\partial \mathbf{v}_a} \right\}. \tag{14.3}$$

Here, $f_a^{\text{scat}}(\mathbf{r}, \mathbf{v}_a, t)$ is the perturbation of the distribution function of the particles of species a corresponding to the scattered wave; $\mathbf{E}(\mathbf{r}, t)$ and $\mathbf{B}(\mathbf{r}, t)$ are the electric and the magnetic field in the scattered wave, and $\mathbf{E}_0(\mathbf{r}, t)$ and $\mathbf{B}_0(\mathbf{r}, t)$ are the fields in the incident wave, which gives rise to the perturbation $f_a^{\text{inc}}(\mathbf{r}, \mathbf{v}_a, t)$ of the plasma; $\delta\mathbf{E}(\mathbf{r}, t)$, $\delta\mathbf{B}(\mathbf{r}, t)$, and $\delta N_a(\mathbf{r}, \mathbf{v}_a, t)$ are, respectively, the fluctuations of the electric and the magnetic field and the number density of particles of species a in the phase space. The distribution functions f_a^{scat} and f_a^{inc} are small compared with the equilibrium, unperturbed distribution function f_{a0} of the plasma particles, which we shall assume below is the Maxwellian distribution (6.14) when we analyze the particular cases of scattering. The smallness of the functions f_a^{scat} and f_a^{inc} corresponds to small fields of the incident and the scattered wave.

We take the field of the incident wave in the form

$$\mathbf{E}_0(\mathbf{r}, t) = \mathbf{E}_0 e^{-i\omega t + i\mathbf{k}\mathbf{r}} \tag{14.4}$$

and substitute this expression into the right-hand side of Eq. (14.3), eliminating the magnetic fields by means of one of the Maxwell equations (1.7). As a result, the current density $\mathbf{j}(\omega', \mathbf{k}')$ producing the scattered radiation:

$$\mathbf{j}(\omega', \mathbf{k}') = \sum_a e_a \int d\mathbf{v}_a f_a^{\text{scat}}(\mathbf{k}', \mathbf{v}_a, \omega') \, \mathbf{v}_a, \tag{14.5}$$

* As in §12, we assume V ≡ 1 in the formulas given below.

† It is assumed that there is no constant external magnetic field $\mathbf{B}_0$. It can be taken into account by adding to the left-hand side of Eq. (14.3) the term $(e/c)[\mathbf{v}_a \mathbf{B}_0](\partial f_a^{\text{scat}}/\partial \mathbf{p}_a)$ ($\mathbf{B}_0$ should not be confused with the magnetic field in the incident wave!).

can be represented after Eq. (14.3) for f_a^{scat} has been solved, by the equation [34]

$$j_i(\omega', \mathbf{k}') = -i\frac{\omega'}{4\pi} E_{0j}\{\varepsilon_{isj}(\omega', \mathbf{k}'; \omega, \mathbf{k})\delta E_s(\omega'', \mathbf{k}'')\ \delta\varepsilon_{ij}(\omega', \mathbf{k}'; \omega'', \mathbf{k}''; \omega, \mathbf{k})\}, \tag{14.6}$$

in which the frequencies and wave vectors are related by the equations

$$\omega'' = \omega' - \omega, \qquad \mathbf{k}'' = \mathbf{k}' - \mathbf{k}, \tag{14.7}$$

which differ from the quantities used previously (7.3) by the opposite sign on the right-hand sides. The tensors ε_{isj} and $\delta\varepsilon_{ij}$ on the right-hand side of (14.6) have a simple physical meaning. Namely, $\{\delta\varepsilon_{ij} + \delta_{ij}\delta(\omega)\delta(\mathbf{k})\}$ is the permittivity tensor of an inhomogeneous plasma in which the distribution function is $\delta N_a(\omega, \mathbf{k}, \mathbf{v}_a)$ [cf. (6.11)]:

$$\delta\varepsilon_{ij}(\omega', \mathbf{k}'; \omega'', \mathbf{k}''; \omega, \mathbf{k}) = \frac{4\pi}{\omega'}\sum_a \frac{e_a^2}{m_a}\int d\mathbf{v}_a v_{ai} g(\omega', \mathbf{k}', \mathbf{v}_a)\Gamma_j(\omega'', \mathbf{k}'', \mathbf{v}_a)\delta N_a(\omega, \mathbf{k}, \mathbf{v}_a). \tag{14.8}$$

The vector Γ_j and the scalar g are the vertex part [cf. (5.29)]

$$\Gamma_j(\omega, \mathbf{k}, \mathbf{v}_a) \equiv \frac{1}{\omega}\{\delta_{ij}(\omega - \mathbf{k}\mathbf{v}_a) + k_i v_{aj}\}\frac{\partial}{\partial v_{ai}} \equiv \alpha_{ij}(\omega, \mathbf{k}, \mathbf{v}_a)\,\partial/\partial v_{ai} \tag{14.9}$$

and the Green's function [cf. (5.28)]

$$g(\omega, \mathbf{k}, \mathbf{v}_a) \equiv \frac{1}{\omega - \mathbf{k}\mathbf{v}_a}, \tag{14.10}$$

that we have used previously. The three-index tensor ε_{isj} has been described in detail in the foregoing chapter (§ 7) and is given by formula (7.1) [with allowance for the difference in the notation of (7.3) and (14.7)]:

$$\varepsilon_{ijs}(\omega', \mathbf{k}'; \omega, \mathbf{k}) = -4\pi i\sum_a \frac{e_a^3}{m_a^2}\int d\mathbf{v}_a \frac{v_{ai}}{\omega'} g(\omega', \mathbf{k}', \mathbf{v}_a)\ \Gamma_j(\omega'', \mathbf{k}'', \mathbf{v}_a) g(\omega, \mathbf{k}, \mathbf{v}_a)\Gamma_s(\omega, \mathbf{k}, \mathbf{v}_a) f_{a0}. \tag{14.11}$$

Substituting the current density (14.6) into (14.1), we obtain the following expression for the cross section of the scattering of waves in the plasma [34]

$$d\sigma(\mathbf{k}, \mathbf{k}') = -\frac{1}{(2\pi)^3}\frac{d\mathbf{k}'}{16\pi^2 W(\mathbf{k})\left|\frac{\partial\omega}{\partial\mathbf{k}}\right|}\operatorname{Im}\Big\{E_{0n}E_{0m}^*\int_{-\infty}^{+\infty}\omega' d\omega' A_{ji}(\omega', \mathbf{k}')\times$$
$$\times[\varepsilon_{isn}(\omega', \mathbf{k}'; \omega, \mathbf{k})\varepsilon_{jpm}^*(\omega', \mathbf{k}'; \omega, \mathbf{k})\langle\delta E_s\delta E_p\rangle_{\omega'', \mathbf{k}''} +$$
$$+ \varepsilon_{jpm}^*(\omega', \mathbf{k}'; \omega, \mathbf{k})\langle\delta\varepsilon_{in}(\omega', \mathbf{k}'; \omega, \mathbf{k}; \omega'', \mathbf{k}'')\delta E_p^*(\omega'', \mathbf{k}'')\rangle +$$
$$+ \varepsilon_{isn}(\omega', \mathbf{k}'; \omega, \mathbf{k})\langle\delta E_s(\omega'', \mathbf{k}'')\delta\varepsilon_{jm}^*(\omega', \mathbf{k}'; \omega, \mathbf{k}; \omega'', \mathbf{k}'')\rangle +$$
$$+ \langle\delta\varepsilon_{in}(\omega', \mathbf{k}'; \omega, \mathbf{k}; \omega'', \mathbf{k}'')\delta\varepsilon_{jm}^*(\omega', \mathbf{k}'; \omega, \mathbf{k}; \omega'', \mathbf{k}'')\rangle]\Big\}. \tag{14.12}$$

For applications, it is convenient to express this formula in terms of the well-known [13] correlation functions

$$\langle\delta E_i\delta E_j\rangle_{\omega, \mathbf{k}};\quad \langle\delta N_a\delta E_i\rangle_{\omega, \mathbf{k}};\quad \langle\delta N_a\delta N_b\rangle_{\omega, \mathbf{k}}, \tag{14.13}$$

if the explicit expression (14.8) for $\delta\varepsilon_{ij}$ is substituted into the right-hand side of (14.12):

$$d\sigma(\mathbf{k}, \mathbf{k}') = -\frac{1}{(2\pi)^3}\frac{d\mathbf{k}'}{16\pi^2 W(\mathbf{k})\left|\frac{\partial\omega}{\partial\mathbf{k}}\right|}\operatorname{Im}\Big\{E_{0n}E_{0m}^*\int_{-\infty}^{+\infty}\omega' d\omega' A_{ji}(\omega', \mathbf{k}')\times$$
$$\times[\varepsilon_{isn}(\omega', \mathbf{k}'; \omega, \mathbf{k})\varepsilon_{jpm}^*(\omega', \mathbf{k}'; \omega, \mathbf{k})\langle\delta E_s\delta E_p\rangle_{\omega'', \mathbf{k}''} -$$

$$-\sum_a \frac{4\pi e_a^2}{m_a} \int d\mathbf{v}_a \alpha_{qi}(\omega', \mathbf{k}', \mathbf{v}_a) g^2(\omega', \mathbf{k}', \mathbf{v}_a) \alpha_{qn}(\omega, \mathbf{k}, \mathbf{v}_a) \times$$

$$\times \langle \delta N_a \delta E_p \rangle_{\omega'', \mathbf{k}''} \varepsilon^*_{jpm}(\omega', \mathbf{k}'; \omega, \mathbf{k}) - \sum_a \frac{4\pi e_a^2}{m_a} \int d\mathbf{v}_a \alpha_{pj}(\omega', \mathbf{k}', \mathbf{v}_a) \times$$

$$\times g^{*2}(\omega', \mathbf{k}', \mathbf{v}_a) \alpha_{pm}(\omega, \mathbf{k}, \mathbf{v}_a) \langle \delta E_s \delta N_a \rangle_{\omega'', \mathbf{k}''} \varepsilon_{isn}(\omega', \mathbf{k}'; \omega, \mathbf{k}) +$$

$$+ \sum_{a,b} \frac{(4\pi e_a e_b)^2}{m_a m_b} \int d\mathbf{v}_a d\mathbf{v}_b \alpha_{pi}(\omega', \mathbf{k}', \mathbf{v}_a) \alpha_{qj}(\omega', \mathbf{k}', \mathbf{v}_b) \alpha_{pn}(\omega, \mathbf{k}, \mathbf{v}_a) \times$$

$$\times \alpha_{qm}(\omega, \mathbf{k}, \mathbf{v}_b) g^2(\omega', \mathbf{k}', \mathbf{v}_a) g^{*2}(\omega', \mathbf{k}', \mathbf{v}_b) \langle \delta N_a \delta N_b \rangle_{\omega'', \mathbf{k}''} \Big] \Big\}. \quad (14.14)$$

A differential cross section like (14.14) was obtained in [40] and analyzed in the case of combination scattering (i.e., the decay and coalescence of waves in the language of the theory of nonlinear interaction). We now show (see the Appendix in [34]) that the expression (14.14) for the cross section is identical with the expression (12.12) that we obtained by a different method in § 12 in the framework of the theory of the nonlinear interaction of waves in a plasma. For this we note that the correlation functions on the right-hand side of (14.14) in the case of an isotropic plasma, which we consider in this chapter, have the form

$$\langle \delta N_a \delta E_i \rangle_{\omega, \mathbf{k}} = -i \frac{e_a}{m_a} g(\omega, \mathbf{k}, \mathbf{v}_a) \frac{\partial f_{a0}}{\partial v_{aj}} \langle \delta E_j \delta E_i \rangle_{\omega, \mathbf{k}} + 4\pi i e_a 2\pi \delta(\omega - \mathbf{k}\mathbf{v}_a) A^*_{ij}(\omega, \mathbf{k}) \alpha_{pj}(\omega, \mathbf{k}, \mathbf{v}_a) \frac{k_p}{k^2} f_{a0}(\mathbf{v}_a); \quad (14.15)$$

$$\langle \delta N_a \delta N_b \rangle_{\omega, \mathbf{k}} = 2\pi \delta_{ab} \delta(\omega - \mathbf{k}\mathbf{v}_a) \delta(\mathbf{v}_a - \mathbf{v}_b) f_{b0}(\mathbf{v}_b) + g(\omega, \mathbf{k}, \mathbf{v}_a) g^*(\omega, \mathbf{k}, \mathbf{v}_b) \frac{e_a e_b}{m_a m_b} \langle \delta E_i \delta E_j \rangle_{\omega, \mathbf{k}} \frac{\partial f_{a0}}{\partial v_{ai}} \frac{\partial f_{b0}}{\partial v_{bj}} -$$

$$- \frac{4\pi}{k^2} \frac{e_a e_b}{m_a} 2\pi \delta(\omega - \mathbf{k}\mathbf{v}_b) g(\omega, \mathbf{k}, \mathbf{v}_a) A_{ij}(\omega, \mathbf{k}) \alpha_{pj}(\omega, \mathbf{k}, \mathbf{v}_b) k_p f_{b0}(\mathbf{v}_b) \frac{\partial f_{a0}}{\partial v_{ai}} -$$

$$- \frac{4\pi}{k^2} \frac{e_a e_b}{m_b} 2\pi \delta(\omega - \mathbf{k}\mathbf{v}_a) g^*(\omega, \mathbf{k}, \mathbf{v}_b) A^*_{ji}(\omega, \mathbf{k}) \alpha_{pj}(\omega, \mathbf{k}, \mathbf{v}_a) k_p f_{a0}(\mathbf{v}_a) \frac{\partial f_{b0}(\mathbf{v}_b)}{\partial v_{bi}}; \quad (14.16)$$

$$\langle \delta E_i \delta E_j \rangle_{\omega, \mathbf{k}} = \sum_a \frac{(4\pi)^2 e_a^2}{k^2} \int d\mathbf{v}_a f_{a0}(\mathbf{v}_a) 2\pi \delta(\omega - \mathbf{k}\mathbf{v}_a) \left\{ \frac{k_i k_j}{k^2} |\varepsilon^l(\omega, \mathbf{k})|^{-2} + \right.$$

$$\left. + \frac{1}{2}\left(\delta_{ij} - \frac{k_i k_j}{k^2}\right) [\mathbf{k}\mathbf{v}_a]^2 |\omega^2 \varepsilon^{tr}(\omega, \mathbf{k}) - c^2 k^2|^{-2} \right\}. \quad (14.17)$$

Here, $\varepsilon^l(\omega, \mathbf{k})$ and $\varepsilon^{tr}(\omega, \mathbf{k})$ are the longitudinal and transverse permittivities of an isotropic plasma with the equilibrium distribution function f_{a0} [cf. (6.19)]

$$\varepsilon^l(\omega, \mathbf{k}) = 1 + \sum_a \frac{4\pi e_a^2}{k^2} \frac{1}{\omega} \int d\mathbf{v}_a \frac{(\mathbf{k}\mathbf{v}_a)}{\omega - \mathbf{k}\mathbf{v}_a} \mathbf{k} \frac{\partial f_{a0}}{\partial \mathbf{v}_a};$$
$$\varepsilon^{tr}(\omega, \mathbf{k}) = 1 + \sum_a \frac{2\pi e_a^2}{k^2} \frac{1}{\omega} \int d\mathbf{v}_a \frac{[[\mathbf{k}\mathbf{v}_a]\,\mathbf{k}]}{\omega - \mathbf{k}\mathbf{v}_a} \frac{\partial f_{a0}}{\partial \mathbf{v}_a}. \quad (14.18)$$

In an equilibrium state with the Maxwell (6.14) distribution function f_{a0}, the correlation function of the fluctuation electric fields $\langle \delta E_i \delta E_j \rangle_{\omega, \mathbf{k}}$ (the spectral function of the fields) can be represented in accordance with (14.17) by the equation

$$\langle \delta E_i \delta E_j \rangle_{\omega, \mathbf{k}} = -\frac{8\pi}{\omega} \varkappa T \, \mathrm{Im}\, A_{ij}(\omega, \mathbf{k}), \quad (14.19)$$

so that the cross section (14.14) with allowance for the explicit expressions (14.15)-(14.17) and (14.19) of the correlation functions that it contains will now depend only on the three-index tensors ε_{ijs} and S_{ijs} [see the definition (3.24)], the inverse tensors A_{ij}, and the tensors α_{ij} (in the last term):

$$d\sigma(\mathbf{k}, \mathbf{k}') = -\frac{1}{(2\pi)^3} \frac{d\mathbf{k}'}{16\pi^2 W(\mathbf{k}) \left| \frac{\partial \omega}{\partial \mathbf{k}} \right|} \mathrm{Im} \left\{ E_{0n} E^*_{0m} \int_{-\infty}^{+\infty} \omega' d\omega' A_{ji}(\omega', \mathbf{k}') \times \right.$$

$$\times\Big[-\frac{8\pi}{\omega''}\varkappa T(\operatorname{Im}A_{sp}(\omega'',\mathbf{k}''))S^*_{jpm}(\omega',\mathbf{k}';\omega,\mathbf{k})S_{isn}(\omega',\mathbf{k}';\omega,\mathbf{k})+$$

$$+8i\pi\frac{\varkappa T}{\omega''}S^*_{jpm}(\omega',\mathbf{k}';\omega,\mathbf{k})\operatorname{Re}\varepsilon_{ins}(\omega',\mathbf{k}';\omega'',\mathbf{k}'')A^*_{sp}(\omega'',\mathbf{k}'')+$$

$$+\operatorname{Re}\varepsilon^*_{jmp}(\omega',\mathbf{k}';\omega'',\mathbf{k}'')S_{ins}(\omega',\mathbf{k}';\omega'',\mathbf{k}'')A_{sp}(\omega'',\mathbf{k}'')+$$

$$+\sum_a\frac{(4\pi)^2\,2\pi e_a^4}{m_a^2}\int d\mathbf{v}_a f_{a0}(\mathbf{v}_a)\alpha_{pi}(\omega',\mathbf{k}',\mathbf{v}_a)\alpha_{qj}(\omega',\mathbf{k}',\mathbf{v}_a)\alpha_{pn}(\omega,\mathbf{k},\mathbf{v}_a)\times$$

$$\times\alpha_{qm}(\omega,\mathbf{k},\mathbf{v}_a)|g(\omega',\mathbf{k}',\mathbf{v}_a)|^4\delta(\omega''-\mathbf{k}''\mathbf{v}_a)\Big]\Big\}. \qquad (14.20)$$

In the real parts Re ε_{ijs} should be taken into account the contribution from only the pole $\omega'' = \mathbf{k}''\mathbf{v}_a$;* then we have the equation [see the definition (3.24) of S_{ijs} and Eqs. (7.1) and (7.5) for ε_{ijs} and ε_{isj}]

$$\operatorname{Re}\varepsilon_{isj}(\omega',\mathbf{k}';\omega'',\mathbf{k}'')=\operatorname{Re}S_{ijs}(\omega',\mathbf{k}';\omega,\mathbf{k}). \qquad (14.21)$$

We shall also assume that Im $A_{ij}(\omega'',\mathbf{k}'')$ and Re S_{ijs}, which are proportional to the number of particles that participate in the scattering, are small. Then, restricting ourselves to the largest of these terms, we find that (14.20) is replaced by

$$d\sigma(\mathbf{k},\mathbf{k}')=\frac{1}{(2\pi)^4}\frac{d\mathbf{k}'}{W(\mathbf{k})\left|\frac{\partial\omega}{\partial\mathbf{k}}\right|}\varkappa T\operatorname{Im}\Big\{E_{0n}E^*_{0m}\int_{-\infty}^{+\infty}\omega' d\omega' A_{ji}(\omega',\mathbf{k}')\times$$

$$\times\operatorname{Im}[A_{lp}(\omega'',\mathbf{k}'')S_{pjm}(\omega'',\mathbf{k}'';\omega',\mathbf{k}')S_{iln}(\omega',\mathbf{k}';\omega,\mathbf{k})-V_{ijnm}(\omega',\mathbf{k}';\omega,\mathbf{k})]\Big\}. \qquad (14.22)$$

Here, Im V_{ijnm} is the last term on the right-hand side of (14.20):

$$\operatorname{Im}V_{ijnm}(\omega',\mathbf{k}';\omega,\mathbf{k})=\frac{\omega''}{\varkappa T}\sum_a\frac{4\pi^2e_a^4}{m_a^2}\int d\mathbf{v}_a\alpha_{pi}(\omega',\mathbf{k}',\mathbf{v}_a)f_{a0}(\mathbf{v}_a)\times$$

$$\times\alpha_{qj}(\omega',\mathbf{k}',\mathbf{v}_a)\alpha_{pn}(\omega,\mathbf{k},\mathbf{v}_a)\alpha_{qm}(\omega,\mathbf{k},\mathbf{v}_a)|g(\omega',\mathbf{k}',\mathbf{v}_a)|^4\,\delta(\omega'-\omega-\mathbf{k}'\mathbf{v}_a+\mathbf{k}\mathbf{v}_a) \qquad (14.23)$$

and we have used the symmetry properties (7.19)-(7.20) of the three-index tensors S_{ijs}. The expression of the cross section in the form (14.22) corresponds completely to the general expression (12.12) used in the theory of nonlinear interaction, since the tensor part of $(E_{0n}E^*_{0m})\operatorname{Im}A_{ji}(\omega',\mathbf{k}')$ is equal† to the product $(E_nE_m)_{\omega,\mathbf{k}}(E_jE_i)_{\omega',\mathbf{k}'}$, of the spectral functions that arise in the equation (3.22b) of the nonlinear interaction (induced scattering on particles) of the characteristic oscillations of the plasma, and the expression (14.23) is equal, in agreement with the notation used for it, to the imaginary part of the four-index tensor V (8.14) in the nonlinear theory of wave interaction [f_{a0} is the Maxwell distribution (6.14)].

We now consider the special case of the differential cross section (14.14), allowing explicitly in it for the fact that the incident and the scattered waves have phase velocities that are appreciably greater than the thermal velocities of both the electrons and the ions. We ex-

---

* As we shall see below, this condition does not, in fact, impose any serious restrictions on the applicability of the theory of wave scattering developed here since it is assumed as a rule in such a theory that one considers the scattering of electromagnetic waves with high phase velocities [$(\omega/k)\gg v_{Te}$, $(\omega'/k')\gg v_{Te}$], i.e., one ignores the contributions of the poles $\omega = \mathbf{k}\mathbf{v}_a$ and $\omega' = \mathbf{k}'\mathbf{v}_a$.

† We recall that the resonance part of the inverse tensor (or the imaginary part, which is the same thing in an isotropic plasma) is equal to the tensor part of the spectral function of the fields of the characteristic oscillations in the plasma.

pand all quantities that depend on ω and $\mathbf{k}$ and ω' and $\mathbf{k}'$ in (14.14), assuming [cf. (7.70)] $(\mathbf{kv}/\omega) \ll 1$ and $(\mathbf{k'v}/\omega) \ll 1$. In addition, in the expression (14.11) for ε_{ijs} we allow for only the electron terms, ignoring quantities of order m/M. Then in the lowest approximation in kv_T/ω obtain from (14.14)

$$d\sigma(\mathbf{k}, \mathbf{k}') = -\frac{1}{(2\pi)^3}\frac{d\mathbf{k}'}{16\pi^2 W(\mathbf{k})\left|\frac{\partial\omega}{\partial\mathbf{k}}\right|}\,\mathrm{Im}\left\{E_{0n}E^*_{0m}\int\limits_{-\infty}^{+\infty}\omega' d\omega' A_{js}(\omega', \mathbf{k}')\times\right.$$

$$\times\Big[\varepsilon_{isn}(\omega', \mathbf{k}'; \omega, \mathbf{k})\,\varepsilon^*_{jpm}(\omega', \mathbf{k}'; \omega, \mathbf{k})\langle\delta E_s\delta E_p\rangle_{\omega'', \mathbf{k}''} - \frac{4\pi e^2}{m\omega'^2}(\varepsilon^*_{jpm}(\omega', \mathbf{k}'; \omega, \mathbf{k})\,\delta_{in}\langle\delta n_e\delta E_p\rangle_{\omega'', \mathbf{k}''} +$$

$$\left.+\,\varepsilon_{isn}(\omega', \mathbf{k}'; \omega, \mathbf{k})\,\delta_{jm}\langle\delta E_s\delta n_e\rangle_{\omega'', \mathbf{k}''}) + \frac{16\pi^2 e^4}{m^2\omega'^4}\delta_{in}\delta_{jm}\langle\delta n_e^2\rangle_{\omega'', \mathbf{k}''}\Big]\right\}. \qquad (14.24)$$

Here the three-index tensors ε_{ijs} have, in accordance with the definition (14.11) [cf. (7.23)], the form

$$\varepsilon_{ijs}(\omega', \mathbf{k}'; \omega, \mathbf{k}) = -i\frac{4\pi N_e e^3}{m^2}\frac{1}{\omega''\omega^2\omega'^3}\{\omega\omega'' k'_j\delta_{si} + \omega\omega' k''_i\delta_{sj} + \delta_{ij}(k_s\omega^2 - k'_s\omega^2)\}, \qquad (14.25)$$

and the correlation functions are given in accordance with Eqs. (14.15)-(14.17) by

$$\langle\delta n_e^2\rangle_{\omega, \mathbf{k}} \equiv \int d\mathbf{v}d\mathbf{v}'\langle\delta N_e(\mathbf{v})\delta N_e(\mathbf{v}')\rangle_{\omega, \mathbf{k}} = \frac{k^2}{2\pi^2 e^2\omega}|\varepsilon^l|^{-2}\{\delta\varepsilon_e^{l''}\varkappa T_e|1 + \delta\varepsilon_i^l|^2 + \varkappa T_i\delta\varepsilon_i^{l''}|\delta\varepsilon_e^l|^2\}; \qquad (14.26)$$

$$\langle\delta n_e\delta E_j\rangle_{\omega, \mathbf{k}} \equiv \int d\mathbf{v}\langle\delta N_e(\mathbf{v})\delta E_j\rangle_{\omega, \mathbf{k}} = 4\pi i\frac{k_j}{k^2}\{e\langle\delta n_e^2\rangle_{\omega, \mathbf{k}} + e_i\langle\delta n_e\delta n_i\rangle_{\omega, \mathbf{k}}\}; \qquad (14.27)$$

$$\langle\delta E_s\delta E_j\rangle_{\omega, \mathbf{k}} = \frac{16\pi}{k^2}\frac{k_s k_j}{k^2}[e^2\langle\delta n_e^2\rangle_{\omega, \mathbf{k}} + e_i^2\langle\delta n_i^2\rangle_{\omega, \mathbf{k}} + e_i e\langle\delta n_e\delta n_i\rangle_{\omega, \mathbf{k}} +$$

$$+\, ee_i\langle\delta n_i\delta n_e\rangle_{\omega, \mathbf{k}}] + \frac{8\pi}{\omega}\frac{k^2}{\omega^2}\left(\delta_{sj} - \frac{k_s k_j}{k^2}\right)\frac{(\varkappa T_e)^2}{m}\delta\varepsilon_e^{l''}\,|\varepsilon^{tr}(\omega, \mathbf{k}) - (c^2k^2/\omega^2)|^{-2}, \qquad (14.28)$$

in which $\delta\varepsilon_e^l$ and $\delta\varepsilon_i^l$ are the partial (electron and ion) parts of the longitudinal permittivity, which depend on the frequency ω and the wave vector $\mathbf{k}$ [see the definition (9.6)]. The second term in the correlation function of the fluctuation fields (14.28), which is proportional to the tensor $[\delta_{sj} - (k_s k_j/k^2)]$, corresponds to fluctuations of the transverse (nonelectrostatic) field. Below, this will be ignored.

For frequencies ω and ω' that are either large compared with the electron plasma frequency ω_{Le} or of order of ω_{Le}, the largest term in (14.24) is the last. And, ignoring all terms except the last in the square brackets in (14.24), we obtain an expression for the cross section of scattering due to fluctuations of the electron number, this corresponding to the expression previously obtained [8, 9, 46]. However, it must be emphasized that this result is by no means exhaustive. Namely, in a considerable number of cases the first nonvanishing term in the expansion of the cross section (14.14) in powers of kv_T/ω is anomalously small. Under these conditions, we must also allow for the higher powers in the expansion of the last term in (14.14) in powers of kv_T/ω. When such an expansion is made, we obtain correlation functions that have the form

$$\langle\delta n_e\delta v_{e,p}\rangle_{\omega, \mathbf{k}} \equiv \int d\mathbf{v}d\mathbf{v}' v'_p\langle\delta N_e(\mathbf{v})\delta N_e(\mathbf{v}')\rangle_{\omega, \mathbf{k}} = \frac{k_p}{k^2}\omega\langle\delta n_e^2\rangle_{\omega, \mathbf{k}}; \qquad (14.29)$$

$$\langle\delta v_{e,p}\delta v_{e,j}\rangle_{\omega, \mathbf{k}} \equiv \int d\mathbf{v}d\mathbf{v}' v_p v'_j\langle\delta N_e(\mathbf{v})\delta N_e(\mathbf{v}')\rangle_{\omega, \mathbf{k}} = \frac{k_p k_j}{k^2}\frac{\omega^2}{k^2}\langle\delta n_e^2\rangle_{\omega, \mathbf{k}} +$$

$$+\frac{\varkappa T_e}{m}\left(\delta_{pj} - \frac{k_p k_j}{k^2}\right)\delta\varepsilon_e^{l''}(\omega, \mathbf{k})\frac{(\varkappa T_e)k^2}{\omega 4\pi e^2}; \qquad (14.30)$$

$$\langle \delta n_e \delta \pi_{e,\,sj} \rangle_{\omega,\,\mathbf{k}} \equiv \int d\mathbf{v} d\mathbf{v}' v'_s v'_j \langle \delta N_e(\mathbf{v}) \delta N_e(\mathbf{v}') \rangle_{\omega,\,\mathbf{k}} =$$

$$= \left[\frac{k_s k_j}{k^2} \frac{\omega^2}{k^2} + \frac{\varkappa T_e}{m} \left(\delta_{sj} - \frac{k_s k_j}{k^2} \right) \right] \langle \delta n_e^2 \rangle_{\omega,\,\mathbf{k}} + \frac{k_s k_j}{k^2} \frac{\varkappa T_e}{m} 2N_e \frac{1}{\omega} |\delta\varepsilon^l|^{-2} [\delta\varepsilon_i^{l''} \delta\varepsilon_e^l - \delta\varepsilon_e^{l''} |1 + \delta\varepsilon_i^l|]. \quad (14.31)$$

The terms in these formulas that contain $\langle \delta n_e^2 \rangle_{\omega,\,\mathbf{k}}$ can be ignored, since they are always small compared with the last term in the square brackets (14.24), if, of course, $\omega'' \ll \omega, \omega'$, i.e., we consider scattering with a small change of frequency [cf. (7.76)]. The last term in the correlation function (14.31) leads in the expression for the cross section to quantities that are of the same order as the terms with $\langle \delta n_e \delta E_i \rangle$, and can also be ignored.

Thus, the only terms that must be taken into account are terms that are transverse with respect to $\mathbf{k}$ and are of the type these corresponding to fluctuations of the solenoidal current. Then the expression for the cross section takes the form

$$\langle \delta v_i \delta v_j \rangle^{tr}_{\omega,\,\mathbf{k}} \equiv \langle \delta v_{e,\,i} \delta v_{e,\,j} \rangle^{tr}_{\omega,\,\mathbf{k}} = \frac{\varkappa T_e}{m} \left(\delta_{ij} - \frac{k_i k_j}{k^2} \right) \frac{(\varkappa T_e) k^2}{\omega 2\pi e^2} \delta\varepsilon_e^{l''}(\omega, \mathbf{k}), \quad (14.30a)$$

these corresponding to fluctuations of the solenoidal current. Then the expression for the cross section takes the form

$$d\sigma(\mathbf{k}, \mathbf{k}') = -\frac{1}{(2\pi)^3} \frac{d\mathbf{k}'}{W(\mathbf{k}) \left| \frac{\partial \omega}{\partial \mathbf{k}} \right|} \frac{e^4}{m^2} \operatorname{Im} \left\{ E_{0n} E^*_{0m} \int_{-\infty}^{+\infty} \frac{d\omega'}{\omega'} A_{ji}(\omega', \mathbf{k}') \times \right.$$

$$\times \left[\delta_{in} \delta_{jm} \langle \delta n_e^2 \rangle_{\omega'',\,\mathbf{k}''} + \frac{1}{\omega^2} \left(k'_n k'_m \langle \delta v_i \delta v_j \rangle^{tr}_{\omega'',\,\mathbf{k}''} + k_i k'_m \langle \delta v_n \delta v_j \rangle^{tr}_{\omega'',\,\mathbf{k}''} + \right. \right.$$

$$\left. \left. \left. + k'_n k_j \langle \delta v_i \delta v_m \rangle^{tr}_{\omega'',\,\mathbf{k}''} + k_i k_j \langle \delta v_n \delta v_m \rangle^{tr}_{\omega'',\,\mathbf{k}''} \right) \right] \right\}. \quad (14.32)$$

The terms $\sim \langle \delta v_p \delta v_q \rangle^{tr}_{\omega'',\,\mathbf{k}''}$ in the cross section (14.32) are important in principle. We must emphasize once more that if only such terms are allowed for, i.e., for fluctuations of the solenoidal current, we obtain exhaustive expressions for the scattering of longitudinal and transverse waves in a plasma. These terms qualitatively change the expressions for,* for example, the cross section of Coulomb scattering of Langmuir waves in an electron plasma, which agrees with the cross section (12.14) that we have obtained earlier in the theory of the nonlinear interaction of waves. In contrast, when the scattering is due to ions [see, for example, the cross section (12.13)] the fluctuations of the solenoidal current are not very important and in the cross section (14.32) it is sufficient to retain only the first term $\sim \langle \delta n_e^2 \rangle_{\omega'',\,\mathbf{k}''}$ [see formula (10.45) in [9]].

In view of the identity proven above of the general expressions (14.22) and (12.12) for the scattering cross sections of waves in a plasma obtained in this section and earlier (§ 12) in the theory of the nonlinear interaction of waves, the expression (14.32) for the cross section of scattering of waves with high phase velocities must be regarded as a different formulation of the results obtained in § 12. It is very useful for a qualitative explanation of the physical meaning of the individual terms that make the most important contribution to the scattering cross section. As an illustration of formula (14.32), let us consider in the conclusion of this section the scattering of a naturally polarized transverse electromagnetic wave with the formation of another wave of this kind. Then, in contrast to the special case considered in § 12 of the scatter-

* Note that in [8, 9, 46-49] the contribution of the fluctuations of the solenoidal current to the cross section is ignored, i.e., allowance is made for only the first term, $\sim \langle \delta n_e^2 \rangle_{\omega'',\,\mathbf{k}''}$, which leads to incomplete expressions for the cross sections for the scattering of waves in an electron plasma.

ing of long-wavelength ($\lambda \gg \lambda_0 = c/\omega_{Le}$) transverse waves, we shall discuss the scattering of short-wavelength (high-frequency) oscillations (6.59), when to a high accuracy we can assume ($\lambda \ll \lambda_0 = c/\omega_{Le}$)

$$\omega = ck, \quad \omega' = ck'. \tag{14.33}$$

Taking into account in the cross section (14.34) the following expressions for the quantities contained in it:

$$E_{0n}E^*_{0m} = |E_0^2|\left(\delta_{nm} - \frac{k_n k_m}{k^2}\right),$$
$$W(\mathbf{k}) = \frac{1}{4\pi}|E_0|^2, \tag{14.34}$$
$$\operatorname{Im} A_{ij}(\omega', \mathbf{k}') = -\pi\left(\delta_{ij} - \frac{k'_i k'_j}{k'^2}\right)\delta\left[1 - \frac{c^2 k'^2}{\omega'^2}\right],$$

we obtain [34]

$$d\sigma^{tt}(\mathbf{k}, \mathbf{k}') = \left(\frac{e^2}{mc^2}\right)^2 \frac{c}{4\pi}\frac{d\mathbf{k}'}{k'^2}\left\{(1+\cos^2\theta)\langle\delta n_e^2\rangle_{\omega'', \mathbf{k}''} + 4\sin^2\theta\,(2-\cos\theta)\frac{N_e}{c}\frac{(k'' r_{De})^2}{k-k'}\delta\varepsilon_e^{l''}(\omega'', \mathbf{k}'')\frac{v_{Te}^2}{c^2}\right\}, \tag{14.35}$$

where θ is the angle between $\mathbf{k}$ and $\mathbf{k}'$.

For scattering on fluctuations of the solenoidal electron velocity to become decisive* it is necessary for the angle θ to be ~ 1 and also

$$\frac{v_{Te}}{c} \gg \frac{m}{M}. \tag{14.36}$$

In a hydrogen plasma this is the case at temperatures above 2000°. In addition, we must have

$$\frac{\omega_{Le}^2}{\omega^2} \gg \sqrt{\frac{v_{Te}}{c}}\frac{\omega_{Le}}{\omega} \gg \frac{\omega''}{\omega} \gg \frac{\omega_{Li}}{\omega}, \quad \frac{v_{Ti}}{c}\sqrt{\ln\left\{\frac{c^2}{v_{Te}v_{Ti}}\frac{T_e}{T_i}\right\}}, \tag{14.37}$$

or

$$\frac{\omega_{Li}}{\omega} \gg \frac{\omega''}{\omega} \gg \sqrt{\frac{m}{M}\frac{v_{Te}}{c}}, \quad \frac{v_{Ti}}{c}\sqrt{\ln\left\{\frac{c^2}{v_{Te}v_{Ti}}\frac{T_e}{T_i}\right\}}. \tag{14.38}$$

It follows in particular that although the frequency of the waves that participate in the scattering can appreciably exceed the electron plasma frequency, it must not be too high. Namely, the frequency of the scattered waves must be less than $\omega_{Le}(c/v_{Te})^{1/2}$.

CHAPTER IV

MULTI-INDEX TENSORS OF THE COMPLEX PERMITTIVITY OF A HOMOGENEOUS MAGNETOACTIVE PLASMA

This chapter is to the nonlinear theory of oscillations of a magnetoactive plasma what Chapter II is to the theory of an isotropic plasma. It consists of seven sections (§ 15-21).

---

*In [8, 9, 46-49] in the study of the scattering of transverse waves allowance is made for only the first term in the square brackets of (14.35), which is proportional to the correlation function $\langle\delta n_e^2\rangle_{\omega'', \mathbf{k}''}$, this determining the fluctuations of the electron number.

The main result of § 15 is the solution (15.53) of Vlasov's equation in the form of a power series in the electric field in a magnetoactive plasma. The procedure for obtaining such a solution is the same as in an isotropic plasma (Ch. II, § 5).

In § 16 we determine explicit expressions for the multi-index tensors $\varepsilon_{ij1\ldots jn}$, which take into account the presence of a constant and homogeneous magnetic field $\mathbf{B}_0$. The greater part of this section is devoted to the permittivity tensor $\varepsilon_{ij}(\omega, \mathbf{k})$ as the simplest representative of the tensor $\varepsilon_{ij1\ldots jn}$ for n = 1. Methodologically, our treatment is characterized by the representation (16.44) of the tensor $\varepsilon_{ij}(\omega, \mathbf{k})$ as a decomposition into six elementary tensors composed of a unit vector $\varkappa$ along the wave vector $\mathbf{k}$, a unit vector $\mathbf{h}$ along the external magnetic field $\mathbf{B}_0$ and the unit tensors δ_{ij} and e_{isj} (see [50]). The notation that we introduce [see, in particular, formulas (16.12) and (16.15) for the tensors W and ρ] goes a long way, in our opinion, to simplify the study of the three- and four-index tensors in nonlinear theory.

In § 17 we discuss expressions for the inverse tensor $A_{ij}(\omega, \mathbf{k})$ and the spectral function of the fields $(E_j E_i)_{\omega, \mathbf{k}}$, which are also represented as decompositions into elementary tensors. The finding of the spectral function is facilitated by its tensor part being equal to the resonance part of the inverse tensor. We consider different special cases (longitudinal and transverse propagation of oscillations along $\mathbf{B}_0$, cold plasma, etc.).

The exposition of the ancillary results from the linear theory of a magnetoactive plasma (see [1], Ch. III) is begun in § 16 and 17 and completed in § 18. Here, we formulate necessary and sufficient conditions for the characteristic oscillations of a magnetoactive plasma to be quasilongitudinal; in general, these conditions differ from those adopted in the literature (cf. [8], Ch. II, § 4, p. 31, 32; [51], Ch. IX, p. 288-289). The simplicity and clarity of these conditions is due to the representation of the spectral function of the fields in the plasma as a decomposition into elementary tensors, this making the relative contribution of each polarization manifest. The major part of this section is devoted to a discussion of the spectra and spectral functions of quasilongitudinal waves in a plasma with an external magnetic field $\mathbf{B}_0$.

In § 19 we study in detail the three-index tensor $S_{ijs}(\omega, \mathbf{k}; \omega', \mathbf{k}')$ and its longitudinal contraction. The need to make a detailed study of the different forms of the first nonlinear tensors S and V in a magnetoactive plasma is dictated, in our opinion, not only by the important role of these tensors in the theory of the nonlinear interaction of waves in a plasma, which is obvious from general considerations, but also by the actual development of this theory and related experiments during the last five or six years. The number of conceivable particular nonlinear processes in which the wave vectors and the magnetic field $\mathbf{B}_0$ have different relative dispositions and the parameters of the magnetoactive plasma vary in a wide range is fairly large. In this connection, despite the large number of cross sections of such processes (see Ch. V) already obtained in the theory it is always necessary, if one wishes to discuss new experiments, to calculate once more some new variant of nonlinear interaction, not studied hitherto. It is in this connection that the importance of the different special cases of the tensor S increases; on the one hand, it is not difficult to make exhaustive studies of these cases, while on the other they allow one to obtain necessary cross sections with comparative ease. This section consists of three parts. First, we analyze the expression (19.18) for the tensor S in a cold magnetoactive plasma. We then consider the longitudinal contraction (19.68) of this tensor in a hot plasma. At the end of the section we give the general symmetric expression (19.97) for the tensor S in a hot magnetoactive plasma.

A characteristic feature of the formulas is their symmetry, which greatly facilitates the calculation of specific cross sections of nonlinear processes.

The four-index tensor V is discussed in § 20. We give the expression (20.3) for the tensor V in a cold magnetoactive plasma and the symmetrized combination (10.7) of the six four-index tensors ε ($v_T = 0$). We consider in more detail the longitudinal contraction of the tensor in a hot magnetoactive plasma; this is used in the next chapter to calculate the induced Coulomb scattering of quasilongitudinal oscillations of such a plasma.

In § 21 the general laws that the equations of the nonlinear interaction of waves in an isotropic and magnetoactive plasma satisfy are discussed. It is shown that in decay processes the total energy and total momentum of the waves is conserved, while the entropy increases. In processes of induced scattering on particles the total number of scattered oscillations is conserved. Deductions concerning the conservation laws in the theory of the nonlinear interaction waves are obtained on the basis of the general symmetry properties of the three- and four-index tensors and in this sense they are valid not only in a plasma but also for a larger class of material media that have this symmetry.

§ 15. Solution of the Transport Equation for a Magnetoactive Plasma in the Form of a Power Series in the Self-Consistent Electric Field

The transport equation is solved in the nonlinear approximation for a plasma in a constant and homogeneous external magnetic field $\mathbf{B}_0$ on the basis of the same physical assumptions as we used for an isotropic plasma in § 5.

Substituting the expansion (5.4) of the distribution function into Vlasov's equation, which takes into account the influence of the external field $\mathbf{B}_0$,

$$\frac{\partial f}{\partial t}+\mathbf{v}\frac{\partial f}{\partial \mathbf{r}}+e\left\{\mathbf{E}+\frac{1}{c}[\mathbf{v},\ \mathbf{B}+\mathbf{B}_0]\right\}\frac{\partial f}{\partial \mathbf{p}}=0, \tag{15.1}$$

we obtain a chain of coupled equations for the nonequilibrium corrections:

$$\frac{\partial f_n}{\partial t}+\mathbf{v}\frac{\partial f_n}{\partial \mathbf{r}}+\frac{e}{c}[\mathbf{v}\mathbf{B}_0]\frac{\partial f_n}{\partial \mathbf{p}}=-e\left\{\mathbf{E}+\frac{1}{c}[\mathbf{v}\mathbf{B}]\right\}\frac{\partial f_{n-1}}{\partial \mathbf{p}},\qquad n=1,\ 2,\ \ldots \tag{15.2}$$

this chain differing from the similar chain (5.6) in an isotropic plasma by the presence of the third term on the left-hand side. It is this term that is responsible for all the specific features of the nonlinear interaction of the characteristic oscillations of a magnetoactive plasma as opposed to an isotropic plasma. The first of the equations of the system (15.2) ($n = 1$),

$$\frac{\partial f_1}{\partial t}+\mathbf{v}\frac{\partial f_1}{\partial \mathbf{r}}+\frac{e}{c}[\mathbf{v}\mathbf{B}_0]\frac{\partial f_1}{\partial \mathbf{p}}=-e\left\{\mathbf{E}+\frac{1}{c}[\mathbf{v}\mathbf{B}]\right\}\frac{\partial f_0}{\partial \mathbf{p}} \tag{15.3}$$

determines the linear (in the electric $\mathbf{E}$ and magnetic $\mathbf{B}$ fields) nonequilibrium distribution function $f_1(\mathbf{p}, \mathbf{r}, t)$ in terms of the unperturbed distribution function f_0. The distribution function f_0 of an unperturbed homogeneous and stationary state of a magnetoactive plasma does not depend on the self-consistent fields $\mathbf{E}$ and $\mathbf{B}$ and is determined by the equation

$$\frac{e}{c}[\mathbf{v}\mathbf{B}_0]\frac{\partial f_0}{\partial \mathbf{p}}=0. \tag{15.4}$$

In the general case of an arbitrary equilibrium state of a magnetoactive plasma, f_0 can be an arbitrary function of $f_0(\mathbf{p}, \mathbf{r}, t)$. In all cases we shall assume that f_0 is given and fixed. In contrast to an isotropic plasma, Eq. (15.4) imposes restrictions on the form of f_0. Namely, it follows from (15.4) that f_0 can be an arbitrary function of the integrals of the equation

$$\frac{d\mathbf{p}}{dt}=\frac{e}{c}[\mathbf{v}\mathbf{B}_0], \tag{15.5}$$

which can be found by successive scalar multiplication of both sides of (15.5) by a unit vector **h** along the external magnetic field $\mathbf{B}_0 = \mathbf{h}B_0$ and by the momentum **p**:

$$(\mathbf{ph}) = \text{const}, \quad [\mathbf{ph}]^2 = \text{const}. \tag{15.6}$$

Physically, the choice of Eq. (15.4) for f_0 amounts to the assumption that the characteristic times and lengths of variation of the self-consistent fields and, therefore, the first nonequilibrium correction $f_1(\mathbf{p}, \mathbf{r}, t)$ are negligibly short and small compared with the characteristic time and length of variation of f_0. To find $f_1(\mathbf{p}, \mathbf{r}, t)$, it is necessary to know the solutions of the equations of the characteristics of the differential operator on the left-hand side of (15.3):

$$\frac{d\mathbf{r}}{dt} = \mathbf{v}, \quad \frac{d\mathbf{p}}{dt} = \frac{e}{c}[\mathbf{v}\mathbf{B}_0] \tag{15.7}$$

Noting that the energy

$$\mathcal{E} = (c^2p^2 + m^2c^4)^{1/2} \tag{15.8}$$

is an integral of Eq. (15.5), and taking into account the relativistic relation between the energy, momentum, and velocity of a particle:

$$\mathbf{p} = \frac{\mathcal{E}}{c^2}\mathbf{v}, \tag{15.9}$$

we rewrite the equations of motion (15.7) differently:

$$\frac{d\mathbf{r}}{dt} = \mathbf{v}, \quad \frac{d\mathbf{v}}{dt} = \Omega[\mathbf{vh}]. \tag{15.10}$$

Here, for the gyroscopic frequency Ω we have used the notation

$$\Omega = \frac{eB_0c}{\mathcal{E}}, \tag{15.11}$$

which goes over in the nonrelativistic approximation (5.30) into the expression

$$\Omega = \frac{eB_0}{mc}, \quad mc^2 \gg cp. \tag{15.12}$$

We solve the second of the equations (15.10) by decomposing the velocity $\mathbf{v}(t)$ with respect to orthogonal vectors,

$$\mathbf{v}(t) = \mathbf{h}A(t) + [\mathbf{v}_0\mathbf{h}]B(t) + [\mathbf{h}[\mathbf{v}_0\mathbf{h}]]C(t), \tag{15.13}$$

which are composed of a unit vector **h** along $\mathbf{B}_0$ and the vector $\mathbf{v}_0$, which is the velocity **v** at the initial time t_0:

$$\mathbf{v}_0 \equiv \mathbf{v}(t_0). \tag{15.14}$$

Multiplying both sides of Eq. (15.13) into the second equation (15.10) and compare terms of vectors with the same directions. This comparison gives two equations for determining B(t) and C(t):

$$(\mathbf{vh}) = A(t) = \text{const} = (\mathbf{v}_0\mathbf{h}). \tag{15.15}$$

We substitute (15.13) into the second equation (15.10) and compare terms of vectors with the same directions. This comparison gives two equations for determining B(t) and C(t):

$$\frac{dB}{dt} = \Omega C, \quad \frac{dC}{dt} = -\Omega B, \tag{15.16}$$

and these have the solutions

$$B = B_1 \sin \Omega t + B_2 \cos \Omega t,$$
$$C = B_1 \cos \Omega t - B_2 \sin \Omega t. \tag{15.17}$$

We find the constants B_1 and B_2 by setting $t = t_0$ in (15.13):

$$B_1 \sin \Omega t_0 + B_2 \cos \Omega t_0 = 0,$$
$$B_1 \cos \Omega t_0 - B_2 \sin \Omega t_0 = 1. \tag{15.18}$$

Namely,

$$B_1 = \cos \Omega t_0, \quad B_2 = -\sin \Omega t_0. \tag{15.19}$$

Collecting Eqs. (15.5), (15.17), and (15.19) in (15.13), we obtain the desired solution (cf. [1], (19.6))

$$\mathbf{v}(t) = (\mathbf{hv}(t_0))\mathbf{h} + [\mathbf{v}(t_0)\mathbf{h}] \sin(\Omega(t - t_0)) + [\mathbf{h}[\mathbf{v}(t_0)\mathbf{h}]] \cos(\Omega(t - t_0)). \tag{15.20}$$

The structure of the right-hand side of (15.20) shows that

$$\mathbf{v}(t) \equiv \mathbf{v}(t - t_0, \mathbf{v}(t_0)). \tag{15.21}$$

Thus, the solution (15.20) is a relation between the velocities at two arbitrary times t and t_0. We find the solution of the first of the equations (15.7) of the characteristics by using (15.20):

$$\mathbf{r}(t) = \mathbf{r}(t_0) + \int_{t_0}^{t} dt' \mathbf{v}(t'), \qquad \mathbf{r}(t_0) = \mathbf{r}(t = t_0). \tag{15.22}$$

For convenience in what follows we use the notation (cf. [1], (19.6))

$$\delta\mathbf{R}(t - t_0, \mathbf{v}(t_0)) \equiv \mathbf{r}(t) - \mathbf{r}(t_0) = \int_{t_0}^{t} dt' \mathbf{v}(t') = (\mathbf{hv}(t_0))\mathbf{h}(t - t_0) +$$
$$+ \frac{[\mathbf{v}(t_0)\mathbf{h}]}{\Omega}\{1 - \cos(\Omega(t - t_0))\} + \sin(\Omega(t - t_0))[\mathbf{h}[\mathbf{v}(t_0)\mathbf{h}]]\frac{1}{\Omega}. \tag{15.23}$$

Equation (15.23) also relates the position **r** of a particle at the two times t and t_0:

$$\mathbf{r}(t) = \mathbf{r}(t_0) + \delta\mathbf{R}(t - t_0, \mathbf{v}(t_0)). \tag{15.24}$$

The characteristics can be used, in conjunction with an equation that follows from (15.3):

$$\frac{df_1}{dt} = -e\left\{\mathbf{E} + \frac{1}{c}[\mathbf{vB}]\right\}\frac{\partial f_0}{\partial \mathbf{p}} \tag{15.25}$$

to write down an expression for the nonequilibrium distribution function as well:

$$f_1(\mathbf{p}, \mathbf{r}, t) = f_1(t_0) - e\int_{t_0}^{t} dt' \left\{\mathbf{E}(\mathbf{r}(t'), t') + \frac{1}{c}[\mathbf{v}(t'), \mathbf{B}(\mathbf{r}(t'), t')]\right\}\frac{\partial f_0}{\partial \mathbf{p}(t')}. \tag{15.26}$$

The first term on the right-hand side of (15.26), which we have denoted by $f_1(t_0)$, is the solution of the homogeneous equation corresponding to (15.3),

$$\frac{\partial f_1}{\partial t} + \mathbf{v}\frac{\partial f_1}{\partial \mathbf{r}} + \frac{e}{c}[\mathbf{vB}_0]\frac{\partial f_1}{\partial \mathbf{p}} = 0, \tag{15.27}$$

this being equal at $t = t_0$ to the "initial" value of the nonequilibrium addition (see [13], p. 111, (9.12)):

$$f_1(t_0) \equiv f_1(\mathbf{p}(t_0 - t, \mathbf{v}), \mathbf{r} + \delta\mathbf{R}(t_0 - t, \mathbf{v}), t_0). \tag{15.28}$$

In the adiabatic approximation (5.8) that we are using in this monograph, this term is omitted everywhere.

The functions $\mathbf{r}(t')$, $\mathbf{v}(t')$, and $\mathbf{p}(t')$ in the second term of (15.26) are determined, respectively, by Eqs. (15.23), (15.20), and (15.9) after the substitution

$$t \to t', \quad t_0 \to t. \tag{15.29}$$

Thus, (cf. [1], (19.5))

$$f_1(\mathbf{p}, \mathbf{r}, t) = -e \int_{-\infty}^{t} dt' \Big\{ \mathbf{E}(\mathbf{r} + \delta\mathbf{R}(t'-t, \mathbf{v}), t') + \\ + \frac{1}{c} [\mathbf{v}(t'-t, \mathbf{v}) \mathbf{B}(\mathbf{r} + \delta\mathbf{R}(t'-t, \mathbf{v}), t')] \Big\} \frac{\partial f_0}{\partial \mathbf{p}(t'-t, \mathbf{v})}. \tag{15.30}$$

The distribution function of the equilibrium state in (15.30) is, in general, an arbitrary function of the integrals (15.6):

$$f_0 = f_0(\mathbf{ph}, [\mathbf{ph}]^2) \tag{15.31}$$

and it does not depend on either the variable of integration t' or the time t. As in Chapters II and III, we shall henceforth take f_0 to be the isotropic nonrelativistic Maxwell distribution (6.14)

In (15.30) we relabel the variable:

$$t' - t = \tau \tag{15.32}$$

and we go over to Fourier transforms, using in the process Eqs. (5.15)-(5.16):

$$f_1(\mathbf{p}, \mathbf{k}, \omega) = -e \int_{-\infty}^{0} d\tau e^{-i\omega\tau + i\mathbf{k}\delta\mathbf{R}(\tau, \mathbf{v})} \alpha_{ij}(\omega, \mathbf{k}, \mathbf{v}(\tau, \mathbf{v})) \frac{\partial f_0}{\partial p_i(\tau, \mathbf{v})} E_j(\mathbf{k}, \omega). \tag{15.33}$$

Equations (15.30) and (15.33) give a complete picture of the solution of the linearized transport equation (15.3) for a magnetoactive plasma in the approximation (5.8) of the adibatic switching on of the field in the infinitely distant past. The method used in this section to calculate the nonequilibrium distribution function f_1 could also be applied in an isotropic plasma (§ 5) (see [1], p. 77; [13], p. 89). In the limit $\mathbf{B}_0 = 0$ it follows from (15.20) and (15.23) that

$$\mathbf{v}(t) = \mathbf{v}(t - t_0, \quad \mathbf{v}(t_0)) = \mathbf{v}(t_0) = \text{const}; \tag{15.34}$$

$$\delta\mathbf{R}(t - t_0, \mathbf{v}(t_0)) = \mathbf{r}(t) - \mathbf{r}(t_0) = \mathbf{v}(t - t_0). \tag{15.35}$$

Therefore, the relation (15.30) takes the form

$$f_1(\mathbf{p}, \mathbf{r}, t) = -e \int_{-\infty}^{t} dt' \Big\{ \mathbf{E}(\mathbf{r} + \mathbf{v}(t'-t), t') + \frac{1}{c} [\mathbf{v}\mathbf{B}(\mathbf{r} + \mathbf{v}(t'-t), t')] \Big\} \frac{\partial f_0}{\partial \mathbf{p}}, \tag{15.36}$$

and the Fourier transform (15.33),

$$f_1(\mathbf{p}, \mathbf{k}, \omega) = -e \int_{-\infty}^{0} d\tau e^{-i\omega\tau + i\mathbf{k}\mathbf{v}\tau} \alpha_{ij}(\omega, \mathbf{k}, \mathbf{v}) \frac{\partial f_0}{\partial p_i} E_j(\mathbf{k}, \omega), \tag{15.37}$$

goes over after integration with respect to τ into the isotropic expression (15.17) if it is noted that [cf. (5.12)]

$$\int_{-\infty}^{0} d\tau e^{-i\omega\tau + i\mathbf{k}\mathbf{v}\tau} = \frac{i}{\omega - \mathbf{k}\mathbf{v}} = i \Big\{ \frac{P}{\omega - \mathbf{k}\mathbf{v}} - i\pi\delta(\omega - \mathbf{k}\mathbf{v}) \Big\}. \tag{15.38}$$

If the unperturbed distribution function f_0 depends arbitrarily on the time and the coordinates, then, as in an isotropic plasma [see (5.20)], one can obtain a generalization of the solution (15.33):

$$f_1(\mathbf{p},\ \mathbf{k},\ \omega) = -e\int d\omega' d\mathbf{k}' d\omega'' d\mathbf{k}'' \delta(\omega - \omega' - \omega'') \times$$

$$\times\,\delta(\mathbf{k} - \mathbf{k}' - \mathbf{k}'') \int_{-\infty}^{0} d\tau_0 e^{-i\omega\tau_0 + i\mathbf{k}\delta\mathbf{R}(\tau_0,\ \mathbf{v})}\, \alpha_{ij}(\omega',\ \mathbf{k}',\ \mathbf{v}(\tau_0,\ \mathbf{v}))\, \frac{\partial f_0(\mathbf{p}(\tau_0,\ \mathbf{v}),\ \omega'',\ \mathbf{k}'')}{\partial p_i(\tau_0,\ \mathbf{v})} E_j(\omega',\ \mathbf{k}'), \quad (15.39)$$

which again goes over under the conditions (5.22) into (15.33). Allowing for the fact that the differential operators on the left-hand sides of (15.2) and (15.3) are equal and that they therefore have the same characteristics, the formal solution of the chain of equations (15.2) can be represented, by means of equations that follow from it:

$$\frac{df_n}{dt} = -e\left\{\mathbf{E} + \frac{1}{c}[\mathbf{vB}]\right\}\frac{\partial f_{n-1}}{\partial \mathbf{p}};\quad n = 1,\ 2,\ \ldots \quad (15.40)$$

in the adiabatic approximation (5.24) in the form of the recursion relation

$$f_n(\mathbf{p},\ \mathbf{r},\ t) = -e\int_{-\infty}^{t} dt'\left\{\mathbf{E}(\mathbf{r} + \delta\mathbf{R}(t' - t,\ \mathbf{v}),\ t') + \frac{1}{c}[\mathbf{v}(t' - t,\ \mathbf{v})\,\mathbf{B}(\mathbf{r} + \delta\mathbf{R}(t' - t,\ \mathbf{v}),\ t')]\right\} \times$$

$$\times\,\frac{\partial}{\partial \mathbf{p}(t' - t,\ \mathbf{v})} f_{n-1}(\mathbf{p}(t' - t,\ \mathbf{v}),\ \mathbf{r} + \delta\mathbf{R}(t' - t,\ \mathbf{v}),\ t'),\ n = 1, 2, \ldots \quad (15.41)$$

Here, the momentum $\mathbf{p}(t' - t, \mathbf{v})$ is related to the velocity $\mathbf{v}(t' - t, \mathbf{v})$ defined in (15.20) by the relation (15.9):

$$\mathbf{p}(t' - t,\ \mathbf{v}) = \frac{\mathcal{E}}{c^2}\mathbf{v}(t' - t,\ \mathbf{v}). \quad (15.42)$$

In the limit of a weak external magnetic field ($\mathbf{B}_0 = 0$), Eqs. (15.41), do not depend on $\mathbf{B}_0$ in accordance with (15.34) and (15.35),

$$f_n(\mathbf{p},\ \mathbf{r},\ t) = -e\int_{-\infty}^{t} dt'\left\{\mathbf{E}(\mathbf{r} + \mathbf{v}(t' - t),\ t') + \right.$$

$$\left. + \frac{1}{c}[\mathbf{vB}(\mathbf{r} + \mathbf{v}(t' - t),\ t')]\right\}\frac{\partial}{\partial \mathbf{p}} f_{n-1}(\mathbf{p},\ \mathbf{r} + \mathbf{v}(t' - t),\ t'),\ n = 1, 2, \ldots \quad (15.43)$$

and can be obtained in such a form by solving the chain of equations (5.6) for an isotropic plasma.

Formula (15.41) can be written in Fourier transforms as follows [cf. (15.39)]:

$$f_n(\mathbf{p},\ \omega,\ \mathbf{k}) = -e\int d\omega' d\mathbf{k}' d\omega'' d\mathbf{k}'' \int_{-\infty}^{0} d\tau e^{-i\omega\tau + i\mathbf{k}\delta\mathbf{R}(\tau,\ \mathbf{v})}\, \alpha_{ij}(\omega',\ \mathbf{k}',\ \mathbf{v}(\tau,\ \mathbf{v}))\,\frac{\partial f_{n-1}(\mathbf{p}(\tau,\ \mathbf{v}),\ \omega'',\ \mathbf{k}'')}{\partial p_i(\tau,\ \mathbf{v})} E_j(\omega',\ \mathbf{k}') \times$$

$$\times\,\delta(\omega - \omega' - \omega'')\,\delta(\mathbf{k} - \mathbf{k}' - \mathbf{k}''),\quad n = 1,\ 2,\ \ldots \quad (15.44)$$

We integrate Eq. (15.44) in order to express the n-th nonequilibrium correction $f_n(\mathbf{p}, \omega, \mathbf{k})$ in terms of f_0 and the electric fields. We substitute f_1 from (15.39) into the right-hand side of (15.44) for n = 2 and we obtain an expression for the Fourier transform of the second correction $f_2(\mathbf{p}, \omega, \mathbf{k})$:

$$f_2(\mathbf{p},\ \omega,\ \mathbf{k}) = (-e)^2\int d\omega_1 d\mathbf{k}_1 d\omega_2 d\mathbf{k}_2 \int_{-\infty}^{0} d\tau_0 \int_{-\infty}^{0} d\tau_1 e^{-i\omega\tau_0 + i\mathbf{k}\delta\mathbf{R}(\tau_0,\ \mathbf{v})} \times$$

$$\times \alpha_{i1j1}(\omega-\omega_1,\ \mathbf{k}-\mathbf{k}_1,\ \mathbf{v}(\tau_0,\ \mathbf{v}))\frac{\partial}{\partial p_{i1}(\tau_0,\ \mathbf{v})}e^{-i\omega_1\tau_1+i\mathbf{k}_1\delta\mathbf{R}(\tau_1,\ \mathbf{v}(\tau_0,\ \mathbf{v}))}\alpha_{i2j2}(\omega_1-\omega_2,\ \mathbf{k}_1-\mathbf{k}_2,\ldots$$

$$\ldots\mathbf{v}(\tau_1,\ \mathbf{v}(\tau_0,\ \mathbf{v})))\frac{\partial f_0(\mathbf{p}(\tau_1,\ \mathbf{v}(\tau_0,\ \mathbf{v})),\ \omega_2,\ \mathbf{k}_2)}{\partial p_{i2}(\tau_1,\ \mathbf{v}(\tau_0,\ \mathbf{v}))}E_{j1}(\omega-\omega_1,\ \mathbf{k}-\mathbf{k}_1)E_{j2}(\omega_1-\omega_2,\ \mathbf{k}_1-\mathbf{k}_2). \quad (15.45)$$

Here, the vectors $\mathbf{v}(\tau_1, \mathbf{v}(\tau_0, \mathbf{v}))$ and $\delta\mathbf{R}(\tau_1, \mathbf{v}(\tau_0, \mathbf{v}))$ can be written in a slightly different form by using the equations

$$v_i(\tau_1,\ \mathbf{v}(\tau_0,\ \mathbf{v}))=v_i(\tau_0+\tau_1,\ \mathbf{v}); \quad (15.46)$$

$$\delta R_i(\tau_1,\ \mathbf{v}(\tau_0,\ \mathbf{v}))=\delta R_i(\tau_0+\tau_1,\ \mathbf{v})-\delta R_i(\tau_0,\ \mathbf{v}). \quad (15.47)$$

In what follows we shall use both forms of expression.

Both the equations (15.46) and (15.47) have a simple physical meaning. The first, (15.46) means that the velocity with which a particle that initially had the velocity $\mathbf{v}$ moves after a time $\tau_0 + \tau_1$ is equal to the velocity of a particle that traverses the time interval $\tau_0 + \tau_1$ in two parts: first, the time τ_0 with initial velocity $\mathbf{v}$ and then the time τ_1 with initial velocity $\mathbf{v}(\tau_0, \mathbf{v})$ equal to the velocity at the end of the first part of the path. Essentially, the proof of (15.46) follows from the definition (15.21) by a simple transformation.

The relation (15.47) can be interpreted as follows: the path traversed by a particle with initial velocity $\mathbf{v}(\tau_0, \mathbf{v})$ during the time τ_1 is equal to the difference of the distances traversed by a particle with initial velocity $\mathbf{v}$ in the times $\tau_0 + \tau_1$ and τ_0. The validity of this relation can be proved by means of the definition (15.23) (for more detail see § 16).

The third nonequilibrium correction $f_3(\mathbf{p}, \omega, \mathbf{k})$ is obtained from (15.44) for n = 3 and (15.45) in exactly the same way as f_2 was obtained from (15.44) for n = 2 and (15.39):

$$f_3(\mathbf{p},\ \omega,\ \mathbf{k})=(-e)^3\int d\omega_1 d\mathbf{k}_1 d\omega_2 d\mathbf{k}_2 d\omega_3 d\mathbf{k}_3\int_{-\infty}^{0}d\tau_0\int_{-\infty}^{0}d\tau_1\int_{-\infty}^{0}d\tau_2\times$$

$$\times\exp[-i\omega\tau_0+i\mathbf{k}\delta\mathbf{R}(\tau_0,\ \mathbf{v})]\,\alpha_{i1j1}(\omega-\omega_1,\ \mathbf{k}-\mathbf{k}_1,\ \mathbf{v}(\tau_0,\ \mathbf{v}))\times$$

$$\times\frac{\partial}{\partial p_{i1}(\tau_0,\ \mathbf{v})}\exp[-i\omega_1\tau_1+i(\mathbf{k}_1,\delta\mathbf{R}(\tau_0+\tau_1,\ \mathbf{v})-\delta\mathbf{R}(\tau_0,\ \mathbf{v}))]\ \alpha_{i2j2}(\omega_1-\omega_2,\ \mathbf{k}_1-\mathbf{k}_2,\ \mathbf{v}(\tau_0+\tau_1,\ \mathbf{v}))\times$$

$$\times\frac{\partial}{\partial p_{i2}(\tau_0+\tau_1,\ \mathbf{v})}\exp[-i\omega_2\tau_2+i(\mathbf{k}_2,\ \delta\mathbf{R}(\tau_0+\tau_1+\tau_2,\ \mathbf{v})-\delta\mathbf{R}(\tau_0+\tau_1,\ \mathbf{v}))]\times$$

$$\times\alpha_{i3j3}(\omega_2-\omega_3,\ \mathbf{k}_2-\mathbf{k}_3,\ \mathbf{v}(\tau_0+\tau_1+\tau_2,\ \mathbf{v}))\ \frac{\partial}{\partial p_{i3}(\tau_0+\tau_1+\tau_2,\ \mathbf{v})}f_0(\mathbf{p}(\tau_0+\tau_1+\tau_2,\ \mathbf{v}),\ \omega_3,\ \mathbf{k}_3)\times$$

$$\times E_{j1}(\omega-\omega_1,\ \mathbf{k}-\mathbf{k}_1)E_{j2}(\omega_1-\omega_2,\mathbf{k}_1-\mathbf{k}_2)E_{j3}(\omega_2-\omega_3,\mathbf{k}_2-\mathbf{k}_3). \quad (15.48)$$

In this expression we have also used equations of the form (15.46) and (15.47).

The expressions (15.45) and (15.48) completely clarify the rule for writing down the n-th nonequilibrium correction. Namely, the result of iterating the recursion formula (15.44) can be represented in the form (this can be proved, for example, by induction) [52]

$$f_n(\mathbf{p},\ \omega,\ \mathbf{k})=(-e)^n\int d\omega_1 d\mathbf{k}_1\ldots d\omega_n d\mathbf{k}_n\int_{-\infty}^{0}d\tau_0\ldots\int_{-\infty}^{0}d\tau_{n-1}G\times$$

$$\times\Gamma_{j1}(\tau_0)G_1\Gamma_{j2}(\tau_0+\tau_1)\ldots G_{n-1}\Gamma_{jn}(\tau_0+\tau_1+\ldots+\tau_{n-1})\times$$

$$\times f_0(\mathbf{p}(\tau_0+\tau_1+\ldots+\tau_{n-1},\ \mathbf{v}),\ \omega_n,\ \mathbf{k}_n)E_{j1}(\omega-\omega_1,\ \mathbf{k}-\mathbf{k}_1)\ldots E_{jn}(\omega_{n-1}-\omega_n,\ \mathbf{k}_{n-1}-\mathbf{k}_n) \quad (15.49)$$

using the notation

$$G_n\equiv\exp\left\{-i\omega_n\tau_n+i\left(\mathbf{k}_n,\delta\mathbf{R}\left(\sum_{l=0}^{n}\tau_l,\mathbf{v}\right)-\delta\mathbf{R}\left(\sum_{l=0}^{n-1}\tau_l,\mathbf{v}\right)\right)\right\},n=1,2,3,\ldots,\quad G\equiv\exp\{-i\omega\tau_0+i\mathbf{k}\delta\mathbf{R}(\tau_0,\ \mathbf{v})\}; \quad (15.50)$$

$$\Gamma_{jn}(\tau_0+\tau_1+\ldots+\tau_{n-1}) \equiv \alpha_{injn}\left(\omega_{n-1}-\omega_n,\ \mathbf{k}_{n-1}-\mathbf{k}_n,\ \mathbf{v}\left(\sum_{l=0}^{n-1}\tau_l,\ \mathbf{v}\right)\right)\frac{\partial}{\partial p_{in}\left(\sum_{l=0}^{n-1}\tau_l,\ \mathbf{v}\right)} =$$

$$= \frac{1}{\omega_{n-1}-\omega_n}\left\{(\mathbf{k}_{n-1}-\mathbf{k}_n)_{in} v_{jn}\left(\sum_{l=0}^{n-1}\tau_l,\ \mathbf{v}\right) + \right.$$

$$\left. + \delta_{in,\,jn}\left(\omega_{n-1}-\omega_n-\left(\mathbf{k}_{n-1}-\mathbf{k}_n,\ \mathbf{v}\left(\sum_{l=0}^{n-1}\tau_l,\ \mathbf{v}\right)\right)\right)\right\}\frac{\partial}{\partial p_{in}\left(\sum_{l=0}^{n-1}\tau_l,\ \mathbf{v}\right)},\quad n\geqslant 2; \tag{15.51}$$

$$\Gamma_{j1}(\tau_0) \equiv \frac{1}{\omega-\omega_1}\{(\mathbf{k}-\mathbf{k}_1)_{i1} v_{j1}(\tau_0,\ \mathbf{v}) + \delta_{i1,\,j1}(\omega-\omega_1-(\mathbf{k}-\mathbf{k}_1,\ \mathbf{v}(\tau_0,\ \mathbf{v})))\}\,\partial/\partial p_{i1}(\tau_0,\ \mathbf{v}),\quad n=1.$$

Note that if the external magnetic field vanishes, $\mathbf{B}_0 = 0$, the vertex parts $\Gamma_{jn}(\tau_0 + \tau_1 + \ldots + \tau_{n-1})$ go over into the isotropic expressions Γ_{jn} in (5.29) in accordance with (15.34). The Green's functions G_n in this limit also simplify, since then

$$\left\{\delta\mathbf{R}\left(\sum_{l=0}^{n}\tau_l,\ \mathbf{v}\right)-\delta\mathbf{R}\left(\sum_{l=0}^{n-1}\tau_l,\ \mathbf{v}\right)\right\}_{\mathbf{B}_0=0} = \tau_n\mathbf{v}. \tag{15.52}$$

Therefore, after integration with respect to the variables $\tau_0, \tau_1, \ldots, \tau_{n-1}$ separated in this manner, the relation (15.49) for f_n goes over into the expression (5.27) when $\mathbf{B}_0 = 0$.

As for the case for an isotropic plasma [see (5.32)], we write down the complete expression for the distribution function $f(\mathbf{p}, \mathbf{r}, t)$, which is the nonlinear solution of Vlasov's equation with allowance for the effect of the external constant and homogeneous magnetic field:

$$f(\mathbf{p},\ \mathbf{r},\ t) = f_0(\mathbf{p},\ \mathbf{r},\ t) + \sum_{n=1}^{\infty}(-e)^n\int d\omega d\mathbf{k}\, e^{-i\omega t+i\mathbf{k}\mathbf{r}}\times$$

$$\times\int d\omega_1 d\mathbf{k}_1\ldots d\omega_n d\mathbf{k}_n\int_{-\infty}^{0} d\tau_0\ldots\int_{-\infty}^{0} d\tau_{n-1} G\Gamma_{j1}(\tau_0)\, G_1\Gamma_{j2}(\tau_0+\tau_1)\ldots$$

$$\ldots G_{n-1}\Gamma_{jn}(\tau_0+\tau_1+\ldots+\tau_{n-1}) f_0(\mathbf{p}(\tau_0+\ldots+\tau_{n-1},\ \mathbf{v}),\ \omega_n,\ \mathbf{k}_n)\times$$

$$\times E_{j1}(\omega-\omega_1,\ \mathbf{k}-\mathbf{k}_1)\ldots E_{jn}(\omega_{n-1}-\omega_n,\ \mathbf{k}_{n-1}-\mathbf{k}_n). \tag{15.53}$$

To conclude this section, we give a slightly different expression for $f(\mathbf{p}, \mathbf{r}, t)$, which is obtained by iterating the recursion relation (15.41) without going over to the Fourier transforms of the fields in the magnetoactive plasma:

$$f(\mathbf{p},\ \mathbf{r},\ t) = f_0(\mathbf{p},\ \mathbf{r},\ t) + \sum_{n=1}^{\infty}(-e)^n\int_{-\infty}^{0} d\tau_0\ldots\int_{-\infty}^{0} d\tau_{n-1}\times$$

$$\times\left\{E_{j1}(\mathbf{r}+\delta\mathbf{R}(\tau_0,\ \mathbf{v}),\ t+\tau_0)+\frac{1}{c}[\mathbf{v}(\tau_0,\ \mathbf{v})\mathbf{B}(\mathbf{r}+\delta\mathbf{R}(\tau_0,\mathbf{v}),\ t+\tau_0)]_{j1}\right\}\times$$

$$\times\frac{\partial}{\partial p_{j1}(\tau_0,\ \mathbf{v})}\cdots\left\{E_{jn}(\mathbf{r}+\delta\mathbf{R}(\tau_0+\ldots+\tau_{n-1},\ \mathbf{v}),\ t+\tau_0+\ldots+\tau_{n-1})+\right.$$

$$\left.+\frac{1}{c}[\mathbf{v}(\tau_0+\ldots+\tau_{n-1},\mathbf{v})\mathbf{B}(\mathbf{r}+\delta\mathbf{R}(\tau_0+\ldots+\tau_{n-1},\mathbf{v}),\ t+\tau_0+\ldots+\tau_{n-1})]_{jn}\right\}\times$$

$$\times\frac{\partial}{\partial p_{jn}(\tau_0+\ldots+\tau_{n-1},\ \mathbf{v})} f_0(p(\tau_0+\ldots+\tau_{n-1},\mathbf{v}),\ \mathbf{r}+\delta\mathbf{R}(\tau_0+\ldots+\tau_{n-1},\ \mathbf{v}),\ t+\tau_0+\ldots+\tau_{n-1}). \tag{15.54}$$

In the limit of an isotropic plasma, $\mathbf{B}_0 = 0$, this formula is identical with (5.33).

§ 16. Permittivity Tensor of a Magnetoactive Plasma as the Simplest Example of the Multi-Index Tensor for n = 1

Our solution of the transport equation enables us to determine the nonlinear induced current density in a magnetoactive plasma:

$$\mathbf{j}=\mathbf{j}_1+\mathbf{j}_2+\ \dots\ +\mathbf{j}_n+\dots; \tag{16.1}$$

$$\begin{aligned}\mathbf{j}_n(\omega,\ \mathbf{k}) = e(-e)^n \int d\mathbf{p}\mathbf{v} \int d\omega_1 d\mathbf{k}_1 \dots d\omega_n d\mathbf{k}_n \int\limits_{-\infty}^{0} d\tau_0 \dots \int\limits_{-\infty}^{0} d\tau_{n-1} G\Gamma_{j1}(\tau_0) \times \\ \times G_1\Gamma_{j2}(\tau_0+\tau_1)\dots G_{n-1}\Gamma_{jn}(\tau_0+\tau_1+\dots+\tau_{n-1}) f_0(\mathbf{p}(\tau_0+\tau_1+\dots \\ \dots+\tau_{n-1},\ \mathbf{v}),\ \omega_n,\ \mathbf{k}_n) E_{j1}(\omega-\omega_1,\ \mathbf{k}-\mathbf{k}_1)\dots E_{jn}(\omega_{n-1}-\omega_n,\ \mathbf{k}_{n-1}-\mathbf{k}_n)\end{aligned} \tag{16.2}$$

[the current density in the ground state vanishes, see (6.4)] and, therefore, the multi-index complex permittivity tensors $\varepsilon_{ij1\dots jn}$ that occur in the nonlinear field equations (2.12) [cf. (6.9)]:

$$\begin{aligned}\varepsilon_{ij1\dots jn}(\omega,\ \mathbf{k};\ \omega_1,\ \mathbf{k}_1;\ \dots;\ \omega_n,\ \mathbf{k}_n) = \delta_{n1}\delta_{i,\,j1}\delta(\omega_1)\delta(\mathbf{k}_1) - 4\pi i(-e)^{n+1} \times \\ \times \int d\mathbf{p}\frac{v_i}{\omega} \int\limits_{-\infty}^{0} d\tau_0 \dots \int\limits_{-\infty}^{0} d\tau_{n-1} G\Gamma_{j1}(\tau_0) G_1\Gamma_{j2}(\tau_0+\tau_1)\dots G_{n-1}\Gamma_{jn}(\tau_0+\dots+\tau_{n-1}) \times \\ \times f_0(\mathbf{p}(\tau_0+\tau_1+\ \dots\ +\tau_{n-1},\ \mathbf{v}),\ \omega_n,\ \mathbf{k}_n).\end{aligned} \tag{16.3}$$

Thus, the field equation (2.12) now enables us to describe different nonlinear effects in an unbounded magnetoactive plasma. In the stationary and homogeneous magnetoactive plasma (5.22), the tensors (16.3) simplify to a certain extent [52]:

$$\varepsilon_{ij1\dots jn}(\omega,\ \mathbf{k};\ \omega_1,\ \mathbf{k}_1;\ \dots;\ \omega_n,\ \mathbf{k}_n) = \delta(\omega_n)\delta(\mathbf{k}_n)\,\varepsilon_{ij1\dots jn}(\omega,\ \mathbf{k};\ \omega_1,\ \mathbf{k}_1;\ \dots;\ \omega_{n-1},\ \mathbf{k}_{n-1}); \tag{16.4}$$

$$\begin{aligned}\varepsilon_{ij1\dots jn}(\omega,\ \mathbf{k};\ \omega_1,\ \mathbf{k}_1;\ \dots;\omega_{n-1},\ \mathbf{k}_{n-1}) = \delta_{n1}\delta_{i,j1} - 4\pi i(-e)^{n+1}\int d\mathbf{p}\frac{v_i}{\omega}\int\limits_{-\infty}^{0} d\tau_0\dots \\ \dots\int\limits_{-\infty}^{0} d\tau_{n-1} G\Gamma_{j1}(\tau_0)\dots G_{n-1}\Gamma_{jn}(\tau_0+\dots+\tau_{n-1}) f_0(\mathbf{p}(\tau_0+\ \dots\ +\tau_{n-1},\ \mathbf{v})),\end{aligned} \tag{16.5}$$

since they depend on a smaller number of pairs of arguments in this case.

It is not difficult to show that in the limit of an infinitesimally small external magnetic field $\mathbf{B}_0$, the tensors (16.5) go over into their isotropic analogs (16.10). Note also that, as in the case of an isotropic plasma, the Fourier transform of the n-th nonequilibrium correction $\rho_n(\omega, \mathbf{k})$ in the charge density ρ [see (6.1)]:

$$\begin{aligned}\rho_n(\omega, \mathbf{k}) = -(-e)^{n+1}\int d\mathbf{p}\int d\omega_1 d\mathbf{k}_1\dots d\omega_n d\mathbf{k}_n \int\limits_{-\infty}^{0} d\tau_0\dots\int\limits_{-\infty}^{0} d\tau_{n-1} G\Gamma_{j1}(\tau_0)\times \\ \times G_1\Gamma_{j2}(\tau_0+\tau_1)\dots G_{n-1}\Gamma_{jn}(\tau_0+\tau_1+\ \dots\ +\tau_{n-1}) f_0(\mathbf{p}(\tau_0+\ \dots \\ \dots+\tau_{n-1},\ \mathbf{v}),\ \omega_n,\ \mathbf{k}_n) E_{j1}(\omega-\omega_1,\ \mathbf{k}-\mathbf{k}_1)\dots E_{jn}(\omega_{n-1}-\omega_n,\ \mathbf{k}_{n-1}-\mathbf{k}_n)\end{aligned} \tag{16.6}$$

is related to the corresponding correction $\mathbf{j}_n(\omega, \mathbf{k})$ in the current density (16.2) by the continuity equation (6.5).

For n = 1 and a constant homogeneous equilibrium function f_0 [see (5.22)], Eqs. (16.3)-(16.5) determine the permittivity tensor $\varepsilon_{ij}(\omega, k)$ of the magnetoactive plasma:

$$\varepsilon_{ij1}(\omega, \mathbf{k}; \omega_1, \mathbf{k}_1) = \delta(\omega_1)\delta(\mathbf{k}_1)\varepsilon_{ij1}(\omega, \mathbf{k}),$$
$$\varepsilon_{ij}(\omega, \mathbf{k}) = \delta_{ij} - 4\pi i e^2 \int d\mathbf{p} \frac{v_i}{\omega} \int_{-\infty}^{0} d\tau_0 G\Gamma_j(\tau_0) f_0(\mathbf{p}(\tau_0, \mathbf{v})). \tag{16.7}$$

We devote this section to a systematic study of the tensor $\varepsilon_{ij}(\omega, \mathbf{k})$ in a magnetoactive plasma, since we shall need the approach used here and the results we obtain in the nonlinear theory of oscillations.

Using the explicit expressions (15.50) and (15.51) for the Green's function G and the vertex part $\Gamma_{j1}(\tau_0)$, we rewrite the tensor (16.7) in the more usual form (see [1], § 19):

$$\varepsilon_{ij}(\omega, \mathbf{k}) = \delta_{ij} - i\frac{4\pi e^2}{\omega}\int d\mathbf{p} v_i \int_{-\infty}^{0} d\tau e^{-i\omega\tau + i\mathbf{k}\delta\mathbf{R}(\tau, \mathbf{v})} \alpha_{nj}(\omega, \mathbf{k}; \mathbf{v}(\tau, \mathbf{v})) \frac{\partial f_0}{\partial p_n(\tau, \mathbf{v})}. \tag{16.8}$$

For an isotropic distribution function of the ground state, an equation of the type (5.18) also holds in a magnetoactive plasma:

$$\alpha_{ij}(\omega, \mathbf{k}; \mathbf{v}(\tau, \mathbf{v})) \frac{\partial f_0(\mathcal{E})}{\partial p_i(\tau, \mathbf{v})} = \frac{\partial f_0(\mathcal{E})}{\partial p_j(\tau, \mathbf{v})}. \tag{16.9}$$

Therefore, (16.8) simplifies (cf. [1], (19.11), p. 137)

$$\varepsilon_{ij}(\omega, \mathbf{k}) = \delta_{ij} - i\frac{4\pi e^2}{\omega}\int d\mathbf{p} v_i \int_{-\infty}^{0} d\tau e^{-i\omega\tau + i\mathbf{k}\delta\mathbf{R}(\tau, \mathbf{v})} \frac{\partial f_0(\mathcal{E})}{\partial p_j(\tau, \mathbf{v})}. \tag{16.10}$$

The vectors $\delta\mathbf{R}(\tau, \mathbf{v})$ and $\mathbf{v}(\tau, \mathbf{v})$ in (16.10) are specified by Eqs. (15.23) and (15.20). In view of the importance of these equations for the following exposition, we rewrite them in a notation that is somewhat different and more convenient; that is, we introduce the tensors $W_{ij}(\tau)$ and $\rho_{ij}(\tau)$:

$$v_i(\tau, \mathbf{v}) = W_{ij}(\tau) v_j, \qquad v_i(0, \mathbf{v}) \equiv v_i; \tag{16.11}$$

$$W_{ij}(\tau) \equiv h_i h_j - e_{isj} h_s \sin\Omega\tau + (\delta_{ij} - h_i h_j)\cos\Omega\tau = h_i h_j + \frac{1}{2} a_{ij} e^{i\Omega\tau} + \frac{1}{2} a_{ji} e^{-i\Omega\tau}; \tag{16.12}$$

$$a_{ij} \equiv \delta_{ij} - h_i h_j + i e_{isj} h_s. \tag{16.13}$$

Here e_{isj} is the tensor we have already used [in Ch. II, § 6, Eqs. (6.73), (6.74), and (6.76)]: the unit, completely antisymmetric pseudotensor of third rank. Similarly, for $\delta\mathbf{R}(\tau, \mathbf{v})$

$$\delta R_i(\tau, \mathbf{v}) = \rho_{ij}(\tau) v_j; \tag{16.14}$$

$$\rho_{ij}(\tau) \equiv h_i h_j \tau + \frac{1}{\Omega} e_{isj} h_s (\cos\Omega\tau - 1) + \frac{\sin\Omega\tau}{\Omega}(\delta_{ij} - h_i h_j) = h_i h_j \tau - \frac{e_{isj} h_s}{\Omega} - \frac{i}{2\Omega} a_{ij} e^{i\Omega\tau} + \frac{i}{2\Omega} a_{ji} e^{-i\Omega\tau}. \tag{16.15}$$

The relationship between the tensors W_{ij} and ρ_{ij} is obvious [see (15.23)]:

$$\rho_{ij}(\tau) = \int_0^{\tau} d\tau' W_{ij}(\tau'), \quad W_{ij}(\tau) = \frac{d\rho_{ij}(\tau)}{d\tau}. \tag{16.16}$$

Using the explicit expressions (16.12) and (16.15) for these tensors, we find some of their properties:

$$W_{ij}(\tau) \equiv W_{ij}(\tau, \mathbf{h}) \equiv W_{ij}(\tau, \mathbf{h}, \Omega) = W_{ji}(\tau, -\mathbf{h}) = W_{ji}(-\tau, \mathbf{h}) =$$

$$= W_{ji}(\tau, \mathbf{h}, -\Omega), \ \rho_{ij}(\tau) \equiv \rho_{ij}(\tau, \mathbf{h}) \equiv \rho_{ij}(\tau, \mathbf{h}, \Omega) = \rho_{ji}(\tau, -\mathbf{h}) = = -\rho_{ji}(-\tau, \mathbf{h}) = \rho_{ji}(\tau, \mathbf{h}, -\Omega). \tag{16.17}$$

In the limit of an infinitesimally weak magnetic field $\mathbf{B}_0$, both tensors reduce to the Kronecker delta:

$$W_{ij}(\tau, \Omega = 0) = \delta_{ij}, \quad \rho_{ij}(\tau, \Omega = 0) = \delta_{ij}\tau, \tag{16.18}$$

and their values for $\tau = 0$ are given by

$$W_{ij}(\tau = 0) = \delta_{ij}, \ \rho_{ij}(\tau = 0) = 0. \tag{16.19}$$

In particular, using the algebraic properties of the tensor a_{ij},

$$a_{ij}h_j = h_i a_{ij} = 0, \ a_{ii} = 2, \ a_{ij}a_{js} = 2a_{is}, \ a_{ij}a_{sj} = a_{ji}a_{js} = 0, \tag{16.20}$$

we can show, using (16.12) and (16.15), that

$$W_{ij}(\tau_0) W_{js}(\tau_1) = W_{ij}(\tau_1) W_{js}(\tau_0) = W_{is}(\tau_0 + \tau_1); \tag{16.21}$$

$$\rho_{ij}(\tau_0) W_{js}(\tau_1) = \rho_{is}(\tau_0 + \tau_1) - \rho_{is}(\tau_1), \tag{16.22}$$

which confirm the relations (15.46)-(15.47) used to derive the general expression for the nonlinear (in the field $\mathbf{E}$) nonequilibrium distribution function in a magnetoactive plasma.

The relations we have so far deduced are also valid for relativistic velocities of the particles of a magnetoactive plasma, since the gyroscopic frequency Ω in these relations is determined by Eq. (15.11). Turning to a nonrelativistic plasma, we go over in (16.10) from the momenta $\mathbf{p}$ to the velocities $\mathbf{v}$ [see (15.30) instead of (15.9)], replacing at the same time the gyroscopic frequency (15.11) by its nonrelativistic limit (15.12):

$$\varepsilon_{ij}(\omega, \mathbf{k}) = \delta_{ij} - i\frac{4\pi e^2}{m}\int d\mathbf{v}\frac{v_i}{\omega}\int_{-\infty}^{0} d\tau e^{-i\omega\tau + i\mathbf{k}\delta\mathbf{R}(\tau, \mathbf{v})}\frac{\partial f_0(\mathscr{E})}{\partial v_j(\tau, \mathbf{v})}. \tag{16.23}$$

Here, we integrate by parts with respect to the velocities $\mathbf{v}$:

$$\varepsilon_{ij}(\omega, \mathbf{k}) = \delta_{ij} + i\frac{4\pi e^2}{m}\int d\mathbf{v}\frac{1}{\omega}f_0\int_{-\infty}^{0} d\tau e^{-i\omega\tau + i\mathbf{k}\delta\mathbf{R}(\tau, \mathbf{v})}\{W_{ij}(-\tau) - iv_i k_n \rho_{nj}(-\tau)\} \tag{16.24}$$

by means of the "differentiation rules"

$$\frac{\partial v_i}{\partial v_j(\tau, \mathbf{v})} = W_{ij}(-\tau), \quad \frac{\partial \delta R_i(\tau, \mathbf{v})}{\partial v_j(\tau, \mathbf{v})} = -\rho_{ij}(-\tau), \tag{16.25}$$

which follow directly from the definitions (16.11) and (16.14) on account of the identities [see (16.21), (16.22)]

$$\begin{gathered} v_i = v_i(-\tau, \mathbf{v}(\tau, \mathbf{v})) = W_{ij}(-\tau) v_j(\tau, \mathbf{v}), \\ \delta R_i(\tau, \mathbf{v}) = \rho_{ij}(\tau) v_j = \rho_{ij}(\tau) W_{js}(-\tau) v_s(\tau, \mathbf{v}) = -\rho_{ij}(-\tau) v_j(\tau, \mathbf{v}). \end{gathered} \tag{16.26}$$

We substitute into (16.24) the Maxwell distribution (6.14) and integrate with respect to the velocities by means of the integrals

$$\frac{1}{N}\int d\mathbf{v} f_0 e^{i\mathbf{a}\mathbf{v}} = \exp\left\{-\frac{1}{2}\mathbf{a}^2 v_T^2\right\}; \tag{16.27}$$

$$\frac{1}{N}\int d\mathbf{v} f_0 v_i e^{i\mathbf{a}\mathbf{v}} = i v_T^2 a_i \exp\left\{-\frac{1}{2}\mathbf{a}^2 v_T^2\right\}, \tag{16.28}$$

where **a** is a real vector that does not depend on the velocity. The permittivity tensor takes the form

$$\varepsilon_{ij}(\omega, \mathbf{k}) = \delta_{ij} + i\frac{\omega_L^2}{\omega}\int_{-\infty}^{0} d\tau e^{-i\omega\tau - \frac{1}{2}v_T^2[k_n\rho_{nm}(\tau)]^2} \{W_{ij}(-\tau) + v_T^2 k_n \rho_{ni}(\tau) k_m \rho_{mj}(-\tau)\}. \quad (16.29)$$

It follows from this that the tensor structure of $\varepsilon_{ij}(\omega, \mathbf{k})$ is completely determined by the tensors W_{ij} and ρ_{ij}. In the limit of an isotropic plasma

$$\mathbf{B}_0 = 0$$

the tensors W_{ij} and ρ_{ij} reduce to the Kronecker deltas (16.18), as a result of which the tensor (16.29) greatly simplifies:

$$\varepsilon_{ij}(\omega, \mathbf{k}) = \delta_{ij} + i\frac{\omega_L^2}{\omega}\int_{-\infty}^{0} d\tau e^{-i\omega\tau - \frac{1}{2}k^2 v_T^2 \tau^2} \{\delta_{ij} - k_i k_j \tau^2 v_T^2\}. \quad (16.30)$$

Hence, using the integrals

$$\int_{-\infty}^{0} d\tau e^{-i\omega\tau - \frac{1}{2}k^2 v_T^2 \tau^2} = \frac{i}{\omega} J_+\left(\frac{\omega}{kv_T}\right); \quad (16.31)$$

$$\int_{-\infty}^{0} d\tau \tau^2 e^{-i\omega\tau - \frac{1}{2}k^2 v_T^2 \tau^2} = \frac{i}{(kv_T)^3}\left\{\frac{1}{\beta} J_+(\beta) - \beta J_+(\beta) + \beta\right\}, \quad \beta \equiv (\omega/kv_T) \quad (16.32)$$

we obtain the expressions (6.17), (6.21), and (6.22) for the permittivity tensor of an isotropic plasma. For a cold magnetoactive plasma,

$$v_T = 0, \quad \mathbf{B}_0 = \text{const} \neq 0, \quad (16.33)$$

we obtain from (16.29)

$$\varepsilon_{ij}(\omega, 0) = \delta_{ij} + i\frac{\omega_L^2}{\omega}\int_{-\infty}^{0} d\tau e^{-i\omega\tau} W_{ij}(-\tau) = \delta_{ij} - \frac{\omega_L^2}{\omega}\Gamma_{ij}(\omega), \quad (16.34)$$

where

$$\Gamma_{ij}(\omega) \equiv \frac{h_i h_j}{\omega} + \frac{1}{2}\frac{a_{ij}}{\omega + \Omega} + \frac{1}{2}\frac{a_{ji}}{\omega - \Omega}. \quad (16.35)$$

Going over to the elementary tensors

$$\delta_{ij}, \quad h_i h_j, \quad e_{isj} h_s, \quad (16.36)$$

we obtain the expression

$$\varepsilon_{ij}(\omega, \mathbf{k}) = \varepsilon_1 \delta_{ij} + (\varepsilon_2 - \varepsilon_1) h_i h_j - ig e_{isj} h_s, \quad (16.37)$$

in which (cf. [1], (20.4))

$$\varepsilon_1 = 1 - \frac{\omega_L^2}{\omega^2 - \Omega^2}, \quad \varepsilon_2 = 1 - \frac{\omega_L^2}{\omega^2}, \quad g = -\frac{\Omega}{\omega}\frac{\omega_L^2}{\omega^2 - \Omega^2}. \quad (16.38)$$

In a right-handed coordinate system with z axis along the external magnetic field,

$$h_z = 1, \quad h_x = h_y = 0, \quad (16.39)$$

the elementary tensors (16.36) can be written in the form of the matrices

$$\delta_{ij} = \begin{pmatrix} 1 & 0 & 0 \\ 0 & 1 & 0 \\ 0 & 0 & 1 \end{pmatrix}, \quad h_i h_j = \begin{pmatrix} 0 & 0 & 0 \\ 0 & 0 & 0 \\ 0 & 0 & 1 \end{pmatrix}, \quad e_{isj} h_s = \begin{pmatrix} 0 & -1 & 0 \\ 1 & 0 & 0 \\ 0 & 0 & 0 \end{pmatrix}, \quad (16.40)$$

and when these are used it is obvious that the permittivity tensor (16.37) in a cold plasma is equal to the expression (20.4) of [1] (p. 140).

We now go over in the general expression (16.29) to tensors that are simpler than ρ_{ij} and W_{ij}. To do this we use (16.15), obtaining

$$\varkappa_{n}\rho_{ni}(\tau)\varkappa_{m}\rho_{mj}(-\tau)=-h_i h_j(\varkappa\mathbf{h})^2\left(\frac{\sin\Omega\tau}{\Omega}-\tau\right)^2-\varkappa_i\varkappa_j\left(\frac{\sin\Omega\tau}{\Omega}\right)^2+$$
$$+[\varkappa\mathbf{h}]_i[\varkappa\mathbf{h}]_j\left(\frac{1-\cos\Omega\tau}{\Omega}\right)^2-\frac{\sin\Omega\tau}{\Omega}\frac{1-\cos\Omega\tau}{\Omega}(\varkappa_i[\varkappa\mathbf{h}]_j-\varkappa_j[\varkappa\mathbf{h}]_i)+$$
$$+\frac{\sin\Omega\tau}{\Omega}\left(\frac{\sin\Omega\tau}{\Omega}-\tau\right)(\varkappa\mathbf{h})(\varkappa_i h_j+\varkappa_j h_i)+\frac{1-\cos\Omega\tau}{\Omega}\left(\frac{\sin\Omega\tau}{\Omega}-\tau\right)(\varkappa\mathbf{h})(h_i[\varkappa\mathbf{h}]_j-h_j[\varkappa\mathbf{h}]_i), \quad (16.41)$$

and, eliminating from it the last two tensors by means of the identities*

$$(\varkappa\mathbf{h})(\varkappa_i h_j+\varkappa_j h_i)=\varkappa_i\varkappa_j+h_i h_j-[\varkappa\mathbf{h}]^2\delta_{ij}+[\varkappa\mathbf{h}]_i[\varkappa\mathbf{h}]_j, \quad (16.42)$$

$$(\varkappa\mathbf{h})(h_i[\varkappa\mathbf{h}]_j-h_j[\varkappa\mathbf{h}]_i)=(\varkappa_i[\varkappa\mathbf{h}]_j-\varkappa_j[\varkappa\mathbf{h}]_i)-[\varkappa\mathbf{h}]^2 e_{isj}h_s, \quad (16.43)$$

we substitute it into (16.29). As a result, the permittivity tensor $\varepsilon_{ij}(\omega, \mathbf{k})$ can be represented in accordance with [53] as a linear combination of six elementary tensors:

$$\varepsilon_{ij}(\omega,\mathbf{k})=a\delta_{ij}+b\varkappa_i\varkappa_j+ch_i h_j+de_{isj}h_s+e(\varkappa_i[\varkappa\mathbf{h}]_j-\varkappa_j[\varkappa\mathbf{h}]_i)+f[\varkappa\mathbf{h}]_i[\varkappa\mathbf{h}]_j \quad (16.44)$$

with the coefficients

$$\begin{aligned}
a&=1+i\frac{\omega_L^2}{\omega}\left\langle\cos\Omega\tau+[\mathbf{kh}]^2v_T^2\frac{\sin\Omega\tau}{\Omega}\left(\tau-\frac{\sin\Omega\tau}{\Omega}\right)\right\rangle,\\
b&=-i\frac{\omega_L^2}{\omega}\left\langle\tau\frac{\sin\Omega\tau}{\Omega}k^2v_T^2\right\rangle,\\
c&=i\frac{\omega_L^2}{\omega}\left\langle 1-\cos\Omega\tau+v_T^2\left\{[\mathbf{kh}]^2\frac{\sin^2\Omega\tau}{\Omega^2}-\tau^2(\mathbf{kh})^2-\tau\frac{\sin\Omega\tau}{\Omega}([\mathbf{kh}]^2-(\mathbf{kh})^2)\right\}\right\rangle,\\
d&=i\frac{\omega_L^2}{\omega}\left\langle\sin\Omega\tau-v_T^2[\mathbf{kh}]^2\frac{1-\cos\Omega\tau}{\Omega}\left(\frac{\sin\Omega\tau}{\Omega}-\tau\right)\right\rangle,\\
e&=-i\frac{\omega_L^2}{\omega}k^2v_T^2\left\langle\tau\frac{1-\cos\Omega\tau}{\Omega}\right\rangle,\\
f&=i\frac{\omega_L^2}{\omega\Omega}k^2v_T^2\left\langle 2\frac{1-\cos\Omega\tau}{\Omega}-\tau\sin\Omega\tau\right\rangle.
\end{aligned} \quad (16.45)$$

Here, the angular brackets denote the integration with respect to τ in accordance with the rule

$$\langle\alpha(\tau)\rangle\equiv\int_{\infty}^{0}d\tau\,\alpha(\tau)\,e^{-i\omega\tau-\frac{1}{2}v_T^2[k_n\rho_{nm}(\tau)]^2}. \quad (16.46)$$

---

* These identities can be proved by rewriting them in matrix form in a coordinate system suitable for this purpose. We note also that they are a direct consequence of the properties of the unit, completely antisymmetric pseudotensor of third rank e_{isj} (see [27], p. 34, footnote). In particular, (16.42) arises from scalar multiplication of both sides of the equation

$$e_{isa}e_{jcb}=\begin{vmatrix}\delta_{ij}&\delta_{ic}&\delta_{ib}\\ \delta_{sj}&\delta_{sc}&\delta_{sb}\\ \delta_{aj}&\delta_{ac}&\delta_{ab}\end{vmatrix}\ \text{by the tensor}\ h_s h_c\varkappa_a\varkappa_b,$$

which is composed of the unit vectors $\varkappa$ and $\mathbf{h}$, and the tensor (16.43) follows from the more general $(\delta_{ab}-h_a h_b)e_{isj}h_s=(\delta_{ib}-h_i h_b)e_{asj}h_s-(\delta_{jb}-h_j h_b)e_{asi}h_s$ by contraction with the tensor $\varkappa_a\varkappa_b$.

The expressions for the coefficients (16.45) agree with those given in [53]. Sometimes (for example, in the limit of frequencies ω that are low compared with the gyroscopic frequencies Ω) it is more convenient to express these coefficients differently. The identity

$$e^{-i\omega\tau-\frac{1}{2}v_T^2[k_n\rho_{nm}(\tau)]^2}\equiv e^{-i\omega\tau-\frac{1}{2}k_z^2v_T^2\tau^2}e^{-\frac{k_\perp^2v_T^2}{\Omega^2}(1-\cos\Omega\tau)}, \tag{16.47}$$

$$k_z\equiv(\mathbf{kh}),\qquad k_\perp^2=[\mathbf{kh}]^2 \tag{16.48}$$

and expansion of the exponential in a series in modified Bessel functions,

$$e^{z\cos\Omega\tau}=\sum_{n=-\infty}^{+\infty}e^{in\Omega\tau}I_n(z); \tag{16.49}$$

$$(I_n(z)=i^{-n}J_n(iz), \tag{16.50}$$

where J_n is a Bessel function of the first-kind, enables us to rewrite the integral (16.46) in a slightly different form:

$$\langle a(\tau)\rangle=\sum_{n=-\infty}^{+\infty}A_n(z)\int_{-\infty}^{0}d\tau a(\tau)e^{-i(\omega-n\Omega)\tau-\frac{1}{2}k_z^2v_T^2\tau^2}, \tag{16.51}$$

$$A_n(z)\equiv e^{-z}I_n(z), \tag{16.52}$$

$$z\equiv\frac{k_\perp^2v_T^2}{\Omega^2}\equiv k_\perp^2\rho^2. \tag{16.53}$$

Then, using the integrals (16.31) and (16.32), and also

$$\int_{-\infty}^{0}d\tau\tau e^{-i\omega\tau-\frac{1}{2}k^2v_T^2\tau^2}=\frac{1}{(kv_T)^2}\left\{J_+\left(\frac{\omega}{kv_T}\right)-1\right\}, \tag{16.54}$$

we readily obtain expressions for the coefficients (16.45) as infinite series:

$$\begin{aligned}
a&=1-\frac{\omega_L^2}{k_z^2v_T^2}\sum_{n=-\infty}^{+\infty}J_+(\beta_n)A_n(z)\frac{k^2v_T^2}{\omega(\omega-n\Omega)}\frac{n}{z}\left(n-\frac{\omega}{\Omega}\frac{k_\perp^2}{k^2}\right),\\
b&=-\frac{\omega_L^2}{k_z^2v_T^2}\sum_{n=-\infty}^{+\infty}J_+(\beta_n)A_n(z)\frac{n}{z}\frac{k^2v_T^2}{\omega\Omega},\\
c&=\frac{\omega_L^2}{k_z^2v_T^2}\left\{1+\sum_{n=-\infty}^{+\infty}J_+(\beta_n)A_n(z)\frac{k^2}{k_\perp^2}\frac{\Omega}{\omega-n\Omega}\left(n-\frac{\omega}{\Omega}\frac{k_\perp^2}{k^2}\right)\right\},\\
d&=i\frac{\omega_L^2}{\omega}\frac{k^2}{k_z^2}\sum_{n=-\infty}^{+\infty}J_+(\beta_n)\frac{dA_n(z)}{dz}\frac{1}{\omega-n\Omega}\left(n-\frac{\omega}{\Omega}\frac{k_\perp^2}{k^2}\right),\\
e&=i\frac{\omega_L^2}{\omega\Omega}\frac{k^2}{k_z^2}\sum_{n=-\infty}^{+\infty}J_+(\beta_n)\frac{dA_n(z)}{dz},\\
f&=\frac{\omega_L^2}{\omega\Omega}\frac{k^2}{k_z^2}\sum_{n=-\infty}^{+\infty}J_+(\beta_n)\left\{2z\frac{k_z^2}{k_\perp^2}\frac{\Omega}{\omega-n\Omega}\frac{dA_n}{dz}-\frac{n}{z}A_n(z)\right\},
\end{aligned} \tag{16.55}$$

$$\beta_n\equiv\frac{\omega-n\Omega}{|k_z|v_T}. \tag{16.56}$$

Formulas of the type (16.44) and (16.55) were first obtained in [53] and they agree with formula

(A.2) of [53] if allowance is made for the relation between the Kramp function W and $J_+(\beta)$:

$$J_+(\beta) = -i\sqrt{\pi}\,\frac{\beta}{\sqrt{2}}\,W\left(\frac{\beta}{\sqrt{2}}\right). \tag{16.57}$$

Note that the expressions (16.55) for the coefficients a, b, c, d, e, and f of the tensor (16.44) can be obtained differently, directly from the general result (16.24) for a plasma with an isotropic distribution function of the ground state without calculating (16.45). We shall here sketch this method, since it will be used in the nonlinear theory of oscillations.

For the Maxwell distribution (6.14)

$$v_i f_0 = -v_T^2\,\frac{\partial f_0}{\partial v_i}. \tag{16.58}$$

Substituting this into the last term on the right-hand side of (16.24) and integrating with respect to the velocity by parts, we obtain

$$\varepsilon_{ij}(\omega,\mathbf{k}) = \delta_{ij} + \frac{i}{\omega}\frac{4\pi e^2}{m}\int d\mathbf{v} f_0 \int_{-\infty}^{0} d\tau e^{-i\omega\tau + ik_n\rho_{nm}(\tau)v_m}\,\{W_{ij}(-\tau) + v_T^2 k_n\rho_{ni}(\tau)\,k_m\rho_{mj}(-\tau)\}. \tag{16.59}$$

The curly brackets on the right-hand side of (16.59) are identical to the corresponding expression in (16.29). If we here integrate with respect to the velocities by means of (16.27), we obtain exact agreement with (16.29). We shall proceed differently. Namely, we go over to a cylindrical coordinate system (in the velocity space)

$$d\mathbf{v} = dv_z v_\perp dv_\perp d\varphi,\quad v_z \equiv \mathbf{vh} \tag{16.60}$$

with z axis along the external magnetic field, "radius vector" $v_\perp$, and polar angle φ, the angle between the x axis and the component $\mathbf{v}_\perp$ of the velocity $\mathbf{v}$ in the plane perpendicular to the external magnetic field. Introducing the angle θ between the component $\mathbf{k}_\perp$ of the wave vector:

$$\mathbf{k}_\perp = \mathbf{k} - \mathbf{h}(\mathbf{kh}) \tag{16.61}$$

and the x axis, we write the contraction $k_n\rho_{nm}(\tau)v_m$ in the expanded form

$$k_n\rho_{nm}(\tau)v_m = k_z v_z\tau + \frac{k_\perp v_\perp}{\Omega}\sin(\Omega\tau + \theta - \varphi) - \frac{k_\perp v_\perp}{\Omega}\sin(\theta - \varphi). \tag{16.62}$$

Then, with allowance for an expansion in series in Bessel functions:

$$e^{ix\sin\varphi} = \sum_{n=-\infty}^{+\infty} J_n(x)e^{in\varphi} \tag{16.63}$$

we write the tensor (16.59) in the form

$$\varepsilon_{ij}(\omega,\mathbf{k}) = \delta_{ij} + i\frac{\omega_L^2}{\omega}\int_{-\infty}^{+\infty}\frac{dv_z}{\sqrt{2\pi}v_T}e^{-\frac{v_z^2}{2v_T^2}}\int_0^\infty \frac{v_\perp dv_\perp}{v_T^2}e^{-\frac{v_\perp^2}{2v_T^2}}\times$$
$$\times\sum_{n,\,l=-\infty}^{+\infty} J_n\left(\frac{k_\perp v_\perp}{\Omega}\right)J_l\left(\frac{k_\perp v_\perp}{\Omega}\right)\frac{1}{2\pi}\int_0^{2\pi} d\varphi\int_{-\infty}^{0} d\tau e^{-i(\omega - n\Omega - k_z v_z)\tau}e^{i(n-l)(\theta-\varphi)}\,\{W_{ij}(-\tau) + v_T^2 k_n\rho_{ni}(\tau)\,k_m\rho_{mj}(-\tau)\}. \tag{16.64}$$

Averaging over the azimuth

$$\frac{1}{2\pi}\int_0^{2\pi} d\varphi e^{i(l-n)\varphi} = \delta_{l-n,\,0} \tag{16.65}$$

and integrating with respect to the transverse velocities,

$$\int_0^\infty \frac{v_\perp dv_\perp}{v_T^2} e^{-\frac{v_\perp^2}{2v_T^2}} J_n^2\left(\frac{k_\perp v_\perp}{\Omega}\right) = A_n\left(\frac{k_\perp^2 v_T^2}{\Omega^2}\right), \tag{16.66}$$

and also replacing the last term in the curly brackets by its expanded expression (16.41) with allowance for (16.42) and (16.43), we obtain the previous expansion (16.44) of the permittivity tensor $\varepsilon_{ij}(\omega, \mathbf{k})$ with the coefficients (16.55) if we use the integrals that arise after integration with respect to τ [cf. (7.43)]:

$$\begin{gathered}
\int_{-\infty}^{+\infty} \frac{dv_z}{\sqrt{2\pi}\, v_T} e^{-\frac{v_z^2}{2v_T^2}} \frac{1}{\omega - n\Omega - k_z v_z} = \frac{1}{\omega - n\Omega} J_+(\beta_n), \\
\int_{-\infty}^{+\infty} \frac{dv_z}{\sqrt{2\pi}\, v_T} e^{-\frac{v_z^2}{2v_T^2}} \frac{1}{(\omega - n\Omega - k_z v_z)^2} = \frac{1}{(k_z v_T)^2} \{J_+(\beta_n) - 1\}, \\
\int_{-\infty}^{+\infty} \frac{dv_z}{\sqrt{2\pi}\, v_T} e^{-\frac{v_z^2}{2v_T^2}} \frac{1}{(\omega - n\Omega - k_z v_z)^3} - \frac{1}{2\,|k_z v_T|^3} \left\{\frac{1}{\beta_n} J_+(\beta_n) - \beta_n J_+(\beta_n) + \beta_n\right\}.
\end{gathered} \tag{16.67}$$

To conclude this section we give two expressions for the permittivity tensor in the two limiting cases of purely transverse,

$$k_z = (\mathbf{kh}) = 0, \tag{16.68}$$

and purely longitudinal,

$$[\mathbf{kh}] = 0, \quad \varkappa_i \varkappa_j = h_i h_j, \tag{16.69}$$

propagation of oscillations relative to the external magnetic field $\mathbf{B}_0$. If the wave vector of a plasma oscillation is strictly perpendicular, (16.68), to the external magnetic field, it follows from the identities (16.42) and (16.43) that the six elementary tensors that determine the structure (16.44) of the permittivity reduce to four:

$$\delta_{ij} = \varkappa_i \varkappa_j + h_i h_j + [\varkappa\mathbf{h}]_i [\varkappa\mathbf{h}]_j; \tag{16.70}$$

$$e_{isj} h_s = \varkappa_i [\varkappa\mathbf{h}]_j - \varkappa_j [\varkappa\mathbf{h}]_i, \quad \varkappa\mathbf{h} = 0. \tag{16.71}$$

Using this fact and (16.44)-(16.46), we obtain for purely transverse propagation

$$\varepsilon_{ij}(\omega, \mathbf{k}) = a_\perp \delta_{ij} + b_\perp \varkappa_i \varkappa_j + c_\perp h_i h_j + d_\perp e_{isj} h_s; \tag{16.72}$$

$$\begin{aligned}
a_\perp &= 1 + i\frac{\omega_L^2}{\omega}\left\langle \cos\Omega\tau + \frac{k^2 v_T^2}{\Omega^2}(1 - \cos\Omega\tau)^2 \right\rangle_\perp, \\
b_\perp &= -i\frac{\omega_L^2}{\omega}\left\langle 2\frac{k^2 v_T^2}{\Omega^2}(1 - \cos\Omega\tau) \right\rangle_\perp, \\
c_\perp &= i\frac{\omega_L^2}{\omega}\left\langle 1 - \cos\Omega\tau - \frac{k^2 v_T^2}{\Omega^2}(1 - \cos\Omega\tau)^2 \right\rangle_\perp, \\
d_\perp &= i\frac{\omega_L^2}{\omega}\left\langle \sin\Omega\tau - \frac{k^2 v_T^2}{\Omega^2}\sin\Omega\tau\,(1 - \cos\Omega\tau) \right\rangle_\perp,
\end{aligned} \tag{16.73}$$

where the angular brackets with the subscript $\perp$ stand for the "averaging"

$$\langle a(\tau)\rangle_\perp \equiv \int_{-\infty}^{\infty} d\tau a(\tau) \exp\left\{-i\omega\tau - \frac{k^2 v_T^2}{\Omega^2}(1 - \cos\Omega\tau)\right\}. \tag{16.74}$$

Similarly, for purely transverse, (16.69), propagation

$$\varepsilon_{ij}(\omega, \mathbf{k}) = \delta_{ij} + i\frac{\omega_L^2}{\omega}\langle\cos\Omega\tau\rangle_{\parallel}(\delta_{ij} - h_i h_j) + i\frac{\omega_L^2}{\omega}h_i h_j\langle 1 - k^2 v_T^2\tau^2\rangle_{\parallel} + i\frac{\omega_L^2}{\omega}e_{isj}h_s\langle\sin\Omega\tau\rangle_{\parallel}. \quad (16.75)$$

Here the angular brackets with the subscript $\parallel$ stand for the integration

$$\langle a(\tau)\rangle_{\parallel} \equiv \int_{-\infty}^{0} d\tau\, a(\tau)\, e^{-i\omega\tau - \frac{1}{2}k^2 v_T^2\tau^2}. \quad (16.76)$$

Carrying out this integration by means of (16.31) and (16.32) and the integrals

$$\int_{-\infty}^{0} d\tau \cos\Omega\tau e^{-i\omega\tau - \frac{1}{2}k^2 v_T^2\tau^2} = \frac{i}{2}\left\{\frac{1}{\omega-\Omega}J_+\left(\frac{\omega-\Omega}{kv_T}\right) + \frac{1}{\omega+\Omega}J_+\left(\frac{\omega+\Omega}{kv_T}\right)\right\}, \quad (16.77)$$

$$\int_{-\infty}^{0} d\tau \sin\Omega\tau e^{-i\omega\tau - \frac{1}{2}k^2 v_T^2\tau^2} = \frac{1}{2}\left\{\frac{1}{\omega-\Omega}J_+\left(\frac{\omega-\Omega}{kv_T}\right) - \frac{1}{\omega-\Omega}J_+\left(\frac{\omega+\Omega}{kv_T}\right)\right\}, \quad (16.78)$$

we obtain the slightly different expression [cf. (16.55)]

$$\varepsilon_{ij}(\omega, \mathbf{k}) = \delta_{ij} - \frac{\omega_L^2}{2\omega}(\delta_{ij} - h_i h_j)\left\{\frac{1}{\omega-\Omega}J_+\left(\frac{\omega-\Omega}{kv_T}\right) + \frac{1}{\omega+\Omega}J_+\left(\frac{\omega+\Omega}{kv_T}\right)\right\} +$$
$$+ \frac{\omega_L^2}{k^2 v_T^2}\left\{1 - J_+\left(\frac{\omega}{kv_T}\right)\right\}h_i h_j + i\frac{\omega_L^2}{2\omega}e_{isj}h_s\left\{\frac{1}{\omega-\Omega}J_+\left(\frac{\omega-\Omega}{kv_T}\right) - \frac{1}{\omega+\Omega}J_+\left(\frac{\omega+\Omega}{kv_T}\right)\right\}. \quad (16.79)$$

We also give an expression for the anti-Hermitian part $\varepsilon_{ij}^a(\omega, \mathbf{k})$ of the permittivity tensor of a magnetoactive plasma with Maxwell distribution (6.14); this follows directly from formulas (16.44), (16.55), (16.56) after separation of the imaginary part of the function $J_+(\beta_n)$ in accordance with (6.20):

$$\varepsilon_{ij}^a(\omega, \mathbf{k}) = a''\delta_{ij} + b''\varkappa_i\varkappa_j + c''h_i h_j + d'e_{isj}h_s + e'(\varkappa_i[\varkappa\mathbf{h}]_j - \varkappa_j[\varkappa\mathbf{h}]_i) + f''[\varkappa\mathbf{h}]_i[\varkappa\mathbf{h}]_j; \quad (16.80)$$

$$a'' = i\sqrt{\frac{\pi}{2}}\frac{\omega_L^2}{\omega|k_z|v_T}\sum_{n=-\infty}^{+\infty}A_n(z)\frac{n}{z(\varkappa\mathbf{h})^2}\left(n - \frac{\omega}{\Omega}[\varkappa\mathbf{h}]^2\right)\exp\left\{-\frac{1}{2}\left(\frac{\omega-n\Omega}{k_z v_T}\right)^2\right\},$$

$$b'' = i\sqrt{\frac{\pi}{2}}\frac{\omega_L^2}{\omega|k_z|v_T}\sum_{n=-\infty}^{+\infty}A_n(z)\frac{n}{z(\varkappa\mathbf{h})^2}\left(n - \frac{\omega}{\Omega}\right)\exp\left\{-\frac{1}{2}\left(\frac{\omega-n\Omega}{k_z v_T}\right)^2\right\},$$

$$c'' = -i\sqrt{\frac{\pi}{2}}\frac{\omega_L^2}{\Omega|k_z|v_T}\sum_{n=-\infty}^{+\infty}A_n(z)\frac{n - \frac{\omega}{\Omega}[\varkappa\mathbf{h}]^2}{z(\varkappa\mathbf{h})^2}\exp\left\{-\frac{1}{2}\left(\frac{\omega-n\Omega}{k_z v_T}\right)^2\right\},$$

$$d' = \sqrt{\frac{\pi}{2}}\frac{\omega_L^2}{\omega|k_z|v_T}\sum_{n=-\infty}^{+\infty}\frac{dA_n(z)}{dz}\frac{n - \frac{\omega}{\Omega}[\varkappa\mathbf{h}]^2}{(\varkappa\mathbf{h})^2}\exp\left\{-\frac{1}{2}\left(\frac{\omega-n\Omega}{k_z v_T}\right)^2\right\}, \quad (16.81)$$

$$e' = \sqrt{\frac{\pi}{2}}\frac{\omega_L^2}{\omega|k_z|v_T}\sum_{n=-\infty}^{+\infty}\frac{dA_n(z)}{dz}\frac{n - \left(\frac{\omega}{\Omega}\right)}{(\varkappa\mathbf{h})^2}\exp\left\{-\frac{1}{2}\left(\frac{\omega-n\Omega}{k_z v_T}\right)^2\right\},$$

$$f'' = -i\sqrt{\frac{\pi}{2}}\frac{\omega_L^2}{\omega|k_z|v_T}\sum_{n=-\infty}^{+\infty}\left\{2\frac{z}{[\varkappa\mathbf{h}]^2}\frac{dA_n(z)}{dz} + \frac{n}{z(\varkappa\mathbf{h})^2}\left(n - \frac{\omega}{\Omega}\right)A_n(z)\right\}\exp\left\{-\frac{1}{2}\left(\frac{\omega-n\Omega}{k_z v_T}\right)^2\right\}.$$

The anti-Hermitian part ε_{ij}^a of the permittivity tensor, which occurs in the generalized kinetic equation (3.22), determines the linear damping of the oscillations of a magnetoactive plasma that participate in a nonlinear interaction.

§ 17. The Inverse Tensor (3.17) and the Spectral Function of the Electric Fields of the Characteristic Oscillations of a Magnetoactive Plasma

The inverse tensor $A_{ij}(\omega, \mathbf{k})$ and the spectral functions of the fields $(E_j E_i)_{\omega,\mathbf{k}}$, which occur in the generalized kinetic equation (3.22), have a more complicated tensor structure in a magnetoactive plasma than in an isotropic plasma. At the same time, the relations obtained in § 16 for the permittivity tensor enable us to find the inverse tensor (3.17) in a magnetoactive plasma (see § 6). At the same time we shall find the explicit form of the spectral function $(E_j E_i)_{\omega,\mathbf{k}}$, since its tensor structure is determined, as we shall show below, by the anti-Hermitian part of the inverse tensor $A_{ji}(\omega, \mathbf{k})$ in the transparency region (4.7) of the plasma.

Every second rank tensor that describes a homogeneous magnetoactive plasma with distribution function f_0 satisfying Eq. (15.4) can be represented as a linear combination of the six independent elementary tensors [cf. (16.44)]

$$\delta_{ij},\ \varkappa_i\varkappa_j,\ h_ih_j,\ e_{isj}h_s,\ \varkappa_i[\varkappa\mathbf{h}]_j-\varkappa_j[\varkappa\mathbf{h}]_i,\ [\varkappa\mathbf{h}]_i[\varkappa\mathbf{h}]_j, \tag{17.1}$$

which are composed of the two unit vectors $\varkappa$ and $\mathbf{h}$ and the two unit tensors δ_{ij} and e_{isj}. The tensors (17.1) are linearly independent, i.e., their linear combination

$$a_1\delta_{ij}+a_2\varkappa_i\varkappa_j+a_3h_ih_j+a_4e_{isj}h_s+a_5(\varkappa_i[\varkappa\mathbf{h}]_j-\varkappa_j[\varkappa\mathbf{h}]_i)+a_6[\varkappa\mathbf{h}]_i[\varkappa\mathbf{h}]_j \tag{17.2}$$

vanishes if and only if all the coefficients $a_1, \ldots, a_6$ of the elementary tensors vanish identically. However, this assertion is only true for "oblique" propagation, when the wave vector $\mathbf{k} = k\varkappa$ of the plasma oscillations is directed at an angle not equal to 0 or $\pi/2$ to the external magnetic field [see the sets of tensors (17.7) and (17.8) in these cases]. The other conceivable elementary tensors that can be composed of the two vectors $\varkappa$ and $\mathbf{h}$ are either a linear combination* already written down in (17.1):

$$\begin{aligned}
&(\varkappa\mathbf{h})(\varkappa_ih_j+\varkappa_jh_i)=\varkappa_i\varkappa_j+h_ih_j-[\varkappa\mathbf{h}]^2\delta_{ij}+[\varkappa\mathbf{h}]_i[\varkappa\mathbf{h}]_j,\\
&(\varkappa\mathbf{h})(h_i[\varkappa\mathbf{h}]_j-h_j[\varkappa\mathbf{h}]_i)=(\varkappa_i[\varkappa\mathbf{h}]_j-\varkappa_j[\varkappa\mathbf{h}]_i)-[\varkappa\mathbf{h}]^2e_{isj}h_s,\\
&(\varkappa\mathbf{h})e_{isj}\varkappa_s=e_{isj}h_s-(\varkappa_i[\varkappa\mathbf{h}]_j-\varkappa_j[\varkappa\mathbf{h}]_i),
\end{aligned} \tag{17.3}$$

or can be rejected on account of the symmetry property of the tensor under consideration. Such tensors are

$$(h_i[\varkappa\mathbf{h}]_j+h_j[\varkappa\mathbf{h}]_i),\ (\varkappa_i[\varkappa\mathbf{h}]_j+\varkappa_j[\varkappa\mathbf{h}]_i),\ h_i\varkappa_j-h_j\varkappa_i=e_{isj}[\varkappa\mathbf{h}]_s. \tag{17.4}$$

In the given case the elementary tensors (17.4) drop out on account of the symmetry (see [1], p. 71, (9.50)) of the permittivity tensor

$$\varepsilon_{ij}(\omega,\ \mathbf{k},\ \mathbf{h})=\varepsilon_{ji}(\omega,\ -\mathbf{k},\ -\mathbf{h}), \tag{17.5}$$

and therefore of the inverse tensor:

$$A_{ij}(\omega,\ \mathbf{k},\ \mathbf{h})=A_{ji}(\omega,\ -\mathbf{k},\ -\mathbf{h}). \tag{17.6}$$

---

* The first two identities (17.3) have already been used to derive $\varepsilon_{ij}(\omega, \mathbf{k})$ in § 16 [see (16.42), (16.43) and the remark in the footnote]. The last, third identity (17.3) follows from the more general equation

$$\delta_{ia}e_{bsj}h_s-\delta_{aj}e_{bsi}h_s+\delta_{ab}e_{jsi}h_s=h_ae_{jbi}$$

by contraction with the tensor $\varkappa_a\varkappa_b$.

In the case of purely longitudinal propagation of the oscillations (16.68), the linearly independent tensors (17.1) reduce to the three tensors [cf. (16.79)]

$$\delta_{ij}, \quad h_i h_j, \quad e_{isj} h_s, \tag{17.7}$$

and in the case of purely transverse propagation (16.68) of the oscillations the number of elementary tensors is four,* since in accordance with the consequences of (17.3) Eqs.(16.70)-(16.71) then hold. Therefore, in the case of transverse propagation of oscillations the independent elementary tensors can be taken to be the first four in (17.1):

$$\delta_{ij}, \ \varkappa_i \varkappa_j, \ h_i h_j, \ e_{isj} h_s. \tag{17.8}$$

Then, knowing the tensor structure of $A_{ij}(\omega, \mathbf{k})$:

$$A_{ij}(\omega, \mathbf{k}) = \frac{1}{\Delta}\{a_1 \delta_{ij} + a_2 \varkappa_i \varkappa_j + a_3 h_i h_j + a_4 e_{isj} h_s + a_5 (\varkappa_i [\varkappa \mathbf{h}]_j - \varkappa_j [\varkappa \mathbf{h}]_i) + a_6 [\varkappa \mathbf{h}]_i [\varkappa \mathbf{h}]_j\}, \tag{17.9}$$

where Δ is determined by the left-hand side of the dispersion relation (2.14),

$$\Delta \equiv \Delta(\omega, \mathbf{k}) = \left| \varepsilon_{ij}(\omega, \mathbf{k}) - \frac{c^2 k^2}{\omega^2}(\delta_{ij} - \varkappa_i \varkappa_j) \right|, \tag{17.10}$$

we can find explicit expressions for the coefficients $a_1, a_2, \ldots, a_6$ and the determinant Δ in terms of the coefficients a, b, c, d, e, f of the elementary tensors in the permittivity tensor (16.44).

It is obvious that this problem is completely analogous to deriving $A_{ij}(\omega, \mathbf{k})$ in an isotropic plasma in the form (6.77) by means of (6.17). To do this, we use a very simple coordinate system, taking the z axis along the external magnetic field, (16.39), and the unit vector $\varkappa$ in the zx plane (ϑ is the angle between $\varkappa$ and **h**):

$$\varkappa_y = 0, \quad \varkappa_x = \sin\vartheta, \quad \varkappa_z = \cos\vartheta. \tag{17.11}$$

In this coordinate system, the elementary tensors (17.7) have the form (16.40), and the other three remaining in (17.1) are given by

$$\begin{gathered}
\varkappa_i \varkappa_j = \begin{pmatrix} \sin^2\vartheta & 0 & \sin\vartheta\cos\vartheta \\ 0 & 0 & 0 \\ \sin\vartheta\cos\vartheta & 0 & \cos^2\vartheta \end{pmatrix}, \\
(\varkappa_i [\varkappa \mathbf{h}]_j - \varkappa_j [\varkappa \mathbf{h}]_i) = \begin{pmatrix} 0 & -\sin^2\vartheta & 0 \\ \sin^2\vartheta & 0 & \sin\vartheta\cos\vartheta \\ 0 & -\sin\vartheta\cos\vartheta & 0 \end{pmatrix}, \\
[\varkappa \mathbf{h}]_i [\varkappa \mathbf{h}]_j = \begin{pmatrix} 0 & 0 & 0 \\ 0 & \sin^2\vartheta & 0 \\ 0 & 0 & 0 \end{pmatrix}.
\end{gathered} \tag{17.12}$$

The matrices (16.40) and (17.12) give the following matrix expression for the Maxwell tensor:

$$M_{ij}(\omega, \mathbf{k}) = \begin{pmatrix} \bar{a} + \bar{b}\sin^2\vartheta & -d - e\sin^2\vartheta & \bar{b}\sin\vartheta\cos\vartheta \\ d + e\sin^2\vartheta & \bar{a} + f\sin^2\vartheta & e\sin\vartheta\cos\vartheta \\ \bar{b}\sin\vartheta\cos\vartheta & -e\sin\vartheta\cos\vartheta & \bar{a} + c + b\cos^2\vartheta \end{pmatrix}, \tag{17.13}$$

$$\bar{a} \equiv a - n^2, \quad \bar{b} \equiv b + n^2, \quad n^2 \equiv (c^2 k^2/\omega^2) \tag{17.14}$$

---

* This assertion is not precise. In general, in the case of purely transverse propagation there exist six elementary tensors: $\varkappa_i h_j + \varkappa_j h_i$, $e_{isj}\varkappa_s = -(h_i[\varkappa\mathbf{h}]_j - h_j[\varkappa\mathbf{h}]_i)$ and (17.8) that satisfy the symmetry condition (17.6). However, the first two tensors ($\varkappa_i h_j + \varkappa_j h_i$) and $e_{isj}\varkappa_s$ do not occur in A_{ij}: the coefficients of these tensors vanish. This is obvious from continuity considerations in the limit ($\varkappa \mathbf{h} \to 0$) and be confirmed by a direct calculation.

and the inverse tensors

$$A_{ij}(\omega, \mathbf{k}) = \frac{1}{\Delta}\begin{pmatrix} a_1 + a_2\sin^2\vartheta & -a_4 - a_5\sin^2\vartheta & a_2\sin\vartheta\cos\vartheta \\ a_1 + a_5\sin^2\vartheta & a_1 + a_6\sin^2\vartheta & a_5\sin\vartheta\cos\vartheta \\ a_2\sin\vartheta\cos\vartheta & -a_5\sin\vartheta\cos\vartheta & a_1 + a_3 + a_2\cos^2\vartheta \end{pmatrix} \tag{17.15}$$

Writing down the elements of the matrices (17.13) and (17.15) by means of the usual definition of the inverse matrix as cofactor, we find the coefficients $a_1, \ldots, a_6$ of the tensor (17.9):

$$\begin{aligned} a_1 &= \bar{a}(a+b+c) + [\varkappa\mathbf{h}]^2\{f(a+b+c) + e(e+d)\},\\ a_2 &= -ed - \bar{a}\bar{b} - [\varkappa\mathbf{h}]^2(e^2 + \bar{b}f),\\ a_3 &= d(e+d) - \bar{a}c - cf[\varkappa\mathbf{h}]^2,\\ a_4 &= -d(a+b+c) - ec[\varkappa\mathbf{h}]^2,\\ a_5 &= \bar{b}d - \bar{a}e,\\ a_6 &= -e(e+d) - f(a+b+c) + \bar{b}c. \end{aligned} \tag{17.16}$$

The quantities $\bar{a}$ and $\bar{b}$ are determined by Eq. (17.14) and the determinant Δ can be found directly by means of (17.13):

$$\begin{aligned} \Delta &= An^4 + Bn^2 + C,\\ A &= a + b + c(\varkappa\mathbf{h})^2,\\ B &= -2a(a+b+c) - [\varkappa\mathbf{h}]^2\{f(a+b+c(\varkappa\mathbf{h})^2) + (e+d)^2 + c(b-a)\},\\ C &= (a^2+d^2)(a+b+c) + [\varkappa\mathbf{h}]^2\{af(a+b+c) + abc - bd^2 + c[\varkappa\mathbf{h}]^2(bf+e^2) + e(ae + 2ad + 2cd)\}. \end{aligned} \tag{17.17}$$

The expression (17.9) for the inverse tensor with the coefficients (17.16) and the determinant (17.17) were first obtained in a similar form in [54]. The coefficients (17.16) are equal to the coefficients (7) in [55] if the identities (16.42) and (16.43) are used to eliminate the tensors

$$\varkappa_i h_j + \varkappa_j h_i, \qquad h_i[\varkappa\mathbf{h}]_j - h_j[\varkappa\mathbf{h}]_i, \tag{17.18}$$

which are used in [55] alongside (17.1).*

We now consider the inverse tensor $A_{ij}(\omega, \mathbf{k})$ in different special cases.

In the case of purely longitudinal propagation (16.69), the tensor structure of $A_{ij}(\omega, \mathbf{k})$ is determined in accordance with the above general arguments by the three elementary tensors (17.7):

$$\begin{aligned} A_{ij}(\omega, \mathbf{k}) &= \frac{1}{\Delta_\parallel}\{\alpha_1\delta_{ij} + \alpha_2\varkappa_i\varkappa_j + \alpha_3 e_{isj}h_s\},\\ \Delta_\parallel &= (\alpha+\beta)[(\alpha-n^2)^2 + \gamma^2],\\ \alpha_1 &= (\alpha - n^2)(\alpha+\beta), \qquad \alpha_2 = \gamma^2 - (\alpha-n^2)(\beta+n^2),\\ \alpha_3 &= -\gamma(\alpha+\beta). \end{aligned} \tag{17.19}$$

Here the coefficients α, β, and γ occur (successively) in front of the elementary tensors (17.7) in the permittivity tensor and they are given in accordance with (16.79) by

$$\alpha = 1 - \frac{\omega_L^2}{2\omega}\left\{\frac{1}{\omega-\Omega}J_+\left(\frac{\omega-\Omega}{kv_T}\right) + \frac{1}{\omega+\Omega}J_+\left(\frac{\omega+\Omega}{kv_T}\right)\right\},$$

---

* In [54, 55] the inverse tensor is decomposed with respect to the eight elementary tensors (17.1) and (17.18). Obviously, the elimination of the last two of (17.18) by means of (16.42) and (16.43) simplifies the expression for $A_{ij}(\omega, \mathbf{k})$.

$$\beta=\frac{\omega_L^2}{k^2v_T^2}\left\{1-J_+\left(\frac{\omega}{kv_T}\right)\right\}+\frac{\omega_L^2}{2\omega}\left\{\frac{1}{\omega-\Omega}J_+\left(\frac{\omega-\Omega}{kv_T}\right)+\frac{1}{\omega+\Omega}J_+\left(\frac{\omega+\Omega}{kv_T}\right)\right\},$$
$$\gamma=\frac{i}{2}\frac{\omega_L^2}{\omega}\left\{\frac{1}{\omega-\Omega}J_+\left(\frac{\omega-\Omega}{kv_T}\right)-\frac{1}{\omega+\Omega}J_+\left(\frac{\omega+\Omega}{kv_T}\right)\right\}. \tag{17.20}$$

To the purely transverse (16.68) propagation of oscillations there corresponds the inverse tensor

$$A_{ij}(\omega,\mathbf{k})=\frac{1}{\Delta_\perp}\{A_1\delta_{ij}+A_2\varkappa_i\varkappa_j+A_3h_ih_j+A_4e_{isj}h_s\},$$
$$A_1=(a_\perp+b_\perp)(a_\perp-n^2+c_\perp),$$
$$A_2=-(b_\perp+n^2)(a_\perp-n^2+c_\perp), \tag{17.21}$$
$$A_3=d_\perp^2-c_\perp(a_\perp+b_\perp),\quad A_4=-d_\perp(a_\perp-n^2+c_\perp),$$
$$\Delta_\perp=n^4(a_\perp+b_\perp)-n^2[(a_\perp+b_\perp)(2a_\perp+c_\perp)+d_\perp^2]+(a_\perp+c_\perp)[d_\perp^2+a_\perp(a_\perp+b_\perp)].$$

Here, the coefficients $a_\perp$, $b_\perp$, $c_\perp$, and $d_\perp$ are determined by Eqs. (16.73). In the limit of a cold plasma ($v_T=0$), the inverse tensor simplifies considerably and (for arbitrary angles between the direction of propagation $\varkappa$ of the oscillation and the external magnetic field $\mathbf{B}_0$) has the form

$$A_{ij}(\omega,\mathbf{k})=\frac{1}{\Delta_0}\{\varepsilon_2(\varepsilon_1-n^2)\delta_{ij}-n^2(\varepsilon_1-n^2)\varkappa_i\varkappa_j-[g^2+(\varepsilon_2-\varepsilon_1)(\varepsilon_1-n^2)]h_ih_j+ig\varepsilon_2e_{isj}h_s-$$
$$-ign^2(\varkappa_i[\varkappa\mathbf{h}]_j-\varkappa_j[\varkappa\mathbf{h}]_i)+n^2(\varepsilon_2-\varepsilon_1)[\varkappa\mathbf{h}]_i[\varkappa\mathbf{h}]_j\}. \tag{17.22}$$

Here, ε_1, ε_2, and g are given by (16.38) and the determinant Δ_0 is given by the well-known expression (see [1], p. 140, (20.5))

$$\Delta_0=n^4(\varepsilon_1\sin^2\vartheta+\varepsilon_2\cos^2\vartheta)-n^2\{(\varepsilon_1^2-g^2-\varepsilon_1\varepsilon_2)\sin^2\vartheta+2\varepsilon_1\varepsilon_2\}+(\varepsilon_1^2-g^2)\varepsilon_2. \tag{17.23}$$

Going over to a cold plasma ($v_T=0$) in formulas (17.19) and (17.21) [or restricting our treatment to purely longitudinal or purely transverse propagation of the oscillations in (17.22)-(17.23)], we obtain expressions for the inverse tensor $A_{ij}(\omega,\mathbf{k})$ in a cold magnetoactive plasma for longitudinal ($[\varkappa\mathbf{h}]=0$)

$$A_{ij}(\omega,\mathbf{k})=\frac{1}{\Delta_\parallel^0}\{\varepsilon_2(\varepsilon_1-n^2)\delta_{ij}+\varkappa_i\varkappa_j[n^4-n^2(2\varepsilon_1-\varepsilon_2)-g^2+\varepsilon_1(\varepsilon_1-\varepsilon_2)]+ig\varepsilon_2e_{isj}h_s\}, \tag{17.24}$$

$$\Delta_\parallel^0=\varepsilon_2n^4-2\varepsilon_1\varepsilon_2n^2+\varepsilon_2(\varepsilon_1^2-g^2) \tag{17.25}$$

and transverse ($\varkappa\mathbf{h}=0$)

$$A_{ij}(\omega,\mathbf{k})=\frac{1}{\Delta_\perp^0}\{\varepsilon_1(\varepsilon_2-n^2)\delta_{ij}+n^2(n^2-\varepsilon_2)\varkappa_i\varkappa_j-[g^2+\varepsilon_1(\varepsilon_2-\varepsilon_1)]h_ih_j+ig(\varepsilon_2-n^2)e_{isj}h_s\}, \tag{17.26}$$

$$\Delta_\perp^0=\varepsilon_1n^4-n^2(\varepsilon_1^2-g^2+\varepsilon_1\varepsilon_2)+\varepsilon_2(\varepsilon_1^2-g^2) \tag{17.27}$$

propagation of the oscillations, respectively.

In the case of a vanishingly weak magnetic field ($\mathbf{B}_0=0$), only two of the six coefficients (16.45) of the permittivity tensor (16.44) are nonvanishing [see (16.30)-(16.32)]:

$$a=1+i\frac{\omega_L^2}{\omega}\int_{-\infty}^{0}d\tau e^{-i\omega\tau-\frac{1}{2}k^2v_T^2\tau^2},$$
$$b=-i\frac{\omega_L^2}{\omega}k^2v_T^2\int_{-\infty}^{0}d\tau\tau^2e^{-i\omega\tau-\frac{1}{2}k^2v_T^2\tau^2}, \tag{17.28}$$
$$c=d=e=f=0,\quad \mathbf{B}_0=0.$$

This, in its turn, simplifies the coefficients (17.16) of the inverse tensor:

$$a_1=\left(a-\frac{c^2k^2}{\omega^2}\right)(a+b),\quad a_2=-\left(a-\frac{c^2k^2}{\omega^2}\right)\left(b+\frac{c^2k^2}{\omega^2}\right),$$
$$a_3=a_4=a_5=a_6=0 \tag{17.29}$$

and also the determinant (17.17):

$$\Delta=(a+b)\left(a-\frac{c^2k^2}{\omega^2}\right)^2 \tag{17.30}$$

and takes us from (17.9) to the expression (6.79) (calculated in an isotropic plasma) if allowance is made for the equations [cf. (17.28), (16.31), (16.32), and (6.21)-(6.22)]

$$a+b=\varepsilon^l(\omega,\ \mathbf{k}),\qquad a=\varepsilon^{tr}(\omega,\ \mathbf{k}),\qquad \mathbf{B}_0=0. \tag{17.31}$$

In the transparency regions of a magnetoactive plasma, when the decay constants of the oscillations are small compared with the frequencies of the oscillations, the inverse tensor $A_{ij}(\omega, \mathbf{k})$ can be split naturally, as in an isotropic plasma, into a resonant and a nonresonant part [cf. (6.82)-(6.83)]:

$$A_{ij}^{\text{nonres}}(\omega,\ \mathbf{k})=\frac{P}{\Delta'(\omega,\ \mathbf{k})}\{a_1\delta_{ij}+a_2\varkappa_i\varkappa_j+a_3h_ih_j+a_4e_{isj}h_s+a_5(\varkappa_i[\varkappa\mathbf{h}]_j-\varkappa_j[\varkappa\mathbf{h}]_i)+a_6[\varkappa\mathbf{h}]_i[\varkappa\mathbf{h}]_j\}, \tag{17.32}$$

$$A_{ij}^{\text{res}}(\omega,\ \mathbf{k})=-i\pi\,\text{sgn}\,\Delta''(\omega,\ \mathbf{k})\,\delta[\Delta'(\omega,\ \mathbf{k})]\{a_1\delta_{ij}+a_2\varkappa_i\varkappa_j+$$
$$+a_3h_ih_j+a_4e_{isj}h_s+a_5(\varkappa_i[\varkappa\mathbf{h}]_j-\varkappa_j[\varkappa\mathbf{h}]_i)+a_6[\varkappa\mathbf{h}]_i[\varkappa\mathbf{h}]_j\} \tag{17.33}$$

Here we have used the approximate relation [cf. (6.80)-(6.81)]

$$\frac{1}{\Delta(\omega,\ \mathbf{k})}\simeq\frac{P}{\Delta'(\omega,\ \mathbf{k})}-i\pi\,\text{sgn}\,\Delta''(\omega,\ \mathbf{k})\,\delta[\Delta'(\omega,\ \mathbf{k})], \tag{17.34}$$

which is the more accurate, the smaller is the imaginary part $\Delta''(\omega, \mathbf{k})$ of the determinant (17.10) compared with the real part $\Delta'(\omega, \mathbf{k})$, which we have introduced in Ch. I [see (4.6)] in deriving the nonlinear dispersion relation (4.12). In its turn, the smallness of $\Delta''(\omega, \mathbf{k})$ is related to the smallness of the damping constant $\gamma(\mathbf{k})$ of the oscillations of a magnetoactive plasma on account of the dispersion relation (2.14) ($\omega \equiv \omega' + i\omega''$):

$$\Delta(\omega,\ \mathbf{k})=\Delta'+i\Delta''\simeq\Delta'(\omega',\ \mathbf{k})+i\omega''\frac{\partial\Delta'(\omega',\ \mathbf{k})}{\partial\omega'}+i\Delta''(\omega',\ \mathbf{k})=0. \tag{17.35}$$

Namely, the oscillation spectrum $\omega(\mathbf{k}) \equiv \omega'$ is then determined by (4.7) and the damping has the form [cf. (6.25a)]

$$\gamma(\mathbf{k})=-\omega''=\Delta''(\omega',\ \mathbf{k})\left\{\frac{\partial\Delta'(\omega',\ \mathbf{k})}{\partial\omega'}\right\}^{-1}. \tag{17.36}$$

As we have already pointed out in Ch. II [see Eq. (3.22a)], the resonance parts (17.33) of the inverse tensors are essential for the description of decay processes. We now show that the resonant part (17.33) of the inverse tensor $A_{ij}^{\text{res}}(\omega, \mathbf{k})$ completely determines the tensor structure of the spectral function $(E_i E_j)_{\omega,\mathbf{k}}$ of the field in the magnetoactive plasma [we have already given this assertion in the form of formula (4.3) without proof]. In accordance with (4.5), the expansion of the determinant $\Delta'(\omega, \mathbf{k})$ with respect to the elements of the i-th row can be written in the form (see [56], p. 27, (2.6)]

$$M_{ij}^H e_{js}=\Delta'\delta_{is}. \tag{17.37}$$

In the case $\Delta' = 0$ in which we are interested, we have

$$M_{ij}^H e_{js}=0, \tag{17.38}$$

and after comparison with the field equations (2.13)

$$M^H_{ij} E_j = 0 \tag{17.39}$$

this implies the relation (cf. [57], § 1.7, p. 20)

$$E_j(\omega, \mathbf{k}) \sim e_{js}(\omega, \mathbf{k}), \tag{17.40}$$

in which the ~ must be understood as a vector equation in j accurate to within a scalar factor; the subscript s is fixed but arbitrary. Thus, for the spectral function

$$(E_j E_i)_{\omega, \mathbf{k}} \sim e_{js} e^*_{ir} \quad \text{(s and r fixed)}, \tag{17.41}$$

which takes in accordance with the condition of transparency (Hermiticity)

$$M^H_{ij} = M^{H*}_{ji}, \quad e_{ij} = e^*_{ji} \tag{17.42}$$

the form

$$(E_j E_i)_{\omega, \mathbf{k}} \sim e_{js} e_{ri}. \tag{17.43}$$

Using the equation ([58], p. 39)

$$e_{js} e_{ri} - e_{rs} e_{ji} = \Delta' (-1)^{i+j+r+s} M^H_{\alpha\beta}, \quad \alpha \neq j, \ \alpha \neq r, \ \beta \neq i, \ \beta \neq s, \tag{17.44}$$

which simplifies in our case

$$e_{js} e_{ri} - e_{rs} e_{ji} = 0 \tag{17.45}$$

on account of $\Delta' = 0$, we obtain with allowance for (17.43) the desired assertion

$$(E_j E_i)_{\omega, \mathbf{k}} \sim e_{ji}(\omega, \mathbf{k}) \sim A^{\mathrm{res}}_{ji}(\omega, \mathbf{k}). \tag{17.46}$$

The tensor $e_{ij}(\omega, \mathbf{k})$ has in fact been calculated above in a general form and in different special cases in our study of the inverse tensor. For the complete determination of the spectral function, we find the scalar coefficient in (17.46), using the relation (4.4) for the spectral energy density $W(\mathbf{k})$ and a consequence of (17.46) [cf. (4.3)]:

$$(E_j E_i)_{\omega, \mathbf{k}} = \mathrm{const}\, e_{ji}(\omega, \mathbf{k}) \, \delta(\Delta'). \tag{17.47}$$

Namely, we substitute (17.47) into (4.4), taking into account the expansion

$$\delta(\Delta') = \sum \left| \frac{\partial \Delta'}{\partial \omega} \right|^{-1} \delta(\omega - \omega(\mathbf{k})), \tag{17.48}$$

in which the summation sign denotes summation over all roots $\omega(\mathbf{k})$ of the dispersion relation (4.7) and the equation for the first derivative of the determinant is ([56], p. 33, Exercise 9)

$$\frac{\partial \Delta'}{\partial \omega} = e_{ji}(\omega, \mathbf{k}) \frac{\partial}{\partial \omega} M^H_{ij}(\omega, \mathbf{k}). \tag{17.49}$$

As a result, for the spectral function of the fields of this characteristic oscillation of the plasma we obtain the final formula

$$(E_j E_i)_{\omega, \mathbf{k}} = \frac{W(\mathbf{k})}{2\pi^2} \sum \frac{1}{\omega} \left\{ \frac{\partial \Delta'}{\partial \omega} \right\}^{-1} e_{ji}(\omega, \mathbf{k}) \, \delta(\omega - \omega(\mathbf{k})), \tag{17.50}$$

in which the summation sign stands for summation over the $\omega(\mathbf{k})$ that are roots of the dispersion equation (4.7) corresponding to the given oscillation [for example, in a cold isotropic plasma for the electron Langmuir oscillation $\omega(\mathbf{k}) = \pm\omega_{Le}$]. The spectral function of the fields in the plasma is the sum of the spectral functions of the fields of the characteristic oscillations of the plasma.

We now consider all possible special cases of our relation (17.50). Let us first consider an isotropic plasma in order to relate Eq. (6.62) for the spectral function introduced in § 6 to (17.50). In an isotropic plasma [see (6.29)]

$$\Delta' = \varepsilon^{l'}(\omega, \mathbf{k})\{\varepsilon^{tr'}(\omega, \mathbf{k}) - n^2\}^2, \qquad \mathbf{B}_0 = 0, \tag{17.51}$$

and the tensor $e_{ij}(\omega, \mathbf{k})$ has the form

$$e_{ji}(\omega, \mathbf{k}) = \varepsilon^{l'}(\varepsilon^{tr'} - n^2)(\delta_{ij} - \varkappa_i\varkappa_j) + (\varepsilon^{tr'} - n^2)^2 \varkappa_i\varkappa_j. \tag{17.52}$$

Substituting (17.51) and (17.52) into the definition (17.50), we obtain the spectral functions for the fields of longitudinal:

$$(E_jE_i)_{\omega, \mathbf{k}} = \frac{W_l(\mathbf{k})}{2\pi^2}\sum\left\{\frac{\partial\omega\varepsilon^{l'}}{\partial\omega}\right\}^{-1}\varkappa_i\varkappa_j\delta(\omega - \omega(\mathbf{k})) \tag{17.53}$$

and transverse:

$$(E_jE_i)_{\omega, \mathbf{k}} = \frac{W_{tr}(\mathbf{k})}{2\pi^2}\sum\frac{1}{2}\left\{\frac{\partial\omega(\varepsilon^{tr'} - n^2)}{\partial\omega}\right\}^{-1}(\delta_{ij} - \varkappa_i\varkappa_j)\delta(\omega - \omega(\mathbf{k})) \tag{17.54}$$

oscillation, respectively, in complete agreement with (6.62), (6.64), and (6.65).

We emphasize that we have obtained from the general relation (17.50) the spectral function (17.54) of the fields of a naturally polarized transverse electromagnetic wave [see (6.68)]. Thus, the method we have used to determine the spectral function of the fields in a plasma by means of the inverse tensor $A_{ij}(\omega, \mathbf{k})$, i.e., in fact by means of the permittivity tensor of the plasma, is not suitable for describing the spectral function (6.76) of a transverse electromagnetic wave with a polarization that is more complicated than natural polarization. This restriction also follows from purely geometrical considerations that arise naturally in the framework of the approach to the construction of the tensor characteristics of a plasma by means of elementary tensors. Namely, the tensors (6.73) in (6.76) are not, in general, covered by the set of elementary tensors (17.1) (for $\mathbf{e} \neq \mathbf{h}$), which form the basis of the arguments in this section. It should however be noted that in a magnetoactive plasma ($\mathbf{B}_0 \neq 0$) this restriction of the applicability of formula (17.50) is in fact lifted since the external magnetic field $\mathbf{B}_0$ eliminates the polarization degeneracy, "transforming" transverse oscillations with the spectral function (6.76) into an ordinary and an extraordinary wave. Indeed, even in a cold magnetoactive plasma, the dispersion relation [see Δ in (17.23)]

$$An^4 + Bn^2 + C = 0 \tag{17.55}$$

has three solutions. Two of them correspond to finite values of the square of the refractive index, $n^2 = c^2k^2/\omega^2$, and are called the extraordinary and the ordinary wave with refractive indices (see [1], p. 141, (20.9))

$$n_{1,2}^2 = \frac{1}{2}(\varepsilon_1\sin^2\vartheta + \varepsilon_2\cos^2\vartheta)^{-1}\left\{(\varepsilon_1^2 - g^2 - \varepsilon_1\varepsilon_2)\sin^2\vartheta + 2\varepsilon_1\varepsilon_2 \pm \sqrt{(\varepsilon_1^2 - g^2 - \varepsilon_1\varepsilon_2)^2\sin^4\vartheta + 4\varepsilon_2^2g^2\cos^2\vartheta}\right\}. \tag{17.56}$$

The third solution of the quadratic equation (17.55) corresponds to an infinite value of the refractive index:

$$n^2 \to \infty. \tag{17.57}$$

To the infinite value of the root of the quadratic equation there corresponds a zero value of the coefficient A of the highest power of n:

$$A \equiv \varepsilon_1\sin^2\vartheta + \varepsilon_2\cos^2\vartheta = 0. \tag{17.58}$$

Geometrically, this limit in the (Δ, n^2) plane amounts to a transformation of the parabola

$\Delta = An^4 + Bn^2 + C$ into the straight line $\Delta = Bn^2 + C$. Of course, the splitting of the solutions of the quadratic equation into three is somewhat arbitrary.* However, this splitting is very convenient and, what is most important, is expedient on account of the exceptional importance of oscillations with low phase velocity

$$\frac{\omega}{k} \ll c \tag{17.59}$$

for the description of a weakly turbulent state of a plasma. Oscillations with infinitely large refractive index are sometimes called plasma resonances ([8], p. 34); they have been investigated in considerable detail in the literature [8, 59].

In accordance with what we have said, we write down the spectral functions of the fields of the characteristic oscillations of a cold magnetoactive plasma. For longitudinal oscillations, we obtain from the general formula (17.50) with allowance for (17.22), (17.23), and (17.57)

$$(E_j E_i)_{\omega,\mathbf{k}} = \frac{W_l(\mathbf{k})}{2\pi^2} \sum \left\{ \frac{\partial \omega \varepsilon}{\partial \omega} \right\}^{-1} \varkappa_i \varkappa_j \delta(\omega - \omega(\mathbf{k})), \tag{17.60}$$

which differs from Eq. (17.53) in a isotropic plasma only by the replacement of $\varepsilon^{l'}$ by ε. Here, ε is the longitudinal contraction of the permittivity tensor (16.37) of a cold magnetoactive plasma, and it is identically equal to the highest coefficient A in Eq. (17.55):

$$\varepsilon \equiv \varkappa_i \varepsilon_{ij} \varkappa_j \equiv A = \varepsilon_1 \sin^2 \vartheta + \varepsilon_2 \cos^2 \vartheta. \tag{17.61}$$

The spectral function preserves its form (17.60) in a hot magnetoactive plasma as well if ε is understood as the real part of the longitudinal contraction of the tensor (16.44) with the coefficients (16.45) [or (16.55)], which take into account the thermal motion of the particles. This question will be analyzed in more detail in the next section.

The spectral functions of the electric fields of the extraordinary and ordinary waves are described by the general relation (17.50) if $e_{ji}(\omega, \mathbf{k})$ in this relation is replaced by $\Delta_0 A_{ji}(\omega, \mathbf{k})$ in accordance with (17.22), the determinant Δ' is replaced by Δ_0 from (17.23), and a summation is carried out over the frequencies of these oscillations. However, such a description is not entirely convenient, since the expression for the frequencies of the extraordinary and ordinary waves cannot be written down analytically without difficulty. It is customary to describe these waves by their refractive indices (17.56). Therefore, we write down the spectral functions of the extraordinary and the ordinary wave, using n_1^2 and n_2^2. Implicit expressions for the frequencies are given by the equations

$$\omega_{1,2}^2 = \frac{c^2 k^2}{n_{1,2}^2}, \tag{17.62}$$

and, taking into account these by means of the relations

$$\frac{1}{2|\omega|} \{\delta(\omega - \omega_{1,2}) + \delta(\omega + \omega_{1,2})\} = \delta(\omega^2 - \omega_{1,2}^2) = \frac{n^2}{\omega^2} \delta(n^2 - n_{1,2}^2) \tag{17.63}$$

we obtain from (17.50)

$$\begin{aligned}(E_j E_i)_{\omega,\mathbf{k}}^{1,2} = {} & \frac{W_{1,2}}{2\pi^2} \frac{n^2}{|\omega|} \left\{ \frac{1}{2} \frac{\partial \omega \Delta_0}{\partial \omega} \right\}^{-1} \delta(n^2 - n_{1,2}^2) \{\varepsilon_2 (\varepsilon_1 - n^2) \delta_{ij} - \\ & - n^2 (\varepsilon_1 - n^2) \varkappa_i \varkappa_j - [g^2 + (\varepsilon_2 - \varepsilon_1)(\varepsilon_1 - n^2)] h_i h_j + ig\varepsilon_2 e_{jsi} h_s - \\ & - ign^2 (\varkappa_j [\varkappa \mathbf{h}]_i - \varkappa_i [\varkappa \mathbf{h}]_j) + n^2 (\varepsilon_2 - \varepsilon_1) [\varkappa \mathbf{h}]_i [\varkappa \mathbf{h}]_j \}. \end{aligned} \tag{17.64}$$

---

* This arbitrariness disappears if allowance is made for even a weak thermal motion of the plasma particles ([1], p. 142, 143) since the dispersion relation (17.55) then becomes an equation of third degree in n^2, one of whose roots is (17.58) ([1], p. 143, formulas (20.16) and (20.17)).

Note that the spectral functions of the electric fields of the extraordinary (1) and the ordinary wave (2) in the limit of a vanishingly small external magnetic field ($\mathbf{B}_0 = 0$) go over into the spectral function (6.68) of a naturally polarized transverse electromagnetic wave in an isotropic plasma.

We give a general expression for the spectral function of the field of an arbitrary characteristic oscillation of a hot ($v_T \neq 0$) magnetoactive plasma:

$$(E_jE_i)_{\omega,\mathbf{k}} = \frac{W(\mathbf{k})}{2\pi^2}\sum\left\{\frac{\partial\omega\Delta'}{\partial\omega}\right\}^{-1}\delta(\omega-\omega(\mathbf{k}))\{[(a-n^2)(a+b+c)+$$
$$+\sin^2\vartheta f(a+b+c)+\sin^2\vartheta e(e+d)]\delta_{ij}-[ed+(a-n^2)(b+n^2)+$$
$$+\sin^2\vartheta e^2+\sin^2\vartheta f(b+n^2)]\varkappa_i\varkappa_j+[d(e+d)-c(a-n^2)-$$
$$-cf\sin^2\vartheta]h_ih_j-[d(a+b+c)+ec\sin^2\vartheta]e_{jsi}h_s+[(b+n^2)d-$$
$$-e(a-n^2)](\varkappa_j[\varkappa\mathbf{h}]_i-\varkappa_i[\varkappa\mathbf{h}]_j)+[(b+n^2)c-e(e+d)-f(a+b+c)][\varkappa\mathbf{h}]_i[\varkappa\mathbf{h}]_j\}. \quad (17.65)$$

Here Δ' is the real part of the determinant (17.17). In this equation a, b, c, d, e, f are the coefficients of the corresponding elementary tensors of the Hermitian part of the permittivity tensor (16.44), i.e., they are determined by Eqs. (16.45) or (16.55), but with allowance for only the principal parts in the corresponding integrals [real parts of the functions $J_+(\beta_n)$ in (16.55)]. We have retained for them the previous notation.

From formula (17.65) one can obtain a set of different special cases of spectral functions of oscillations in a hot magnetoactive plasma that propagate, for example, along or at right angles to the external magnetic field by using the corresponding simplified forms of the coefficients a, b, c, d, e, f of the permittivity tensor (16.44). In particular, in a cold magnetoactive plasma we obtain from (17.65) the relation

$$(E_jE_i)_{\omega,\mathbf{k}} = \frac{W(\mathbf{k})}{2\pi^2}\sum\left\{\frac{\partial\Delta_0\omega}{\partial\omega}\right\}^{-1}\delta(\omega-\omega(\mathbf{k}))\{\varepsilon_2(\varepsilon_1-n^2)\delta_{ij}-n^2(\varepsilon_1-n^2)\varkappa_i\varkappa_j-$$
$$-[g^2+(\varepsilon_2-\varepsilon_1)(\varepsilon_1-n^2)]h_ih_j+ig\varepsilon_2e_{jsi}h_s-ign^2(\varkappa_j[\varkappa\mathbf{h}]_i-\varkappa_i[\varkappa\mathbf{h}]_j)+n^2(\varepsilon_2-\varepsilon_1)[\varkappa\mathbf{h}]_i[\varkappa\mathbf{h}]_j\}, \quad (17.64a)$$

which is identical with (17.64) if the frequencies are replaced by the refractive indices in accordance with (17.63). The following section is devoted to the study of different special examples of the spectral function (17.65) for almost electrostatic oscillations of a magnetoactive plasma.

§ 18. Quasilongitudinal Oscillations of a Magnetoactive Plasma. Spectra and Spectral Functions

We consider first in more detail the spectra and spectral functions of longitudinal (electrostatic) oscillations in a cold magnetoactive plasma. As we have already pointed out in § 17, the condition for waves to be longitudinal in such a plasma is an infinitely large value of the refractive index (17.57). Then of the six terms of the tensor $e_{ji}(\omega, \mathbf{k})$, which determines the polarization of an oscillation [see (17.64a)], only one is distinguished, namely, the tensor $n^4\varkappa_i\varkappa_j$, and this corresponds to longitudinal oscillations. In the determinant Δ_0, the principal term is $n^4A \equiv n^4\varepsilon$. The fourth powers of the refractive index in the tensor in the curly brackets (17.64a) and the factor $\{\partial\Delta_0\omega/\partial\omega\}^{-1}$ cancel, which reduces the expression (17.64a) to the spectral function of electrostatic oscillations (17.60). In fact, of course, one is dealing with waves whose refractive index is large but not infinite. We shall say that a wave with a large refractive index is quasilongitudinal. The extent to which a wave is longitudinal is determined by the relative weight of the five terms in the tensor $e_{ji}(\omega, \mathbf{k})$ compared with the term containing the longitudinal elementary tensor $\varkappa_i\varkappa_j$. It is natural to call the conditions for the coefficients of the five elementary tensors

$$\delta_{ij},\quad h_ih_j,\quad e_{isj}h_s,\quad \varkappa_i[\varkappa\mathbf{h}]_j-\varkappa_j[\varkappa\mathbf{h}]_i,\quad [\varkappa\mathbf{h}]_i[\varkappa\mathbf{h}]_j \quad (18.1)$$

to be small compared with the coefficient of the longitudinal elementary tensor $\varkappa_i\varkappa_j$ necessary and sufficient conditions for the oscillation to be longitudinal. It is obvious that a large value of the refractive index is a necessary condition for an oscillation to be longitudinal in not only a cold but also in a hot magnetoactive plasma. Because the elementary tensors that characterize the polarization of an oscillation in a hot plasma are the same as the elementary tensors in a cold plasma, the necessary and sufficient condition for an oscillation in a hot plasma to be longitudinal is the same as that formulated above.

Let us now consider a cold plasma. The spectra of quasilongitudinal oscillations of a cold magnetoactive plasma are determined by the dispersion equation (17.58), which, with allowance for the explicit expressions (16.38) for ε_1 and ε_2 in an electron–ion plasma with one ion species takes the following form (the subscripts e and i are appended to the electron and ion variables, respectively):

$$\varepsilon = 1 - \frac{\omega_{L_e}^2}{\omega^2 - \Omega_e^2}\frac{k_\perp^2}{k^2} - \frac{\omega_{L_e}^2}{\omega^2}\frac{k_z^2}{k^2} - \frac{\omega_{L_i}^2}{\omega^2 - \Omega_i^2}\frac{k_\perp^2}{k^2} - \frac{\omega_{L_i}^2}{\omega^2}\frac{k_z^2}{k^2} = 0. \tag{18.2}$$

The spectra of quasilongitudinal oscillations of a cold plasma have been studied in detail in the literature (see [8, 54], and [59, 60]) on the basis of this equation.

At frequencies ω that are low compared with the ion gyroscopic frequency Ω_i

$$\omega \ll \Omega_i, \tag{18.3}$$

we obtain from (18.2) the spectrum

$$(p) \qquad \omega^2 = \frac{k_z^2}{k^2}\omega_{L_e}^2\left\{1 + \frac{k_\perp^2}{k^2}\frac{\omega_{L_i}^2}{\Omega_i^2}\right\}^{-1}. \tag{18.4}$$

We shall indicate oscillations with such a spectrum by the letter p, which is placed on the left of Eq. (18.4) for the frequency. Frequently, the electron plasma frequency is much higher than the ion gyroscopic frequency. Indeed, the square of the ratio of these frequencies in a hydrogen plasma is given approximately by

$$\frac{\omega_{L_e}^2}{\Omega_i^2} \approx 3.5\,\frac{10 N_e}{B_0^2}, \qquad e_i = |e|, \tag{18.5}$$

in which the electron density N_e is measured in cm^{-3}, and the magnetic field B_0 is measured in gauss. The strong inequality for $B_0 = 10^2$–10^4 G

$$\omega_{L_e}^2 \gg \Omega_i^2 \tag{18.6}$$

is already satisfied at very low densities: $N_e = 10^4$–10^8 cm^{-3}. Therefore, the condition (18.3) imposes definite restrictions on oscillations with the spectrum (18.4). Namely, in a rarefied plasma,

$$\omega_{L_i}^2 \lessdot \Omega_i^2, \tag{18.7}$$

the angle ϑ between the direction of the wave vector of the oscillation and B_0, the external magnetic field, must be fairly close to $\pi/2$ in accordance with the inequality

$$\cos^2\vartheta \ll \frac{\Omega_i^2}{\omega_{L_e}^2} \ll 1. \tag{18.8}$$

In a dense plasma,

$$\omega_{L_i}^2 \gg \Omega_i^2, \tag{18.9}$$

the restriction on the angles is less stringent (cf. [8], p. 36, (5.9)):

$$\cos^2\vartheta \ll \frac{m}{M}, \qquad \left|\frac{\pi}{2} - \vartheta\right| \ll \left(\frac{m}{M}\right)^{1/2}. \tag{18.8a}$$

The spectral function of the fields of the p oscillation can be described by means of the general relation (17.60) for longitudinal oscillations in a cold plasma. Namely, substituting into (17.60) the expression for the longitudinal permittivity in the low-frequency limit (18.3)

$$\varepsilon = 1 + \frac{\omega_{Li}^2}{\Omega_i^2} - \frac{\omega_{Le}^2}{\omega^2}\frac{k_z^2}{k^2}, \tag{18.10}$$

we obtain the spectral function of the electric fields of this oscillation:

$$(E_j E_i)^p_{\omega, \mathbf{k}} = \frac{W_p(\mathbf{k})}{4\pi^2} \varkappa_i \varkappa_j \left(1 + \frac{\omega_{Li}^2}{\Omega_i^2}\right)^{-1} \left\{\delta\left(\omega - \frac{k_z}{k}\omega_{Le}\left[1 + \frac{\omega_{Li}^2}{\Omega_i^2}\right]^{-1/2}\right) + \delta\left(\omega + \frac{k_z}{k}\omega_{Le}\left[1 + \frac{\omega_{Li}^2}{\Omega_i^2}\right]^{-1/2}\right)\right\}. \tag{18.11}$$

Note the approximate nature of this equation in the sense of the description of the polarization of the p oscillation: we have omitted the elementary tensors that differ from the longitudinal tensor $\varkappa_i\varkappa_j$ and occur in the general expression (17.64a) with albeit small but nonvanishing coefficients. Waves with frequency near the ion gyroscopic frequency:

$$|\omega - \Omega_i| \ll \Omega_i, \tag{18.12}$$

which also exist in a cold plasma, are a direct continuation of this oscillation branch to higher frequencies. Separating out the highest terms in (18.2) in accordance with the inequality (18.12), we obtain an approximate expression for ε:

$$\varepsilon = 1 - \frac{\omega_{Le}^2}{\omega^2}\frac{k_z^2}{k^2} - \frac{\omega_{Li}^2}{\omega^2 - \Omega_i^2}\frac{k_\perp^2}{k^2}, \tag{18.13}$$

in which for angles $\vartheta = \angle \mathbf{k}\,\mathbf{B}_0$ not too near $\pi/2$ the unity can be ignored on account of (18.6). We then obtain the spectrum

$$(\tilde{c}_i') \qquad \omega^2 = \Omega_i^2\left\{1 - \frac{m}{M}\frac{k_\perp^2}{k_z^2}\right\}. \tag{18.14}$$

Proximity of the frequency to the ion gyroscopic frequency (18.12) imposes a condition on the angles [cf. (18.8a)]:

$$\frac{k_\perp^2}{k_z^2} \ll \frac{M}{m}, \quad \vartheta < 88.5°. \tag{18.15}$$

Note that the spectra (18.4) and (18.14) of the oscillations p and $\tilde{c}_i'$ are an analytic form of expression of the dispersion relation for one and the same branch in a cold plasma but at different regions of angles ϑ: near $\pi/2$ [see (18.8), (18.8a)] and not too near this angle: (18.5).

Although the frequency of the $\tilde{c}_i'$ oscillation differs little from the ion gyroscopic frequency (18.12), ion cyclotron damping is exponentially small,* since in a cold plasma

$$|k_z v_{T_i}| \ll |\omega - \Omega_i|. \tag{18.16}$$

The spectral function of the electric fields of this oscillation has the form

$$(E_j E_i)^{\tilde{c}_i'}_{\omega, \mathbf{k}} = \frac{1}{4\pi^2} W_{\tilde{c}_i'}(\mathbf{k}) \left(\frac{m}{M}\frac{k_\perp^2}{k_z^2}\right)\left(\frac{\omega_{Le}^2 k_z^2}{\Omega_i^2 k^2}\right)^{-1} \varkappa_i\varkappa_j \left\{\delta\left(\omega - \Omega_i\left[1 - \frac{1}{2}\frac{k_\perp^2}{k_z^2}\frac{m}{M}\right]^{1/2}\right) + \delta\left(\omega + \Omega_i\left[1 - \frac{1}{2}\frac{k_\perp^2}{k_z^2}\frac{m}{M}\right]^{1/2}\right)\right\}, \tag{18.17}$$

if the contribution of tensors that are nonlongitudinal with respect to the wave vector is ignored in the general expression (17.64a). Taking the example of this oscillation, let us estimate the

* Strictly speaking, in a cold ($T_e = T_i = 0$) collisionless magnetoactive plasma the damping of oscillations with frequencies that are not exactly equal to the gyroscopic frequency is equal to zero. In this sense, the inequality (18.16) determines the ion temperature below which the plasma can be regarded as cold.

order of magnitude of the omitted nonlongitudinal terms. The quantities ε_1, ε_2, and g in the coefficients of the elementary tensors in $e_{ji}(\omega, \mathbf{k})$ are given approximately by the equations

$$g=\varepsilon_1=x\cot^2\vartheta, \qquad \varepsilon_2=-x\equiv-\frac{\omega_{L_e}^2}{\Omega_i^2}, \tag{18.18}$$

and the tensor $e_{ji}(\omega, \mathbf{k})$ with allowance for all the polarization corrections is given by

$$e_{ji}(\omega, \mathbf{k})=x(n^2-x\cot^2\vartheta)\delta_{ij}+n^2(n^2-x\cot^2\vartheta)\varkappa_i\varkappa_j-$$
$$-\frac{x}{\sin^2\vartheta}(n^2-x\cos^2\vartheta)h_ih_j-ix^2\cot^2\vartheta e_{jsi}h_s+ixn^2\cot^2\vartheta(\varkappa_i[\varkappa\mathbf{h}]_j-\varkappa_j[\varkappa\mathbf{h}]_i)-\frac{xn^2}{\sin^2\vartheta}[\varkappa\mathbf{h}]_i[\varkappa\mathbf{h}]_j. \tag{18.19}$$

At angles $\vartheta \sim 1$, all the trigonometric functions in (18.19) can be assumed to be of order unity; then the ratio of the coefficients of the five elementary tensors (18.1) to the coefficient of the longitudinal tensor $\varkappa_i\varkappa_j$ is determined by the fraction x/n^2. Thus, the condition for the oscillation (18.14) to be quasilongitudinal and, therefore, the condition for Eq. (18.17) to hold for the spectral function of the electric fields can be written in the form

$$n^2\gg\frac{\omega_{L_e}^2}{\Omega_i^2}\gg1 \quad \text{or} \quad c^2k^2\gg\omega_{L_e}^2\gg\Omega_i^2. \tag{18.20}$$

In accordance with (18.5), this condition relates the "limiting" values of the plasma density, the external magnetic field, and the refractive index (hydrogen plasma):

$$n^2\gg35\frac{N_e}{B_0^2}\gg1. \tag{18.21}$$

At frequencies much higher than the ion gyroscopic frequency,

$$\omega\gg\left(\frac{M}{m}\right)^{1/2}\Omega_i, \tag{18.22}$$

the motion of the ions can be completely ignored (see [1], p. 158), so that the longitudinal permittivity of a cold magnetoactive plasma is entirely determined by the electrons:

$$\varepsilon=1-\frac{\omega_{L_e}^2}{\omega^2-\Omega_e^2}\frac{k_\perp^2}{k^2}-\frac{\omega_{L_e}^2}{\omega^2}\frac{k_z^2}{k^2}, \tag{18.23}$$

and the corresponding oscillation spectra have the form ([8], p. 35, (5.3) and (5.4))

$$(l_\pm) \qquad \omega_\pm^2=\frac{1}{2}(\omega_{L_e}^2+\Omega_e^2)\pm\frac{1}{2}\left[(\omega_{L_e}^2+\Omega_e^2)^2-4\omega_{L_e}^2\Omega_e^2\frac{k_z^2}{k^2}\right]^{1/2}. \tag{18.24}$$

The oscillation l_+ has the spectrum ω_+^2 in accordance with (18.24) in the complete range of variation of the angle ϑ between the direction of propagation and the external magnetic field: $0 \le (k_z^2/k^2) \le 1$.

At angles ϑ fairly near but still not equal to $\pi/2$,

$$1\gg(k_z^2/k^2)\gg\frac{m}{M}, \tag{18.25}$$

the spectrum ω_-^2 of the oscillations l_- can be represented by the equation

$$(l_-) \qquad \omega_-^2=\omega_{L_e}^2\frac{k_z^2}{k^2}\left(1+\frac{k_\perp^2}{k^2}\frac{\omega_{L_e}^2}{\Omega_e^2}\right)^{-1}, \tag{18.26}$$

which follows directly from the vanishing of (18.23) when allowance is made for the left-hand inequality in (18.25). But if the angle ϑ is very near $\pi/2$, so that the right-hand side of the inequality (18.25) is replaced by the opposite inequality (18.8a), the frequency of the oscillation that is a direct continuation of l_- can be found from the general equation (18.2) with allowance for the frequency ω being small compared with the electron gyroscopic frequency Ω_e ([8],

p. 36, (5.8)):

$$(l_-) \qquad \omega_-^2 = (\omega_{L_i}^2 + \Omega_i^2)\left(1 + \frac{\omega_{L_e}^2}{\Omega_e^2}\right)^{-1}, \qquad \omega_- \ll \Omega_e. \tag{18.27}$$

Thus, in this range of angles the expression ω_-^2 in (18.24) is not valid and is replaced by (18.27).

The polarization of the oscillations $l_\pm$ is determined by the spectral function of the electric fields in accordance with (17.60) and (18.23):

$$(E_jE_i)^{l\pm}_{\omega,\mathbf{k}} = \frac{1}{4\pi^2} W_{l\pm}(\mathbf{k}) \frac{k^2}{\omega_{L_e}^2}\left\{\frac{k_z^2}{\omega_\pm^2} + \frac{k_\perp^2 \omega_\pm^2}{(\omega_\pm^2 - \Omega_e^2)^2}\right\}^{-1} \varkappa_i\varkappa_j\{\delta(\omega - \omega_\pm) + \delta(\omega + \omega_\pm)\}. \tag{18.28}$$

Let us summarize our investigation of the spectra of quasilongitudinal oscillations in a cold magnetoactive plasma with one ion species. The general equation (18.2) is an algebraic equation of third degree in ω^2. Accordingly, there exist altogether three branches of longitudinal oscillations in the limit $n^2 \to \infty$. The low-frequency branch ($\omega \lessgtr \Omega_i$) is described by the spectra (18.4) and (18.14) with polarization given by (18.11) and (18.17), respectively. Comparatively high-frequency oscillations are represented by the two branches $l_\pm$, so that the l_+ spectrum is given by one and the same expression ω_+^2 from (18.24) and the dispersion relation for oscillations of the branch l_- can be determined three equations for ω_-^2 depending on the proximity of the direction of propagation of the oscillation to the exactly transverse direction relative to the external magnetic field $\mathbf{B}_0$.

The general conditions for the approximation of a cold plasma to hold have the form [59, 60]

$$z_{e,i} \ll 1, \qquad |\omega - n\Omega_{e,i}| \gg |k_z v_{T_{e,i}}|, \qquad n = 0, \pm 1, \ldots \tag{18.29}$$

Applying these conditions to the longitudinal contraction of the permittivity tensor that arises from (16.44) and (16.55) (cf. [60], p. 78 (12.23))

$$\varepsilon(\omega, \mathbf{k}) = \varkappa_i \varepsilon_{ij}(\omega, \mathbf{k})\varkappa_j = 1 + \frac{\omega_L^2}{k^2 v_T^2}\left\{1 - \sum_{n=-\infty}^{+\infty} \frac{\omega}{\omega - n\Omega} A_n(z) J_+\left(\frac{\omega - n\Omega}{|k_z| v_T}\right)\right\}, \tag{18.30}$$

by means of the expansion

$$A_n(z) = \frac{z^n}{2^n n!}\left\{1 - z + \frac{z^2}{2}\frac{2n+3}{2n+2}\right\}, \qquad z \ll 1, \ n > 0, \tag{18.31}$$

and the asymptotic behavior (6.32) of the function $J_+(\beta)$, we obtain the longitudinal permittivity ε of a cold plasma ($v_T = 0$) which is on the left-hand side of Eq. (18.2), and all the spectra described above. At the same time, the terms in $\varepsilon(\omega, \mathbf{k})$ that depend on the thermal velocity of the particles are omitted, as small and unimportant corrections. The opposite case, when in the dispersion relation with $\varepsilon(\omega, \mathbf{k})$ from (18.30)

$$\varepsilon(\omega, \mathbf{k}) = 0 \tag{18.32}$$

one must allow for the temperature-dependent terms in order to determine the oscillation spectrum, corresponds to the approximation of a hot plasma. In this approximation there exist, i.e., weakly damped, some further quasilongitudinal oscillations [8, 54, 59]. The damping constants of such oscillations are completely determined in accordance with (17.36) [see also formula (6.25), § 6, Ch. II, for the damping of electrostatic oscillations in a isotropic plasma] by the longitudinal complex permittivity (18.30):

$$\gamma(\mathbf{k}) = \varepsilon''(\omega, \mathbf{k})\left\{\frac{\partial \varepsilon'(\omega, \mathbf{k})}{\partial \omega}\right\}^{-1}. \tag{18.33}$$

In this formula, the frequencies ω are equal to the frequencies $\omega(\mathbf{k})$ of the electrostatic oscillations of a hot magnetoactive plasma found as solutions of the dispersion equation [cf. (18.32)]:

$$\varepsilon'(\omega, \mathbf{k}) \equiv \operatorname{Re}\varepsilon(\omega, \mathbf{k}) = 1 + \frac{\omega_L^2}{k^2 v_T^2}\left\{1 - \frac{\omega}{|k_z| v_T} \sum_{n=-\infty}^{+\infty} A_n(z) \exp(-\beta_n^2/2) \int_0^{\beta_n} dt \exp(t^2/2)\right\} = 0. \quad (18.34)$$

Summing over the species of charged particles for a plasma with one ion species, we can express Eq. (18.34) in the expanded form

$$1 + \frac{\omega_{Le}^2}{k^2 v_{T_e}^2}\left\{1 - \frac{\omega}{|k_z| v_{T_e}} \sum_{n=-\infty}^{+\infty} A_n(z_e) \exp[-(\beta_n^e)^2/2] \int_0^{\beta_n^e} dt e^{\frac{t^2}{2}}\right\} +$$

$$+ \frac{\omega_{Li}^2}{k^2 v_{T_i}^2}\left\{1 - \frac{\omega}{|k_z| v_{T_i}} \sum_{n=-\infty}^{+\infty} A_n(z_i) \exp[-(\beta_n^i)^2/2] \int_0^{\beta_n^i} dt e^{t^2/2}\right\} = 0. \quad (18.35)$$

We recall that here, in accordance with (16.52), (16.53), and (16.56),

$$z_{e,i} \equiv (k_\perp \rho_{e,i})^2, \qquad A_n(z) \equiv I_n(z) e^{-z}, \qquad \beta_n^{e,i} \equiv \frac{\omega - n\Omega_{e,i}}{|k_z| v_{T_{e,i}}}.$$

The imaginary part of the complex permittivity $\varepsilon(\omega, \mathbf{k})$ also arises from (18.30) by separating the imaginary part of the function J_+:

$$\varepsilon''(\omega, \mathbf{k}) = \sqrt{\frac{\pi}{2}} \frac{\omega}{|k_z| v_T} \frac{\omega_L^2}{k^2 v_T^2} \sum_{n=-\infty}^{+\infty} A_n(z) e^{-\frac{1}{2}\left(\frac{\omega - n\Omega}{k_z v_T}\right)^2} \quad (18.36)$$

and it consists of the electron and the ion term, which determine the damping (18.33) of quasilongitudinal oscillations on electrons and ions, respectively:

$$\varepsilon''(\omega, \mathbf{k}) = \sqrt{\frac{\pi}{2}} \frac{\omega}{|k_z| v_{T_e}} \frac{\omega_{Le}^2}{k^2 v_{T_e}^2} \sum_{n=-\infty}^{+\infty} A_n(z_e) \exp\left\{-\frac{1}{2}\left(\frac{\omega - n\Omega_e}{k_z v_{T_e}}\right)^2\right\} +$$

$$+ \sqrt{\frac{\pi}{2}} \frac{\omega}{|k_z| v_{T_i}} \frac{\omega_{Li}^2}{k^2 v_{T_i}^2} \sum_{n=-\infty}^{+\infty} A_n(z_i) \exp\left\{-\frac{1}{2}\left(\frac{\omega - n\Omega_i}{k_z v_{T_i}}\right)^2\right\}. \quad (18.37)$$

The derivative $(\partial\varepsilon'/\partial\omega)$ in the decay constant (18.33) can, on account of (18.34), also be represented in the explicit form

$$\frac{\partial \varepsilon'(\omega, \mathbf{k})}{\partial \omega} = -\frac{1}{|k_z| v_T} \frac{\omega_L^2}{k^2 v_T^2}\left\{\beta + \sum_{n=-\infty}^{+\infty} A_n(z)(1 - \beta\beta_n) e^{-\beta_n^2/2} \int_0^{\beta_n} dt e^{t^2/2}\right\},$$

$$\beta \equiv (\beta_n)_{n=0} = \frac{\omega}{|k_z v_T|}. \quad (18.38)$$

We emphasize that the conditions deduced below for quasilongitudinal oscillations of a magnetoactive plasma to exist are first and foremost conditions for these oscillations to be weakly damped, i.e., for their decay constants to be small compared with the frequencies. At the same time, such conditions enable us to obtain explicit analytic expressions for the spectra.

Suppose that the oscillation frequency ω is near a multiple $n\Omega_i$ of the ion gyroscopic frequency, so that the longitudinal thermal motion of the ions is weak:

$$\Omega_i \gg |\omega - n\Omega_i| \gg |k_z| v_{T_i}, \qquad n = \pm 1, \ldots, \quad (18.39)$$

the transverse motion is arbitrary,

$$z_i \gtrsim 1, \tag{18.40}$$

and the electron thermal velocity satisfies the inequalities

$$z_e \ll 1, \quad |\Omega_e| \gg \omega \gg |k_z| v_{T_e}. \tag{18.41}$$

Using these inequalities in (18.35), we obtain the spectrum of such an oscillation:

$$(\tilde{c}_i) \quad \omega(\mathbf{k}) = n\Omega_i \left\{1 + A_n(z_i) \left[1 + (kr_{D_i})^2 \left(1 + \frac{k_\perp^2 \omega_{L_i}^2}{k^2 \Omega_i^2} - \frac{k_z^2 \omega_{L_e}^2}{k^2 (n\Omega_i)^2}\right)\right]^{-1}\right\}, \tag{18.42}$$

which simplifies somewhat in the region of angles (18.15) to

$$(\tilde{c}_i) \quad \omega(\mathbf{k}) = n\Omega_i \left\{1 - \frac{m}{M} \left(\frac{n\Omega_i}{k_z v_{T_i}}\right)^2 A_n(z_i)\right\}. \tag{18.43}$$

Conversely, in the case of almost perpendicular propagation (18.8a) of ion-cyclotron waves in a dense plasma (18.9) under the conditions (18.39)-(18.41), we can obtain the following expression for the spectrum ([28], p. 32, (2.17)):

$$(\tilde{c}_i) \quad \omega(\mathbf{k}) = n\Omega_i \left\{1 - \frac{n^2 - 1}{z_i} A_n(z_i)\right\}, \tag{18.44}$$

which holds for $n \geq 2$. The spectral function of the oscillation $\tilde{c}_i$ is determined by the general expression (17.60). For example, for the spectrum (18.43) it can be represented in the form

$$(E_j E_i)^{\tilde{c}_i}_{\omega, \mathbf{k}} = \frac{1}{2\pi^2} W_{\tilde{c}_i}(\mathbf{k}) \varkappa_i \varkappa_j \left\{\frac{\partial \omega \varepsilon'}{\partial \omega}\right\}^{-1} \left\{\delta\left(\omega - n\Omega_i \left[1 - \frac{m}{M}\left(\frac{n\Omega_i}{k_z v_{T_i}}\right)^2 A_n(z_i)\right]\right) + \delta\left(\omega + n\Omega_i \left[1 - \frac{m}{M}\left(\frac{n\Omega_i}{k_z v_{T_i}}\right)^2 A_n(z_i)\right]\right)\right\}. \tag{18.45}$$

Note that for $n = 1$ and weak thermal motion of the ions ($z_i \ll 1$), the spectrum of the ion-cyclotron oscillation (18.43) goes over into the long-wave limit (18.14). Allowance for thermal motion of the electrons ($|k_z v_{Te}| \gg \omega$) enables us to write down an expression for the spectrum of the first ion-cyclotron harmonic ([28], p. 34, (2.19)) that depends on the electron temperature T_e:

$$(c_i) \quad \omega(\mathbf{k}) = \Omega_i \left\{1 + A_1(z_i) \left[\frac{T_i}{T_e} - \frac{k_z^2 v_{T_i}^2}{\Omega_i^2}\right]^{-1}\right\}, \quad kr_{D_e} \ll 1. \tag{18.46}$$

Such an oscillation is weakly damped in a nonisothermal ($T_e \gg T_i$) magnetoactive plasma under the conditions (18.39)-(18.40). At the same time, the electron thermal velocity does not satisfy (18.41) but the inequalities

$$z_e \ll 1, \quad |\Omega_e| \gg |k_z| v_{T_e} \gg \Omega_i. \tag{18.47}$$

In the limit

$$\left(\frac{k_z v_{T_i}}{\Omega_i}\right)^2 \ll \frac{T_i}{T_e}, \quad z_i \ll 1, \tag{18.48}$$

the spectrum (18.46) takes the form

$$(c_i) \quad \omega(\mathbf{k}) = \Omega_i \left\{1 + \frac{1}{2} \frac{k_\perp^2 v_s^2}{\Omega_i^2}\right\}, \quad kr_{D_e} \ll 1, \tag{18.49}$$

which is identical with (8.5) of [8] (p. 55). This spectrum can also be obtained directly from the dispersion equation (18.34), which becomes much simpler in the given case:

$$\varepsilon'(\omega, \mathbf{k}) = \frac{\omega_{Le}^2}{k^2 v_{T_e}^2} - \frac{\omega_{Li}^2}{\Omega_i^2}\frac{k_\perp^2}{k^2}\frac{\omega^2}{\omega^2 - \Omega_i^2} = 0. \tag{18.50}$$

The total permittivity $\varepsilon(\omega, \mathbf{k})$ corresponding to the oscillations (18.49) at resonance $\omega \simeq \Omega_i$ is determined by

$$\varepsilon(\omega, \mathbf{k}) = \frac{\omega_{Le}^2}{k^2 v_{T_e}^2} - \frac{1}{2}\frac{\omega_{Li}^2}{\Omega_i^2}\frac{k_\perp^2}{k^2}\frac{\omega}{\omega - \Omega_i} +$$

$$+ i\sqrt{\frac{\pi}{2}}\left\{\frac{\omega_{Le}^2}{k^2 v_{T_e}^2}\frac{\omega}{|k_z| v_{T_e}} + \frac{\omega_{Li}^2}{k^2 v_{T_i}^2}\frac{\Omega_i}{|k_z| v_{T_i}}\exp\left(-\frac{1}{2}\frac{\Omega_i^2}{k_z^2 v_{T_i}^2}\right) + \frac{1}{2}\frac{k_\perp^2}{k^2}\frac{\omega_{Li}^2}{\Omega_i^2}\frac{\Omega_i}{|k_z v_{T_i}|}\exp\left(-\frac{1}{8}\frac{T_e}{T_i}\frac{k_\perp^2}{k_z^2}\frac{k_\perp^2 v_s^2}{\Omega_i^2}\right)\right\}, \tag{18.51}$$

from which, in accordance with the general definition (18.33), we find the damping constant of the ion-cyclotron harmonic (18.49) on electrons,

$$\gamma_{c_i}^e(\mathbf{k}) = \sqrt{\frac{\pi}{8}\frac{m}{M}}\frac{k_\perp}{|k_z|}(k_\perp v_s) \equiv \sqrt{\frac{\pi}{8}}\frac{\Omega_i}{|k_z| v_{T_e}}\frac{k_\perp^2 v_s^2}{\Omega_i^2}\omega \tag{18.52}$$

and on ions,

$$\gamma_{c_i}^i(\mathbf{k}) = \sqrt{\frac{\pi}{8}}\frac{\Omega_i}{|k_z| v_{T_i}}\frac{k_\perp^2 v_s^2}{\Omega_i^2}\omega\left\{\frac{T_e}{T_i}\exp\left(-\frac{1}{2}\frac{\Omega_i^2}{k_z^2 v_{T_i}^2}\right) + \frac{1}{2}(k_\perp r_{D_e})^2\frac{\omega_{Li}^2}{\Omega_i^2}\exp\left(-\frac{1}{8}\frac{T_e}{T_i}\frac{k_\perp^2}{k_z^2}\frac{k_\perp^2 v_s^2}{\Omega_i^2}\right)\right\}. \tag{18.53}$$

It can be seen from these formulas that the oscillation (18.49) is indeed weakly damped [the decay constants (18.52)-(18.53) are small compared with the frequency (18.49)] under the conditions (18.39), (18.47), and (18.48). Similarly, we can show that the ion-cyclotron oscillations (18.42)-(18.43) have, for example, when the inequalities (18.39) and (18.41) are satisfied, exponentially small damping:

$$\gamma_{\tilde{c}_i}^e(\mathbf{k}) \sim \omega(\mathbf{k})\exp\left\{-\frac{1}{2}\left(\frac{n\Omega_i}{k_z v_{T_e}}\right)^2\right\}, \quad \gamma_{\tilde{c}_i}^i(\mathbf{k}) \sim \omega(\mathbf{k})\, e^{-\frac{1}{2}\left(\frac{\omega(\mathbf{k}) - n\Omega_i}{k_z v_{T_i}}\right)^2}. \tag{18.54}$$

To conclude our discussion of the ion-cyclotron harmonics, we give an expression for the spectral function of the electric fields of the oscillation (18.49):

$$(E_j E_i)_{\omega, \mathbf{k}}^{c_i} = \frac{1}{4\pi^2} W_{c_i}(\mathbf{k})(kr_{D_e})^2\left(\frac{k_\perp v_s}{\Omega_i}\right)^2 \varkappa_i\varkappa_j\left\{\delta\left(\omega - \Omega_i - \frac{1}{2}\frac{k_\perp^2 v_s^2}{\Omega_i}\right) + \delta\left(\omega + \Omega_i + \frac{1}{2}\frac{k_\perp^2 v_s^2}{\Omega_i}\right)\right\}. \tag{18.55}$$

Apart from the ion-cyclotron harmonics, there exist in a magnetoactive plasma oscillations with a frequency $n\Omega_e$ that is a multiple of the electron gyroscopic frequency. For the derivation of the spectra of such waves, the ion motion can be completely ignored and the plasma regarded as an electron plasma. In such an electron plasma with weak thermal motion:

$$z_e \ll 1, \quad |\Omega_e| \gg |\omega - n\Omega_e| \gg |k_z| v_{T_e} \tag{18.56}$$

the spectrum of the higher ($n \geq 2$) electron cyclotron quasilongitudinal oscillations has the form ([28], p. 34, (2.20))

$$(c_e) \qquad \omega(\mathbf{k}) = n\Omega_e\left\{1 + A_n(z_e)\left[(kr_{D_e})^2 - \frac{k_\perp^2\rho_e^2}{n^2 - 1} - \frac{k_z^2\rho_e^2}{n^2}\right]^{-1}\right\}, \tag{18.57}$$

and the spectral function is given by

$$(E_j E_i)_{\omega, \mathbf{k}}^{c_e} = \frac{1}{2\pi^2} W_{c_e}(\mathbf{k})\left\{\frac{\partial\omega\varepsilon'}{\partial\omega}\right\}^{-1}\varkappa_i\varkappa_j\left\{\delta\left(\omega - n\Omega_e - n\Omega_e A_n(z_e)\left[(kr_{D_e})^2 - \frac{k_\perp^2\rho_e^2}{n^2 - 1} - \frac{k_z^2\rho_e^2}{n^2}\right]^{-1}\right) +\right.$$

$$\left. + \delta\left(\omega + n\Omega_e + n\Omega_e A_n(z_e)\left[(kr_{D_e})^2 - \frac{k_\perp^2\rho_e^2}{n^2 - 1} - \frac{k_z^2\rho_e^2}{n^2}\right]^{-1}\right)\right\}. \tag{18.58}$$

For the first harmonic under the same conditions the spectrum is much simpler:

$$(c_e) \qquad \omega(\mathbf{k}) = \Omega_e\left\{1 + \frac{1}{2}\frac{k_\perp^2}{k^2}\left(\frac{\Omega_e^2}{\omega_{L_e}^2} - \frac{k_z^2}{k^2}\right)^{-1}\right\}, \qquad \omega \simeq \Omega_e, \tag{18.59}$$

and can be obtained directly from the expression (18.23) for the longitudinal permittivity of a cold electron plasma. Obviously, (18.59) is a special case of the spectra (18.24) of the Langmuir oscillations $l_\pm$ modified by the magnetic field. The fact that the oscillation (18.59), which is a special case of (18.24), occurs at the frequency Ω_e, and not at the electron plasma frequency, indicates that the analogy between the waves (18.24) and electron Langmuir oscillations in an isotropic plasma is of limited validity. Proximity of the frequency ω to the electron gyroscopic frequency in (18.59) can be achieved in the case of almost longitudinal propagation of the oscillation ($k_\perp \ll k$) or when the electrons are magnetized ($\Omega_e \gg \omega_{Le}$). But if the transverse wave number $k_\perp$ of the electron-cyclotron oscillation at the gyroscopic frequency Ω_e is not too small compared with the reciprocal Larmor radius of the electrons:

$$z_e \gtrsim 1, \qquad |\Omega_e| \gg |\omega - \Omega_e| \gg |k_z| v_{T_e}, \tag{18.60}$$

the corrections to the spectrum $\omega \simeq \Omega_e$ begin to depend on the electron thermal velocity ([28], p. 35, (2.22))

$$(c_e) \qquad \omega(\mathbf{k}) = \Omega_e\left\{1 + A_1(z_e)\left[1 + (kr_{D_e})^2 - \frac{1 - A_0(z_e)}{z_e}\right]^{-1}\right\}. \tag{18.57a}$$

The damping constants of the waves (18.57), (18.59), and (18.57a) are exponentially small.

As in an isotropic plasma, weakly damped (characteristic) oscillations of ion-acoustic type exist in a nonisothermal magnetoactive plasma with hot electrons and cold ions (6.36). The general feature of such oscillations is the small value of their phase velocity ω/k_z compared with the electron thermal velocity [cf. (6.37)]. At frequencies ω that are low compared with Ω_i, their spectrum can be represented by the equation

$$(s_i') \qquad \omega^2(\mathbf{k}) = \omega_{L_i}^2 \frac{k_z^2}{k^2}\left\{1 + \frac{k_\perp^2}{k^2}\frac{\omega_{L_i}^2}{\Omega_i^2} + \frac{\omega_{L_e}^2}{k^2 v_{T_e}^2}\right\}^{-1}, \tag{18.61}$$

which holds if the following inequalities do:

$$|\Omega_e| \gg |k_z| v_{T_e} \gg \omega \gg |k_z| v_{T_i}, \qquad \Omega_i \gg |k_z| v_{T_i}, \qquad z_{e,i} \ll 1, \tag{18.62}$$

which also ensure that the constants of damping on electrons and ions are small, as can be seen from the following expression for the longitudinal permittivity of the plasma, this arising directly from (18.30) under the conditions (18.62):

$$\varepsilon(\omega, \mathbf{k}) = 1 + \frac{k_\perp^2 \omega_{L_i}^2}{k^2 \Omega_i^2} + \frac{\omega_{L_e}^2}{k^2 v_{T_e}^2} - \frac{k_z^2}{k^2}\frac{\omega_{L_i}^2}{\omega^2} + i\sqrt{\frac{\pi}{2}}\frac{\omega_{L_e}^2}{k^2 v_{T_e}^2}\frac{\omega}{|k_z| v_{T_e}} + i\sqrt{\frac{\pi}{2}}\frac{\omega_{L_i}^2}{k^2 v_{T_i}^2}\frac{\omega}{|k_z| v_{T_i}} e^{-\frac{1}{2}\frac{\omega^2}{k_z^2 v_{T_i}^2}}. \tag{18.63}$$

Indeed, from this and the general relation (18.33) we obtain [cf. (6.40b)]

$$\gamma_{s_i'}^e(\mathbf{k}) = \sqrt{\frac{\pi}{8}}\,\omega_{L_i}\frac{k_z}{k}\left(\frac{\omega_{L_i}}{kv_{T_e}}\right)\frac{\omega_{L_e}^2}{k^2 v_{T_e}^2}\left\{1 + \frac{k_\perp^2}{k^2}\frac{\omega_{L_i}^2}{\Omega_i^2} + \frac{\omega_{L_e}^2}{k^2 v_{T_e}^2}\right\}^{-2}; \tag{18.64}$$

$$\gamma_{s_i'}^i(\mathbf{k}) = \sqrt{\frac{\pi}{8}}\,\omega_{L_i}\frac{k_z}{k}\left(\frac{\omega_{L_i}}{kv_{T_i}}\right)^3\left\{1 + \frac{k_\perp^2}{k_2}\frac{\omega_{L_i}^2}{\Omega_i^2} + \frac{\omega_{L_e}^2}{k^2 v_{T_e}^2}\right\}^{-2}\exp\left\{-\frac{1}{2}\frac{\omega_{L_i}^2}{k^2 v_{T_i}^2}\left[1 + \frac{k_\perp^2}{k^2}\frac{\omega_{L_i}^2}{\Omega_i^2} + \frac{\omega_{L_e}^2}{k^2 v_{T_e}^2}\right]^{-1}\right\}, \tag{18.65}$$

which are small compared with the frequency (18.61) of low-frequency ion-acoustic oscillations.

In accordance with (17.60), the spectral function of the fields of such oscillations has the form

$$(E_jE_i)^{s_i'}_{\omega,\mathbf{k}}=\frac{1}{4\pi^2}W_{s_i'}(\mathbf{k})\left\{1+\frac{k_\perp^2}{k^2}\frac{\omega_{Li}^2}{\Omega_i^2}+\frac{\omega_{Le}^2}{k^2v_{T_e}^2}\right\}^{-1}\varkappa_i\varkappa_j\left\{\delta\left(\omega-\omega_{Li}\frac{k_z}{k}\left[1+\frac{k_\perp^2}{k^2}\frac{\omega_{Li}^2}{\Omega_i^2}+\frac{\omega_{Le}^2}{k^2v_{T_e}^2}\right]^{-1}\right)+\right.$$
$$\left.+\delta\left(\omega+\omega_{Li}\frac{k_z}{k}\left[1+\frac{k_\perp^2}{k^2}\frac{\omega_{Li}^2}{\Omega_i^2}+\frac{\omega_{Le}^2}{k^2v_{T_e}^2}\right]^{-1}\right)\right\}. \qquad (18.66)$$

But if the frequency of the ion-acoustic oscillations of a magnetoactive plasma is greater than the ion gyroscopic frequency, the spectrum of the oscillations is identical with the expression (6.40a) for the spectrum of sound in an isotropic plasma ($\mathbf{B}_0 = 0$), and the small damping is determined by the constant (6.40b) with the factor (ω_{Li}/kv_{Te}), replaced by $(\omega_{Li}/|k_z|v_{Te})$ in the electron term. The corresponding spectral function of the electric fields is determined by the right-hand side of Eq. (6.67). We shall denote by the symbol s_i this comparatively high-frequency ($\omega \gg \Omega_i$) ion sound in a magnetoactive plasma.

In the opposite case of a nonisothermal magnetoactive plasma with hot ions and cold electrons:

$$T_i \gg T_e \qquad (18.67)$$

at low frequencies ω and high phase velocities

$$\Omega_i \gg \omega \gg |k_z|v_{T_i}, \qquad |\Omega_e| \gg \omega \gg |k_z|v_{T_e} \qquad (18.68)$$

there exist oscillations that can be naturally called electron sound:

$$(s_e) \qquad \omega^2=\frac{k_z^2}{k^2}\omega_{Le}^2\left\{1+\frac{k_\perp^2}{k^2}\frac{\omega_{Le}^2}{\Omega_e^2}+\frac{\omega_{Li}^2}{k^2v_{T_i}^2}[1-A_0(z_i)]\right\}^{-1}. \qquad (18.69)$$

Apart from the conditions (18.68), it is necessary for the existence of these oscillations that their transverse (relative to the direction of $\mathbf{B}_0$) wave number $k_\perp$ be small compared with the reciprocal electron Larmor radius but not small compared with the reciprocal ion Larmor radius:

$$z_e \ll 1, \quad z_i \gtrsim 1. \qquad (18.70)$$

To such electron-acoustic oscillations there corresponds a real part of the longitudinal permittivity in the form

$$\varepsilon'(\omega,\mathbf{k})=1+\frac{k_\perp^2}{k^2}\frac{\omega_{Le}^2}{\Omega_e^2}+\frac{\omega_{Li}^2}{k^2v_{T_i}^2}[1-A_0(z_i)]-\frac{k_z^2}{k^2}\frac{\omega_{Le}^2}{\omega^2} \qquad (18.71)$$

and an exponentially small [in accordance with (18.68)] imaginary part ε''. In this connection, the decay constants $\gamma^e_{s_e}$ and $\gamma^i_{s_e}$ are exponentially small compared with the frequency (18.69). Note that in the limit of long waves, the spectrum (18.69) goes over into the spectrum (18.4) of the low-frequency oscillation p in a cold magnetoactive plasma. The designation of the oscillation s_e can be justified to a certain extent by noting that in a plasma with magnetized electrons ($\omega_{Le}^2 < \Omega_e^2$) and not too hot ions ($\omega_{Li}^2 \gg k^2v_{T_i}^2$) the spectrum (18.69) simplifies to

$$(s_e) \qquad \omega^2=k_z^2\frac{\varkappa T_i}{m}[1-A_0(z_i)]^{-1}, \qquad T_i \gg T_e, \qquad (18.72)$$

from which it follows that the electron and ion variables in the electron-sound oscillation s_e play, roughly speaking, the same role as the ion and electron variables in the magnetized ($\omega \ll \Omega_i$) long-wave ($\omega_{Le}^2 \gg k^2v_{Te}^2$) ion sound s_i':

$$(s_i') \qquad \omega^2=k_z^2\frac{\varkappa T_e}{M}, \qquad T_e \gg T_i. \qquad (18.73)$$

The electron sound s_e is determined by the spectral function

$$(E_jE_i)^{s_e}_{\omega,\mathbf{k}} = \frac{1}{4\pi^2} W_{s_e}(\mathbf{k}) \left\{1 + \frac{k_\perp^2}{k^2}\frac{\omega_{L_e}^2}{\Omega_e^2} + \frac{\omega_{L_i}^2}{k^2 v_{T_i}^2}[1 - A_0(z_i)]\right\}^{-1} \varkappa_i\varkappa_j \left\{\delta\left(\omega - \omega_{L_e}\frac{k_z}{k}\left[1 + \frac{k_\perp^2}{k^2}\frac{\omega_{L_e}^2}{\Omega_e^2} + \frac{\omega_{L_i}^2}{k^2 v_{T_i}^2}(1 - A_0(z_i))\right]^{-1/2}\right) + \right.$$

$$\left. + \delta\left(\omega + \omega_{L_e}\frac{k_z}{k}\left[1 + \frac{k_\perp^2}{k^2}\frac{\omega_{L_e}^2}{\Omega_e^2} + \frac{\omega_{L_i}^2}{k^2 v_{T_i}^2}(1 - A_0(z_i))\right]^{-1/2}\right)\right\}. \quad (18.74)$$

To conclude this section we note that, as in a cold plasma, the spectral functions we have derived for the electric fields of the characteristic oscillations of a hot magnetoactive plasma are only approximately tensors that are longitudinal with respect to the wave vector **k** (i.e., electrostatic). The corresponding conditions for the oscillations to be longitudinal, which restrict the values of the coefficients of the five elementary tensors (18.1) of the spectral function (17.65) that differ from the longitudinal tensor by the value of the coefficient of the longitudinal tensor $\varkappa_i\varkappa_j$ impose, of course, certain restrictions on not only the plasma density and the external magnetic field [as occurs in a cold plasma: see, for example (18.21)] but also the plasma temperature. However, we have restricted ourselves here to the already given results for quasilongitudinal oscillations of a magnetoactive plasma, assuming that for them the refractive index is sufficiently large (17.59), and we now turn to the study of the first nonlinear tensor S_{ijs} in a magnetoactive plasma.

§ 19. Three-Index Tensor $S_{ijs}(\omega, \mathbf{k}; \omega', \mathbf{k}')$ in a Magnetoactive Plasma

As in an isotropic plasma, the tensors S and V determine the evolution of the nonlinear interaction of characteristic oscillations of a magnetoactive plasma since their contractions with the spectral functions and the inverse tensor give the kernel of the integrodifferential equation that describes the nonlinear process [see (3.22a) for decay processes].

In accordance with the general definition (3.24), the tensor S_{ijs} in a magnetoactive plasma also consists of a sum of two three-index tensors. In accordance with Eq. (16.3), the first of these is given by [61]

$$\varepsilon_{ijs}(\omega, \mathbf{k}; \omega', \mathbf{k}') = i4\pi e^3 \int d\mathbf{p}\frac{v_i}{\omega} \int_{-\infty}^{0} d\tau_0 \int_{-\infty}^{0} d\tau_1 \exp\{-i\omega\tau_0 + i\mathbf{k}\delta\mathbf{R}(\tau_0, \mathbf{v})\} \times$$

$$\times \alpha''_{nj}(\tau_0)\frac{\partial}{\partial p_n(\tau_0, \mathbf{v})} \exp\{-i\omega'\tau_1 + i\mathbf{k}'\delta\mathbf{R}(\tau_0 + \tau_1, \mathbf{v}) - i\mathbf{k}'\delta\mathbf{R}(\tau_0, \mathbf{v})\}\, \alpha'_{ms}(\tau_0 + \tau_1)\frac{df_0}{\partial p_m(\tau_0 + \tau_1, \mathbf{v})}, \quad (19.1)$$

in which for the tensors α_{ij} we have used the abbreviated notation [cf. (7.2)]

$$\begin{gathered} \alpha''_{nj}(\tau_0) \equiv \alpha_{nj}(\omega'', \mathbf{k}''; \mathbf{v}(\tau_0, \mathbf{v})), \\ \alpha'_{ms}(\tau_0 + \tau_1) \equiv \alpha_{ms}(\omega', \mathbf{k}'; \mathbf{v}(\tau_0 + \tau_1, \mathbf{v})), \end{gathered} \quad (19.2)$$

and the vectors $\delta\mathbf{R}(\tau, \mathbf{v})$ and $\mathbf{v}(\tau, \mathbf{v})$ are defined in § 15-16 of this chapter. The second term of the tensor S is obtained from the first by transposing the subscripts $s \rightleftharpoons j$ and the arguments $\omega, \mathbf{k} \rightleftharpoons \omega'', \mathbf{k}''$ in the tensors $\alpha_{nj}(\omega'', \mathbf{k}''; \mathbf{v}(\tau_0, \mathbf{v}))$ and $\alpha_{ms}(\omega', \mathbf{k}'; \mathbf{v}(\tau_0 + \tau_1; \mathbf{v}))$ and simultaneously replacing $\exp\{-i\omega'\tau_1 + i\mathbf{k}'\delta\mathbf{R}(\tau_0 + \tau_1, \mathbf{v}) - i\mathbf{k}'\delta\mathbf{R}(\tau_0, \mathbf{v})\}$ by $\exp\{-i\omega''\tau_1 + i\mathbf{k}''\delta\mathbf{R}(\tau_0 + \tau_1, \mathbf{v}) - i\mathbf{k}''\delta\mathbf{R}(\tau_0, \mathbf{v})\}$:

$$\varepsilon_{isj}(\omega, \mathbf{k}; \omega'', \mathbf{k}'') = i4\pi e^3 \int d\mathbf{p}\frac{v_i}{\omega} \int_{-\infty}^{0} d\tau_0 \int_{-\infty}^{0} d\tau_1 \exp\{-i\omega\tau_0 + i\mathbf{k}\delta\mathbf{R}(\tau_0, \mathbf{v})\} \times$$

$$\times \alpha'_{ns}(\tau_0)\frac{\partial}{\partial p_n(\tau_0, \mathbf{v})} \exp\{-i\omega''\tau_1 + i\mathbf{k}''\delta\mathbf{R}(\tau_0 + \tau_1, \mathbf{v}) - i\mathbf{k}''\delta\mathbf{R}(\tau_0, \mathbf{v})\}\, \alpha''_{mj}(\tau_0 + \tau_1)\frac{\partial f_0}{\partial p_m(\tau_0 + \tau_1, \mathbf{v})}. \quad (19.3)$$

In both terms Eqs. (7.3) are satisfied for the frequencies and wave vectors $\omega = \omega' + \omega''$, $\mathbf{k} = \mathbf{k}' + \mathbf{k}''$. It already follows from the definition of the tensor S that it is symmetric with respect to the last two subscripts [cf. (3.24)]:

$$S_{ijs}(\omega, \mathbf{k}; \omega', \mathbf{k}') = S_{isj}(\omega, \mathbf{k}; \omega'', \mathbf{k}''). \tag{19.4}$$

In a nonrelativistic plasma it is natural to use the velocities $\mathbf{v}$ instead of the momenta p:

$$\varepsilon_{ijs}(\omega, \mathbf{k}; \omega', \mathbf{k}') = i\frac{4\pi e^3}{m^2}\int d\mathbf{v}\,\frac{v_i}{\omega}\int\limits_{-\infty}^{0} d\tau_0 \int\limits_{-\infty}^{0} d\tau_1 \exp\{-i\omega\tau_0 + i\mathbf{k}\delta\mathbf{R}(\tau_0, \mathbf{v})\}\times$$
$$\times\, \alpha''_{nj}(\tau_0)\frac{\partial}{\partial v_n(\tau_0, \mathbf{v})}\exp\{-i\omega'\tau_1 + i\mathbf{k}'\delta\mathbf{R}(\tau_0+\tau_1, \mathbf{v}) - i\mathbf{k}'\delta\mathbf{R}(\tau_0, \mathbf{v})\}\,\alpha'_{ms}(\tau_0+\tau_1)\frac{\partial f_0}{\partial v_m(\tau_0+\tau_1, \mathbf{v})}. \tag{19.5}$$

In the limit of a weak external magnetic field ($\mathbf{B}_0 = 0$), the three-index tensors ε and, therefore, the tensor S go over into their analogs in an isotropic plasma. Let us follow this transition for the example (19.1). When $\mathbf{B}_0 = 0$ we have [cf. (16.18) and (15.52)]

$$\mathbf{k}\delta\mathbf{R}(\tau_0, \mathbf{v}) = \mathbf{k}\mathbf{v}\tau_0, \qquad \mathbf{v}(\tau_0, \mathbf{v}) = \mathbf{v}, \tag{19.6}$$
$$\mathbf{k}'\delta\mathbf{R}(\tau_0+\tau_1, \mathbf{v}) - \mathbf{k}'\delta\mathbf{R}(\tau_0, \mathbf{v}) = \mathbf{k}'\mathbf{v}\tau_1,$$

so that the tensor (19.1) becomes

$$\varepsilon_{ijs}(\omega, \mathbf{k}; \omega', \mathbf{k}') = i4\pi e^3\int d\mathbf{p}\frac{v_i}{\omega}\int\limits_{-\infty}^{0} d\tau_0\int\limits_{-\infty}^{0} d\tau_1 e^{-i(\omega-\mathbf{k}\mathbf{v})\tau_0}\,\alpha''_{nj}\frac{\partial}{\partial p_n}e^{-i(\omega'-\mathbf{k}'\mathbf{v})\tau_1}\alpha'_{ms}\frac{\partial f_0}{\partial p_m}, \quad \mathbf{B}_0 = 0, \tag{19.7}$$

and after integration with respect to τ_0 and τ_1 in accordance with (15.38) it is equal to (7.1).

In (19.5) we integrate twice by parts with respect to dv, using the differentiation rules [cf. (16.25)]

$$\frac{\partial v_i}{\partial v_n(\tau_0, \mathbf{v})} = W_{in}(-\tau_0), \quad \frac{\partial \alpha''_{nj}(\tau_0)}{\partial v_n(\tau_0, \mathbf{v})} = 0, \quad \frac{\partial \delta R_i(\tau_0, \mathbf{v})}{\partial v_n(\tau_0, \mathbf{v})} = -\rho_{in}(-\tau_0) \tag{19.8}$$

in the first integration and

$$\frac{\partial v_i}{\partial v_m(\tau_0+\tau_1, \mathbf{v})} = W_{im}(-\tau_0-\tau_1), \quad \frac{\partial \alpha'_{ms}(\tau_0+\tau_1)}{\partial v_m(\tau_0+\tau_1, \mathbf{v})} = 0, \quad \frac{\partial \alpha''_{nj}(\tau_0)}{\partial v_m(\tau_0+\tau_1, \mathbf{v})} = \left(\frac{k''_n}{\omega''}\delta_{bj} - \delta_{nj}\frac{k''_b}{\omega''}\right)W_{bm}(-\tau_1),$$
$$\frac{\partial \delta R_s(\tau_0+\tau_1, \mathbf{v})}{\partial v_m(\tau_0+\tau_1, \mathbf{v})} = -\rho_{sm}(-\tau_0-\tau_1), \quad \frac{\partial \delta R_i(\tau_0, \mathbf{v})}{\partial v_m(\tau_0+\tau_1, \mathbf{v})} = \rho_{im}(-\tau_1) - \rho_{im}(-\tau_0-\tau_1) \tag{19.9}$$

in the second. These rules are a direct consequence of the definitions of the vectors $\mathbf{v}(\tau, \mathbf{v})$ and $\delta\mathbf{R}(\tau, \mathbf{v})$ in terms of the tensors $W_{ij}(\tau)$ and $\rho_{ij}(\tau)$; in particular, we have used the identity (16.22). After integration, we obtain the following expression for ε_{ijs}:

$$\varepsilon_{ijs}(\omega, \mathbf{k}; \omega', \mathbf{k}') = i\frac{4\pi e^3}{m^2}\int d\mathbf{v} f_0\frac{1}{\omega}\int\limits_{-\infty}^{0} d\tau_0\int\limits_{-\infty}^{0} d\tau_1\exp\{-i\omega\tau_0 - i\omega'\tau_1 +$$
$$+ i\mathbf{k}'\delta\mathbf{R}(\tau_0+\tau_1, \mathbf{v}) + i\mathbf{k}''\delta\mathbf{R}(\tau_0, \mathbf{v})\}\,\alpha'_{ms}(\tau_0+\tau_1)\Big\{i\alpha''_{nj}(\tau_0)\{-k_l\rho_{ln}(-\tau_0)W_{im}(-\tau_0-\tau_1) + [W_{in}(-\tau_0) -$$
$$- iv_ik_l\rho_{ln}(-\tau_0)][k''_l\rho_{lm}(-\tau_1) - k_l\rho_{lm}(-\tau_0-\tau_1)]\} +$$
$$+ \{W_{in}(-\tau_0) - iv_ik_l\rho_{ln}(-\tau_0)\}\left[\frac{k''_n}{\omega''}\delta_{bj} - \delta_{nj}\frac{k''_b}{\omega''}\right]W_{bm}(-\tau_1)\Big\}. \tag{19.10}$$

Similarly, we obtain the expression for the second tensor ε_{ijs}, which we omit, since it is readily obtained by replacing in (19.10) all the subscripts j by s and s by j and the arguments ω'', $\mathbf{k}''$ by ω', $\mathbf{k}'$ and ω', $\mathbf{k}'$ by ω'', $\mathbf{k}''$.

We now turn to the case of a cold plasma, setting $v_T = 0$ in both the tensors ε. We then obtain an expression for the tensor S_{ijs} for a cold magnetoactive plasma:

$$S_{ijs}(\omega, \mathbf{k}; \omega', \mathbf{k}') = i\frac{e}{m}\frac{\omega_L^2}{\omega}\int_{-\infty}^{0} d\tau_0 \int_{-\infty}^{0} d\tau_1 \Big\{e^{-i\omega\tau_0 - i\omega'\tau_1}\Big[-ik_n\rho_{nj}(-\tau_0)\times$$
$$\times W_{is}(-\tau_0-\tau_1) + iW_{ij}(-\tau_0)k''_n\rho_{ns}(-\tau_1) - iW_{ij}(-\tau_0)k_n\rho_{ns}(-\tau_0-\tau_1) +$$
$$+ W_{in}(-\tau_0)W_{bs}(-\tau_1)\Big(\frac{k''_n}{\omega''}\delta_{bj} - \delta_{nj}\frac{k''_b}{\omega''}\Big)\Big] + e^{-i\omega\tau_0 - i\omega''\tau_1}\Big[-ik_n\rho_{ns}(-\tau_0)\times$$
$$\times W_{ij}(-\tau_0-\tau_1) + iW_{is}(-\tau_0)k'_n\rho_{nj}(-\tau_1) - iW_{is}(-\tau_0)k_n\rho_{nj}(-\tau_0-\tau_1) +$$
$$+ W_{in}(-\tau_0)W_{cj}(-\tau_1)\Big(\frac{k'_n}{\omega'}\delta_{cs} - \delta_{ns}\frac{k'_c}{\omega'}\Big)\Big]\Big\}. \tag{19.11}$$

Before we integrate here with respect to τ_0 and τ_1, we make a very important remark. When the tensor S is expressed in this manner we have two kinds of term: terms with and without explicitly separated variables of integration τ_0 and τ_1. Of the second kind are the first and third terms of the square brackets. However, in them the variables can be separated by using the conditional identity

$$\int_{-\infty}^{0} d\tau_0 e^{-i\omega\tau_0}\int_{-\infty}^{0} d\tau_1 e^{-i\omega'\tau_1}A_{ij}(-\tau_0)B_{as}(-\tau_0-\tau_1) + \int_{-\infty}^{0} d\tau_0 e^{-i\omega\tau_0}\int_{-\infty}^{0} d\tau_1 e^{-i\omega''\tau_1}A_{ij}(-\tau_0-\tau_1)B_{as}(-\tau_0) =$$
$$= \int_{-\infty}^{0} d\tau_1 e^{-i\omega'\tau_1}\int_{-\infty}^{0} d\tau_2 e^{-i\omega''\tau_2}A_{ij}(-\tau_2)B_{as}(-\tau_1), \tag{19.12}$$

which holds if the first of Eqs. (7.3) for the frequencies is fulfilled. The geometrical meaning of this identity is trivial: on the right-hand side we have an integral over the complete third quadrant of the τ_1, τ_2 plane and on the left-hand side the sum of the integrals (of the same integrand as on the right-hand side) over the half-quadrants divided by the bisectrix $\tau_1 = \tau_2$. In an isotropic plasma, the analog of the identity (19.12) is the equation

$$\frac{1}{\omega\omega'} + \frac{1}{\omega\omega''} = \frac{1}{\omega'\omega''}, \tag{19.13}$$

which is an elementary modification of the relation (7.3) for the frequencies. Application of the identity (19.12) to the terms in (19.11) with nonseparated variables gives the relations

$$\int_{-\infty}^{0} d\tau_0 e^{-i\omega\tau_0}\int_{-\infty}^{0} d\tau_1 \{e^{-i\omega'\tau_1}k_n\rho_{nj}(-\tau_0)W_{is}(-\tau_0-\tau_1) + e^{-i\omega''\tau_1}k_n\rho_{nj}(-\tau_0-\tau_1)W_{is}(-\tau_0)\} =$$
$$= \int_{-\infty}^{0} d\tau_1 e^{-i\omega'\tau_1}\int_{-\infty}^{0} d\tau_2 e^{-i\omega''\tau_2}k_n\rho_{nj}(-\tau_2)W_{is}(-\tau_1),$$
$$\int_{-\infty}^{0} d\tau_0 e^{-i\omega\tau_0}\int_{-\infty}^{0} d\tau_1 \{e^{-i\omega'\tau_1}W_{ij}(-\tau_0)k_n\rho_{ns}(-\tau_0-\tau_1) + e^{-i\omega''\tau_1}W_{ij}(-\tau_0-\tau_1)k_n\rho_{ns}(-\tau_0)\} =$$
$$= \int_{-\infty}^{0} d\tau_1 e^{-i\omega'\tau_1}\int_{-\infty}^{0} d\tau_2 e^{-i\omega''\tau_2}W_{ij}(-\tau_2)k_n\rho_{ns}(-\tau_1), \tag{19.14}$$

which leads to complete separation of the variables in all terms of the tensor S:

$$S_{ijs}(\omega, \mathbf{k}; \omega', \mathbf{k}') = i\frac{e}{m}\frac{\omega_L^2}{\omega}\int_{-\infty}^{0} d\tau_0 \int_{-\infty}^{0} d\tau_1 \Big\{-ie^{-i\omega'\tau_0}e^{-i\omega''\tau_1}[W_{is}(-\tau_0)k_n\rho_{nj}(-\tau_1) +$$
$$+ k_n\rho_{ns}(-\tau_0)W_{ij}(-\tau_1)] + e^{-i\omega\tau_0 - i\omega'\tau_1}\Big[iW_{ij}(-\tau_0)k''_n\rho_{ns}(-\tau_1) -$$

$$- W_{ij}(-\tau_0)\frac{k''_n}{\omega''}W_{ns}(-\tau_1) + W_{js}(-\tau_1)\frac{k''_n}{\omega''}W_{in}(-\tau_0)\Big] +$$

$$+ e^{-i\omega\tau_0 - i\omega''\tau_1}\Big[iW_{is}(-\tau_0)k'_n W_{nj}(-\tau_1) - W_{is}(-\tau_0)\frac{k'_n}{\omega'}W_{nj}(-\tau_1) + W_{sj}(-\tau_1)\frac{k'_n}{\omega'}W_{in}(-\tau_0)\Big]\Big\}. \quad (19.15)$$

Integrating with respect to τ_0 and τ_1, we obtain an expression for the tensor S_{ijs} in a cold magnetoactive plasma in terms of the frequencies and wave vectors of the interacting oscillations:

$$S_{ijs}(\omega, \mathbf{k}; \omega', \mathbf{k}') = i\frac{e}{m}\frac{\omega_L^2}{\omega\omega'\omega''}\{\omega\Gamma_{is}(\omega)\Gamma_{aj}(\omega'')k'_a + \omega\Gamma_{ij}(\omega)\Gamma_{as}(\omega')k''_a -$$

$$- \omega''\Gamma_{ij}(\omega'')\Gamma_{as}(\omega')k_a - \omega''\Gamma_{sj}(\omega'')\Gamma_{ia}(\omega)k'_a - \omega'\Gamma_{is}(\omega')\Gamma_{aj}(\omega'')k_a - \omega'\Gamma_{js}(\omega')\Gamma_{ia}(\omega)k''_a\}. \quad (19.16)$$

Here, the tensors Γ_{ij} are determined by Eq. (16.35).

Going over in accordance with the relation (16.34) from the tensors Γ_{ij} to the tensor $\delta\varepsilon_{ij}$, which is related to the permittivity tensor of a cold magnetoactive plasma by the definition

$$\varepsilon_{ij}(\omega, 0) = \delta_{ij} + \sum \delta\varepsilon_{ij}(\omega), \quad (19.17)$$

we represent (19.16) in a slightly different form:†

$$S_{ijs}(\omega, \mathbf{k}; \omega', \mathbf{k}') = i\frac{e}{m}\frac{1}{\omega_L^2}\Big\{\omega\delta\varepsilon_{ij}(\omega)\delta\varepsilon_{bs}(\omega')\frac{k''_b}{\omega''} + \omega\delta\varepsilon_{is}(\omega)\delta\varepsilon_{cj}(\omega'')\frac{k'_c}{\omega'} -$$

$$- \omega'\delta\varepsilon_{is}(\omega')\delta\varepsilon_{aj}(\omega'')\frac{k_a}{\omega} - \omega'\delta\varepsilon_{js}(\omega')\delta\varepsilon_{ib}(\omega)\frac{k''_b}{\omega''} - \omega''\delta\varepsilon_{ij}(\omega'')\delta\varepsilon_{as}(\omega')\frac{k_a}{\omega} - \omega''\delta\varepsilon_{sj}(\omega'')\delta\varepsilon_{ic}(\omega)\frac{k'_c}{\omega'}\Big\}. \quad (19.18)$$

Thus, the tensor S of a cold magnetoactive plasma is a symmetric bilinear combination of the polarizability tensors of one component of such a plasma. Note that on the right-hand side of Eqs. (19.16) and (19.18) summation over the particle species is understood in accordance with the general condition adopted in this monograph. Therefore, $\delta\varepsilon_{ij}$ are the partial polarizability tensors of a cold magnetoactive plasma.

These equations enable us to find the tensor $S_{jsi}(\omega'', \mathbf{k}''; \omega, \mathbf{k})$ by a simple transposition of the arguments and the indices in (19.18), which yields the symmetry property

$$S_{ijs}(\omega, \mathbf{k}; \omega', \mathbf{k}'; \mathbf{B}_0) = -S_{jsi}(\omega'', \mathbf{k}''; \omega, \mathbf{k}; -\mathbf{B}_0). \quad (19.19)$$

If we restrict ourselves to the case of a nonvanishing frequency difference, $\omega'' \neq 0$,

$$\frac{1}{\omega''} = \frac{P}{\omega''} - i\pi\delta(\omega'') = \frac{P}{\omega''}, \quad (19.20)$$

then this property can be rewritten as

$$S_{ijs}(\omega, \mathbf{k}; \omega', \mathbf{k}') = S^*_{jsi}(\omega'', \mathbf{k}''; \omega, \mathbf{k}). \quad (19.21)$$

Note that for (19.19) and (19.21) to be identical, it is necessary for not only (19.20) to hold but also the analogous equations for ω^{-1} and ω'^{-1}, which follow automatically from the assumption $\omega \neq 0$, $\omega' \neq 0$ (static perturbations of the plasma are not considered). Conversely, if $\omega \simeq \omega'$, i.e., $\omega'' \approx 0$, the term $-i\pi\delta(\omega'')$ in (19.18) is very important and only the first symmetry property (19.19) holds, this generalizing (7.10) to the case of a strong field $\mathbf{B}_0 = \text{const} \neq 0$ in accordance with the well-known rule for modifying the symmetry principle of transport coefficients when an external magnetic field is added.

† Formulas (19.16) and (19.18) for the tensor S_{ijs} in a cold magnetoactive plasma could also be obtained by a hydrodynamic description (see the Appendix). Note that, as in an isotropic plasma, the tensor S_{ijs} in (19.18) is linear in the wave vectors $\mathbf{k}$, $\mathbf{k}'$, $\mathbf{k}''$.

In the limit of a weak magnetic field ($\mathbf{B}_0 = 0$), the tensor (19.18) goes over into the isotropic expression (7.23). The purely longitudinal contraction has the form

$$S(\omega, \mathbf{k}; \omega', \mathbf{k}') \equiv S_{ijs}(\omega, \mathbf{k}; \omega', \mathbf{k}') \frac{k_i k''_j k'_s}{kk'k''} = -i \frac{e}{m} \frac{\omega_L^2}{kk'k''} \left\{ \left[k_i \frac{\Gamma_{ij}(\omega'')}{\omega''} k''_j \right] \left[k_a \frac{\Gamma_{as}(\omega')}{\omega'} k'_s \right] - \right.$$
$$\left. - \left[k_i \frac{\Gamma_{is}(\omega)}{\omega} k'_s \right] \left[k'_c \frac{\Gamma_{cj}(\omega'')}{\omega''} k''_j \right] - \left[k_i \frac{\Gamma_{ij}(\omega)}{\omega} k''_j \right] \left[k''_b \frac{\Gamma_{bs}(\omega')}{\omega'} k'_s \right] \right\} \tag{19.22}$$

or, in the notation $\delta\varepsilon_{ij}$,

$$S(\omega, \mathbf{k}; \omega', \mathbf{k}') = -i \frac{e}{m} \frac{\omega_L^{-2}}{kk'k''} \{ [k_i \delta\varepsilon_{ij}(\omega'') k''_j] [k_a \delta\varepsilon_{as}(\omega') k'_s] -$$
$$- [k_i \delta\varepsilon_{is}(\omega) k'_s] [k'_c \delta\varepsilon_{cj}(\omega'') k''_j] - [k_i \delta\varepsilon_{ij}(\omega) k''_j] [k''_b \delta\varepsilon_{bs}(\omega') k'_s] \}. \tag{19.23}$$

If the magnetic field is very strong (in the limit $B_0 \to \infty$), then from the tensors Γ_{ij} and $\delta\varepsilon_{ij}$ one can separate out only tensors that are purely longitudinal with respect to a unit vector $\mathbf{h}$ along $\mathbf{B}_0$:

$$B_0 = \infty, \quad \Gamma_{ij}(\omega) = \frac{h_i h_j}{\omega}, \quad \delta\varepsilon_{ij} = -\frac{\omega_L^2}{\omega^2} h_i h_j. \tag{19.24}$$

Physically, this case corresponds to low frequencies of the interacting oscillations compared with the gyroscopic frequencies of the plasma particles (more precisely, to the zeroth approximation in the small ratio of the frequencies):

$$\omega, \omega', \omega'' \ll \Omega \tag{19.25}$$

in contrast to the limit of a weak magnetic field, when the opposite inequalities hold:

$$\omega, \omega', \omega'' \gg \Omega. \tag{19.26}$$

Then (the subscript z denotes the projection of a vector onto the z axis along $\mathbf{B}_0$)

$$S_{ijs}(\omega, \mathbf{k}; \omega', \mathbf{k}'; B_0 \to \infty) = -i \frac{e}{m} \omega_L^2 \frac{h_i h_j h_s}{\omega\omega'\omega''} \left\{ \frac{k_z}{\omega} + \frac{k''_z}{\omega''} + \frac{k'_z}{\omega'} \right\}, \tag{19.27}$$

and the longitudinal contraction has the form

$$S(\omega, \mathbf{k}; \omega', \mathbf{k}'; B_0 \to \infty) = -i \frac{e}{m} \frac{\omega_L^2}{\omega\omega'\omega''} \frac{k_z k'_z k''_z}{kk'k''} \left\{ \frac{k_z}{\omega} + \frac{k''_z}{\omega''} + \frac{k'_z}{\omega'} \right\}, \tag{19.28}$$

which corresponds to a one-dimensional nonlinear interaction.

A very strong magnetic field "draws out" all processes along the characteristic direction, so that all variables that characterize the process depend only on the longitudinal components of the wave vectors of the interacting oscillations. On account of the identity (7.58), the tensor can be represented in the different form

$$S_{ijs}(\omega, \mathbf{k}; \omega', \mathbf{k}'; B_0 \to \infty) = -i \frac{e}{m} \omega_L^2 h_i h_j h_s \left\{ \frac{k_z}{(\omega'\omega'')^2} - \frac{k''_z}{(\omega\omega')^2} - \frac{k'_z}{(\omega\omega'')^2} \right\}. \tag{19.29}$$

Note that study of different special cases of the nonlinear tensor S_{ijs} and its contractions in a cold magnetoactive plasma is made easier by the fact that it reduces essentially to studying the well-known linear tensors $\delta\varepsilon_{ij}$ of the partial polarizability of the plasma. In particular, if two of the three frequencies, for example, ω and ω', are much higher than the gyroscopic frequency, and the third, ω'', is arbitrary:

$$\omega, \omega' \gg \Omega, \quad \omega'' \lesseqgtr \Omega, \tag{19.30}$$

then S_{ijs} can be expressed in terms of the partial polarizability $\delta\varepsilon_{ij}(\omega'')$ of the plasma at the frequency ω'':

$$S_{ijs}(\omega,\ \mathbf{k};\ \omega',\ \mathbf{k}') = i\frac{e}{m}\frac{1}{\omega\omega'\omega''}\left\{\omega''^2\delta\varepsilon_{aj}(\omega'')\left[\frac{k_i'}{\omega}\delta_{as} + \frac{k_s}{\omega'}\delta_{ai} + \frac{k_a''}{\omega''}\delta_{is}\right] + \omega_L^2\left[\delta_{ij}\frac{k_s''}{\omega'} - \frac{k_i''}{\omega}\delta_{js}\right]\right\}. \qquad (19.31)$$

We give some concrete forms of the longitudinal contraction (19.23) corresponding to different geometrical dispositions of the wave vectors $\mathbf{k}$, $\mathbf{k}'$, $\mathbf{k}''$ of the interacting oscillations and the unit vector $\mathbf{h}$ along the external magnetic field.

If all three wave vectors lie in the plane perpendicular to the external magnetic field (two-dimensional interaction) i.e.,

$$\mathbf{kh} = \mathbf{k}'\mathbf{h} = \mathbf{k}''\mathbf{h} = 0, \qquad (19.32)$$

then

$$S(\omega,\ \mathbf{k};\ \omega',\ \mathbf{k}') = i\frac{e}{m}\frac{\omega_L^2}{kk'k''}\left\{\frac{(\mathbf{kk}'')(\mathbf{k}'\mathbf{k}'')}{(\omega^2-\Omega^2)(\omega'^2-\Omega^2)} + \frac{(\mathbf{kk}')(\mathbf{k}''\mathbf{k}')}{(\omega^2-\Omega^2)(\omega''^2-\Omega^2)} - \right.$$

$$\left. - \frac{(\mathbf{k}'\mathbf{k})(\mathbf{k}''\mathbf{k})}{(\omega'^2-\Omega^2)(\omega''^2-\Omega^2)} + \Omega^2\frac{(\mathbf{k}\,[\mathbf{k}'\mathbf{h}])^2}{\omega\omega'\omega''}\left[\frac{\omega''}{(\omega^2-\Omega^2)(\omega'^2-\Omega^2)} + \frac{\omega'}{(\omega''^2-\Omega^2)(\omega^2-\Omega^2)} - \frac{\omega}{(\omega'^2-\Omega^2)(\omega''^2-\Omega^2)}\right]\right\}. \qquad (19.33)$$

The case when all three vectors $\mathbf{k}$, $\mathbf{k}'$, $\mathbf{k}''$ are parallel to the external magnetic field:

$$[\mathbf{kh}] = [\mathbf{k}'\mathbf{h}] = [\mathbf{k}''\mathbf{h}] = 0, \qquad (19.34)$$

can be described by the contraction (19.28) obtained in the approximation of an infinitely strong magnetic field. This fact is a geometric illustration of the limit of an infinitely strong magnetic field $\mathbf{B}_0$. In principle, a one-dimensional nonlinear interaction ($[\mathbf{kk}'] = 0$) is also possible at an arbitrary angle to the external magnetic field. Then

$$S(\omega,\ \mathbf{k};\ \omega',\ \mathbf{k}') = i\frac{e}{m}\frac{\omega_L^2}{kk'k''}\left\{-\frac{k_zk_z'k_z''}{\omega\omega'\omega''}\left(\frac{k_z}{\omega} + \frac{k_z'}{\omega'} + \frac{k_z''}{\omega''}\right) + \right.$$

$$\left. + \frac{([\mathbf{kh}]\,[\mathbf{k}''\mathbf{h}])([\mathbf{k}'\mathbf{h}]\,[\mathbf{k}''\mathbf{h}])}{(\omega^2-\Omega^2)(\omega'^2-\Omega^2)} + \frac{([\mathbf{kh}]\,[\mathbf{k}'\mathbf{h}])([\mathbf{k}''\mathbf{h}]\,[\mathbf{k}'\mathbf{h}])}{(\omega^2-\Omega^2)(\omega''^2-\Omega^2)} - \frac{([\mathbf{k}'\mathbf{h}]\,[\mathbf{kh}])([\mathbf{k}''\mathbf{h}]\,[\mathbf{k}'\mathbf{h}])}{(\omega'^2-\Omega^2)(\omega''^2-\Omega^2)}\right\}. \qquad (19.35)$$

This formula is also valid in the more general case of an arbitrary mutual disposition of the wave vectors with the only restriction that $\mathbf{h}$ lies in the plane of the triangle of the wave vectors ($\mathbf{k} = \mathbf{k}' + \mathbf{k}''$), i.e.,

$$(\mathbf{k}\,[\mathbf{k}'\mathbf{h}]) = 0. \qquad (19.36)$$

In particular, if the one-dimensional interaction occurs at right angles to $\mathbf{B}_0$:

$$(\mathbf{kh}) = (\mathbf{k}'\mathbf{h}) = (\mathbf{k}''\mathbf{h}) = 0, \qquad [\mathbf{kk}'] = 0, \qquad (19.37)$$

we then obtain with allowance for the identity (7.54a)

$$S(\omega,\ \mathbf{k};\ \omega',\ \mathbf{k}') = -i\frac{e}{m}\frac{1}{kk'k''}\frac{\omega_L^2}{(\omega^2-\Omega^2)(\omega'^2-\Omega^2)(\omega''^2-\Omega^2)}\{(\mathbf{kk}'')(\mathbf{k}'\mathbf{k}'')(\omega^2-\omega''^2) + (\mathbf{kk}')(\mathbf{k}'\mathbf{k}'')(\omega^2-\omega'^2)\}. \qquad (19.38)$$

But if as a result of the nonlinear interaction with arbitrary mutual disposition of the wave vectors an oscillation is formed with wave vector parallel to $\mathbf{h}$, we then obtain from (19.22) with allowance for the condition $[\mathbf{kh}] = 0$

$$S(\omega,\ \mathbf{k};\ \omega',\ \mathbf{k}') = -i\frac{e}{m}\frac{\omega_L^2}{kk'k''}\left\{\frac{k_zk_z'k_z''}{\omega\omega'\omega''}\left(\frac{k_z}{\omega} + \frac{k_z'}{\omega'} + \frac{k_z''}{\omega''}\right) + \frac{k_z}{\omega^2}\left[k_z''\frac{[\mathbf{k}'\mathbf{h}]^2}{\omega'^2-\Omega^2} + k_z'\frac{[\mathbf{k}''\mathbf{h}]^2}{\omega''^2-\Omega^2}\right]\right\}. \qquad (19.39)$$

In the opposite case, when the interaction gives rise to an oscillation with wave vector perpendicular to the external magnetic field:

$$\mathbf{kh}=0,$$

the contraction of S is determined by the equation

$$\begin{aligned}S(\omega,\ \mathbf{k};\ \omega',\ \mathbf{k}') = i\frac{e}{m}\frac{\omega_L^2}{kk'k''}\Big\{\frac{[\mathbf{k}'\mathbf{h}]\,[\mathbf{k}''\mathbf{h}]}{\omega^2-\Omega^2}\Big(\frac{\mathbf{kk}''}{\omega'^2-\Omega^2}+\frac{\mathbf{kk}'}{\omega''^2-\Omega^2}\Big)- \\ -\frac{(\mathbf{kk}')\,(\mathbf{kk}'')}{(\omega'^2-\Omega^2)\,(\omega''^2-\Omega^2)}+\Omega^2\frac{(\mathbf{k}\,[\mathbf{k}'\mathbf{h}])^2}{\omega\omega'\omega''}\Big(\frac{\omega''}{(\omega^2-\Omega^2)\,(\omega'-\Omega^2)}+ \\ +\frac{\omega'}{(\omega^2-\Omega^2)\,(\omega''^2-\Omega^2)}-\frac{\omega}{(\omega'^2-\Omega^2)\,(\omega''^2-\Omega^2)}\Big)+\frac{k_z'k_z''}{\omega^2-\Omega^2}\Big[\Big(\frac{[\mathbf{kh}]\,[\mathbf{k}'\mathbf{h}]}{\omega'^2}+\frac{[\mathbf{kh}]\,[\mathbf{k}'\mathbf{h}]}{\omega''^2}\Big)+i\frac{\Omega}{\omega}(\mathbf{k}\,[\mathbf{k}'\mathbf{h}])\Big(\frac{1}{\omega''^2}-\frac{1}{\omega'^2}\Big)\Big]\Big\}.\end{aligned} \tag{19.40}$$

It is readily seen that all these concrete forms of the longitudinal contraction of the tensor S go over into each other when the restrictions imposed on the wave vectors and the direction of B_0 are identical.

Let us now consider some special cases of the tensor S_{ijs} (19.16) corresponding to different geometries of the four vectors **k**, **k'**, **k"**, **h** in the general case of nonlinear interaction of characteristic oscillations of a magnetoactive plasma of arbitrary polarization.

If all the wave vectors are parallel to $\mathbf{B}_0$ (19.34), then on account of the equation

$$k_i\Gamma_{ij}(\omega) = k_z h_i \Gamma_{ij}(\omega) = h_j\frac{k_z}{\omega}, \quad [\mathbf{kh}]=0, \tag{19.41}$$

the tensors S_{ijs} simplifies considerably:

$$\begin{aligned}S_{ijs}(\omega,\ \mathbf{k};\ \omega',\ \mathbf{k}') = i\frac{e}{m}\frac{\omega_L^2}{\omega\omega'\omega''}\Big\{\frac{h_j}{\omega''}[\omega k_z'\Gamma_{is}(\omega)-\omega' k_z\Gamma_{is}(\omega')]+ \\ +\frac{h_s}{\omega'}[\omega k_z''\Gamma_{ij}(\omega)-\omega'' k_z\Gamma_{ij}(\omega'')]-\frac{h_i}{\omega}[\omega' k_z''\Gamma_{js}(\omega')+\omega'' k_z'\Gamma_{sj}(\omega'')]\Big\}.\end{aligned} \tag{19.42}$$

Hence, in particular, we obtain (19.28). In terms of $\delta\varepsilon_{ij}$ this equation can be rewritten as

$$\begin{aligned}S_{ijs}(\omega,\ \mathbf{k};\ \omega',\ \mathbf{k}') = -i\frac{e}{m}\frac{1}{\omega\omega'\omega''}\Big\{\frac{h_j}{\omega''}[\omega^2 k_z'\delta\varepsilon_{is}(\omega)-\omega'^2 k_z\delta\varepsilon_{is}(\omega')]+ \\ +\frac{h_s}{\omega'}[\omega^2 k_z''\delta\varepsilon_{ij}(\omega)-\omega''^2 k_z\delta\varepsilon_{ij}(\omega'')]-\frac{h_i}{\omega}[\omega'^2 k_z''\delta\varepsilon_{js}(\omega')+\omega''^2 k_z'\delta\varepsilon_{sj}(\omega'')]\Big\}.\end{aligned} \tag{19.43}$$

When oscillations with arbitrary polarization but wave vectors in the plane perpendicular to $\mathbf{B}_0$ interact, we obtain, using relations of the type

$$\Gamma_{aj}(\omega'')\,k_a = \frac{\omega''}{\omega''^2-\Omega^2}k_j - i\frac{\Omega}{\omega''^2-\Omega^2}[\mathbf{kh}]_j, \quad \mathbf{kh}=0, \tag{19.44}$$

the following equation from (19.16):

$$\begin{aligned}S_{ijs}(\omega,\ \mathbf{k};\ \omega',\ \mathbf{k}') = i\frac{e}{m}\omega_L^2\Big\{\frac{1}{\omega''^2-\Omega^2}\Big[\frac{k_j'}{\omega'}\Gamma_{is}(\omega)-\frac{k_j}{\omega}\Gamma_{is}(\omega')\Big]+ \\ +\frac{1}{\omega'^2-\Omega^2}\Big[\frac{k_s''}{\omega''}\Gamma_{ij}(\omega)-\frac{k_s}{\omega}\Gamma_{ij}(\omega'')\Big]-\frac{1}{\omega^2-\Omega^2}\Big[\frac{k_i'}{\omega'}\Gamma_{sj}(\omega'')+ \\ +\frac{k_i''}{\omega''}\Gamma_{js}(\omega')\Big]-i\frac{\Omega}{\omega''}\frac{1}{\omega''^2-\Omega^2}\Big(\frac{[\mathbf{k}'\mathbf{h}]_j}{\omega'}\Gamma_{is}(\omega)-\frac{[\mathbf{kh}]_j}{\omega}\Gamma_{is}(\omega')\Big)- \\ -i\frac{\Omega}{\omega'}\frac{1}{\omega'^2-\Omega^2}\Big(\frac{[\mathbf{k}''\mathbf{h}]_s}{\omega''}\Gamma_{ij}(\omega)-\frac{[\mathbf{kh}]_s}{\omega}\Gamma_{ij}(\omega'')\Big)-i\frac{\Omega}{\omega}\frac{1}{\omega^2-\Omega^2}\Big(\frac{[\mathbf{k}'\mathbf{h}]_i}{\omega'}\Gamma_{sj}(\omega'')+\frac{[\mathbf{k}''\mathbf{h}]_i}{\omega''}\Gamma_{js}(\omega')\Big)\Big\}.\end{aligned} \tag{19.45}$$

When studying the nonlinear interaction of longitudinal oscillations with frequencies ω', ω'' and wave vectors **k'**, **k"** that gives rise to an oscillation of an arbitrary polarization, frequency ω, and wave vector $\mathbf{k}$, we must use the incomplete longitudinal contraction

$$S_{ijs}(\omega,\ \mathbf{k};\ \omega',\ \mathbf{k}')\frac{k''_j k'_s}{k'k''}=-i\frac{e}{m}\frac{1}{k'k''}\frac{\omega_L^2}{\omega\omega'\omega''}\{\omega'[k_a\Gamma_{aj}(\omega'')k''_j\Gamma_{is}(\omega')k'_s-$$
$$-\Gamma_{is}(\omega)k'_s k'_c\Gamma_{cj}(\omega'')k''_j]+\omega''[k_a\Gamma_{as}(\omega')k'_s\Gamma_{ij}(\omega'')k''_j-\Gamma_{ij}(\omega)k''_j k''_b\Gamma_{bs}(\omega')k'_s]\}. \tag{19.46}$$

If the resulting oscillation then propagates along $\mathbf{B}_0$:

$$[\mathbf{kh}]=0,$$

then (19.46) can be reduced by means of the equations

$$k_a\Gamma_{aj}(\omega'')k''_j=\frac{k_z k''_z}{\omega''},\quad k'_c\Gamma_{cj}(\omega'')k''_j=\frac{k_z k''_z}{\omega''}-k''_b\Gamma_{bj}(\omega'')k''_j \tag{19.47}$$

to the form

$$S_{ijs}(\omega,\ \mathbf{k};\ \omega',\ \mathbf{k}')\frac{k''_j k'_s}{k'k''}=-i\frac{e}{m}\frac{\omega_L^2}{k'k''}\left\{\frac{k_z k''_z}{\omega\omega''}\frac{k'_s}{\omega''}[\Gamma_{is}(\omega')-\Gamma_{is}(\omega)]+\right.$$
$$\left.+\frac{k_z k'_z}{\omega\omega'}\frac{k''_j}{\omega'}[\Gamma_{ij}(\omega'')-\Gamma_{ij}(\omega)]+\frac{k''_b\Gamma_{bj}(\omega'')k''_j}{\omega''}\frac{\Gamma_{is}(\omega)k'_s}{\omega}+\frac{k'_c\Gamma_{cs}(\omega')k'_s}{\omega'}\frac{\Gamma_{ij}(\omega)k''_j}{\omega}\right\}. \tag{19.48}$$

In the last two terms of the contraction, we can give a more detailed expression by means of the relation

$$\frac{k_i\Gamma_{ij}(\omega)k_j}{\omega}=\frac{k_z^2}{\omega^2}+\frac{k_\perp^2}{\omega^2-\Omega^2}, \tag{19.49}$$

where $k_\perp$ is the projection of $\mathbf{k}$ onto the plane perpendicular to $\mathbf{h}$. Similarly, in the case $\mathbf{kh}=0$ we obtain from (19.46)

$$S_{ijs}(\omega,\ \mathbf{k};\ \omega',\ \mathbf{k}')\frac{k''_j k'_s}{k'k''}=-i\frac{e}{m}\frac{\omega_L^2}{kk'k''}\left\{\left[\frac{\omega''}{\omega''^2-\Omega^2}\frac{\mathbf{kk}''}{\omega\omega''}-i\frac{\Omega}{\omega''^2-\Omega^2}\frac{(\mathbf{k}''[\mathbf{kh}])}{\omega\omega''}\right]\times\right.$$
$$\times[\Gamma_{is}(\omega')-\Gamma_{is}(\omega)]k'_s+\left[\frac{\omega'}{\omega'^2-\Omega^2}\frac{\mathbf{kk}'}{\omega\omega'}-i\frac{\Omega}{\omega'^2-\Omega^2}\frac{(\mathbf{k}'[\mathbf{kh}])}{\omega\omega'}\right][\Gamma_{ij}(\omega'')-$$
$$\left.-\Gamma_{ij}(\omega)]k''_j+\frac{k''_b\Gamma_{bj}(\omega'')k''_j}{\omega''}\frac{\Gamma_{is}(\omega)k'_s}{\omega}+\frac{k'_c\Gamma_{cs}(\omega')k'_s}{\omega'}\frac{\Gamma_{ij}(\omega)k''_j}{\omega}\right\}. \tag{19.50}$$

We restrict our investigation of the structure of the tensor S in a cold magnetoactive plasma to these examples, and we now turn to the problem of allowing for thermal motion.

Before we consider the general expression for the tensor S_{ijs} in a hot magnetoactive plasma, let us consider its longitudinal contraction. We contract the nonrelativistic expression (19.5) for the tensor ε_{ijs} with the longitudinal tensor $\varkappa_i\varkappa''_j\varkappa'_s$ in (7.66) and add it to the similar contraction of the tensor ε_{isj}. We then obtain an expression for the longitudinal contraction of S in a hot ($v_T\neq 0$) magnetoactive plasma:

$$S(\omega,\ \mathbf{k};\ \omega',\ \mathbf{k}')=i\frac{4\pi e^3}{m^2}\frac{1}{kk'k''}\int d\mathbf{v}\int_{-\infty}^{0}d\tau_0 e^{-i\omega\tau_0+ik_i\rho_{in}(\tau_0)v_n}\int_{-\infty}^{0}d\tau_1\left\{k'_s\frac{\partial}{\partial v_s(\tau_0,\ \mathbf{v})}\exp[-i\omega''\tau_1+ik''_j\rho_{jn}(\tau_1)v_n(\tau_0,\ \mathbf{v})]\times\right.$$
$$\left.\times k''_j\frac{\partial f_0}{\partial v_j(\tau_1,\ \mathbf{v}(\tau_0,\ \mathbf{v}))}+k''_j\frac{\partial}{\partial v_j(\tau_0,\ \mathbf{v})}\exp[-i\omega'\tau_1+ik'_s\rho_{sn}(\tau_1)v_n(\tau_0,\ \mathbf{v})]k'_s\frac{\partial f_0}{\partial v_s(\tau_1,\ \mathbf{v}(\tau_0,\ \mathbf{v}))}\right\}. \tag{19.51}$$

Here, we have gone over from the vectors $\delta\mathbf{R}$ to the tensors ρ_{ij} in accordance with the relating definition (16.14) and we have used equations of the form

$$\alpha''_{nj}(\tau_0)k''_j=k''_n \tag{19.52}$$

and the fact that in accordance with the equations of continuity of the nonlinear additions to the current and the charge (6.5) the fraction $\mathbf{kv}/\omega$ in the integrand of the integral with respect to the velocity $\mathbf{v}$ in the contractions of the tensors ε can be equated formally to unity. Integrating twice by parts with respect to the $d\mathbf{v}$ in (18.51) and also making the change of variables $\mathbf{v}(\tau_0, \mathbf{v}) \rightleftharpoons \mathbf{v}$ (the Jacobian of the transformation is equal to unity), we obtain

$$S(\omega, \mathbf{k}; \omega', \mathbf{k}') = -i\frac{4\pi e^3}{m^2}\frac{1}{kk'k''}\int d\mathbf{v} f_0 \int_{-\infty}^{0} d\tau_0 \int_{-\infty}^{0} d\tau_1 e^{-i\omega\tau_0 + ik_i\rho_{ni}(\tau_0)v_n} \times$$
$$\times \{e^{-i\omega''\tau_1 + ik''_j\rho_{jn}(\tau_1)v_n}[k'_s\rho_{si}(\tau_0)k_i][k''_j\rho_{ja}(\tau_0+\tau_1)k_a - k''_j\rho_{jc}(\tau_1)k'_c] +$$
$$+ e^{-i\omega'\tau_1 + ik'_s\rho_{sn}(\tau_1)v_n}[k''_j\rho_{ji}(\tau_0)k_i][k'_s\rho_{sa}(\tau_0+\tau_1)k_a - k'_s\rho_{sb}(\tau_1)k''_b]\}. \qquad (19.53)$$

The same equation can be obtained if, instead of the general expression (19.5) one uses the contraction of the tensor ε_{ijs} integrated by parts in (19.10). As in the case of the tensor S_{ijs} in a cold magnetoactive plasma (19.11), there are terms of two kinds on the right-hand side of (19.53) with separated and nonseparated variables of integration τ_0 and τ_1. We transform the latter by means of the identity

$$\int d\mathbf{v} f_0 \int_{-\infty}^{0} d\tau_0 \int_{-\infty}^{0} d\tau_1 e^{-i\omega\tau_0 + ik_i\rho_{ni}(\tau_0)v_n}\{A_{ij}(-\tau_0)B_{as}(-\tau_0-\tau_1) \times$$
$$\times e^{-i\omega''\tau_1 + ik''_j\rho_{jn}(\tau_1)v_n} + A_{ij}(-\tau_0-\tau_1)B_{as}(-\tau_0)e^{-i\omega'\tau_1 + ik'_s\rho_{sn}(\tau_1)v_n}\} \equiv$$
$$\equiv \int d\mathbf{v} f_0 \int_{-\infty}^{0} d\tau_1 e^{-i\omega'\tau_1 + ik'_s\rho_{sn}(\tau_1)v_n} \int_{-\infty}^{0} d\tau_2 e^{-i\omega''\tau_2 + ik''_j\rho_{jn}(\tau_2)v_n} A_{ij}(-\tau_2)B_{as}(-\tau_1), \qquad (19.54)$$

which directly generalizes (19.12) to the case of a hot plasma. In proving this we have used the equation (7.3) for the frequencies and wave vectors, the identities

$$-i\omega\tau_0 - i\omega''\tau_1 + ik_i\rho_{ni}(\tau_0)v_n + ik''_j\rho_{jn}(\tau_1)v_n = -i\omega''(\tau_0+\tau_1) +$$
$$+ ik''_j\rho_{jm}(\tau_0+\tau_1)v_m(-\tau_0, \mathbf{v}) - i\omega'\tau_0 + ik'_s\rho_{sm}(\tau_0)v_m(-\tau_0, \mathbf{v}),$$
$$-i\omega\tau_0 - i\omega'\tau_1 + ik_i\rho_{ni}(\tau_0)v_n + ik'_s\rho_{sn}(\tau_1)v_n = -i\omega'(\tau_0+\tau_1) + \qquad (19.55)$$
$$+ ik'_s\rho_{sm}(\tau_0+\tau_1)v_m(-\tau_0, \mathbf{v}) - i\omega''\tau_0 + ik''_j\rho_{jm}(\tau_0)v_m(-\tau_0, \mathbf{v})$$

and the condition that the particle distribution function f_0 in the unperturbed state of the plasma be invariant under the substitution $\mathbf{v}(\tau_0, \mathbf{v}) \leftrightharpoons \mathbf{v} \leftrightharpoons \mathbf{v}(-\tau_0, \mathbf{v})$ of the variables of integration with respect to the velocity. This last condition is satisfied, for example, for arbitrary f_0 that depends on the integrals $\mathbf{v}_z$ and $\mathbf{v}_\perp$ and, in particular, for the isotropic Maxwell distribution, which is used throughout in all specific special cases.

Using the identity (19.54) to simplify the first and the third terms on the right-hand side of (19.53), we obtain a symmetric representation for the longitudinal contraction of the tensor S in a hot magnetoactive plasma:

$$S(\omega, \mathbf{k}; \omega', \mathbf{k}') = -i\frac{4\pi e^3}{m^2}\frac{1}{kk'k''}\int d\mathbf{v} f_0 \int_{-\infty}^{0} d\tau_0 \int_{-\infty}^{0} d\tau_1 \{e^{-i\omega'\tau_1 + ik'_s\rho_{sn}(\tau_1)v_n} e^{-i\omega''\tau_0 + ik''_j\rho_{jn}(\tau_0)v_n}[k'_s\rho_{si}(\tau_1)k_i][k''_j\rho_{ja}(\tau_0)k_a] -$$
$$- e^{-i\omega\tau_0 + ik_i\rho_{ni}(\tau_0)v_n} e^{-i\omega''\tau_1 + ik''_j\rho_{jn}(\tau_1)v_n}[k'_s\rho_{si}(\tau_0)k_i][k''_j\rho_{jc}(\tau_1)k'_c] -$$
$$- e^{-i\omega\tau_0 + ik_i\rho_{ni}(\tau_0)v_n} e^{-i\omega'\tau_1 + ik'_s\rho_{sn}(\tau_1)v_n}[k''_j\rho_{ji}(\tau_0)k_i][k'_s\rho_{sb}(\tau_1)k''_b]\}. \qquad (19.56)$$

Going over here to the limit of a weak external magnetic field ($\mathbf{B}_0 = 0$) and integrating with respect to τ_0 and τ_1, we arrive at a symmetric longitudinal contraction (7.31) in an isotropic plasma, since [see (6.18)]

$$k_i\rho_{ni}(\tau_0) = k_n\tau_0, \quad \mathbf{B}_0 = 0. \qquad (19.57)$$

In a very strong magnetic field ($\mathbf{B}_0 \to \infty$) it follows in accordance with the auxiliary relation

$$k_i \rho_{ni}(\tau_0) = k_z h_n \tau_0, \qquad \mathbf{B}_0 \to \infty, \tag{19.58}$$

that the contraction (19.56) is identical with the one-dimensional expression (7.35) in a hot isotropic plasma, which reduces after integration with respect to $\mathbf{v}$ in the case of the Maxwell distribution to (7.46) and (7.48). Physically, this case corresponds to the zeroth approximation in the Larmor radii $\rho = v_T/\Omega$:

$$k\rho, \quad k'\rho, \quad k''\rho \ll 1 \tag{19.59}$$

and low frequencies ω, ω', ω'' compared with the gyroscopic frequencies (19.25). In a cold magnetoactive plasma ($v_T = 0$, $\mathbf{B}_0 = \mathrm{const} \neq 0$) the contraction (19.56) is identical with (19.22).

The contraction (19.56) can be represented in a different, no less useful form by using equations of the type

$$\begin{aligned} -i\omega\tau + ik_i\rho_{in}(\tau)v_n &= -i(\omega - k_z v_z)\tau + i\frac{k_\perp v_\perp}{\Omega}\sin(\Omega\tau + \theta - \varphi) - i\frac{k_\perp v_\perp}{\Omega}\sin(\theta - \varphi), \\ -i\omega\tau + ik_i\rho_{ni}(\tau)v_n &= -i(\omega - k_z v_z)\tau + i\frac{k_\perp v_\perp}{\Omega}\sin(\Omega\tau - \theta + \varphi) + i\frac{k_\perp v_\perp}{\Omega}\sin(\theta - \varphi), \end{aligned} \tag{19.60}$$

obtained directly from the definition (16.15) of the tensor ρ on the transition to a right-handed coordinate system with z axis along $\mathbf{B}_0$ and x and y axes in the plane perpendicular to $\mathbf{h}$. As before, the subscript z is appended to the projections of the vectors $\mathbf{k}$, $\mathbf{k}'$, $\mathbf{k}''$, $\mathbf{v}$ onto the z axis (i.e., along the direction of $\mathbf{B}_0$) and the subscript $\perp$ corresponds to the projections of these vectors onto the xy plane, and θ, θ', θ'', and φ are the angles between the x axis and the components $\mathbf{k}_\perp$, $\mathbf{k}'_\perp$, $\mathbf{k}''_\perp$, and $\mathbf{v}_\perp$ of the vectors $\mathbf{k}$, $\mathbf{k}'$, $\mathbf{k}''$, and $\mathbf{v}$ in the xy plane. Substituting Eqs. (19.60) into (19.56) and using the decomposition (16.63), we obtain for the longitudinal contraction of S in a hot magnetoactive plasma the representation

$$\begin{aligned} S(\omega, \mathbf{k}; \omega', \mathbf{k}') = &-i\frac{e}{m}\frac{4\pi e^2}{m}\frac{k_i k''_j k'_s}{kk'k''}\int d\mathbf{v} f_0 \sum_{n,\,l,\,m=-\infty}^{+\infty} J_l\left(\frac{k''_\perp v_\perp}{\Omega}\right) J_n\left(\frac{k'_\perp v_\perp}{\Omega}\right) \times \\ &\times J_m\left(\frac{k_\perp v_\perp}{\Omega}\right)\exp\{il\theta'' + in\theta' - im\theta\}\exp\{-il\varphi - in\varphi + im\varphi\} \times \\ &\times \int_{-\infty}^{0} d\tau_0 \int_{-\infty}^{0} d\tau_1 \{e^{-i\omega_2\tau_0 - i\omega_1\tau_1}\rho_{ij}(-\tau_0)k_a\rho_{as}(-\tau_1) - e^{-i\omega_0\tau_0 - i\omega_2\tau_1}\rho_{is}(-\tau_0) \times \\ &\times k'_c\rho_{cj}(-\tau_1) - e^{-i\omega_0\tau_0 - i\omega_1\tau_1}\rho_{ij}(-\tau_0)k''_b\rho_{bs}(-\tau_1)\}. \end{aligned} \tag{19.61}$$

Here we have used the notation [cf. (7.17) and (7.36)]

$$\omega_0 = \omega - k_z v_z - m\Omega, \qquad \omega_1 = \omega' - k'_z v_z - n\Omega, \qquad \omega_2 = \omega'' - k''_z v_z - l\Omega \tag{19.62}$$

and the equation

$$k_\perp v_\perp \sin(\theta - \varphi) = k'_\perp v_\perp \sin(\theta' - \varphi) + k''_\perp v_\perp \sin(\theta'' - \varphi), \tag{19.63}$$

by means of which the number of Bessel functions in the expansion of the Green's function in (19.56) in accordance with (19.60) and (16.63) in a general term in the series n, l, m is reduced to three. Indeed, (19.63) is equivalent to the equation

$$\exp\left\{i\frac{k_\perp v_\perp}{\Omega}\sin(\theta - \varphi)\right\} = \exp\left\{i\frac{k'_\perp v_\perp}{\Omega}\sin(\theta' - \varphi)\right\}\exp\left\{i\frac{k''_\perp v_\perp}{\Omega}\sin(\theta'' - \varphi)\right\},$$

from which, using (16.63), we obtain the composition theorem for the Bessel functions:

$$J_m\left(\frac{k_\perp v_\perp}{\Omega}\right)e^{im(\theta-\varphi)} = \sum_{n=-\infty}^{+\infty} J_n\left(\frac{k'_\perp v_\perp}{\Omega}\right) J_{m-n}\left(\frac{k''_\perp v_\perp}{\Omega}\right)e^{in(\theta'-\varphi)+i(m-n)(\theta''-\varphi)}. \tag{19.64}$$

In the special case of an isotropic Maxwell distribution f_0 of the particles we can integrate with respect to the azimuthal angles φ of the velocity $\mathbf{v}$ on the right-hand side of the contraction (19.61) and represent it in the simpler form

$$S(\omega, \mathbf{k}; \omega', \mathbf{k}') = -i\frac{e}{m}\omega_L^2 \frac{k_i k''_j k'_s}{kk'k''} \int\limits_{-\infty}^{+\infty} \frac{dv_z}{\sqrt{2\pi}v_T} e^{-\frac{v_z^2}{2v_T^2}} \sum_{n,\, l=-\infty}^{+\infty} A_{n,\, l} \times$$
$$\times \left\{ \frac{\Gamma_{ij}(\omega_2)}{\omega_2} k_a \frac{\Gamma_{as}(\omega_1)}{\omega_1} - \frac{\Gamma_{is}(\omega_0)}{\omega_0} k'_c \frac{\Gamma_{cj}(\omega_2)}{\omega_2} - \frac{\Gamma_{ij}(\omega_0)}{\omega_0} k''_b \frac{\Gamma_{bs}(\omega_1)}{\omega_1} \right\} \qquad (19.65)$$

by means of the integrals

$$\frac{1}{2\pi}\int\limits_0^{2\pi} d\varphi e^{i(m-l-n)\varphi} = \delta_{m,\, l+n}, \qquad \int\limits_{-\infty}^{0} d\tau e^{-i\omega\tau}\rho_{ij}(-\tau) = -\frac{\Gamma_{ij}(\omega)}{\omega}. \qquad (19.66)$$

Here, in contrast to (19.61), the denominators ω_0, ω_1, ω_2 are defined in accordance with Eqs. (19.62), but already for $m = l + n$ in accordance with the first integral of (19.66); $A_{n,l}$ is given by the equation

$$A_{n,\, l} \equiv \int\limits_0^{\infty} \frac{v_\perp dv_\perp}{v_T^2} e^{-\frac{v_\perp^2}{2v_T^2}} J_n\left(\frac{k'_\perp v_\perp}{\Omega}\right) J_l\left(\frac{k''_\perp v_\perp}{\Omega}\right) J_{l+n}\left(\frac{k_\perp v_\perp}{\Omega}\right) \exp\{il\theta'' + in\theta' - i(l+n)\theta\}. \qquad (19.67)$$

Note that a longitudinal contraction of this kind can be obtained differently by expanding with respect to Bessel functions already in the asymmetric representation (19.53), integrating with respect to φ, and applying Eq. (19.12) with the frequencies ω, ω', ω'', replaced by ω_0, ω_1, ω_2 (we recall that $\omega_0 = \omega_1 + \omega_2$) for symmetrization. On account of the importance of the representation (19.65), we rewrite it in the terms of $\delta\varepsilon_{ij}$ [cf. (19.23)]:

$$S(\omega, \mathbf{k}; \omega', \mathbf{k}') = -i\frac{e}{m}\frac{\omega_L^{-2}}{kk'k''} \int\limits_{-\infty}^{+\infty} \frac{dv_z}{\sqrt{2\pi}v_T} e^{-\frac{v_z^2}{2v_T^2}} \sum_{n,\, l=-\infty}^{+\infty} A_{n,\, l} \{[k_i \delta\varepsilon_{ij}(\omega_2) k''_j][k_a \delta\varepsilon_{as}(\omega_1) k'_s] -$$
$$- [k_i \delta\varepsilon_{is}(\omega_0) k'_s][k'_c \delta\varepsilon_{cj}(\omega_2) k''_j] - [k_i \delta\varepsilon_{ij}(\omega_0) k''_j][k''_b \delta\varepsilon_{bs}(\omega_1) k'_s]\}. \qquad (19.68)$$

In the limit of a cold plasma, the contractions (19.65) and (19.68) go over into (19.22) and (19.23) respectively; for this it is sufficient to set $v_T \equiv 0$ in them and integrate over the velocities by means of the δ function that then arises:

$$\delta(\mathbf{v}) = \lim_{v_T \to 0} \frac{1}{(2\pi)^{3/2} v_T^3} e^{-\frac{\mathbf{v}^2}{2v_T^2}}. \qquad (19.69)$$

Of course, formulas of the type (19.65) and (19.68) could also be obtained not only for an isotropic Maxwell distribution f_0 but also for any $f_0(v_\perp, v_z)$:

$$S(\omega, \mathbf{k}; \omega', \mathbf{k}') = -i\frac{e}{m}\omega_L^2 \frac{k_i k''_j k'_s}{kk'k''} \int d\mathbf{v} f_0(v_\perp, v_z) \sum_{n,\, l=-\infty}^{+\infty} J_n\left(\frac{k'_\perp v_\perp}{\Omega}\right) \times$$
$$\times J_l\left(\frac{k''_\perp v_\perp}{\Omega}\right) J_{l+n}\left(\frac{k_\perp v_\perp}{\Omega}\right) \exp\{il\theta'' + in\theta' - i(l+n)\theta\} \times$$
$$\times \left\{ \frac{\Gamma_{ij}(\omega_2)}{\omega_2} k_a \frac{\Gamma_{as}(\omega_1)}{\omega_1} - \frac{\Gamma_{is}(\omega_0)}{\omega_0} k'_c \frac{\Gamma_{cj}(\omega_2)}{\omega_2} - \frac{\Gamma_{ij}(\omega_0)}{\omega_0} k''_b \frac{\Gamma_{bs}(\omega_1)}{\omega_1} \right\}, \qquad (19.70)$$

if one integrates with respect to the azimuth φ, τ_0, and τ_1 in (19.61).

As in the case of the linear theory of a magnetoactive plasma (§ 16), for an isotropic Maxwell distribution f_0 we can integrate on the right-hand side of Eq. (19.56) with respect to the velocities by means of the integral (16.27), which gives a further form of expression of the longitudinal contraction in a hot plasma in the form of double integrals with respect to τ_0 and τ_1 [compare the expression for the permittivity tensor (16.29)]:

$$
\begin{aligned}
S(\omega,\ \mathbf{k};\ \omega',\ \mathbf{k}') = & -i\frac{e}{m}\omega_L^2\frac{k_i k''_j k'_s}{kk'k''}\int_{-\infty}^{0} d\tau_0 \int_{-\infty}^{0} d\tau_1 \times \\
& \times\Big\{\exp\Big[-i\omega'\tau_1 - i\omega''\tau_0 - \frac{1}{2}v_T^2\left(k'_s\rho_{sm}(\tau_1) + k''_j\rho_{jm}(\tau_0)\right)^2\Big]\rho_{ij}(-\tau_0)\,k_a\rho_{as}(-\tau_1) - \\
& - \exp\Big[-i\omega\tau_0 - i\omega''\tau_1 - \frac{1}{2}v_T^2\left(k_i\rho_{mi}(\tau_0) + k''_j\rho_{jm}(\tau_1)\right)^2\Big]\rho_{is}(-\tau_0)\,k'_c\rho_{cj}(-\tau_1) - \\
& - \exp\Big[i\omega\tau_0 - i\omega'\tau_1 - \frac{1}{2}v_T^2\left(k_i\rho_{mi}(\tau_0) + k'_s\rho_{sm}(\tau_1)\right)^2\Big]\rho_{ij}(-\tau_0)\,k''_b\rho_{bs}(-\tau_1)\Big\}. \qquad (19.71)
\end{aligned}
$$

If we here take $v_T = 0$, then after integration with respect to τ_0 and τ_1 we obtain the contraction of S for a cold magnetoactive plasma, but if $\mathbf{B}_0 = 0$, then we obtain an expression for the isotropic contraction in a representation that is slightly different from that used in § 7, which was devoted to the study of the tensor S_{ijs} in an isotropic plasma:

$$
\begin{aligned}
S(\omega,\ \mathbf{k};\ \omega',\ \mathbf{k}') = & -i\frac{e}{m}\frac{\omega_L^2}{kk'k''}\int_{-\infty}^{0} d\tau_0\tau_0 \int_{-\infty}^{0} d\tau_1\tau_1\ \Big\{(\mathbf{k}\mathbf{k}')(\mathbf{k}\mathbf{k}'')\exp\Big[-i\omega'\tau_1 - i\omega''\tau_0 - \frac{1}{2}v_T^2(\mathbf{k}'\tau_1 + \mathbf{k}''\tau_0)^2\Big] - \\
& - (\mathbf{k}\mathbf{k}')(\mathbf{k}'\mathbf{k}'')\ \mathrm{epx}\Big[-i\omega\tau_0 - i\omega''\tau_1 - \frac{1}{2}v_T^2(\mathbf{k}\tau_0 + \mathbf{k}''\tau_1)^2\Big] - \\
& - (\mathbf{k}\mathbf{k}'')(\mathbf{k}'\mathbf{k}'')\exp\Big[-i\omega\tau_0 - i\omega'\tau_1 - \frac{1}{2}v_T^2(\mathbf{k}\tau_0 + \mathbf{k}'\tau_1)^2\Big]\Big\}, \quad \mathbf{B}_0 = 0. \qquad (19.72)
\end{aligned}
$$

Obviously, this additional form of expression of the relations (7.28) and (7.31) from § 7 is identical with those obtained by means of (15.38) and (16.27).

For studying different special cases of the longitudinal contraction by means of (19.65) it is helpful to use the modification

$$
\begin{aligned}
S(\omega,\ \mathbf{k};\ \omega',\ \mathbf{k}') = & -i\frac{e}{m}\frac{\omega_L^2}{kk'k''}\int_{-\infty}^{+\infty}\frac{dv_z}{\sqrt{2\pi}v_T}e^{-\frac{v_z^2}{2v_T^2}}\sum_{n,\ l=-\infty}^{+\infty}\Big\{\Big\{k_z k'_z k''_z - \\
& - i(\mathbf{k}[\mathbf{k}'\mathbf{h}])\Big(\frac{k_z\omega' - k'_z\omega}{\Omega} + k'_z l - k''_z n\Big)\Big\}\Big\{\frac{k_z}{(\omega_1\omega_2)^2} - \frac{k'_z}{(\omega_0\omega_2)^2} - \frac{k''_z}{(\omega_0\omega_1)^2}\Big\}A_{n,\ l} - \\
& - \frac{1}{2\Omega}\sum_{\lambda}\Big\{\frac{k''_z}{\omega_2^2}\Big(\frac{k_z}{\omega_1} - \frac{k'_z}{\omega_0}\Big)(k_i a_{is} k'_s \delta_{\lambda,\ 1} - k'_s a_{si} k_i \delta_{\lambda,\ -1})A_{n+\lambda,\ l} + \\
& + \frac{k'_z}{\omega_1^2}\Big(\frac{k_z}{\omega_2} - \frac{k''_z}{\omega_0}\Big)(k_i a_{ij} k''_j \delta_{\lambda,\ 1} - k''_j a_{ji} k_i \delta_{\lambda,\ -1})A_{n,\ l+\lambda} - \frac{k_z}{\omega_0^2}\Big(\frac{k'_z}{\omega_2} + \frac{k''_z}{\omega_1}\Big)\times \\
& \times(k'_s a_{sj} k''_j \delta_{\lambda,\ 1} - k''_j a_{js} k'_s \delta_{\lambda,\ -1})A_{n-\lambda,\ l+\lambda}\Big\} + \frac{1}{4\Omega^2}\sum_{\lambda,\ \nu}A_{n+\nu,\ l+\lambda}\Big\{\frac{1}{\omega_1\omega_2}(k_i a_{ij} k''_j \delta_{\lambda,\ 1} - \\
& - k''_j a_{ji} k_i \delta_{\lambda,\ -1})(k_a a_{as} k'_s \delta_{\nu,\ 1} - k'_s a_{sa} k_a \delta_{\nu,\ -1}) - \frac{1}{\omega_0\omega_2}(k_i a_{is} k'_s \delta_{\lambda+\nu,\ 1} - k'_s a_{si} k_i \delta_{\lambda+\nu,\ -1})(k'_c a_{cj} k''_j \delta_{\lambda,\ 1} - k''_j a_{jc} k'_c \delta_{\lambda,\ -1}) - \\
& - \frac{1}{\omega_0\omega_1}(k_i a_{ij} k''_j \delta_{\lambda+\nu,\ 1} - k''_j a_{ji} k_i \delta_{\lambda+\nu,\ -1})(k''_b a_{bs} k'_s \delta_{\nu,\ 1} - k'_s a_{sb} k''_b \delta_{\nu,\ -1})\Big\}\Big\}. \qquad (19.73)
\end{aligned}
$$

It is readily obtained by doing without the quantities Γ_{ij} by using the relations

$$
\begin{aligned}
\Gamma_{ij}(\omega) &= \omega\sum_{\nu} O^{\nu}_{ij}\frac{1}{\omega + \nu\Omega}, \\
O^{\nu}_{ij} &\equiv \frac{h_i h_j}{\omega}\delta_{\nu,\ 0} + i\frac{e_{isj}h_s}{\Omega}\delta_{\nu,\ 0} - \frac{1}{2}\frac{a_{ij}}{\Omega}\delta_{\nu,\ 1} + \frac{1}{2}\frac{a_{ji}}{\Omega}\delta_{\nu,\ -1}, \qquad (19.74) \\
a_{ij} &\equiv \delta_{ij} - h_i h_j + ie_{isj}h_s
\end{aligned}
$$

and shifting the summation indices n and l. The contractions of the tensors a_{ij} in (19.73) can be represented in more detail by the equations

$$\begin{aligned} k_i a_{is} k'_s &= [\mathbf{kh}][\mathbf{k'h}] + i(\mathbf{k'}[\mathbf{kh}]), \\ k_i a_{ij} k''_j &= [\mathbf{kh}][\mathbf{k''h}] - i(\mathbf{k'}[\mathbf{kh}]), \\ k'_s a_{sj} k''_j &= [\mathbf{k'h}][\mathbf{k''h}] - i(\mathbf{k'}[\mathbf{kh}]). \end{aligned} \tag{19.75}$$

Summarizing our analysis of the different symmetric representations of the longitudinal contraction of S in a hot magnetoactive plasma, we give a symmetry relation for this contraction that follows formally from the equation (19.19) for the tensor S_{ijs} when there is no thermal motion:

$$S(\omega, \mathbf{k};\ \omega', \mathbf{k}';\ \mathbf{B}_0) = -S(\omega'', \mathbf{k}'';\ \omega, \mathbf{k};\ -\mathbf{B}_0). \tag{19.76}$$

For the study of decay processes of quasilongitudinal waves, the contributions from the resonance parts of the expressions ω_0^{-1}, ω_1^{-1}, ω_2^{-1} are omitted by definition, so that the corresponding Cauchy type integrals are taken with respect to the velocity projection v_z in the sense of the principal value. In this sense, we have the different symmetry relation

$$PS(\omega, \mathbf{k};\ \omega', \mathbf{k}') = PS^*(\omega'', \mathbf{k}'';\ \omega, \mathbf{k}), \tag{19.77}$$

in which the letter P in front of the contraction reflects the fact that we have separated out from the contraction only those terms that contain the principal parts of the Cauchy-type integrals.

We now turn to the study of all possible exppressions for the longitudinal contraction of S_{ijs} under different restrictions on the thermal motion of the particles, the magnetic field $\mathbf{B}_0$, and the frequencies and the wave lengths of the interacting oscillations. These restrictions arise naturally in the nonlinear theory of oscillations when one considers concrete nonlinear processes and are none other than the conditions for the existence of (weakly damped) characteristic oscillations given in the foregoing section. To shorten the formulas, we denote

$$S \equiv S(\omega, \mathbf{k};\ \omega', \mathbf{k}') = S_{ijs}(\omega, \mathbf{k};\ \omega', \mathbf{k}')\frac{k_i k''_j k'_s}{k k' k''}. \tag{19.78}$$

In the limit of low frequencies (19.25) of the interacting oscillations (ω, ω', $\omega'' \ll \Omega$) and long wavelengths in the direction of $\mathbf{B}_0$ compared with the Larmor radii

$$k_z\rho,\ k'_z\rho,\ k''_z\rho \ll 1 \tag{19.79}$$

we obtain from the representation (19.73) the following expression for the longitudinal contraction ($\omega_0 \equiv \omega - k_z v_z$, $\omega_1 \equiv \omega' - k'_z v_z$, $\omega_2 \equiv \omega'' - k''_z v_z$):

$$S = -i\frac{e}{m}\frac{\omega_L^2}{kk'k''}\int\limits_{-\infty}^{+\infty}\frac{dv_z}{\sqrt{2\pi}v_T}e^{-\frac{v_z^2}{2v_T^2}}A_{0,0}\left\{k_z k'_z k''_z + i(\mathbf{k}[\mathbf{k'h}])\frac{k'_z\omega - k_z\omega'}{\Omega}\right\}\left\{\frac{k_z}{(\omega_1\omega_2)^2} - \frac{k'_z}{(\omega_0\omega_2)^2} - \frac{k''_z}{(\omega_0\omega_1)^2}\right\} \tag{19.80}$$

or, after integration with respect to the longitudinal velocities,

$$S = i\frac{e}{m}\frac{\omega_L^2}{kk'k''}\frac{1}{v_T^2}\frac{A_{0,0}}{(\omega' k_z - \omega k'_z)}\left\{\frac{k_z k'_z k''_z}{\omega' k_z - \omega k'_z} - i\frac{(\mathbf{k}[\mathbf{k'h}])}{\Omega}\right\}\left\{k_z J_+\left(\frac{\omega}{|k_z| v_T}\right) - k'_z J_+\left(\frac{\omega'}{|k'_z| v_T}\right) - k''_z J_+\left(\frac{\omega''}{|k''_z| v_T}\right)\right\}. \tag{19.81}$$

In the limit of a very strong magnetic field ($B_0 \to \infty$), this expression goes over into Eq. (7.46) which arises in the case of a one-dimensional interaction, since the second term in the first curly brackets of (19.81) is then equal to zero and the integral $A_{0,0}$ [see the definition (19.67)] is equal to unity. In the integration of (19.80) we have assumed that the phase velocities ω/k_z, ω'/k'_z, ω''/k''_z of the oscillations are different, i.e. [see (7.37)],

$$\omega' k_z - \omega k'_z \neq 0. \tag{19.82}$$

In the opposite case (7.39), the contraction (19.80) can be reduced to the different expression

$$S = \frac{i}{2}\frac{e}{m}\frac{\omega_L^2}{kk'k''}\frac{A_{0,0}}{v_T^4}\left\{\left(3 - \frac{\omega^2}{k_z^2 v_T^2}\right) J_+\left(\frac{\omega}{|k_z| v_T}\right) + \left(\frac{\omega^2}{k_z^2 v_T^2} - 2\right)\right\}, \tag{19.83}$$

which differs from (7.48) only by the factor $A_{0,0}$ and goes over into (7.48) in the case of long wavelengths of the oscillations at right angles to the magnetic compared with the Larmor radii:

$$k_\perp \rho,\ k'_\perp \rho,\ k''_\perp \rho \ll 1. \tag{19.84}$$

From (19.81) in the case of weak thermal motion:

$$\frac{\omega}{k_z},\ \frac{\omega'}{k'_z},\ \frac{\omega'}{k''_z} \gg v_T \tag{19.85}$$

we obtain, using the asymptotic behavior (6.32) of the function J_+, a more specialized form of the longitudinal contraction:

$$S = i\frac{e}{m}\frac{\omega_L^2}{kk'k''}\frac{A_{0,0}}{(\omega' k_z - \omega k'_z)^2}\left\{k_z k'_z k''_z + i(\mathbf{k}[\mathbf{k}'\mathbf{h}])\frac{\omega k'_z - \omega' k_z}{\Omega}\right\}\left\{\left[\frac{k_z^3}{\omega^2} - \frac{k_z'^3}{\omega'^2} - \frac{k_z''^3}{\omega''^2}\right] - \right.$$
$$\left. - i\sqrt{\frac{\pi}{2}}\frac{1}{v_T^3}\left[\omega \operatorname{sgn} k_z \exp\left(-\frac{1}{2}\frac{\omega^2}{k_z^2 v_T^2}\right) - \omega' \operatorname{sgn} k'_z \exp\left(-\frac{1}{2}\frac{\omega'^2}{k_z'^2 v_T^2}\right) - \omega'' \operatorname{sgn} k''_z \exp\left(-\frac{1}{2}\frac{\omega''^2}{k_z''^2 v_T^2}\right)\right]\right\}. \tag{19.86}$$

Here the first term in the second curly brackets can be expressed differently, by using the identities (7.58). In particular, ignoring the imaginary terms in the second curly brackets in (19.86) which is of interest in the case of decay processes, we obtain a simpler equation for S:

$$S = -i\frac{e}{m}\frac{\omega_L^2}{\omega\omega'\omega''}\frac{A_{0,0}}{kk'k''}\left\{k_z k'_z k''_z + i(\mathbf{k}[\mathbf{k}'\mathbf{h}])\frac{\omega k'_z - \omega' k_z}{\Omega}\right\}\left\{\frac{k_z}{\omega} + \frac{k'_z}{\omega'} + \frac{k''_z}{\omega''}\right\}, \tag{19.87}$$

this holding – by continuity considerations – not only for different but also equal phase velocities (7.39). Here, the second term in the first curly brackets can be ignored when, for example, the wave numbers of the oscillations along and at right angles to the external magnetic field are of the same order of magnitude. Note that the right-hand side of (19.87) without the factor $A_{0,0}$ can also be obtained from the longitudinal contraction (19.22) in a cold plasma.

We also give some longitudinal contractions of S that ignore the contribution from the poles of the Green's functions, i.e., we calculate the integrals in S in the sense of the principal value. Such expressions, as we have already mentioned on more than one occassion, are very helpful for the study of decay processes.

If the oscillation frequencies are near the gyroscopic:

$$\omega \simeq (l+n)\Omega, \qquad \omega' \simeq n\Omega, \qquad \omega'' \simeq l\Omega \tag{19.88}$$

and cyclotron damping is weak:

$$\begin{gathered}\Omega \gg \omega - (l+n)\Omega \gg k_z v_T, \qquad \Omega \gg \omega' - n\Omega \gg k'_z v_T, \\ \Omega \gg \omega'' - l\Omega \gg k''_z v_T,\end{gathered} \tag{19.89}$$

then in accordance with (19.73) the longitudinal contraction is

$$S = -i\frac{e}{m}\frac{\omega_L^2}{kk'k''}\frac{A_{n,l}}{[\omega - (l+n)\Omega][\omega' - n\Omega][\omega'' - l\Omega]}\left\{\frac{k_z}{\omega - (l+n)\Omega} + \right.$$
$$\left. + \frac{k'_z}{\omega' - n\Omega} + \frac{k''_z}{\omega'' - l\Omega}\right\}\left\{k_z k'_z k''_z - i\frac{(\mathbf{k}[\mathbf{k}'\mathbf{h}])}{\Omega}[k_z(\omega' - n\Omega) - k'_z(\omega - l\Omega - n\Omega)]\right\} \tag{19.90}$$

and it differs from its low-frequency analog (19.87) by the substitution $\omega \to \omega - (l + n)\Omega$, $\omega' \to \omega' - n\Omega$, $\omega'' \to \omega'' - l\Omega$ and the factor $A_{n,l}$. Note that the exponential containing the angles θ, θ', θ'' in $A_{n,l}$ does not occur in the kernel of the integrodifferential equation that describes decay since the kernel is determined by the squared modulus of the contraction: $|S^2|$ (see § 21).

The right-frequency limit (19.26) of the longitudinal contraction in the case of weak thermal motion [cf. (19.85)]:

$$\frac{\omega}{k}, \frac{\omega'}{k'}, \frac{\omega''}{k''} \gg v_T \tag{19.91}$$

for decay processes is equal to the isotropic expression (7.30) in a cold plasma ($v_T = 0$). This conclusion follows readily from the representations (19.56) and (19.71) of the longitudinal contraction. Formulas (19.65) and (19.73) are less convenient for this purpose, since, on the transition to the isotropic limit, one must use in them a different asymptotic behavior for the Bessel functions depending on whether the index or the argument is larger (cf. [60], p. 75). To conclude our treatment of the various properties of the longitudinal contraction, we give a further relation for S in a cold plasma for the case when the motion of the particles of all the plasma components except one can be ignored. Such a situation is realized, for example in an electron plasma, for which the great mass of the ions ($M = \infty$) enables one to ignore the ion motion, so that the ion component represents a homogeneous positive background that neutralizes the electron component. Then S in this case is in accordance with (19.23) a sum of three terms each of which is proportional to the product of two contractions of the tensor $\delta\varepsilon_{ij}$ of the electron polarizability with the wave vectors of the interacting oscillations. In the case of the decay of longitudinal oscillations we obtain in accordance with the field equations (2.13)

$$-\delta\varepsilon^e_{ij}(\omega) k_j = k_i, \qquad -\delta\varepsilon^e_{ij}(\omega') k'_j = k'_i, \qquad -\delta\varepsilon^e_{ij}(\omega'') k''_j = k''_i, \tag{19.92}$$

which means that the magnetic field drops out of S:

$$S_e = -i \frac{e}{m} \frac{\omega_L^{-2}}{kk'k''} \{(\mathbf{kk}'')(\mathbf{kk}') - (\mathbf{kk}')(\mathbf{k}'\mathbf{k}'') - (\mathbf{kk}'')(\mathbf{k}'\mathbf{k}'')\}. \tag{19.93}$$

Then, applying the identity (7.54a), we reduce the contraction S_e in an electron plasma for decay processes to the simple form

$$S_e = -i \frac{e}{m} \frac{1}{\omega^2_{L_e}} \frac{[\mathbf{kk}']^2}{kk'k''} \tag{19.94}$$

and it ceases to depend on the electron mass:

$$S_e = -\frac{i}{4\pi e N_e} \frac{[\mathbf{kk}']^2}{kk'k''}. \tag{19.95}$$

In an isotropic electron plasma, this relation does not hold, since the first of Eqs. (7.3) for the frequencies cannot be satisfied because the only longitudinal oscillation ($v_T = 0$) in such a plasma is the electron Langmuir oscillation ($\omega = \omega_{Le}$), which has a nondecay spectrum.

Note that in a cold electron plasma we can, using the tensor (19.18), obtain an expression similar to the longitudinal contraction (19.93) in the case of the decay of oscillations of arbitrary polarization if we use the expressions for the polarization vectors of the interacting oscillations. Then $|S|^2$ in the kernel of the integrodifferential equation must be replaced by the contraction $S_{ijs}e^*_i e''_j e'_s$ of the tensor S_{ijs} with the polarization vectors $e^*_i \equiv e_i(\omega)$, $e''_j \equiv e''_j(\omega'')$, $e'_s \equiv e'_s(\omega')$, so that in accordance with equations of the form [see (2.13)]

$$\delta\varepsilon_{ij}(\omega) e_j(\omega) = (n^2 - 1) e_i - n^2 \varkappa_i (\boldsymbol{\varkappa}\mathbf{e}), \quad n^2 \equiv \frac{c^2k^2}{\omega^2}, \quad \varkappa_i \equiv \frac{k_i}{k} \tag{19.96}$$

the contraction $S_{ijs} e_i^* e_s' e_j''$ depends on the magnetic field and the electron mass only through the polarization vectors and the expressions for the spectra.

To conclude this exposition we give a more symmetric expression for the tensor $S_{ijs}(\omega, \mathbf{k}; \omega', \mathbf{k}')$ in a hot nonrelativistic magnetoactive plasma:

$$S_{ijs}(\omega, \mathbf{k}; \omega', \mathbf{k}') = \frac{4\pi e^3}{m^2} \int d\mathbf{v} f_0 \left\{ \frac{1}{\omega} \int_{-\infty}^{0} d\tau' \int_{-\infty}^{0} d\tau'' G' G'' \alpha''_{bj}(\tau'') \alpha'_{cs}(\tau') \times \right.$$
$$\times [W_{ic}(-\tau') k_a \rho_{ab}(-\tau'') + W_{ib}(-\tau'') k_a \rho_{ac}(-\tau') - i v_i k_a \rho_{ac}(-\tau') k_n \rho_{nb}(-\tau'')] -$$
$$- \frac{1}{\omega''} \int_{-\infty}^{0} d\tau \int_{-\infty}^{0} d\tau' G G' \alpha_{ai}(-\tau) \alpha'_{cs}(\tau') [W_{aj}(-\tau) k''_b \rho_{bc}(-\tau') - W_{jc}(-\tau') \rho_{ab}(-\tau) k''_b - i v_j k''_b \rho_{bc}(-\tau') k''_n \rho_{an}(-\tau)] -$$
$$\left. - \frac{1}{\omega'} \int_{-\infty}^{0} d\tau \int_{-\infty}^{0} d\tau'' G G'' \alpha_{ai}(-\tau) \alpha''_{bj}(\tau'') [W_{as}(-\tau) k'_c \rho_{cb}(-\tau'') - W_{sb}(-\tau'') \rho_{ac}(-\tau) k'_c - i v_s k'_c \rho_{cb}(-\tau'') k'_n \rho_{an}(-\tau)] \right\}. \tag{19.97}$$

Here, the Green's functions G, G', G'' have the form

$$G \equiv \exp\{-i\omega\tau - i k_a \rho_{an}(-\tau) v_n\}, \quad G' \equiv \exp\{-i\omega'\tau' + i k'_c \rho_{cn}(\tau') v_n\},$$
$$G'' \equiv \exp\{-i\omega''\tau'' + i k''_b \rho_{bn}(\tau'') v_n\}, \tag{19.98}$$

and the remaining notation has been frequently explained. The formula (19.97) can be obtained from the expressions for the three-index tensors ε [see (19.10) for $\varepsilon_{ijs}(\omega, \mathbf{k}; \omega', \mathbf{k}')$]. For the sake of symmetrizing the expressions, it is convenient to use Eq. (19.54) and the two identities

$$k''_b \int d\mathbf{v} f_0 \int_{-\infty}^{0} d\tau \int_{-\infty}^{0} d\tau' G G' \alpha_{ai}(-\tau) \alpha'_{ns}(\tau') \{W_{aj}(-\tau) \rho_{bn}(-\tau') -$$
$$- \rho_{aj}(-\tau) W_{bn}(-\tau') + i(\omega'' - \mathbf{k}''\mathbf{v}) \rho_{aj}(-\tau) \rho_{bn}(-\tau')\} \equiv 0,$$

$$k'_c \int d\mathbf{v} f_0 \int_{-\infty}^{0} d\tau \int_{-\infty}^{0} d\tau'' G G'' \alpha_{ai}(-\tau) \alpha''_{mj}(\tau'') \{W_{as}(-\tau) \rho_{cm}(-\tau'') -$$
$$- \rho_{as}(-\tau) W_{cm}(-\tau'') + i(\omega' - \mathbf{k}'\mathbf{v}) \rho_{as}(-\tau) \rho_{cm}(-\tau'')\} \equiv 0. \tag{19.99}$$

We emphasize that the relation (19.97) for S_{ijs} holds for any unperturbed distribution function f_0 that satisfies Eq. (15.4), i.e., [see (15.31)] for an arbitrary function $f_0(v_\perp, v_z)$. From (19.97) we obtain all the expressions so far considered in the present paper for the tensor S and its contractions. For example, forming the longitudinal contraction in (19.97) with the tensor (7.66), we obtain Eq. (19.56). In the case of the one-dimensional interaction (19.34) along the magnetic field $\mathbf{B}_0$, the tensor (19.97) simplifies appreciably [ω_0, ω_1, ω_2 are defined by (7.36)]:

$$S_{ijs}(\omega, \mathbf{k}; \omega', \mathbf{k}') = -i \frac{4\pi e^3}{m^2} \int d\mathbf{v} f_0 \left\{ \frac{h_i h_j h_s}{\omega_0 \omega_1 \omega_2} \left[\frac{k_z}{\omega_0} + \frac{k'_z}{\omega_1} + \frac{k''_z}{\omega_2} \right] + \right.$$
$$+ \frac{1}{2} \frac{h_i}{\omega_0^2} \left[\frac{k''_z}{\omega''} \left(\frac{a_{sj}}{\omega_1 - \Omega} + \frac{a_{js}}{\omega_1 + \Omega} \right) + \frac{k'_z}{\omega'} \left(\frac{a_{js}}{\omega_2 - \Omega} + \frac{a_{sj}}{\omega_2 + \Omega} \right) \right] +$$
$$+ \frac{1}{2} \frac{h_j}{\omega_2^2} \left[\frac{k_z}{\omega} \left(\frac{a_{si}}{\omega_1 - \Omega} + \frac{a_{is}}{\omega_1 + \Omega} \right) - \frac{k'_z}{\omega'} \left(\frac{a_{si}}{\omega_0 - \Omega} + \frac{a_{is}}{\omega_0 + \Omega} \right) \right] +$$
$$\left. + \frac{1}{2} \frac{h_s}{\omega_1^2} \left[\frac{k_z}{\omega} \left(\frac{a_{ji}}{\omega_2 - \Omega} + \frac{a_{ij}}{\omega_2 + \Omega} \right) - \frac{k''_z}{\omega''} \left(\frac{a_{ji}}{\omega_0 - \Omega} + \frac{a_{ij}}{\omega_0 + \Omega} \right) \right] \right\}. \tag{19.100}$$

Setting here $\Omega = 0$, we arrive at the tensor (7.64), which arises in the case of a one-dimensional interaction in an isotropic plasma. In the case of an isotropic Maxwell distribution of the plasma particles (6.14), the integration over the velocities on the right-hand side of (19.100) can be eliminated on account of (7.28) and (7.46) and the relation

$$(1/N)\int d\mathbf{v} f_0 \frac{1}{(\omega - k_z v_z)^2 (\omega' - k'_z v_z)} = \frac{1}{(\omega' k_z - \omega k'_z)} \frac{1}{k_z v_T^2} \left\{ J_+\left(\frac{\omega}{|k_z| v_T}\right) - 1 \right\} - \\ - \frac{k'_z}{(\omega' k_z - \omega k'_z)^2} \left\{ \frac{k_z}{\omega} J_+\left(\frac{\omega}{|k_z| v_T}\right) - \frac{k'_z}{\omega'} J_+\left(\frac{\omega'}{|k'_z| v_T}\right) \right\}. \tag{19.101}$$

Then the tensor (19.100) takes the form ($a \equiv \omega'_z k_z - \omega k'_z = \omega k''_z - \omega'' k_z = \omega' k''_z - \omega'' k'_z$)

$$S_{ijs}(\omega, \mathbf{k}; \omega', \mathbf{k}') = i \frac{e}{m} \frac{\omega_L^2}{v_T^2} \frac{h_i h_j h_s}{a^2} \left\{ k_z J_+\left(\frac{\omega}{|k_z| v_T}\right) - k'_z J_+\left(\frac{\omega'}{|k'_z| v_T}\right) - \right. \\ \left. - k''_z J_+\left(\frac{\omega''}{|k''_z| v_T}\right) \right\} - i \frac{e}{m} \frac{\omega_L^2}{2\omega'\omega''} h_i \left\{ \frac{a}{k_z v_T^2} \left(\frac{a_{sj}}{a - \Omega k_z} + \frac{a_{js}}{a + \Omega k_z} \right) \left[J_+\left(\frac{\omega}{|k_z| v_T}\right) - 1 \right] - \right. \\ - k_z k'_z k''_z \left[\frac{a_{sj}}{(a - \Omega k_z)^2} + \frac{a_{js}}{(a + \Omega k_z)^2} \right] J_+\left(\frac{\omega}{|k_z| v_T}\right) + \frac{k'_z k''_z}{(a - \Omega k_z)^2} a_{sj} \times \\ \times \left[k'_z \frac{\omega'}{\omega' - \Omega} J_+\left(\frac{\omega' - \Omega}{|k'_z| v_T}\right) + k''_z \frac{\omega''}{\omega'' + \Omega} J_+\left(\frac{\omega'' + \Omega}{|k''_z| v_T}\right) \right] + \\ \left. + \frac{k'_z k''_z}{(a + \Omega k_z)^2} a_{js} \left[k'_z \frac{\omega'}{\omega' + \Omega} J_+\left(\frac{\omega' + \Omega}{|k'_z| v_T}\right) + k''_z \frac{\omega''}{\omega'' - \Omega} J_+\left(\frac{\omega'' - \Omega}{|k''_z| v_T}\right) \right] \right\} - \\ - i \frac{e}{m} \frac{\omega_L^2}{2\omega\omega'} h_j \left\{ \frac{a}{k''_z v_T^2} \left(\frac{a_{si}}{a - \Omega k''_z} + \frac{a_{is}}{a + \Omega k''_z} \right) \left[J_+\left(\frac{\omega''}{|k''_z| v_T}\right) - 1 \right] + \right. \\ + k_z k'_z k''_z \left[\frac{a_{si}}{(a - \Omega k''_z)^2} + \frac{a_{is}}{(a + \Omega k''_z)^2} \right] J_+\left(\frac{\omega''}{|k''_z| v_T}\right) + \\ + \frac{k_z k'_z}{(a - \Omega k''_z)^2} a_{si} \left[k'_z \frac{\omega'}{\omega' - \Omega} J_+\left(\frac{\omega' - \Omega}{|k'_z| v_T}\right) - k_z \frac{\omega}{\omega - \Omega} J_+\left(\frac{\omega - \Omega}{|k_z| v_T}\right) \right] + \\ \left. + \frac{k_z k'_z}{(a + \Omega k''_z)^2} a_{is} \left[k'_z \frac{\omega'}{\omega' + \Omega} J_+\left(\frac{\omega' + \Omega}{|k'_z| v_T}\right) - k_z \frac{\omega}{\omega + \Omega} J_+\left(\frac{\omega + \Omega}{|k_z| v_T}\right) \right] \right\} - \\ - i \frac{e}{m} \frac{\omega_L^2}{2\omega\omega''} h_s \left\{ \frac{a}{k'_z v_T^2} \left(\frac{a_{ji}}{a + \Omega k'_z} + \frac{a_{ij}}{a - \Omega k'_z} \right) \left[J_+\left(\frac{\omega'}{|k'_z| v_T}\right) - 1 \right] + \right. \\ + k_z k'_z k''_z \left[\frac{a_{ji}}{(a + \Omega k'_z)^2} + \frac{a_{ij}}{(a - \Omega k'_z)^2} \right] J_+\left(\frac{\omega'}{|k'_z| v_T}\right) + \\ + \frac{k_z k''_z}{(a + \Omega k'_z)^2} a_{ji} \left[k''_z \frac{\omega''}{\omega'' - \Omega} J_+\left(\frac{\omega'' - \Omega}{|k''_z| v_T}\right) - k_z \frac{\omega}{\omega - \Omega} J_+\left(\frac{\omega - \Omega}{|k_z| v_T}\right) \right] + \\ \left. + \frac{k_z k''_z}{(a - \Omega k'_z)^2} a_{ij} \left[k''_z \frac{\omega''}{\omega'' + \Omega} J_+\left(\frac{\omega'' + \Omega}{|k''_z| v_T}\right) - k_z \frac{\omega}{\omega + \Omega} J_+\left(\frac{\omega + \Omega}{|k_z| v_T}\right) \right] \right\}. \tag{19.102}$$

In the limit of a vanishing magnetic field ($\mathbf{B}_0 = 0$), this expression coincides with the right-hand side of Eq. (7.65) calculated in an isotropic plasma. In the general case of a three-dimensional nonlinear interaction in the same limit ($\mathbf{B}_0 = 0$), the tensor (19.97) yields a further modification of the three-index tensor S_{ijs} in a hot isotropic plasma [see (7.16) and (7.17)]:

$$S_{ijs}(\omega, \mathbf{k}; \omega', \mathbf{k}') = -i \frac{4\pi e^3}{m^2} \frac{1}{\omega\omega'\omega''} \int d\mathbf{v} f_0 \left\{ \beta''_{bj} \beta'_{cs} \left(\frac{k_b}{\omega_2} \delta_{ic} + \frac{k_c}{\omega_1} \delta_{ib} + v_i \frac{k_c k_b}{\omega_1 \omega_2} \right) + \right. \\ \left. + \beta_{ai} \beta'_{cs} \left(\frac{k''_a}{\omega_0} \delta_{jc} - \frac{k''_c}{\omega_1} \delta_{aj} - v_j \frac{k''_c k''_a}{\omega_0 \omega_1} \right) + \beta_{ai} \beta''_{bj} \left(\frac{k'_a}{\omega_0} \delta_{bs} - \frac{k'_b}{\omega_2} \delta_{as} - v_s \frac{k'_a k'_b}{\omega_0 \omega_2} \right) \right\}. \tag{19.103}$$

In a hot magnetoactive plasma with the Maxwell (6.14) distribution function f_0, one can integrate on the right-hand side of (19.97) with respect to the velocities using (16.27) and (16.28) and the equations

$$(1/N)\int d\mathbf{v} f_0 e^{i\mathbf{av}} v_i v_j = v_T^2 (\delta_{ij} - v_T^2 a_i a_j) \exp\left(-\frac{1}{2} a^2 v_T^2\right), \tag{19.104}$$

$$(1/N)\int d\mathbf{v} f_0 e^{i\mathbf{av}} v_i v_j v_s = i v_T^4 (\delta_{ij} a_s + \delta_{is} a_j + \delta_{js} a_i - v_T^2 a_i a_j a_s) e^{-\frac{1}{2} a^2 v_T^2}, \tag{19.105}$$

which follow from (16.27). We then obtain an expression for the tensor S_{ijs} in the form of double integrals with respect to τ [cf. formula (19.71) for the longitudinal contraction]:

$$\begin{aligned}
&S_{ijs}(\omega,\ \mathbf{k};\ \omega',\ \mathbf{k}') = \frac{e}{m}\frac{\omega_L^2}{\omega}\int_{-\infty}^{0} d\tau' \int_{-\infty}^{0} d\tau'' \times \\
&\times \exp\Big\{-i\omega'\tau' - i\omega''\tau'' - \frac{1}{2} v_T^2 [k'_c \rho_{cn}(\tau') + k''_b \rho_{bn}(\tau'')]^2\Big\} \Big\{ W_{is}(-\tau') k_a \rho_{aj}(-\tau'') + \\
&+ W_{ij}(-\tau'') k_a \rho_{as}(-\tau') + v_T^2 [k'_q \rho_{qp}(\tau') + k''_q \rho_{qp}(\tau'')] \Big\{ \delta_{pi} k_a \rho_{as}(-\tau') k_n \rho_{nj}(-\tau'') + \\
&+ i W_{mp}(\tau') [W_{ic}(-\tau') k_a \rho_{aj}(-\tau'') + W_{ij}(-\tau'') k_a \rho_{ac}(-\tau')] \left(\frac{k'_c}{\omega'}\delta_{sm} - \frac{k'_m}{\omega'}\delta_{cs}\right) + \\
&+ i W_{np}(\tau'') [W_{is}(-\tau') k_a \rho_{ab}(-\tau'') + W_{ib}(-\tau'') k_a \rho_{as}(-\tau')] \left(\frac{k''_b}{\omega''}\delta_{jn} - \delta_{bj}\frac{k''_n}{\omega''}\right)\Big\} + \\
&+ v_T^2 \Big\{ W_{np}(\tau'') W_{mp}(\tau') [W_{ic}(-\tau') k_a \rho_{ab}(-\tau'') + W_{ib}(-\tau'') k_a \rho_{ac}(-\tau')] \Big(\frac{k'_c}{\omega'}\delta_{sm} - \\
&- \delta_{cs}\frac{k'_m}{\omega'}\Big) \left(\frac{k''_b}{\omega''}\delta_{jn} - \delta_{bj}\frac{k''_n}{\omega''}\right) - i W_{mi}(\tau') k_a \rho_{ac}(-\tau') k_n \rho_{nj}(-\tau'') \left(\frac{k'_c}{\omega'}\delta_{sm} - \delta_{cs}\frac{k'_m}{\omega'}\right) - \\
&- i W_{ni}(\tau'') k_a \rho_{as}(-\tau') k_m \rho_{mb}(-\tau'') \left(\frac{k''_b}{\omega''}\delta_{jn} - \delta_{bj}\frac{k''_n}{\omega''}\right)\Big\} + v_T^4 [k'_p \rho_{pr}(\tau') + \\
&+ k''_p \rho_{pr}(\tau'')] [k'_q \rho_{qk}(\tau') + k''_q \rho_{qk}(\tau'')] \Big\{ -W_{nr}(\tau'') W_{mk}(\tau') [W_{ic}(-\tau') k_a \rho_{ab}(-\tau'') + \\
&+ W_{ib}(-\tau'') k_a \rho_{ac}(-\tau')] \left(\frac{k'_c}{\omega'}\delta_{sm} - \delta_{cs}\frac{k'_m}{\omega'}\right) \left(\frac{k''_b}{\omega''}\delta_{jn} - \delta_{bj}\frac{k''_n}{\omega''}\right) + \\
&+ i\delta_{ir} W_{mk}(\tau') k_a \rho_{ac}(-\tau') k_l \rho_{lj}(-\tau'') \left(\frac{k'_c}{\omega'}\delta_{sm} - \delta_{cs}\frac{k'_m}{\omega'}\right) + \\
&+ i\delta_{ik} W_{nr}(\tau'') k_a \rho_{as}(-\tau') k_l \rho_{lb}(-\tau'') \left(\frac{k''_b}{\omega''}\delta_{jn} - \delta_{bj}\frac{k''_n}{\omega''}\right)\Big\} + \\
&+ v_T^4 W_{mk}(\tau') W_{nr}(\tau'') k_a \rho_{ac}(-\tau') k_p \rho_{pb}(-\tau'') [k'_q \rho_{ql}(\tau') + k''_q \rho_{ql}(\tau'')] (\delta_{ik}\delta_{lr} + \\
&+ \delta_{ir}\delta_{kl} + \delta_{kr}\delta_{il}) \left(\frac{k'_c}{\omega'}\delta_{sm} - \delta_{cs}\frac{k'_m}{\omega'}\right) \left(\frac{k''_b}{\omega''}\delta_{jn} - \delta_{bj}\frac{k''_n}{\omega''}\right) - v_T^6 W_{mk}(\tau') W_{nr}(\tau'') \times \\
&\times k_a \rho_{ac}(-\tau') k_p \rho_{pb}(-\tau'') [k'_q \rho_{qi}(\tau') + k''_q \rho_{qi}(\tau'')] [k'_l \rho_{lk}(\tau') + k''_l \rho_{lk}(\tau'')] [k'_t \rho_{tr}(\tau') + \\
&+ k''_t \rho_{tr}(\tau'')] \left(\frac{k'_c}{\omega'}\delta_{sm} - \delta_{cs}\frac{k'_m}{\omega'}\right) \left(\frac{k''_b}{\omega''}\delta_{jn} - \delta_{bj}\frac{k''_n}{\omega''}\right)\Big\} + (\mathbf{k}' \rightleftarrows \mathbf{k}'' \rightleftarrows \mathbf{k}).
\end{aligned} \tag{19.106}$$

Here, we have written out in full the group of terms that correspond to the first term (first square brackets) in the curly brackets on the right-hand side of Eq. (19.97). The symbol $(\mathbf{k}' \rightleftarrows \mathbf{k}'' \rightleftarrows \mathbf{k})$ denotes the two other groups of terms corresponding to the last two terms of the tensor (19.97). They arise from the terms given explicitly in (19.106) by cyclic permutation of the primes and indices in accordance with the rule (7.18). As an illustration of formula (19.106) for S_{ijs} in a hot magnetoactive plasma with Maxwell distribution function of the particles, we give two special cases of the formula. In an isotropic plasma ($\mathbf{B}_0 = 0$) it follows from (19.106) that

$$\begin{aligned}
&S_{ijs}(\omega,\ \mathbf{k};\ \omega',\ \mathbf{k}') = -\frac{e}{m}\frac{\omega_L^2}{\omega}\int_{-\infty}^{0} d\tau' \int_{-\infty}^{0} d\tau'' \times \\
&\times \exp\Big\{-i\omega'\tau' - i\omega''\tau'' - \frac{1}{2} v_T^2 (\mathbf{k}'\tau' + \mathbf{k}''\tau'')^2\Big\} \Big\{(k_j \delta_{is}\tau'' + k_s \delta_{ij}\tau') + v_T^2 (k'_n\tau' + k''_n\tau'') \times \\
&\times \Big[-\delta_{in} k_s k_j \tau'\tau'' + i(k_b \delta_{is}\tau'' + k_s \delta_{ib}\tau') \left(\frac{k''_b}{\omega''}\delta_{nj} - \delta_{jb}\frac{k''_n}{\omega''}\right) + i(k_j \delta_{ic}\tau'' + k_c \delta_{ij}\tau') \times \\
&\times \left(\frac{k'_c}{\omega'}\delta_{sn} - \frac{k'_n}{\omega'}\delta_{cs}\right)\Big] + v_T^2 \Big[(k_b \delta_{ic}\tau'' + k_c \delta_{ib}\tau') \left(\frac{k''_b}{\omega''}\delta_{nj} - \delta_{jb}\frac{k''_n}{\omega''}\right) \left(\frac{k'_c}{\omega'}\delta_{sn} - \delta_{cs}\frac{k'_n}{\omega'}\right) +
\end{aligned}$$

$$
\begin{aligned}
&+ik_sk_b\tau'\tau''\left(\frac{k''_b}{\omega''}\delta_{ij}-\frac{k''_i}{\omega''}\delta_{bj}\right)+ik_ck_j\tau'\tau''\left(\frac{k'_c}{\omega'}\delta_{is}-\frac{k'_i}{\omega'}\delta_{cs}\right)\Big]-v_T^4\,(k'_n\tau'+k''_n\tau'')\times\\
&\times(k'_m\tau'+k''_m\tau'')\Big[(k_b\delta_{ic}\tau''+k_c\delta_{ib}\tau')\left(\frac{k''_b}{\omega''}\delta_{jn}-\delta_{bj}\frac{k''_n}{\omega''}\right)\left(\frac{k'_c}{\omega'}\delta_{sm}-\delta_{cs}\frac{k'_m}{\omega'}\right)+\\
&+i\delta_{im}k_sk_b\tau'\tau''\left(\frac{k''_b}{\omega''}\delta_{nj}-\delta_{bj}\frac{k''_n}{\omega''}\right)+i\delta_{in}k_jk_c\tau'\tau''\left(\frac{k'_c}{\omega'}\delta_{sm}-\delta_{cs}\frac{k'_m}{\omega'}\right)\Big]-\\
&-v_T^4k_ck_b\tau'\tau''\,(k'_p\tau'+k''_p\tau'')\,(\delta_{in}\delta_{mp}+\delta_{im}\delta_{np}+\delta_{mn}\delta_{ip})\left(\frac{k''_b}{\omega''}\delta_{jn}-\frac{k''_n}{\omega''}\delta_{bj}\right)\times\\
&\times\left(\frac{k'_c}{\omega'}\delta_{sm}-\frac{k'_m}{\omega'}\delta_{cs}\right)+v_T^6k_ck_b\tau'\tau''\,(k'_m\tau'+k''_m\tau'')\,(k'_n\tau'+k''_n\tau'')\,(k'_i\tau'+k''_i\tau'')\times\\
&\times\left(\frac{k''_b}{\omega''}\delta_{jn}-\delta_{bj}\frac{k''_n}{\omega''}\right)\left(\frac{k'_c}{\omega'}\delta_{sm}-\delta_{cs}\frac{k'_m}{\omega'}\right)\Big\}-\frac{e}{m}\frac{\omega_L^2}{\omega''}\int_{-\infty}^{0}d\tau\int_{-\infty}^{0}d\tau'\times\\
&\times\exp\left\{-i\omega\tau-i\omega'\tau'-\frac{1}{2}v_T^2\,(\mathbf{k}\tau+\mathbf{k}'\tau')^2\right\}\Big\{(k''_i\delta_{js}\tau-k''_s\delta_{ij}\tau')+v_T^2\,(k_n\tau+k'_n\tau')\times\\
&\times\Big[\delta_{jn}k''_ik''_s\tau\tau'+i\,(k''_i\delta_{jc}\tau-k''_c\delta_{ij}\tau')\left(\frac{k'_c}{\omega'}\delta_{ns}-\frac{k'_n}{\omega'}\delta_{cs}\right)+i\,(k''_a\delta_{sj}\tau-k''_s\delta_{aj}\tau')\times\\
&\times\left(\frac{k_a}{\omega}\delta_{im}-\frac{k_m}{\omega}\delta_{ai}\right)\Big]+v_T^2\Big[(k''_a\delta_{jc}\tau-k''_c\delta_{aj}\tau')\left(\frac{k_a}{\omega}\delta_{in}-\delta_{ai}\frac{k_n}{\omega}\right)\left(\frac{k'_c}{\omega'}\delta_{ns}-\frac{k'_n}{\omega'}\delta_{cs}\right)-\\
&-ik''_ik''_c\tau\tau'\left(\frac{k'_c}{\omega'}\delta_{sj}-\frac{k'_j}{\omega'}\delta_{cs}\right)-ik''_ak''_s\tau\tau'\left(\frac{k_a}{\omega}\delta_{ij}-\frac{k_j}{\omega}\delta_{ai}\right)\Big]-v_T^4\,(k_n\tau+k'_n\tau')\,(k_m\tau+\\
&+k'_m\tau')\Big[(k''_a\delta_{jc}\tau-k''_c\delta_{aj}\tau')\left(\frac{k_a}{\omega}\delta_{im}-\delta_{ai}\frac{k_m}{\omega}\right)\left(\frac{k'_c}{\omega'}\delta_{ns}-\frac{k'_n}{\omega'}\delta_{cs}\right)-\\
&-i\delta_{jm}k''_ik''_c\tau\tau'\left(\frac{k'_c}{\omega'}\delta_{ns}-\frac{k'_n}{\omega'}\delta_{cs}\right)-i\delta_{jn}k''_ak''_s\tau\tau'\left(\frac{k_a}{\omega}\delta_{im}-\delta_{ai}\frac{k_m}{\omega}\right)\Big]+\\
&+v_T^4k''_ak''_c\tau\tau'\,(k_p\tau+k'_p\tau')\,(\delta_{jn}\delta_{mp}+\delta_{nm}\delta_{jp}+\delta_{mj}\delta_{np})\left(\frac{k_a}{\omega}\delta_{im}-\frac{k_m}{\omega}\delta_{ai}\right)\left(\frac{k'_c}{\omega'}\delta_{ns}-\right.\\
&\left.-\frac{k'_n}{\omega'}\delta_{cs}\right)-v_T^6k''_ak''_c\tau\tau'\,(k_j\tau+k'_j\tau')\,(k_n\tau+k'_n\tau')\,(k_m\tau+k'_m\tau')\left(\frac{k_a}{\omega}\delta_{im}-\frac{k_m}{\omega}\delta_{ai}\right)\times\\
&\times\left(\frac{k'_c}{\omega'}\delta_{ns}-\delta_{cs}\frac{k'_n}{\omega'}\right)\Big\}-\frac{e}{m}\frac{\omega_L^2}{\omega'}\int_{-\infty}^{0}d\tau\int_{-\infty}^{0}d\tau''\times\\
&\times\exp\left\{-i\omega\tau-i\omega''\tau''-\frac{1}{2}v_T^2\,(\mathbf{k}\tau+\mathbf{k}''\tau'')^2\right\}\Big\{(k'_i\delta_{sj}\tau-k'_j\delta_{is}\tau'')+v_T^2\,(k_n\tau+k''_n\tau'')\times\\
&\times\Big[\delta_{sn}k'_ik'_j\tau\tau''+i\,(k'_i\delta_{bs}\tau-k'_b\delta_{is}\tau'')\left(\frac{k''_b}{\omega''}\delta_{nj}-\delta_{bj}\frac{k''_n}{\omega''}\right)+i\,(k'_a\delta_{sj}\tau-k'_j\delta_{as}\tau'')\times\\
&\times\left(\frac{k_a}{\omega}\delta_{in}-\frac{k_n}{\omega}\delta_{ai}\right)\Big]+v_T^2\Big[(k'_a\delta_{sb}\tau-k'_b\delta_{as}\tau'')\left(\frac{k_a}{\omega}\delta_{in}-\frac{k_n}{\omega}\delta_{ai}\right)\left(\frac{k''_b}{\omega''}\delta_{nj}-\frac{k''_n}{\omega''}\delta_{bj}\right)-\\
&-ik'_ik'_b\tau\tau''\left(\frac{k''_b}{\omega''}\delta_{js}-\frac{k''_s}{\omega''}\delta_{bj}\right)-ik'_ak'_j\tau\tau''\left(\frac{k_a}{\omega}\delta_{is}-\frac{k_s}{\omega}\delta_{ai}\right)\Big]-v_T^4\,(k_n\tau+k''_n\tau'')\,(k_m\tau+\\
&+k''_m\tau'')\Big[(k'_a\delta_{sb}\tau-k'_b\delta_{as}\tau'')\left(\frac{k_a}{\omega}\delta_{im}-\delta_{ai}\frac{k_m}{\omega}\right)\left(\frac{k''_b}{\omega''}\delta_{nj}-\frac{k''_n}{\omega''}\delta_{bj}\right)-\\
&-i\delta_{ns}k'_ak'_j\tau\tau''\left(\frac{k_a}{\omega}\delta_{im}-\frac{k_m}{\omega}\delta_{ai}\right)-i\delta_{ms}k'_ik'_b\tau\tau''\left(\frac{k''_b}{\omega''}\delta_{nj}-\frac{k''_n}{\omega''}\delta_{bj}\right)\Big]+\\
&+v_T^4k'_ak'_b\tau\tau''\,(k_p\tau+k''_p\tau'')\,(\delta_{sn}\delta_{mp}+\delta_{nm}\delta_{sp}+\delta_{sm}\delta_{np})\left(\frac{k_a}{\omega}\delta_{im}-\frac{k_m}{\omega}\delta_{ai}\right)\left(\frac{k''_b}{\omega''}\delta_{nj}-\right.\\
&\left.-\frac{k''_n}{\omega''}\delta_{bj}\right)-v_T^6k'_ak'_b\tau\tau''\,(k_m\tau+k''_m\tau'')\,(k_n\tau+k''_n\tau'')\,(k_s\tau+k''_s\tau'')\left(\frac{k_a}{\omega}\delta_{im}-\right.\\
&\left.-\delta_{ai}\frac{k_m}{\omega}\right)\left(\frac{k''_b}{\omega''}\delta_{nj}-\frac{k''_n}{\omega''}\delta_{bj}\right)\Big\}. \qquad (19.107)
\end{aligned}
$$

Hence, in particular, for the longitudinal contraction with the tensor (7.66) in a hot isotropic plasma we obtain the expression

$$
\begin{aligned}
&S(\omega,\mathbf{k};\omega',\mathbf{k}')=-\frac{e}{m}\frac{\omega_L^2}{kk'k''}\Big\{\frac{(\mathbf{kk}')(\mathbf{kk}'')}{\omega}\int_{-\infty}^{0}d\tau'\int_{-\infty}^{0}d\tau''\times\\
&\times\exp\Big[-i\omega'\tau'-i\omega''\tau''-\frac{1}{2}v_T^2\,(\mathbf{k}'\tau'+\mathbf{k}''\tau'')^2\Big][\tau'+\tau''-v_T^2\tau'\tau''\,(\mathbf{kk}'\tau'+\mathbf{kk}''\tau'')]+\\
&+\frac{(\mathbf{k}''\mathbf{k})(\mathbf{k}''\mathbf{k}')}{\omega''}\int_{-\infty}^{0}d\tau\int_{-\infty}^{0}d\tau'\exp\Big[-i\omega\tau-i\omega'\tau'-\frac{1}{2}v_T^2\,(\mathbf{k}\tau+\mathbf{k}'\tau')^2\Big][\tau-\tau'+\\
&+v_T^2\tau\tau'\,(\mathbf{k}''\mathbf{k}\tau+\mathbf{k}''\mathbf{k}'\tau')]+\frac{(\mathbf{k}'\mathbf{k})(\mathbf{k}'\mathbf{k}'')}{\omega'}\int_{-\infty}^{0}d\tau\int_{-\infty}^{0}d\tau''\times
\end{aligned}
$$

$$\times \exp\Big[-i\omega\tau - i\omega''\tau'' - \frac{1}{2}v_T^2(\mathbf{k}\tau+\mathbf{k}''\tau'')^2\Big][\tau-\tau''+v_T^2\tau\tau''(\mathbf{k}'\mathbf{k}\tau+\mathbf{k}'\mathbf{k}''\tau'')]\Big\}, \tag{19.108}$$

which reduces to that previously obtained, (19.72), by means of identities of the form

$$\int_{-\infty}^{0} d\tau \int_{-\infty}^{0} d\tau' \exp\Big[-i\omega\tau - i\omega'\tau' - \frac{1}{2}v_T^2(\mathbf{k}\tau+\mathbf{k}'\tau')^2\Big][\tau-\tau'+$$
$$+v_T^2\tau\tau'(\mathbf{k}''\mathbf{k}\tau+\mathbf{k}''\mathbf{k}'\tau')]+i\omega''\int_{-\infty}^{0} d\tau\tau\int_{-\infty}^{0} d\tau'\tau'\exp\Big[-i\omega\tau-i\omega'\tau'-\frac{1}{2}v_T^2(\mathbf{k}\tau+\mathbf{k}'\tau')^2\Big]\equiv 0. \tag{19.109}$$

As another special case of formula (19.106) we give an expression for S_{ijs} in a magnetoactive plasma with allowance for the first thermal correction:

$$S_{ijs}(\omega,\ \mathbf{k};\ \omega',\ \mathbf{k}') = \frac{e}{m}\frac{\omega_L^2}{\omega}\int_{-\infty}^{0} d\tau' \int_{-\infty}^{0} d\tau'' e^{-i\omega'\tau'-i\omega''\tau''}\Big\{W_{is}(-\tau')k_a\rho_{aj}(-\tau'')+$$
$$+W_{ij}(-\tau'')k_a\rho_{as}(-\tau')+v_T^2[k'_q\rho_{qp}(\tau')+k''_q\rho_{qp}(\tau'')]\Big\{\delta_{pi}k_a\rho_{as}(-\tau')k_n\rho_{nj}(-\tau'')+$$
$$+iW_{mp}(\tau')[W_{ic}(-\tau')k_a\rho_{aj}(-\tau'')+W_{ij}(-\tau'')k_a\rho_{ac}(-\tau')]\Big(\frac{k'_c}{\omega'}\delta_{sm}-\delta_{cs}\frac{k'_m}{\omega'}\Big)+$$
$$+iW_{np}(\tau'')[W_{is}(-\tau')k_a\rho_{ab}(-\tau'')+W_{ib}(-\tau'')k_a\rho_{as}(-\tau')]\Big(\frac{k''_b}{\omega''}\delta_{jn}-\delta_{bj}\frac{k''_n}{\omega''}\Big)\Big\}+$$
$$+v_T^2\Big\{W_{np}(\tau'')W_{mp}(\tau')[W_{ic}(-\tau')k_a\rho_{ab}(-\tau'')+W_{ib}(-\tau'')k_a\rho_{ac}(-\tau')]\times$$
$$\times\Big(\frac{k'_c}{\omega'}\delta_{sm}-\delta_{cs}\frac{k'_m}{\omega'}\Big)\Big(\frac{k''_b}{\omega''}\delta_{jn}-\delta_{bj}\frac{k''_n}{\omega''}\Big)-iW_{mi}(\tau')k_a\rho_{ac}(-\tau')k_n\rho_{nj}(-\tau'')\times$$
$$\times\Big(\frac{k'_c}{\omega'}\delta_{sm}-\delta_{cs}\frac{k'_m}{\omega'}\Big)-iW_{ni}(\tau'')k_a\rho_{as}(-\tau')k_m\rho_{mb}(-\tau'')\Big(\frac{k''_b}{\omega''}\delta_{jn}-\delta_{bj}\frac{k''_n}{\omega''}\Big)-$$
$$-\frac{1}{2}[k'_c\rho_{cn}(\tau')+k''_b\rho_{bn}(\tau'')]^2[W_{is}(-\tau')k_a\rho_{aj}(-\tau'')+W_{ij}(-\tau'')k_a\rho_{as}(-\tau')]\Big\}\Big\}-$$
$$-\frac{e}{m}\frac{\omega_L^2}{\omega''}\int_{-\infty}^{0} d\tau\int_{-\infty}^{0} d\tau' e^{-i\omega\tau-i\omega''\tau''}\Big\{W_{ij}(-\tau)k''_b\rho_{bs}(-\tau')-W_{js}(-\tau')\rho_{ib}(-\tau)k''_b+$$
$$+v_T^2[k'_q\rho_{qp}(\tau')-k_q\rho_{qp}(-\tau)]\Big\{\delta_{pj}k''_b\rho_{bs}(-\tau')k''_n\rho_{in}(-\tau)+$$
$$+iW_{np}(-\tau)[W_{aj}(-\tau)k''_b\rho_{bs}(-\tau')-W_{js}(-\tau')\rho_{ab}(-\tau)k''_b]\Big(\frac{k_a}{\omega}\delta_{in}-\frac{k_n}{\omega}\delta_{ai}\Big)+$$
$$+iW_{mp}(\tau')[W_{ij}(-\tau)k''_b\rho_{bc}(-\tau')-W_{jc}(-\tau')\rho_{ib}(-\tau)k''_b]\Big(\frac{k'_c}{\omega'}\delta_{sm}-\delta_{cs}\frac{k'_m}{\omega'}\Big)\Big\}+$$
$$+v_T^2\Big\{W_{np}(-\tau)W_{mp}(\tau')[W_{aj}(-\tau)k''_b\rho_{bc}(-\tau')-W_{jc}(-\tau')\rho_{ab}(-\tau)k''_b]\times$$
$$\times\Big(\frac{k_a}{\omega}\delta_{in}-\frac{k_n}{\omega}\delta_{ai}\Big)\Big(\frac{k'_c}{\omega'}\delta_{sm}-\frac{k'_m}{\omega'}\delta_{cs}\Big)-iW_{nj}(-\tau)k''_b\rho_{bs}(-\tau')k''_m\rho_{am}(-\tau)\times$$
$$\times\Big(\frac{k_a}{\omega}\delta_{in}-\frac{k_n}{\omega}\delta_{ai}\Big)-iW_{mj}(\tau')k''_b\rho_{bc}(-\tau')k''_n\rho_{in}(-\tau)\Big(\frac{k'_c}{\omega'}\delta_{sm}-\frac{k'_m}{\omega'}\delta_{cs}\Big)-$$
$$-\frac{1}{2}[k'_c\rho_{cn}(\tau')-k_a\rho_{an}(-\tau)]^2[W_{ij}(-\tau)k''_b\rho_{bs}(-\tau')-W_{js}(-\tau')\rho_{ib}(-\tau)k''_b]\Big\}\Big\}-$$
$$-\frac{e}{m}\frac{\omega_L^2}{\omega'}\int_{-\infty}^{0} d\tau\int_{-\infty}^{0} d\tau'' e^{-i\omega\tau-i\omega''\tau''}\Big\{W_{is}(-\tau)k'_c\rho_{cj}(-\tau'')-W_{sj}(-\tau'')\rho_{ic}(-\tau)k'_c+$$
$$+v_T^2[k''_q\rho_{qp}(\tau'')-k_q\rho_{qp}(-\tau)]\Big\{\delta_{ps}k'_c\rho_{cj}(-\tau'')k'_n\rho_{in}(-\tau)+$$
$$+iW_{mp}(-\tau)[W_{as}(-\tau)k'_c\rho_{cj}(-\tau'')-W_{sj}(-\tau'')\rho_{ac}(-\tau)k'_c]\Big(\frac{k_a}{\omega}\delta_{im}-\frac{k_m}{\omega}\delta_{ai}\Big)+$$
$$+iW_{np}(\tau'')[W_{is}(-\tau)k'_c\rho_{cb}(-\tau'')-W_{sb}(-\tau'')\rho_{ic}(-\tau)k'_c]\Big(\frac{k''_b}{\omega''}\delta_{jn}-\delta_{bj}\frac{k''_n}{\omega''}\Big)\Big\}+$$
$$+v_T^2\Big\{W_{mp}(-\tau)W_{np}(\tau'')[W_{as}(-\tau)k'_c\rho_{cb}(-\tau'')-W_{sb}(-\tau'')\rho_{ac}(-\tau)k'_c]\Big(\frac{k_a}{\omega}\delta_{im}-$$
$$-\frac{k_m}{\omega}\delta_{ai}\Big)\Big(\frac{k''_b}{\omega''}\delta_{jn}-\delta_{bj}\frac{k''_n}{\omega''}\Big)-iW_{ns}(\tau'')k'_c\rho_{cb}(-\tau'')k'_m\rho_{im}(-\tau)\Big(\frac{k''_b}{\omega''}\delta_{jn}-\frac{k''_n}{\omega''}\delta_{bj}\Big)-$$
$$-iW_{ms}(-\tau)k'_c\rho_{cj}(-\tau'')k'_n\rho_{an}(-\tau)\Big(\frac{k_a}{\omega}\delta_{im}-\delta_{ai}\frac{k_m}{\omega}\Big)-\frac{1}{2}[k''_b\rho_{bn}(\tau'')-$$
$$-k_a\rho_{an}(-\tau)]^2[W_{is}(-\tau)k'_c\rho_{cj}(-\tau'')-W_{sj}(-\tau'')\rho_{ic}(-\tau')k'_c]\Big\}\Big\}. \tag{19.110}$$

In the limit $v_T = 0$, this takes the form (19.16). In an isotropic plasma, (19.110) yields the more compact expression

$$\begin{aligned}
S_{ijs}(\omega,\ \mathbf{k};\ \omega',\ \mathbf{k}') = &-i\frac{e}{m}\frac{\omega_L^2}{\omega\omega'\omega''}\left\{\frac{k_i}{\omega}\delta_{js}+\frac{k_j''}{\omega''}\delta_{is}+\frac{k_s'}{\omega'}\delta_{ij}\right\}- \\
&-i\frac{e}{m}\frac{\omega_L^2}{\omega\omega'\omega''}v_T^2\left\{\delta_{ij}\left[\left(\frac{k}{\omega}\right)^2\frac{k_s''}{\omega''}+\left(\frac{k''}{\omega''}\right)^2\frac{k_s}{\omega}+\frac{k_s'}{\omega'}\frac{\mathbf{k}\mathbf{k}''}{\omega\omega''}+3\left(\frac{k'}{\omega'}\right)^2\frac{k_s'}{\omega'}\right]+\right. \\
&+\delta_{is}\left[\left(\frac{k}{\omega}\right)^2\frac{k_j'}{\omega'}+\left(\frac{k'}{\omega'}\right)^2\frac{k_j}{\omega}+\frac{k_j''}{\omega''}\frac{\mathbf{k}\mathbf{k}'}{\omega\omega'}+3\left(\frac{k''}{\omega''}\right)^2\frac{k_j''}{\omega''}\right]+\delta_{js}\left[\left(\frac{k''}{\omega''}\right)^2\frac{k_i'}{\omega'}+\right. \\
&\left.+\left(\frac{k'}{\omega'}\right)^2\frac{k_i''}{\omega''}+\frac{k_i}{\omega}\frac{\mathbf{k}'\mathbf{k}''}{\omega'\omega''}+3\left(\frac{k}{\omega}\right)^2\frac{k_i}{\omega}\right]+\frac{k_j''k_s}{\omega''\omega}\left(\frac{k_i}{\omega}+\frac{k_i''}{\omega''}\right)+\frac{k_s'k_j}{\omega'\omega}\left(\frac{k_i}{\omega}+\frac{k_i'}{\omega'}\right)+ \\
&\left.+\frac{k_ik_s''}{\omega\omega''}\left(\frac{k_j''}{\omega''}+\frac{k_j}{\omega}\right)+\frac{k_s'k_i''}{\omega'\omega''}\left(\frac{k_j''}{\omega''}+\frac{k_j'}{\omega'}\right)+\frac{k_ik_j'}{\omega\omega'}\left(\frac{k_s'}{\omega'}+\frac{k_s}{\omega}\right)+\frac{k_i'k_j''}{\omega'\omega''}\left(\frac{k_s'}{\omega'}+\frac{k_s''}{\omega''}\right)\right\}.
\end{aligned} \tag{19.111}$$

In the case of an arbitrary (for example, non-Maxwell) isotropic distribution function f_0 of the unperturbed plasma state the square of the thermal velocity on the right-hand side of (19.111) must be replaced by $(1/3N)\int d\mathbf{v}v^2f_0$. For reference, we give below the following thermal correction to the tensor S_{ijs} in an isotropic plasma with arbitrary isotropic f_0:

$$\begin{aligned}
\delta S_{ijs}(\omega,\ \mathbf{k};\ \omega',\ \mathbf{k}') = &-i\frac{e}{m}\frac{\omega_L^2}{\omega\omega'\omega''}\frac{1}{5N}\int d\mathbf{v}v^4f_0\left\{\left(\frac{k}{\omega}\right)^4\frac{k_i}{\omega}\delta_{js}+\left(\frac{k''}{\omega''}\right)^4\frac{k_j''}{\omega''}\delta_{is}+\right. \\
&\left.+\left(\frac{k'}{\omega'}\right)^4\frac{k_s'}{\omega'}\delta_{ij}\right\}-i\frac{e}{m}\frac{\omega_L^2}{\omega\omega'\omega''}\frac{1}{15N}\int d\mathbf{v}v^4f_0\left\{12\left[\left(\frac{k}{\omega}\right)^4\frac{k_i}{\omega}\delta_{js}+\left(\frac{k''}{\omega''}\right)^4\frac{k_j''}{\omega''}\delta_{is}+\right.\right. \\
&\left.+\left(\frac{k'}{\omega'}\right)^4\frac{k_s'}{\omega'}\delta_{ij}\right]+\frac{k_sk_j''}{\omega\omega''}\left[\left(\frac{k}{\omega}\right)^2\frac{k_i''}{\omega''}+\left(\frac{k''}{\omega''}\right)^2\frac{k_i}{\omega}+2\frac{\mathbf{k}\mathbf{k}''}{\omega\omega''}\left(\frac{k_i}{\omega}+\frac{k_i''}{\omega''}\right)+3\left(\frac{k}{\omega}\right)^2\frac{k_i}{\omega}+\right. \\
&\left.+3\left(\frac{k''}{\omega''}\right)^2\frac{k_i''}{\omega''}\right]+\frac{k_jk_s'}{\omega\omega'}\left[\left(\frac{k}{\omega}\right)^2\left(\frac{k_i'}{\omega'}+3\frac{k_i}{\omega}\right)+\left(\frac{k'}{\omega'}\right)^2\left(\frac{k_i}{\omega}+3\frac{k_i'}{\omega'}\right)+\right. \\
&\left.+2\frac{\mathbf{k}\mathbf{k}'}{\omega\omega'}\left(\frac{k_i}{\omega}+\frac{k_i'}{\omega'}\right)\right]+\frac{k_ik_s''}{\omega\omega''}\left[\left(\frac{k}{\omega}\right)^2\left(\frac{k_j''}{\omega''}+3\frac{k_j}{\omega}\right)+\left(\frac{k''}{\omega''}\right)^2\left(\frac{k_j}{\omega}+3\frac{k_j''}{\omega''}\right)+\right. \\
&\left.+2\frac{\mathbf{k}\mathbf{k}''}{\omega\omega''}\left(\frac{k_j}{\omega}+\frac{k_j''}{\omega''}\right)\right]+\frac{k_s'k_i''}{\omega'\omega''}\left[\left(\frac{k'}{\omega'}\right)^2\left(\frac{k_j''}{\omega''}+3\frac{k_j'}{\omega'}\right)+\left(\frac{k''}{\omega''}\right)^2\left(\frac{k_j'}{\omega'}+3\frac{k_j''}{\omega''}\right)+\right. \\
&\left.+2\frac{\mathbf{k}'\mathbf{k}''}{\omega'\omega''}\left(\frac{k_j'}{\omega'}+\frac{k_j''}{\omega''}\right)\right]+\frac{k_ik_j'}{\omega\omega'}\left[\left(\frac{k}{\omega}\right)^2\left(\frac{k_s'}{\omega'}+3\frac{k_s}{\omega}\right)+\left(\frac{k'}{\omega'}\right)^2\left(\frac{k_s}{\omega}+3\frac{k_s'}{\omega'}\right)+\right. \\
&\left.+2\frac{\mathbf{k}\mathbf{k}'}{\omega\omega'}\left(\frac{k_s}{\omega}+\frac{k_s'}{\omega'}\right)\right]+\frac{k_i'k_j''}{\omega'\omega''}\left[\left(\frac{k'}{\omega'}\right)^2\left(\frac{k_s''}{\omega''}+3\frac{k_s'}{\omega'}\right)+\left(\frac{k''}{\omega''}\right)^2\left(\frac{k_s'}{\omega'}+3\frac{k_s''}{\omega''}\right)+\right. \\
&\left.\left.+2\frac{\mathbf{k}'\mathbf{k}''}{\omega'\omega''}\left(\frac{k_s''}{\omega''}+\frac{k_s'}{\omega'}\right)\right]\right\}-i\frac{e}{m}\frac{\omega_L^2}{\omega\omega'\omega''}\frac{1}{15N}\int d\mathbf{v}v^4f_0\left\{2\left(\frac{k}{\omega}\right)^2\frac{k_s''}{\omega''}\left[3\frac{k_ik_j}{\omega^2}+\frac{k_ik_j''}{\omega\omega''}+\right.\right. \\
&\left.+\frac{k_jk_i''}{\omega\omega''}+\frac{k_i''k_j''}{\omega''^2}\right]+2\left(\frac{k''}{\omega''}\right)^2\frac{k_s}{\omega}\left[3\frac{k_i''k_j''}{\omega''^2}+\frac{k_ik_j''}{\omega\omega''}+\frac{k_i''k_j}{\omega''\omega}+\frac{k_ik_j}{\omega^2}\right]+ \\
&+2\left(\frac{k}{\omega}\right)^2\frac{k_j'}{\omega'}\left[3\frac{k_ik_s}{\omega^2}+\frac{k_ik_s'}{\omega\omega'}+\frac{k_sk_i'}{\omega\omega'}+\frac{k_i'k_s'}{\omega'^2}\right]+2\left(\frac{k'}{\omega'}\right)^2\frac{k_j}{\omega}\left[3\frac{k_i'k_s'}{\omega'^2}+\frac{k_ik_s'}{\omega\omega'}+\right. \\
&\left.+\frac{k_i'k_s}{\omega'\omega}+\frac{k_ik_s}{\omega^2}\right]+2\left(\frac{k''}{\omega''}\right)^2\frac{k_i'}{\omega'}\left[3\frac{k_j''k_s''}{\omega''^2}+\frac{k_j'k_s''}{\omega'\omega''}+\frac{k_j''k_s'}{\omega'\omega''}+\frac{k_j'k_s'}{\omega'^2}\right]+ \\
&+2\left(\frac{k'}{\omega'}\right)^2\frac{k_i''}{\omega''}\left[3\frac{k_j'k_s'}{\omega'^2}+\frac{k_j'k_s''}{\omega'\omega''}+\frac{k_j''k_s'}{\omega''\omega'}+\frac{k_j''k_s''}{\omega''^2}\right]+\frac{\mathbf{k}\mathbf{k}''}{\omega\omega''}\frac{k_s'}{\omega'}\left[2\left(\frac{k_ik_j}{\omega^2}+\frac{k_i'k_j'}{\omega'^2}+\right.\right. \\
&\left.\left.+\frac{k_i''k_j''}{\omega''^2}\right)+\frac{k_i'k_j}{\omega'\omega}+\frac{k_j'k_i}{\omega'\omega}+\frac{k_ik_j''}{\omega\omega''}+\frac{k_i''k_j}{\omega''\omega}+\frac{k_i'k_j''}{\omega'\omega''}+\frac{k_i''k_j'}{\omega''\omega'}\right]+\frac{\mathbf{k}\mathbf{k}'}{\omega\omega'}\frac{k_j''}{\omega''}\times \\
&\times\left[2\left(\frac{k_ik_s}{\omega^2}+\frac{k_i'k_s'}{\omega'^2}+\frac{k_i''k_s''}{\omega''^2}\right)+\frac{k_ik_s'}{\omega\omega'}+\frac{k_i'k_s}{\omega'\omega}+\frac{k_ik_s''}{\omega\omega''}+\frac{k_i''k_s}{\omega''\omega}+\frac{k_i'k_s''}{\omega'\omega''}+\frac{k_i''k_s'}{\omega''\omega'}\right]+ \\
&+\frac{\mathbf{k}'\mathbf{k}''}{\omega'\omega''}\frac{k_i}{\omega}\left[2\left(\frac{k_jk_s}{\omega^2}+\frac{k_j'k_s'}{\omega'^2}+\frac{k_j''k_s''}{\omega''^2}\right)+\frac{k_jk_s'}{\omega\omega'}+\frac{k_j'k_s}{\omega'\omega}+\frac{k_jk_s''}{\omega\omega''}+\frac{k_j''k_s}{\omega''\omega}+\right. \\
&\left.+\frac{k_j'k_s''}{\omega'\omega''}+\frac{k_j''k_s'}{\omega''\omega'}\right]+\left[\left(\frac{k}{\omega}\right)^2+\left(\frac{k'}{\omega'}\right)^2+\left(\frac{k''}{\omega''}\right)^2+\frac{\mathbf{k}\mathbf{k}'}{\omega\omega'}+\frac{\mathbf{k}\mathbf{k}''}{\omega\omega''}+\frac{\mathbf{k}'\mathbf{k}''}{\omega'\omega''}\right]\times \\
&\times\left[\frac{\mathbf{k}\mathbf{k}''}{\omega\omega''}\frac{k_s'}{\omega'}\delta_{ij}+\frac{\mathbf{k}\mathbf{k}'}{\omega\omega'}\frac{k_j''}{\omega''}\delta_{is}+\frac{\mathbf{k}'\mathbf{k}''}{\omega'\omega''}\frac{k_i}{\omega}\delta_{js}\right]+\delta_{ij}\left(\frac{k}{\omega}\right)^2\frac{k_s''}{\omega''}\left[3\left(\frac{k}{\omega}\right)^2+\right. \\
&\left.+2\frac{\mathbf{k}\mathbf{k}''}{\omega\omega''}+\left(\frac{k''}{\omega''}\right)^2\right]+\delta_{ij}\left(\frac{k''}{\omega''}\right)^2\frac{k_s}{\omega}\left[3\left(\frac{k''}{\omega''}\right)^2+2\frac{\mathbf{k}\mathbf{k}''}{\omega\omega''}+\left(\frac{k}{\omega}\right)^2\right]+
\end{aligned} \tag{19.112}$$

$$+\delta_{is}\left(\frac{k}{\omega}\right)^2\frac{k'_j}{\omega'}\left[3\left(\frac{k}{\omega}\right)^2+2\frac{\mathbf{kk}'}{\omega\omega'}+\left(\frac{k'}{\omega'}\right)^2\right]+\delta_{is}\left(\frac{k'}{\omega'}\right)^2\frac{k_j}{\omega}\left[3\left(\frac{k'}{\omega'}\right)^2+\right.$$
$$\left.+2\frac{\mathbf{kk}'}{\omega\omega'}+\left(\frac{k}{\omega}\right)^2\right]+\delta_{js}\left(\frac{k''}{\omega''}\right)^2\frac{k'_i}{\omega'}\left[3\left(\frac{k''}{\omega''}\right)^2+2\frac{\mathbf{k}'\mathbf{k}''}{\omega'\omega''}+\left(\frac{k'}{\omega'}\right)^2\right]+$$
$$+\delta_{js}\left(\frac{k'}{\omega'}\right)^2\frac{k''_i}{\omega''}\left[3\left(\frac{k'}{\omega'}\right)^2+2\frac{\mathbf{k}'\mathbf{k}''}{\omega'\omega''}+\left(\frac{k''}{\omega''}\right)^2\right]\Bigg\}-i\frac{e}{m}\frac{\omega_L^2}{\omega\omega'\omega''}\frac{1}{15N}\int d\mathbf{v}v^4 f_0\times$$
$$\times\left\{\left(\frac{k_i}{\omega}\delta_{js}+\frac{k_j}{\omega}\delta_{is}+\frac{k_s}{\omega}\delta_{ij}\right)\left[2\left(\frac{k}{\omega}\right)^2\frac{\mathbf{k}'\mathbf{k}''}{\omega'\omega''}+\left(\frac{k''}{\omega''}\right)^2\frac{\mathbf{kk}'}{\omega\omega'}+\left(\frac{k'}{\omega'}\right)^2\frac{\mathbf{kk}''}{\omega\omega''}\right]+\right.$$
$$+\left(\frac{k'_i}{\omega'}\delta_{js}+\frac{k'_j}{\omega'}\delta_{is}+\frac{k'_s}{\omega'}\delta_{ij}\right)\left[2\left(\frac{k'}{\omega'}\right)^2\frac{\mathbf{kk}''}{\omega\omega''}+\left(\frac{k''}{\omega''}\right)^2\frac{\mathbf{kk}'}{\omega\omega'}+\left(\frac{k}{\omega}\right)^2\frac{\mathbf{k}'\mathbf{k}''}{\omega'\omega''}\right]+$$
$$\left.+\left(\frac{k''_i}{\omega''}\delta_{js}+\frac{k''_j}{\omega''}\delta_{is}+\frac{k''_s}{\omega''}\delta_{ij}\right)\left[2\left(\frac{k''}{\omega''}\right)^2\frac{\mathbf{kk}'}{\omega\omega'}+\left(\frac{k}{\omega}\right)^2\frac{\mathbf{k}'\mathbf{k}''}{\omega'\omega''}+\left(\frac{k'}{\omega'}\right)^2\frac{\mathbf{kk}''}{\omega\omega''}\right]\right\}.$$

In the case of the Maxwell distribution function (6.14) we must here bear in mind the equation

$$(1/15N)\int d\mathbf{v}v^4 f_0 = v_T^4.$$

In the theory of the parametric effect of powerful fluxes of electromagnetic waves on a plasma one is interested in the special case of the tensor S_{ijs} when one of the wave vectors of the interacting oscillations is equal to zero:

$$\mathbf{k}'=0,\quad \mathbf{k}''=\mathbf{k};\quad \omega'\equiv\omega_0,\quad \omega''=\omega-\omega_0. \tag{19.113}$$

Under the conditions (19.113) the general expression (19.97) for the tensor S_{ijs} in a hot magnetoactive plasma gives

$$S_{ijs}(\omega,\ \mathbf{k};\ \omega_0,\ 0)=i\frac{e}{m}\frac{1}{\omega_L^2}\left\{\frac{\omega}{\omega-\omega_0}k_a\delta\varepsilon_{as}(\omega_0,\ 0)\,\delta\varepsilon_{ij}(\omega,\ -\mathbf{k})-\frac{\omega-\omega_0}{\omega}k_a\delta\varepsilon_{as}(\omega_0,\ 0)\,\delta\varepsilon_{ij}(\omega-\omega_0,\ \mathbf{k})-\right.$$
$$\left.-\frac{\omega_0}{\omega-\omega_0}k_a\delta\varepsilon_{ia}(\omega,\ -\mathbf{k})\,\delta\varepsilon_{js}(\omega_0,\ 0)-\frac{\omega_0}{\omega}k_a\delta\varepsilon_{aj}(\omega-\omega_0,\ \mathbf{k})\,\delta\varepsilon_{is}(\omega_0,\ 0)\right\}. \tag{19.114}$$

Here, $\delta\varepsilon_{ij}(\omega, \mathbf{k})$ is the partial polarizability of a hot magnetoactive plasma, $\delta\varepsilon_{ij} = (\varepsilon_{ij} - \delta_{ij})$, which has been studied in detail in § 16.

§ 20. Four-Index Tensor $V_{iajb}(\omega,\ \mathbf{k};\ \omega',\ \mathbf{k}')$ in a Magnetoactive Plasma

In accordance with the general definition (3.23), the tensor $V_{iajb}(\omega, \mathbf{k}; \omega', \mathbf{k}')$ in a magnetoactive plasma consists of a sum of two four-index tensors ε. The first term is given in accordance with Eq. (16.5) by [52]

$$\varepsilon_{iajb}(\omega,\ \mathbf{k};\ \omega+\omega',\ \mathbf{k}+\mathbf{k}';\ \omega',\ \mathbf{k}')=-i4\pi e^4\int d\mathbf{p}\frac{v_i}{\omega}\int_{-\infty}^{0}d\tau_0\int_{-\infty}^{0}d\tau_1\int_{-\infty}^{0}d\tau_2\times$$
$$\times\exp\{-i\omega\tau_0+i\mathbf{k}\delta\mathbf{R}(\tau_0,\ \mathbf{v})\}\,\alpha'_{na}(\tau_0)\frac{\partial}{\partial p_n(\tau_0,\ \mathbf{v})}\exp\{-i(\omega+\omega')\tau_1+$$
$$+i(\mathbf{k}+\mathbf{k}',\delta\mathbf{R}(\tau_0+\tau_1,\mathbf{v})-\delta\mathbf{R}(\tau_0,\mathbf{v}))\}\,\alpha_{mj}(\tau_0+\tau_1)\Big|\frac{\partial}{\partial p_m(\tau_0+\tau_1,\mathbf{v})}\exp\{-i\omega'\tau_2+i\times$$
$$\times\left|(\mathbf{k}',\ \delta\mathbf{R}(\tau_0+\tau_1+\tau_2,\ \mathbf{v})-\delta\mathbf{R}(\tau_0+\tau_1,\ \mathbf{v}))\right\}\alpha'_{lb}(\tau_0+\tau_1+\tau_2)\frac{\partial f_0}{\partial p_l(\tau_0+\tau_1+\tau_2,\ \mathbf{v})}, \tag{20.1}$$

where we have used the usual abbreviated notation (19.2) for the tensors α. The second term of the tensor V is obtained from this by transposing the subscripts j $\leftrightarrows$ b and the arguments $\omega, \mathbf{k} \leftrightarrows \omega', \mathbf{k}'$ in the tensors $\alpha_{mj}(\omega,\ \mathbf{k};\ \mathbf{v}(\tau_0+\tau_1,\mathbf{v}))$ and $\alpha_{lb}(\omega',\ \mathbf{k}';\ \mathbf{v}(\tau_0+\tau_1+\tau_2,\ \mathbf{v}))$ and at the same time replacing $\exp\{-i\omega'\tau_2+i(\mathbf{k}',\ \delta\mathbf{R}(\tau_0+\tau_1+\tau_2,\ \mathbf{v})-\delta\mathbf{R}(\tau_0+\tau_1,\ \mathbf{v}))\}$ by $\exp\{-i\omega\tau_2+i(\mathbf{k},\ \delta\mathbf{R}(\tau_0+$

$$+\tau_1+\tau_2, \mathbf{v}) - \delta\mathbf{R}(\tau_0+\tau_1, \mathbf{v}))\}:$$

$$\varepsilon_{iabj}(\omega, \mathbf{k}; \omega+\omega', \mathbf{k}+\mathbf{k}'; \omega, \mathbf{k}) = -i4\pi e^4 \int d\mathbf{p} \frac{v_i}{\omega} \int_{-\infty}^{0} d\tau_0 \int_{-\infty}^{0} d\tau_1 \int_{-\infty}^{0} d\tau_2 \times$$
$$\times \exp\{-i\omega\tau_0 + i\mathbf{k}\delta\mathbf{R}(\tau_0, \mathbf{v})\}\, \alpha'_{na}(\tau_0) \frac{\partial}{\partial p_n(\tau_0, \mathbf{v})} \exp\{-i(\omega+\omega')\tau_1 +$$
$$+ i(\mathbf{k}+\mathbf{k}', \delta\mathbf{R}(\tau_0+\tau_1, \mathbf{v}) - \delta\mathbf{R}(\tau_0, \mathbf{v}))\}\, \alpha'_{mb}(\tau_0+\tau_1) \times$$
$$\times \frac{\partial}{\partial p_m(\tau_0+\tau_1, \mathbf{v})} \exp\{-i\omega\tau_2 + i(\mathbf{k}, \delta\mathbf{R}(\tau_0+\tau_1+\tau_2, \mathbf{v}) - \delta\mathbf{R}(\tau_0+\tau_1, \mathbf{v}))\} \times \alpha_{lj}(\tau_0+\tau_1+\tau_2) \frac{\partial f_0}{\partial p_l(\tau_0+\tau_1+\tau_2, \mathbf{v})}. \tag{20.2}$$

For $\mathbf{B}_0 = 0$ these formulas go over into expressions (8.1) and (8.2) of § 8. As in the case of an isotropic plasma, the tensor $V_{iajb}(\omega, \mathbf{k}, \omega', \mathbf{k}')$, in contrast to $S_{ijs}(\omega, \mathbf{k}; \omega', \mathbf{k}')$, is not equal to the corresponding symmetrized combination composed of six four-index tensors ε. Of course, this greatly complicates its symmetry properties.

On the right-hand sides of Eqs. (20.1) and (20.2) we integrate thrice with respect to the momenta, freeing ourselves in this way from the differentiation operators in the vertex parts of (15.51), and we then go over in the resulting expressions to the limit $v_T = 0$ of a cold magnetoactive plasma ($\mathbf{B}_0 = \text{const} \neq 0$). Then the tensor V takes the form

$$V_{iajb}(\omega, \mathbf{k}; \omega', \mathbf{k}') = \{\varepsilon_{iajb}(\omega, \mathbf{k}; \omega+\omega', \mathbf{k}+\mathbf{k}'; \omega', \mathbf{k}') +$$
$$+ \varepsilon_{iabj}(\omega, \mathbf{k}; \omega+\omega', \mathbf{k}+\mathbf{k}'; \omega, \mathbf{k})\} = -i\frac{e^2}{m^2}\frac{\omega_L^2}{\omega} \int_{-\infty}^{0} d\tau_0 e^{-i\omega\tau_0} \times$$
$$\times \int_{-\infty}^{0} d\tau e^{-i\omega\tau} \int_{-\infty}^{0} d\tau_1 e^{-i\omega'\tau_1} \Big\{\frac{i}{\omega'}(k'_s\delta_{ca} - k'_c\delta_{sa})\, k_n\rho_{ns}(-\tau_0) \times$$
$$\times [W_{cj}(-\tau) W_{ib}(-\tau_0-\tau_1) + W_{cb}(-\tau_1) W_{ij}(-\tau_0-\tau)] + W_{ia}(-\tau_0) \times$$
$$\times [k_n\rho_{nb}(-\tau_0-\tau_1) k_m\rho_{mj}(-\tau_0-\tau) + k'_s\rho_{sj}(-\tau) k_n\rho_{nb}(-\tau_0-\tau_1) +$$
$$+ k_m\rho_{mj}(-\tau_0-\tau) k'_c\rho_{cb}(-\tau_1) + k'_c\rho_{cb}(-\tau_1) k'_s\rho_{sj}(-\tau)] +$$
$$+ i\frac{k'_s}{\omega'}[W_{is}(-\tau_0) W_{aj}(-\tau) - W_{sj}(-\tau) W_{ia}(-\tau_0)][k'_c\rho_{cb}(-\tau_1) +$$
$$+ k_n\rho_{nb}(-\tau_0-\tau_1)] + i\frac{k'_c}{\omega'}[W_{ic}(-\tau_0) W_{ab}(-\tau_1) - W_{cb}(-\tau_1) W_{ia}(-\tau_0)] \times$$
$$\times [k'_s\rho_{sj}(-\tau) + k_n\rho_{nj}(-\tau_0-\tau)] + k_n\rho_{na}(-\tau_0) W_{ij}(-\tau_0-\tau) [k'_c\rho_{cb}(-\tau_1) +$$
$$+ k_n\rho_{nb}(-\tau_0-\tau_1)] + k_n\rho_{na}(-\tau_0) W_{ib}(-\tau_0-\tau_1) [k'_s\rho_{sj}(-\tau) +$$
$$+ k_n\rho_{nj}(-\tau_0-\tau)]\Big\} + i\frac{e^2}{m^2}\frac{\omega_L^2}{\omega} \int_{-\infty}^{0} d\tau_0 \int_{-\infty}^{0} d\tau_1 e^{-i\omega\tau_0} e^{-i(\omega+\omega')\tau_1} \times$$
$$\times \Big\{ \int_{-\infty}^{0} d\tau_2 e^{-i\omega'\tau_2} \Big[\frac{k_n}{\omega} W_{jb}(-\tau_2) + i\frac{\omega+\omega'}{\omega}\delta_{nj} k_m\rho_{mb}(-\tau_2)\Big] \times$$
$$\times \Big\{\frac{k'_s}{\omega'}[W_{is}(-\tau_0) W_{an}(-\tau_1) + i\omega\rho_{sn}(-\tau_1) W_{ia}(-\tau_0)] - i[k_m\rho_{ma}(-\tau_0) \times$$
$$\times W_{in}(-\tau_0-\tau_1) + k_m\rho_{mn}(-\tau_0-\tau_1) W_{ia}(-\tau_0)]\Big\} + \int_{-\infty}^{0} d\tau_2 e^{-i\omega\tau_2} \times$$
$$\times \Big[\frac{k'_c}{\omega'} W_{bj}(-\tau_2) + i\frac{\omega+\omega'}{\omega'}\delta_{cb} k'_s\rho_{sj}(-\tau_2)\Big] \Big\{\frac{k'_s}{\omega'}[W_{is}(-\tau_0) \times$$
$$\times W_{ac}(-\tau_1) + i\omega\rho_{sc}(-\tau_1) W_{ia}(-\tau_0)] - i[k_m\rho_{ma}(-\tau_0) \times$$
$$\times W_{ic}(-\tau_0-\tau_1) + k_m\rho_{mc}(-\tau_0-\tau_1) W_{ia}(-\tau_0)]\Big\}\Big\}. \tag{20.3}$$

In this expression there are two groups of terms. In the first of them the sum frequencies $\omega + \omega'$ do not occur in the denominators after integration with respect to the variables τ_0, τ, τ_1. In contrast, the second group of terms contains $\omega + \omega'$. This splitting of terms into two groups can be made if one adds the tensors ε by means of equations similar to (19.14). We regard the splitting as important, because, if the oscillations of a magnetoactive plasma have frequencies ω and ω' that are nearly equal in magnitude but opposite in sign, the terms of the second group are much larger than those of the first group, so that the latter can be ignored [compare with the condition (7.76)]. In the limit $\mathbf{B}_0 = 0$, we obtain from formula (20.3) the corresponding isotropic expressions, which can be obtained independently in the nonlinear theory of an isotropic plasma. Note that, as in an isotropic plasma, in the limit of weak thermal motion ($v_T = 0$) the tensor V is in accordance with (20.3) a bilinear form in the wave vectors of the oscillations. For reference purposes we also give the symmetrized combination composed of the six four-index tensors ε in the limit of a cold magnetoactive plasma. In contrast to the tensors (20.1) and (20.2), the tensors ε that occur in this combination do not depend on two but on three pairs of independent arguments, so that the frequencies and wave vectors in them are related by

$$\begin{aligned} \omega &= \omega' + \omega'' + \omega''', \\ \mathbf{k} &= \mathbf{k}' + \mathbf{k}'' + \mathbf{k}''', \end{aligned} \tag{20.4}$$

which holds in processes of nonlinear interaction of four oscillations. Thus the sum of the six four-index tensors ε

$$\begin{aligned} &\varepsilon_{i(jsr)}(\omega, \mathbf{k}; \omega', \mathbf{k}'; -\omega'', \mathbf{k}''; \omega''', \mathbf{k}''') \equiv \frac{1}{3!}\{\varepsilon_{ijsr}(\omega, \mathbf{k}; \omega'' + \omega''', \mathbf{k}'' + \mathbf{k}'''; \omega''', \mathbf{k}''') + \\ &+ \varepsilon_{irjs}(\omega, \mathbf{k}; \omega' + \omega'', \mathbf{k}' + \mathbf{k}''; \omega'', \mathbf{k}'') + \varepsilon_{isrj}(\omega, \mathbf{k}; \omega''' + \omega', \mathbf{k}''' + \mathbf{k}'; \omega', \mathbf{k}') + \\ &+ \varepsilon_{isjr}(\omega, \mathbf{k}; \omega' + \omega''', \mathbf{k}' + \mathbf{k}'''; \omega''', \mathbf{k}''') + \varepsilon_{irsj}(\omega, \mathbf{k}; \omega'' + \omega', \mathbf{k}'' + \mathbf{k}'; \omega', \mathbf{k}') + \\ &+ \varepsilon_{ijrs}(\omega, \mathbf{k}; \omega''' + \omega'', \mathbf{k}''' + \mathbf{k}''; \omega'', \mathbf{k}'')\}, \end{aligned} \tag{20.5}$$

in which the first term, for example, has the form

$$\begin{aligned} &\varepsilon_{ijsr}(\omega, \mathbf{k}; \omega'' + \omega''', \mathbf{k}'' + \mathbf{k}'''; \omega''', \mathbf{k}''') = -4\pi i e^4 \int d\mathbf{p}\, \frac{v_i}{\omega} \int_{-\infty}^{0} d\tau_1 \int_{-\infty}^{0} d\tau_2 \int_{-\infty}^{0} d\tau_3 \times \\ &\times \exp\{-i\omega\tau_1 + i\mathbf{k}\delta\mathbf{R}(\tau_1, \mathbf{v})\}\, \alpha_{nj}(\omega', \mathbf{k}'; \mathbf{v}(\tau_1, \mathbf{v})) \times \\ &\times \frac{\partial}{\partial p_n(\tau_1, \mathbf{v})} \exp\{-i(\omega'' + \omega''')\tau_2 + i(\mathbf{k}'' + \mathbf{k}''', \delta\mathbf{R}(\tau_1 + \tau_2, \mathbf{v}) - \delta\mathbf{R}(\tau_1, \mathbf{v}))\} \times \\ &\times \alpha_{ms}(\omega'', \mathbf{k}''; \mathbf{v}(\tau_1 + \tau_2, \mathbf{v})) \frac{\partial}{\partial p_m(\tau_1 + \tau_2, \mathbf{v})} \exp\{-i\omega'''\tau_3 + i(\mathbf{k}''', \delta\mathbf{R}(\tau_1 + \\ &+ \tau_2 + \tau_3, \mathbf{v}) - \delta\mathbf{R}(\tau_1 + \tau_2, \mathbf{v}))\}\, \alpha_{lr}(\omega''', \mathbf{k}'''; \mathbf{v}(\tau_1 + \tau_2 + \tau_3, \mathbf{v})) \frac{\partial f_0}{\partial p_l(\tau_1 + \tau_2 + \tau_3, \mathbf{v})}, \end{aligned} \tag{20.6}$$

can be represented in the limit of a cold plasma by the equation

$$\begin{aligned} &\varepsilon_{i(jsr)}(\omega, \mathbf{k}; \omega', \mathbf{k}'; \omega'', \mathbf{k}'', \omega''', \mathbf{k}''') = \frac{1}{6}\frac{e^2}{m^2}\frac{\omega_L^2}{\omega}\Big\{\Big[\gamma''_{bj,r}(\omega''') - \delta_{bj}k''_a\frac{\Gamma_{ar}(\omega''')}{\omega'''}\Big] \times \\ &\times \Big[\Gamma_{ic}(\omega)\gamma'_{cs,b}(\omega'' + \omega''') - \Gamma_{is}(\omega)k'_a\frac{\Gamma_{ab}(\omega'' + \omega''')}{\omega'' + \omega'''} + \Gamma_{is}(\omega')k_a\frac{\Gamma_{ab}(\omega'' + \omega''')}{\omega'' + \omega'''} + \\ &+ \Gamma_{ib}(\omega'' + \omega''')k_a\frac{\Gamma_{as}(\omega')}{\omega'}\Big] + \Big[\gamma'''_{dr,j}(\omega'') - \delta_{dr}k'''_a\frac{\Gamma_{aj}(\omega'')}{\omega''}\Big]\Big[\Gamma_{ic}(\omega)\gamma'_{cs,d}(\omega'' + \omega''') - \\ &- \Gamma_{is}(\omega)k'_a\frac{\Gamma_{ad}(\omega'' + \omega''')}{\omega'' + \omega'''} + \Gamma_{is}(\omega')k_a\frac{\Gamma_{ad}(\omega'' + \omega''')}{\omega'' + \omega'''} + \Gamma_{id}(\omega'' + \omega''')k_a\frac{\Gamma_{as}(\omega')}{\omega'}\Big] + \\ &+ \Big[\gamma'_{cs,r}(\omega''') - \delta_{cs}k'_a\frac{\Gamma_{ar}(\omega''')}{\omega'''}\Big]\Big[\Gamma_{ib}(\omega)\gamma''_{bj,c}(\omega' + \omega''') - \Gamma_{ij}(\omega)k''_a\frac{\Gamma_{ac}(\omega' + \omega''')}{\omega' + \omega'''} + \\ &+ \Gamma_{ij}(\omega'')k_a\frac{\Gamma_{ac}(\omega' + \omega''')}{\omega' + \omega'''} + \Gamma_{ic}(\omega' + \omega''')k_a\frac{\Gamma_{aj}(\omega'')}{\omega''}\Big] + \Big[\gamma'''_{dr,s}(\omega') - \delta_{dr}k'''_a\frac{\Gamma_{as}(\omega')}{\omega'}\Big] \times \\ &\times \Big[\Gamma_{ib}(\omega)\gamma''_{bj,d}(\omega' + \omega''') - \Gamma_{ij}(\omega)k''_a\frac{\Gamma_{ad}(\omega' + \omega''')}{\omega' + \omega'''} + \Gamma_{ij}(\omega'')k_a\frac{\Gamma_{ad}(\omega' + \omega''')}{\omega' + \omega'''} + \end{aligned}$$

$$
\begin{aligned}
&+\Gamma_{id}(\omega'+\omega''')\,k_a\frac{\Gamma_{aj}(\omega'')}{\omega''}\Big]+\Big[\gamma'_{cs,j}(\omega'')-\delta_{cs}k'_a\frac{\Gamma_{aj}(\omega'')}{\omega''}\Big]\Big[\Gamma_{id}(\omega)\,\gamma'''_{dr,c}(\omega'+\omega'')-\\
&-\Gamma_{ir}(\omega)\,k'''_a\frac{\Gamma_{ac}(\omega'+\omega'')}{\omega'+\omega''}+\Gamma_{ir}(\omega''')\,k_a\frac{\Gamma_{ac}(\omega'+\omega'')}{\omega'+\omega''}+\Gamma_{ic}(\omega'+\omega'')\,k_a\frac{\Gamma_{ar}(\omega''')}{\omega'''}\Big]+\\
&+\Big[\gamma''_{bj,s}(\omega')-\delta_{bj}k''_a\frac{\Gamma_{as}(\omega')}{\omega'}\Big]\Big[\Gamma_{id}(\omega)\,\gamma'''_{dr,b}(\omega'+\omega'')-\Gamma_{ir}(\omega)\,k'''_a\frac{\Gamma_{ab}(\omega'+\omega'')}{\omega'+\omega''}+\\
&+\Gamma_{ir}(\omega''')\,k_a\frac{\Gamma_{ab}(\omega'+\omega'')}{\omega'+\omega''}+\Gamma_{ib}(\omega'+\omega'')\,k_a\frac{\Gamma_{ar}(\omega''')}{\omega'''}\Big]\Big\}+\frac{1}{6}\,\frac{e^2}{m^2}\,\frac{\omega_L^2}{\omega}\Big\{\Gamma_{is}(\omega)\,k'_a\times\\
&\times\frac{\Gamma_{ar}(\omega''')}{\omega'''}\,k'_c\frac{\Gamma_{cj}(\omega'')}{\omega''}+\Gamma_{ir}(\omega)\,k'''_a\frac{\Gamma_{as}(\omega')}{\omega'}\,k'''_d\frac{\Gamma_{dj}(\omega'')}{\omega''}+\Gamma_{ij}(\omega)\,k''_a\frac{\Gamma_{as}(\omega')}{\omega'}\,k''_b\frac{\Gamma_{br}(\omega''')}{\omega'''}-\\
&-\Gamma_{ic}(\omega)\,\gamma'_{cs,j}(\omega'')\,k'_a\frac{\Gamma_{ar}(\omega''')}{\omega'''}-\Gamma_{ic}(\omega)\,\gamma'_{cs,r}(\omega''')\,k'_a\frac{\Gamma_{aj}(\omega'')}{\omega''}-\\
&-\Gamma_{id}(\omega)\,\gamma'''_{dr,s}(\omega')\,k'''_a\frac{\Gamma_{aj}(\omega'')}{\omega''}-\Gamma_{id}(\omega)\,\gamma'''_{dr,j}(\omega'')\,k'''_a\frac{\Gamma_{as}(\omega')}{\omega'}-\\
&-\Gamma_{ib}(\omega)\,\gamma''_{bj,r}(\omega''')\,k''_a\frac{\Gamma_{as}(\omega')}{\omega'}-\Gamma_{ib}(\omega)\,\gamma''_{bj,s}(\omega')\,k''_a\frac{\Gamma_{ar}(\omega''')}{\omega'''}+\\
&+\Gamma_{is}(\omega')\,k_b\frac{\Gamma_{bj}(\omega'')}{\omega''}\,k_a\frac{\Gamma_{ar}(\omega''')}{\omega'''}+\Gamma_{ij}(\omega'')\,k_c\frac{\Gamma_{cs}(\omega')}{\omega'}\,k_a\frac{\Gamma_{ar}(\omega''')}{\omega'''}+\\
&+\Gamma_{ir}(\omega''')\,k_c\frac{\Gamma_{cs}(\omega')}{\omega'}\,k_a\frac{\Gamma_{aj}(\omega'')}{\omega''}\Big\}.
\end{aligned}
\qquad (20.7)
$$

The three-index tensors γ in this expression are used to denote sums of the form

$$\gamma'_{cs,j}(\omega)\equiv\frac{k'_a}{\omega'}\{\delta_{ca}\Gamma_{sj}(\omega)-\delta_{cs}\Gamma_{aj}(\omega)\}, \qquad (20.8)$$

so that the argument of the tensors $\gamma(\omega)$ is equal to the argument of $\Gamma(\omega)$ in the curly brackets of (20.8) and the prime in γ' indicates the number of primes above the wave vector and the frequency in the factor in front of the curly brackets. For example, the first term in the first square brackets of (20.7) has the form

$$\gamma''_{bj,r}(\omega''')=\frac{k''_a}{\omega''}\{\delta_{ab}\Gamma_{jr}(\omega''')-\delta_{bj}\Gamma_{ar}(\omega''')\}. \qquad (20.8a)$$

The tensors γ are transverse with respect to the wave vector for contraction over the central index:

$$\gamma'_{cs,j}k'_s=\gamma''_{cs,j}k''_s=\gamma'''_{cs,j}k'''_s=0. \qquad (20.9)$$

Recalling Eqs. (16.34) and (19.17), which relate the tensor Γ_{ij} to the partial permittivity tensor $\delta\varepsilon_{ij}$, we note that the symmetrized combination of the four-index tensors (20.7) is a trilinear form in $\delta\varepsilon_{ij}$ [cf. (19.18)].

Apart from the tensor $V_{i\alpha jb}(\omega, \mathbf{k};\ \omega', \mathbf{k}')$, the nonlinear interaction equation contains in accordance with (21.27) the tensor $V_{i\alpha jb}(\omega, \mathbf{k}; -\omega', -\mathbf{k}')$. For all the concrete processes considered below, the contribution of the terms with the last tensor is much greater at positive frequencies than the contribution of the terms with the tensor $V_{i\alpha jb}(\omega, \mathbf{k}; \omega', \mathbf{k}')$. The physical difference between these two types of terms is described in Ch. II (§ 8) in the case of an isotropic plasma and remains in a magnetoactive plasma. Bearing in mind this remark and the fact that in what follows we shall be interested in the nonrelativistic limit $v \ll c$, for the first term of the tensor $V_{i\alpha jb}(\omega, \mathbf{k}; -\omega', -\mathbf{k}')$ we obtain the equation

$$
\begin{aligned}
&\varepsilon_{i\alpha jb}(\omega,\ \mathbf{k};\ \omega'',\ \mathbf{k}'';\ -\omega',\ -\mathbf{k}')=-i\frac{4\pi e^4}{m^3}\int d\mathbf{v}\,\frac{v_i}{\omega}\int_{-\infty}^{0}d\tau_0\int_{-\infty}^{0}d\tau_1\int_{-\infty}^{0}d\tau_2\times\\
&\times\exp\{-i\omega\tau_0+i\mathbf{k}\delta\mathbf{R}(\tau_0,\ \mathbf{v})\}\,\alpha'_{na}(\tau_0)\frac{\partial}{\partial v_n(\tau_0,\ \mathbf{v})}\exp\{-i\omega''\tau_1+i(\mathbf{k}'',\ \delta\mathbf{R}(\tau_0+\tau_1,\ \mathbf{v})-\\
&-\delta\mathbf{R}(\tau_0,\ \mathbf{v}))\}\,\alpha_{mj}(\tau_0+\tau_1)\frac{\partial}{\partial v_m(\tau_0+\tau_1,\ \mathbf{v})}\exp\{i\omega'\tau_2-i(\mathbf{k}',\delta\mathbf{R}(\tau_0+\tau_1+\tau_2,\ \mathbf{v})-\\
&-\delta\mathbf{R}(\tau_0+\tau_1,\ \mathbf{v}))\}\,\alpha'_{lb}(\tau_0+\tau_1+\tau_2)\frac{\partial f_0}{\partial v_l(\tau_0+\tau_1+\tau_2,\ \mathbf{v})}.
\end{aligned}
\qquad (20.10)
$$

Here, ω'' and $\mathbf{k}''$ are related to the frequencies and wave vectors of the scattered oscillations by the usual relations (7.3). The second term of the tensor $V_{i\alpha jb}(\omega, \mathbf{k}; -\omega', -\mathbf{k}')$ is obtained from the first in (20.10) by transposing the arguments and the indices in the same way as (20.2) follows from (20.1); therefore, we omit it here. Note that it can be obtained from (20.2) by the substitution $\omega' \to -\omega'$, $\mathbf{k}' \to -\mathbf{k}'$.

The general expression for the tensor V in a hot magnetoactive plasma is cumbersome, so that we shall restrict ourselves below to considering only the longitudinal contraction

$$V \equiv V_{iajb}(\omega,\ \mathbf{k};\ -\omega',\ -\mathbf{k}')\frac{k_i k'_a k_j k'_b}{(kk')^2}, \tag{20.11}$$

by means of which we shall ellucidate the characteristic features of the term containing the tensor V in the generalized kinetic equation (3.22) in the case of a magnetoactive plasma. Contracting the tensor (20.10) against the longitudinal $k_i k'_a k_j k'_b/(kk')^2$ and adding the result to the similar contraction of the second term of the tensor $V_{i\alpha jb}(\omega, \mathbf{k}; -\omega', -\mathbf{k}')$, we obtain this expression for the longitudinal contraction V:

$$\begin{aligned} V = & -i\frac{4\pi e^4}{m^3}\int d\mathbf{v}\frac{\mathbf{kv}}{\omega}\int_{-\infty}^{0} d\tau_0 \int_{-\infty}^{0} d\tau_1 \int_{-\infty}^{0} d\tau_2 \exp\{-i\omega\tau_0 + i\mathbf{k}\delta\mathbf{R}(\tau_0,\ \mathbf{v})\} \times \\ & \times k'_a \frac{\partial}{\partial v_a(\tau_0,\ \mathbf{v})}\exp\{-i\omega''\tau_1 + i(\mathbf{k}'',\ \delta\mathbf{R}(\tau_0+\tau_1,\ \mathbf{v}) - \delta\mathbf{R}(\tau_0,\ \mathbf{v}))\} \times \\ & \times \Big\{k_j \frac{\partial}{\partial v_j(\tau_0+\tau_1,\ \mathbf{v})}\exp[i\omega'\tau_2 - i(\mathbf{k}',\ \delta\mathbf{R}(\tau_0+\tau_1+\tau_2,\ \mathbf{v}) - \delta\mathbf{R}(\tau_0+\tau_1,\ \mathbf{v}))]\, k'_b \frac{\partial f_0}{\partial v_b(\tau_0+\tau_1+\tau_2,\ \mathbf{v})} + \\ & + k'_b \frac{\partial}{\partial v_b(\tau_0+\tau_1,\ \mathbf{v})}\exp[-i\omega\tau_2 + i(\mathbf{k},\ \delta\mathbf{R}(\tau_0+\tau_1+\tau_2,\ \mathbf{v}) - \delta\mathbf{R}(\tau_0+\tau_1,\ \mathbf{v}))]\, k_j \frac{\partial f_0}{\partial v(\tau_0+\tau_1+\tau_2,\ \mathbf{v})}\Big\}\frac{1}{(kk')^2}. \end{aligned} \tag{20.12}$$

To simplify this, we integrate thrice by parts with respect to $\mathbf{v}$, using differentiation rules of the same type (19.8) and (19.9) as we used in the tensor S_{ijs} and taking into account the fact that in accordance with the continuity equation (6.5) the fraction $\mathbf{kv}/\omega$ in the integrand can be set equal to unity. Then

$$\begin{aligned} V = & -\frac{4\pi e^4}{m^3}\int d\mathbf{v} f_0 \int_{-\infty}^{0} d\tau_0 \exp[-i\omega\tau_0 + ik_i\rho_{ni}(\tau_0)v_n][k_i\rho_{is}(-\tau_0)k'_s]\frac{1}{(kk')^2} \times \\ & \times \int_{-\infty}^{0} d\tau_1 \exp[-i\omega''\tau_1 + ik''_j\rho_{jn}(\tau_1)v_n]\Big\{k_a k'_c \int_{-\infty}^{0} d\tau_2 \exp[i\omega'\tau_2 - ik'_s\rho_{sn}(\tau_1+\tau_2)v_n + ik'_s\rho_{sn}(\tau_1)v_n] + \\ & + k_c k'_a \int_{-\infty}^{0} d\tau_2 \exp[-i\omega\tau_2 + ik_i\rho_{in}(\tau_1+\tau_2)v_n - ik_i\rho_{in}(\tau_1)v_n]\Big\}\{k'_s\rho_{sa}(-\tau_1) - \\ & - k_n\rho_{na}(-\tau_0-\tau_1)\}\{k_n\rho_{nc}(-\tau_2) + k'_s\rho_{sc}(-\tau_1-\tau_2) - k_n\rho_{nc}(-\tau_0-\tau_1-\tau_2)\}. \end{aligned} \tag{20.13}$$

Applying equations of the type (19.54), which we used to symmetrize the longitudinal contraction of S in a magnetoactive plasma, we transform (20.13) to the slightly different form

$$\begin{aligned} V = & -\frac{4\pi e^4}{m^3}\int d\mathbf{v} f_0 \int_{-\infty}^{0} d\tau_0 \exp[-i\omega\tau_0 + ik_i\rho_{ni}(\tau_0)v_n][k_i\rho_{is}(-\tau_0)k'_s]\frac{1}{(kk')^2} \times \\ & \times \Big\{\int_{-\infty}^{0} d\tau \exp[-i\omega\tau + ik_i\rho_{in}(\tau)v_n]\int_{-\infty}^{0} d\tau_1 \exp[i\omega'\tau_1 - ik'_s\rho_{sn}(\tau_1)v_n] \times \\ & \times [k'_a\rho_{aj}(-\tau)k_j - k_i\rho_{ij}(-\tau_0-\tau)k_j][k_a\rho_{ab}(-\tau_0-\tau_1)k'_b - k'_a\rho_{ab}(-\tau_1)k'_b] - \\ & - \int_{-\infty}^{0} d\tau_2 \exp[-i\omega''\tau_2 + ik''_j\rho_{jn}(\tau_2)v_n]\int_{-\infty}^{0} d\tau_1 \exp[i\omega'\tau_1 - ik'_s\rho_{sn}(\tau_1+\tau_2)v_n + \end{aligned}$$

$$+ik'_s\rho_{sn}(\tau_2)v_n][k_i\rho_{ib}(-\tau_1)k'_b][k_a\rho_{aj}(-\tau_0-\tau_2)k_j-k'_a\rho_{aj}(-\tau_2)k_j]-$$
$$-\int_{-\infty}^{0}d\tau_2\exp[-i\omega''\tau_2+ik''_j\rho_{jn}(\tau_2)v_n]\int_{-\infty}^{0}d\tau\exp[-i\omega\tau+ik_i\rho_{in}(\tau_1+\tau_2)v_n-$$
$$-ik_i\rho_{in}(\tau_2)v_n][k'_c\rho_{cj}(-\tau)k_j][k'_a\rho_{ab}(-\tau_2)k'_b-k_a\rho_{ab}(-\tau_0-\tau_2)k'_b]\}. \quad (20.14)$$

It is clear from this form of expression that, in contrast to S, the longitudinal contraction V [like the tensor (20.3)] cannot be symmetrized by means of an equation of the form (19.54). In the curly brackets on the right-hand side of (20.14) there are terms of two types. The first term does not depend on the frequency ω'' and the wave vector $\mathbf{k}''$ of the virtual oscillation. In contrast, the other two terms are essentially determined by the Green's function corresponding to the virtual oscillation. Calculating the imaginary part Im V of the longitudinal contraction in the equation of the nonlinear interaction, we find that the contribution of the first term is nonvanishing. Nevertheless, when calculating the kernel of the integrodifferential equation of a concrete process, we shall ignore such contributions, since the physical nature of the process that corresponds to them differs qualitatively from induced scattering on particles. Indeed, the imaginary part of the first term of the contraction (20.14) is determined by the contribution of the poles with frequencies and wave vectors of the scattered oscillations and corresponds to radiation effects that are small compared with the effect of linear damping on account of the fact that the entire treatment is in the framework of perturbation theory.

Like the longitudinal contraction of S (§ 19), V can be represented as a series in Bessel functions, using the expansion (16.63). Then Eq. (20.14) can be written in the equivalent form

$$V=-\frac{e^2}{m^2}\omega_L^2\int_{-\infty}^{+\infty}\frac{dv_z}{\sqrt{2\pi}\,v_T}e^{-\frac{v_z^2}{2v_T^2}}\sum_{n,\,l,\,m=-\infty}^{+\infty}B_{n,l,m}\int_{-\infty}^{0}d\tau_0e^{-i\omega_0\tau_0}[k_i\rho_{is}(-\tau_0)k'_s]\times$$
$$\times\left\{\int_{-\infty}^{0}d\tau e^{-i\bar\omega\tau}\int_{-\infty}^{0}d\tau_1e^{i\omega_1\tau_1}[k'_a\rho_{aj}(-\tau)k_j-k_a\rho_{aj}(-\tau_0-\tau)k_j]\times\right.$$
$$\times[k_i\rho_{ib}(-\tau_0-\tau_1)k'_b-k'_a\rho_{ab}(-\tau_1)k'_b]-\int_{-\infty}^{0}d\tau_1e^{i\omega_1\tau_1}\int_{-\infty}^{0}d\tau_2e^{-i\omega_2\tau_2}[k_i\rho_{ib}(-\tau_1)k'_b]\times$$
$$\times[k_a\rho_{aj}(-\tau_0-\tau_2)k_j-k'_a\rho_{aj}(-\tau_2)k_j]-\int_{-\infty}^{0}d\tau e^{-i\bar\omega\tau}\int_{-\infty}^{0}d\tau_2e^{-i\omega_2\tau_2}[k'_c\rho_{cj}(-\tau)k_j]\times$$
$$\left.\times[k'_a\rho_{ab}(-\tau_2)k'_b-k_a\rho_{ab}(-\tau_0-\tau_2)k'_b]\right\}\frac{1}{(kk')^2}, \quad (20.15)$$

where we have assumed the isotropic Maxwell distribution (6.14) of the plasma particles in the unperturbed state and we have used the notation

$$B_{n,l,m}\equiv\int_0^\infty\frac{v_\perp dv_\perp}{v_T^2}e^{-\frac{v_\perp^2}{2v_T^2}}J_l\left(\frac{k_\perp v_\perp}{\Omega}\right)J_m\left(\frac{k_\perp v_\perp}{\Omega}\right)J_n\left(\frac{k'_\perp v_\perp}{\Omega}\right)J_{m+n-l}\left(\frac{k'_\perp v_\perp}{\Omega}\right)\exp\{i(l-m)(\theta-\theta')\}; \quad (20.16)$$

$$\omega_0\equiv\omega-m\Omega-k_zv_z,\quad \bar\omega\equiv\omega-l\Omega-k_zv_z,\quad \omega_1\equiv\omega'-n\Omega-k'_zv_z,$$
$$\omega_2\equiv\omega''-(l-n)\Omega-k''_zv_z. \quad (20.17)$$

As usual, θ and θ' are the angles between the components $\mathbf{k}_\perp$ and $\mathbf{k}'_\perp$ of the wave vectors of the scattered oscillations and the x axis in the plane perpendicular to $\mathbf{B}_0$, which is directed along the z axis. The remaining notation is as in § 19. In the case of low frequencies of the scattered oscillations (19.25) and weak thermal motion [cf. (19.85)]:

$$(\omega/k_z)\gg v_T,\quad (\omega'/k'_z)\gg v_T, \quad (20.18)$$

Eq. (20.15) yields the following expression for the imaginary part of the longitudinal contraction:

$$\mathrm{Im}\ V=\sqrt{\frac{\pi}{2}}\frac{e^2}{m^2}\frac{\omega_L^2}{v_T^2}\frac{k_z''^2}{(\omega' k_z-\omega k_z')^2}\left\{\left(\frac{k_z k_z'}{kk'}\right)^2\frac{k_z''^2}{(\omega' k_z-\omega k_z')^2}+\frac{1}{\Omega^2}\frac{(\mathbf{k}\,[\mathbf{k}'\mathbf{h}])^2}{(kk')^2}\right\}B_{0,0,0}\frac{\omega''}{|k_z''|v_T}\exp\left\{-\frac{1}{2}\frac{\omega''^2}{k_z''^2 v_T^2}\right\}, \quad (20.19)$$

provided one retains only those terms in (20.15) that correspond to the interaction of the virtual oscillation with particles.

In the limit of a very strong magnetic field ($B_0 \to \infty$), the right-hand side of this equation goes over into the relation (8.19), which was derived in an isotropic plasma for a one-dimensional interaction. Note that (20.19) holds only in the case of different phase velocities of the interacting oscillations (19.82). The opposite case, when the phase velocities of the interacting oscillations are the same (7.39), is not interesting, since the contribution of the pole with the frequency and wave vector of the virtual oscillation is proportional to $\beta''\exp\{-\beta''^2/2\}$ ($\beta''\equiv\omega''/|k_z''v_T|$), which is of the same order as the contributions from the poles with the frequencies ω' and ω of the scattered oscillations, which are proportional to $\beta'\exp\{-\beta'^2/2\}$ and $\beta\exp\{-\beta^2/2\}$, respectively, ($\beta'\equiv\omega'/|k_z'v_T|$, $\beta\equiv\omega/|k_z v_T|$) and are ignored as small radiation corrections. In this sense, the greatest interest attaches to the case when the phase velocities of the scattered oscillations are large compared with the phase velocity of the virtual oscillation [cf. (7.74)]. Then (20.19) simplifies to

$$\mathrm{Im}\,V=\sqrt{\frac{\pi}{2}}\frac{e^2}{m^2}\frac{\omega_L^2}{v_T^2}\frac{1}{\omega^4}\left\{\left(\frac{k_z k_z'}{kk'}\right)^2+\frac{\omega^2}{\Omega^2}\frac{(\mathbf{k}\,[\mathbf{k}'\mathbf{h}])^2}{(kk')^2}\right\}B_{0,0,0}\frac{\omega''}{|k_z''v_T|}\exp\left\{-\frac{1}{2}\frac{\omega''^2}{(k_2''v_T)^2}\right\}. \quad (20.20)$$

We note a very important property of the imaginary part of the contraction V in the example (20.19). Namely, as also in an isotropic plasma, Im V is antisymmetric under the transposition $(\omega, \mathbf{k}) \rightleftharpoons (\omega', \mathbf{k}')$ of the frequencies and wave vectors of the scattered oscillations, which is in complete agreement with the principle of detailed balance: the induced scattering of an oscillation with frequency ω and wave vector $\mathbf{k}$ into an oscillation with ω' and $\mathbf{k}'$ is just as probable as the inverse process. Of course, if we were to retain in the contraction V the terms responsible for the radiation corrections this symmetry would be lost.

Summarizing our study of the nonlinear tensors S and V in a magnetoactive plasma, we emphasize that the relations for these tensors given in § 19 and 20 in conjunction with the expressions for the linear tensors – the spectral functions, ε_{ij}, and the inverse tensor, which are also obtained here (§ 16-18) – completely describe, in accordance with the generalized kinetic equation (3.22) and the nonlinear dispersion equation (4.12), the evolution of nonlinear processes in a magnetoactive plasma in which three oscillations participate and they also describe the resulting nonlinear corrections to the spectra. The problem of deriving the kernels of the integrodifferential equations that describe the various forms of nonlinear interaction thus reduces to constructing the appropriate contractions of the nonlinear tensors with the linear tensors and simplifying these contractions in the same way as we did in an isotropic plasma.

§ 21. Conservation Laws in the Nonlinear Interaction of Plasma Oscillations as a Consequence of the Symmetry of the Multi-Index Tensors

In this section we shall give some other forms of the nonlinear interaction equations (3.22a) and (3.22b), these following from the generalized kinetic equation (3.22) for waves, by replacing the spectral functions $(E_j E_i)_{\omega,\mathbf{k}}$ by the spectral energy densities $W(\mathbf{k})$ [see (4.4)] and the number $N_{\mathbf{k}}$ of oscillations, which is related to the energy density by [see the remark on formula (9.43) in § 9, Ch. III]

$$N_{\mathbf{k}}=\frac{W(\mathbf{k})}{\hbar\omega(\mathbf{k})}. \quad (21.1)$$

The applicability of the results obtained below is restricted to the region of applicability of the symmetry properties of the nonlinear tensors [see, for example (7.19) and (19.19)] proved earlier for collisionless isotropic and magnetoactive plasmas.

Going over in Eq. (3.22a) of the decay interaction in accordance with equation (17.60) from the spectral functions of the fields to the energy densities W and restricting ourselves in the inverse tensors A to terms that are longitudinal with respect to the corresponding wave vectors, we obtain the following very useful form of the equation that describes the evolution of the energy density $W_l(\mathbf{k})$ of a given longitudinal oscillation due to its nonlinear interaction with other electrostatic oscillations of a magnetoactive [or isotropic, cf. (9.2)] plasma:

$$\frac{dW_l(\mathbf{k})}{dt}=\frac{\omega(\mathbf{k})}{2\pi}\int d\mathbf{k}'d\mathbf{k}''\int_{-\infty}^{+\infty}d\omega'\int_{-\infty}^{+\infty}d\omega''Q\delta(\omega-\omega'-\omega'')\,\delta(\mathbf{k}-\mathbf{k}'-\mathbf{k}'')\{W_l(\mathbf{k}')\,W_l(\mathbf{k}'')\,\omega(\mathbf{k})\operatorname{sign}\varepsilon''(\omega,\mathbf{k})-$$
$$-W_l(\mathbf{k}')\,W_l(\mathbf{k})\,\omega''\operatorname{sign}\varepsilon''(\omega'',\mathbf{k}'')-W_l(\mathbf{k})\,W_l(\mathbf{k}'')\,\omega'\operatorname{sign}\varepsilon''(\omega',\mathbf{k}')\}\,\{\delta(\omega'-\omega'(\mathbf{k}'))\,\delta(\omega''-\omega''(\mathbf{k}''))+$$
$$+\delta(\omega'+\omega'(-\mathbf{k}'))\,\delta(\omega''+\omega''(-\mathbf{k}''))+\delta(\omega'-\omega'(\mathbf{k}'))\,\delta(\omega''+\omega''(-\mathbf{k}''))+\delta(\omega'+\omega'(-\mathbf{k}'))\,\delta(\omega''-\omega''(\mathbf{k}''))\}. \tag{21.2}$$

Here, the kernel Q is determined by the longitudinal contraction of S:

$$Q\equiv\left|S_{ijs}(\omega,\mathbf{k};\omega',\mathbf{k}')\frac{k_ik''_jk'_s}{kk'k''}\right|^2\left\{\frac{\partial\omega\varepsilon'(\omega,\mathbf{k})}{\partial\omega}\right\}^{-1}_{\omega=\omega(\mathbf{k})}\{\partial\omega'\varepsilon'(\omega',\mathbf{k}')/\partial\omega'\}^{-1}\{\partial\omega''\varepsilon'(\omega'',\mathbf{k}'')/\partial\omega''\}^{-1}, \tag{21.3}$$

in which, as usual,

$$\varepsilon'=\varepsilon'(\omega,\mathbf{k})=\operatorname{Re}\varepsilon(\omega,\mathbf{k})=\operatorname{Re}\frac{k_i\varepsilon_{ij}(\omega,\mathbf{k})\,k_j}{k^2}. \tag{21.4}$$

In deriving this equation we have made essential use of the symmetry property (19.77) of the longitudinal contraction of S. Namely, on account of this property, the kernel Q can be represented in the form (21.3) and taken in front of the curly brackets in Eq. (21.2) as a common factor. Note that all the terms of the first curly brackets on the right-hand side of (21.2) contain sign functions of the imaginary part ε'' of the longitudinal permittivity, which describes a definite form of the corresponding longitudinal oscillation that participates in the decay.

In the special case considered here of a homogeneous unbounded magnetoactive (or isotropic) plasma with isotropic Maxwell distribution of the plasma particles in the unperturbed state, the values of the sign functions are sgn ω, sgn ω', sgn ω'', i.e., all the oscillations are stable. The set of δ functions in the second curly brackets in (21.2) corresponds to allowance for the positive and negative values of the frequencies ω' and ω'' with respect to which the integration is made when there are positive expressions for the spectra, $\omega'(\mathbf{k}')=\omega'(-\mathbf{k}')>0$, $\omega(\mathbf{k})>0$, $\omega''(\mathbf{k}'')=\omega''(-\mathbf{k}'')>0$ of the decaying oscillations. In particular, when allowance is made for only the first term in the second curly brackets, Eq. (21.2) takes the form [f_0 is the isotropic distribution (6.14)]

$$\frac{dW_l(\mathbf{k})}{dt}=\frac{\omega(\mathbf{k})}{2\pi}\int d\mathbf{k}'d\mathbf{k}''\delta(\mathbf{k}-\mathbf{k}'-\mathbf{k}'')\,\delta(\omega(\mathbf{k})-\omega'(\mathbf{k}')-\omega''(\mathbf{k}''))\times$$
$$\times Q\{\omega(\mathbf{k})\,W_l(\mathbf{k}')\,W_l(\mathbf{k}'')-\omega''(\mathbf{k}'')\,W_l(\mathbf{k})\,W_l(\mathbf{k}')-\omega'(\mathbf{k}')\,W_l(\mathbf{k})\,W_l(\mathbf{k}'')\}. \tag{21.5}$$

Here, the kernel Q is determined by (21.3) at the frequency values $\omega'\equiv\omega'(\mathbf{k}')$, $\omega''\equiv\omega''(\mathbf{k}'')$, $\omega=\omega(\mathbf{k})$, and the total derivative with respect to the time (d/dt) on the left-hand side [as also on the left-hand side of (21.2)] is the substitutional derivative, which takes into account [see (3.22a)] the variation of $W_l(\mathbf{k})$ in both time and space [cf. (9.8)-(9.9)]:

$$\frac{dW_l(\mathbf{k})}{dt}\equiv\frac{\partial W_l(\mathbf{k})}{\partial t}+\mathbf{v}_{\mathrm{gr}}\frac{\partial W_l(k)}{\partial\mathbf{r}} \tag{21.6}$$

with group velocity

$$\mathbf{v}_{gr}=\frac{\partial\omega(\mathbf{k})}{\partial\mathbf{k}}=-\left\{\frac{\partial\varepsilon'(\omega,\mathbf{k})}{\partial\mathbf{k}}\left[\frac{\partial\varepsilon'(\omega,\mathbf{k})}{\partial\omega}\right]^{-1}\right\}_{\omega=\omega(\mathbf{k})}. \tag{21.7}$$

Equation (21.5) describes the change of the energy density $W_l(\mathbf{k})$ due to three processes of decay type: coalescence of two longitudinal oscillations with spectra $\omega'(\mathbf{k}')$ and $\omega''(\mathbf{k}'')$ and energies $W_l(\mathbf{k}')$ and $W_l(\mathbf{k}'')$ into the oscillation $\omega(\mathbf{k})$ (the first term on the right-hand side) and decay of the oscillation $\omega(\mathbf{k})$ giving rise to the oscillations $\omega''(\mathbf{k}'')$ and $\omega'(\mathbf{k}')$ respectively (the second and the third term on the right-hand side). Using the definition (21.1) to go over to the number $N_\mathbf{k}$ of oscillations in (21.5), we obtain a further form of the equation:

$$\begin{aligned}\frac{dN_\mathbf{k}}{dt}=\frac{\hbar}{2\pi}\int d\mathbf{k}'d\mathbf{k}''\omega(\mathbf{k})\,\omega'(\mathbf{k}')\,\omega''(\mathbf{k}'')\,\delta(\mathbf{k}-\mathbf{k}'-\mathbf{k}'')\times\\ \times\delta(\omega(\mathbf{k})-\omega'(\mathbf{k}')-\omega''(\mathbf{k}''))\,Q\,\{N_{\mathbf{k}'}N_{\mathbf{k}''}-N_\mathbf{k}N_{\mathbf{k}'}-N_\mathbf{k}N_{\mathbf{k}''}\},\end{aligned} \tag{21.8}$$

which describes the evolution of the oscillation number due to the nonlinear interaction of decay type.

Equations (21.2), (21.5), and (21.8) hold for both isotropic and magnetoactive plasmas.* Then, using these equations, we can describe decay processes in which longitudinal oscillations with the same and different expressions for the spectra participate. The important difference between decay processes in a magnetoactive plasma and ones in an isotropic plasma is that in the latter one cannot have decays in which three oscillations with the same spectrum participate. In a magnetoactive plasma such decays are possible.

Taking the example of the decay of longitudinal oscillations, let us ellucidate by means of Eq. (21.8) the basic features (conservation laws) of decay processes in magnetoactive and isotropic plasmas. To do this, we obtain from Eq. (21. 8) two further equations for the numbers $N_{\mathbf{k}'}$ and $N_{\mathbf{k}''}$ of the two other oscillations:

$$\frac{dN_{\mathbf{k}'}}{dt}=\frac{\hbar}{2\pi}\int d\mathbf{k}d\mathbf{k}''\omega(\mathbf{k})\,\omega'(\mathbf{k}')\,\omega''(\mathbf{k}'')\,\delta(\mathbf{k}'+\mathbf{k}''-\mathbf{k})\,\delta(\omega'(\mathbf{k}')+\omega''(\mathbf{k}'')-\omega(\mathbf{k}))\,Q\,\{N_\mathbf{k}N_{\mathbf{k}''}+N_{\mathbf{k}'}N_\mathbf{k}-N_{\mathbf{k}'}N_{\mathbf{k}''}\}; \tag{21.9}$$

$$\frac{dN_{\mathbf{k}''}}{dt}=\frac{\hbar}{2\pi}\int d\mathbf{k}'d\mathbf{k}\omega(\mathbf{k})\,\omega'(\mathbf{k}')\,\omega''(\mathbf{k}'')\,\delta(\mathbf{k}''+\mathbf{k}'-\mathbf{k})\,\delta(\omega''(\mathbf{k}'')+\omega'(\mathbf{k}')-\omega(\mathbf{k}))\,Q\,\{N_\mathbf{k}N_{\mathbf{k}'}+N_\mathbf{k}N_{\mathbf{k}''}-N_{\mathbf{k}'}N_{\mathbf{k}''}\}. \tag{21.10}$$

Here, the kernels Q are the same as in Eq. (21.8), although the contractions of S depend in accordance with the definition (21.3) on different arguments: $S(\omega, \mathbf{k}, \omega', \mathbf{k}')$, $S(\omega'', \mathbf{k}''; \omega, \mathbf{k})$, $S(\omega', \mathbf{k}'; \omega, \mathbf{k})$. It is at this point that we can clearly see the relationship between the symmetry of the contraction of S that follows from the symmetric representations of § 7 and 19 and the conservation laws that characterize the decay. Multiplying both sides of Eqs. (21.8)-(21.10) by $\omega(\mathbf{k})$, $\omega'(\mathbf{k}')$, $\omega''(\mathbf{k}'')$, respectively, and integrating in the space of wave vectors, we obtain the energy conservation law of the decaying oscillations:

$$\frac{d}{dt}\left\{\int d\mathbf{k}\hbar\omega(\mathbf{k})\,N_\mathbf{k}+\int d\mathbf{k}'\hbar\omega'(\mathbf{k}')\,N_{\mathbf{k}'}+\int d\mathbf{k}''\hbar\omega''(\mathbf{k}'')N_{\mathbf{k}''}\right\}=0. \tag{21.11}$$

* It should be noted that in an isotropic plasma the exact division of characteristic oscillations into electrostatic and solenoidal oscillations means that the purely longitudinal terms can be separated out exactly and not approximately, as in a magnetoactive plasma, from the generalized kinetic equation (3.22a) for decays.

Or, in a different form,

$$\int d\mathbf{k} W_l(\mathbf{k}) + \int d\mathbf{k}' W_l(\mathbf{k}') + \int d\mathbf{k}'' W_l(\mathbf{k}'') = \text{const.} \tag{21.12}$$

Similarly, from the same equations, multiplying them by **k**, **k'**, **k"**, we obtain the conservation law for the total momentum:

$$\frac{d}{dt}\left\{\int d\mathbf{k}\hbar\mathbf{k}N_{\mathbf{k}} + \int d\mathbf{k}'\hbar\mathbf{k}'N_{\mathbf{k}'} + \int d\mathbf{k}''\hbar\mathbf{k}''N_{\mathbf{k}''}\right\} = 0. \tag{21.13}$$

Thus, in decay processes the total energy and total momentum of all the decaying oscillations are conserved.

Using the analogy between N_k and the number of quanta, we introduce the entropy ([62], p. 185, (54.8)

$$S = \frac{1}{(2\pi)^3}\int d\mathbf{k} \ln N_{\mathbf{k}} \tag{21.14}$$

of the gas of oscillations, or quasiparticles, and we consider its variation in decay processes, following [63]. We multiply each of the equations (21.8)-(21.10) by N_k^{-1}, $N_{k'}^{-1}$, $N_{k''}^{-1}$, respectively, and integrate both sides with respect to **dk**. Then, for example, from (21.8) we obtain

$$\frac{d}{dt}\int d\mathbf{k} \ln N_{\mathbf{k}} = \frac{\hbar}{2\pi}\int d\mathbf{k}d\mathbf{k}'d\mathbf{k}''\omega(\mathbf{k})\,\omega'(\mathbf{k}')\,\omega''(\mathbf{k}'')\,\delta(\mathbf{k}-\mathbf{k}'-\mathbf{k}'')\,\delta(\omega(\mathbf{k}) -$$

$$-\omega'(\mathbf{k}') - \omega''(\mathbf{k}''))\,QN_{\mathbf{k}}N_{\mathbf{k}'}N_{\mathbf{k}''}\left\{\frac{1}{N_{\mathbf{k}}} - \frac{1}{N_{\mathbf{k}''}} - \frac{1}{N_{\mathbf{k}'}}\right\}\frac{1}{N_{\mathbf{k}}}. \tag{21.15}$$

Adding this equation term by term to the two others that follow from (21.9) and (21.10), we obtain a relation that describes the change of the entropy:

$$\frac{dS}{dt} = \frac{1}{3}\frac{\hbar}{(2\pi)^4}\int d\mathbf{k}'d\mathbf{k}''d\mathbf{k}\omega(\mathbf{k})\,\omega'(\mathbf{k}')\,\omega''(\mathbf{k}'')\,\delta(\mathbf{k}-\mathbf{k}'-\mathbf{k}'')\times$$

$$\times\delta(\omega(\mathbf{k}) - \omega'(\mathbf{k}') - \omega''(\mathbf{k}''))\,QN_{\mathbf{k}}N_{\mathbf{k}'}N_{\mathbf{k}''}\left\{\frac{1}{N_{\mathbf{k}}} - \frac{1}{N_{\mathbf{k}'}} - \frac{1}{N_{\mathbf{k}''}}\right\}^2. \tag{21.16}$$

The integrand on the right-hand side of (21.16) being positive,* we arrive at the following important conclusion: the entropy of oscillations participating in decay increases. The change in the entropy is equal to zero only in the special case when the doubled number N_k of oscillations $\omega(\mathbf{k})$ is the harmonic mean of the numbers $N_{k'}$ and $N_{k''}$ of oscillations that coalesce to form the $\omega(\mathbf{k})$ oscillation:

$$\frac{1}{N_{\mathbf{k}}} = \frac{1}{N_{\mathbf{k}'}} + \frac{1}{N_{\mathbf{k}''}}. \tag{21.17}$$

Bearing in mind Eqs. (7.3) for the frequencies and the wave vectors in the case of decay $[\omega(\mathbf{k}) = \omega'(\mathbf{k}') + \omega''(\mathbf{k}'')]$, we find that the relation (21.17) holds if the number of oscillations N_k is in-

---

* In the case under consideration, the oscillation numbers N_k are positive and the kernel Q is positive if

$$\left[\frac{\partial\omega\varepsilon'(\omega,\,\mathbf{k})}{\partial\omega}\right]_{\omega=\omega(\mathbf{k})} > 0,\quad \left[\frac{\partial\omega'\varepsilon'(\omega'\,\mathbf{k}')}{\partial\omega'}\right]_{\omega'=\omega'(\mathbf{k}')} > 0,\quad \left[\frac{\partial\omega''\varepsilon'(\omega'',\,\mathbf{k}'')}{\partial\omega''}\right]_{\omega''=\omega''(\mathbf{k}'')} > 0$$

[these inequalities hold, for example, if f_0 is the Maxwell distribution (6.14)].

versely proportional to the frequency $\omega(\mathbf{k})$:

$$N_{\mathbf{k}} \sim \frac{1}{\omega(\mathbf{k})}, \quad N_{\mathbf{k}'} \sim \frac{1}{\omega'(\mathbf{k}')}, \quad N_{\mathbf{k}''} \sim \frac{1}{\omega''(\mathbf{k}'')}, \tag{21.17a}$$

i.e., in other words, if the Rayleigh–Jeans law holds. In its turn, (21.17), yields (21.17a).

We emphasize that these conclusions are valid for decay processes in which there participate oscillations with arbitrary and not only longitudinal polarization; for it follows from the general equation (3.22a) of the decay interaction, in the same way as (21.2), by means of Eq. (17.50) that the equation for the change in the energy of an oscillation of arbitrary polarization is

$$\frac{dW(\mathbf{k})}{dt} = \frac{\omega(\mathbf{k})}{2\pi}\int d\mathbf{k}'d\mathbf{k}''\int_{-\infty}^{+\infty} d\omega'\int_{-\infty}^{+\infty} d\omega'' Q\delta(\omega-\omega'-\omega'')\,\delta(\mathbf{k}-\mathbf{k}'-\mathbf{k}'')\times$$
$$\times\{W(\mathbf{k}')W(\mathbf{k}'')\omega(\mathbf{k})\,\mathrm{sign}\,\Delta''(\omega,\mathbf{k}) - W(\mathbf{k})W(\mathbf{k}')\omega''\,\mathrm{sign}\,\Delta''(\omega'',\mathbf{k}'') -$$
$$- W(\mathbf{k})W(\mathbf{k}'')\omega'\,\mathrm{sign}\,\Delta''(\omega',\mathbf{k}')\}\{\delta(\omega'-\omega'(\mathbf{k}'))\,\delta(\omega''-\omega''(\mathbf{k}'')) +$$
$$+\delta(\omega'+\omega'(-\mathbf{k}'))\,\delta(\omega''+\omega''(-\mathbf{k}'')) + \delta(\omega'-\omega'(\mathbf{k}'))\,\delta(\omega''+\omega''(-\mathbf{k}'')) + \delta(\omega'+\omega'(-\mathbf{k}'))\,\delta(\omega''-\omega''(\mathbf{k}''))\}, \tag{21.18}$$

where in contrast to (21.3) the kernel Q is determined by the more general equation

$$Q \equiv S_{ijs}(\omega,\mathbf{k};\omega',\mathbf{k}')\,S^{*}_{abc}(\omega,\mathbf{k};\omega',\mathbf{k}')\,e^{*}_{ia}(\omega,\mathbf{k})\times$$
$$\times e_{jb}(\omega'',\mathbf{k}'')\,e_{sc}(\omega',\mathbf{k}')\left\{\frac{\partial\omega\Delta'(\omega,\mathbf{k})}{\partial\omega}\right\}^{-1}_{\omega=\omega(\mathbf{k})}\left\{\frac{\partial\omega'\Delta'(\omega',\mathbf{k}')}{\partial\omega'}\right\}^{-1}\left\{\frac{\partial\omega''\Delta'(\omega'',\mathbf{k}'')}{\partial\omega''}\right\}^{-1}, \tag{21.19}$$

in which the Hermitian tensors e_{ij} are the corresponding tensor parts of the spectral functions of the decaying oscillations, and Δ' and Δ'' are the real and the imaginary parts of the determinant Δ [see (17.10)], by means of which the spectra and damping constants of the oscillations can be found from the dispersion relation (2.14). The representation of the kernel Q in the form of the expression (21.19), which is taken in front of Eq. (21.18) as a common factor, can be justified by means of the symmetric relations (7.16) and (19.97) for the three-index tensors S in the same way as in the special case of electrostatic oscillations.* Making the same assumptions as in the derivation of Eq. (21.8) from (21.19), we find an equation for the number N_k of oscillations of arbitrary polarization, this having the same superficial form as (21.8). The difference (and it is important) is that the kernel Q is now determined not by (21.3) but by (21.19), in which all the frequencies must be replaced by the corresponding spectra.

We now fulfill the outlined program for a different type of nonlinear process: induced scattering of characteristic oscillations on plasma particles. For this, we use the general equation for nonlinear scattering (3.22b) and obtain first, as in the case of decay, an equation for the evolution of the energy $W_l(\mathbf{k})$ of a longitudinal oscillation, substituting into (3.22b) the expressions (17.60) for the spectral functions in terms of the energy densities and allowing for only Coulomb scattering (we omit the term with linear damping):

$$\frac{dW_l(\mathbf{k})}{dt} = -\frac{\omega(\mathbf{k})}{\pi^2}W_l(\mathbf{k})\int d\mathbf{k}'W_l(\mathbf{k}')\left\{\frac{\partial\omega\varepsilon'(\omega,\mathbf{k})}{\partial\omega}\right\}^{-1}_{\omega=\omega(\mathbf{k})}\left\{\frac{\partial\omega'\varepsilon'(\omega',\mathbf{k}')}{\partial\omega'}\right\}^{-1}_{\omega'=\omega'(\mathbf{k}')}\times$$

---

* As in the case of the longitudinal waves (21.6)-(21.7), the left-hand side of (21.8) contains the substantial derivative $\partial/\partial t = \partial/\partial t + \mathbf{v}_{gr}(\partial/\partial\mathbf{r})$ with group velocity

$$v_{gr} = \frac{d\omega(\mathbf{k})}{d\mathbf{k}} = -\left\{\frac{\partial\Delta'(\omega,\mathbf{k})}{\partial\mathbf{k}}\right\}\left\{\frac{\partial\Delta'(\omega,\mathbf{k})}{\partial\omega}\right\}^{-1}_{\omega=\omega(\mathbf{k})},$$

determined by the spectrum of the corresponding oscillation.

$$\times \operatorname{Im}\{[V(\omega, \mathbf{k}; -\omega', -\mathbf{k}') - S(\omega, \mathbf{k}; \omega', \mathbf{k}')\, S(\omega-\omega', \mathbf{k}-\mathbf{k}'; \omega, \mathbf{k})\, \varepsilon^{-1}(\omega-\omega', \mathbf{k}-\mathbf{k}')] +$$
$$+ [V(\omega, \mathbf{k}; \omega', \mathbf{k}') - S(\omega, \mathbf{k}; -\omega', -\mathbf{k}')\, S(\omega+\omega', \mathbf{k}+\mathbf{k}'; \omega, \mathbf{k})\, \varepsilon^{-1}(\omega+\omega', \mathbf{k}+\mathbf{k}')]\}. \tag{21.20}$$

In this equation, as also in the equations for decays, the spectra $\omega(\mathbf{k})$ and $\omega'(\mathbf{k}')$ of the scattered oscillations are given by positive-definite expressions. The quantities ε, S, and V are the contractions of the corresponding tensors with longitudinal sets of wave vectors of the interacting and the virtual oscillations (see § 7, 8, 19, 20). In the curly brackets in (21.20) we have separated out two groups of terms represented by the two square brackets. In the first brackets, the longitudinal contraction $V(\omega, \mathbf{k}; -\omega', -\mathbf{k}')$ depends on the differences of the frequencies and the wave vectors, $\omega - \omega'$ and $\mathbf{k} - \mathbf{k}'$, of the scattered oscillations (to simplify the notation we have replaced the spectra $\omega(\mathbf{k})$ and $\omega'(\mathbf{k}')$ in the arguments of the functions in the curly brackets of (21.20) by the frequencies ω and ω' that represent them]. The second term in the first square brackets also contains the contractions S and ε, which depend on the differences $\omega - \omega'$ and $\mathbf{k} + \mathbf{k}'$ of the frequencies and wave vectors of the scattered oscillations. These two groups of terms correspond to two essentially different scattering mechanisms. The first square brackets corresponds to the process of re-emission of an oscillation: an incident oscillation is "absorbed" by the plasma, momentum being exchanged with the plasma particles through the virtual oscillation with frequency $\omega - \omega'$ and $\mathbf{k} - \mathbf{k}'$, and is then re-emitted in the form of the second of the interacting oscillations. Similarly, the second mechanism which corresponds to the second square brackets in (21.20), can be represented as a process of simultaneous emission (or absorption) of two interacting oscillations. All the scattering processes considered in this monograph belong to the first group* and occur with a much greater probability than processes of the other type.

To establish the characteristic features of induced scattering of waves on particles, we go over in Eq. (21.20) to the oscillation number in accordance with the definition (21.1):

$$\frac{dN_{\mathbf{k}}}{dt} = -\frac{\hbar}{\pi^2}\int d\mathbf{k}' N_{\mathbf{k}} N_{\mathbf{k}'} \left\{\frac{\partial \varepsilon'(\omega, \mathbf{k})}{\partial \omega}\right\}^{-1}_{\omega=\omega(\mathbf{k})} \left\{\frac{\partial \varepsilon'(\omega', \mathbf{k}')}{\partial \omega'}\right\}^{-1}_{\omega'=\omega'(\mathbf{k}')} \times$$
$$\times \operatorname{Im}\{[V(\omega, \mathbf{k}; -\omega', -\mathbf{k}') - S(\omega, \mathbf{k}; \omega', \mathbf{k}')\, S(\omega-\omega', \mathbf{k}-\mathbf{k}'; \omega, \mathbf{k})\, \varepsilon^{-1}(\omega-\omega', \mathbf{k}-\mathbf{k}')]$$
$$+ [V(\omega, \mathbf{k}; \omega', \mathbf{k}') - S(\omega, \mathbf{k}; -\omega', -\mathbf{k}')\, S(\omega+\omega', \mathbf{k}+\mathbf{k}'; \omega, \mathbf{k})\, \varepsilon^{-1}(\omega+\omega', \mathbf{k}+\mathbf{k}')]\}. \tag{21.21}$$

Using the symmetric representations for the contractions S and V (§ 7, 8, 19, 20), we find that the second square brackets (21.21) is symmetric under the transposition $\mathbf{k} \rightleftharpoons \mathbf{k}'$ while the first is antisymmetric. This assertion directly yields a conservation law of the total number of oscillations for the induced scattering described by the terms in the first square brackets (first mechanism):

$$\frac{d}{dt}\int d\mathbf{k} N_{\mathbf{k}} = 0, \tag{21.22}$$

if the interacting oscillations have one and the same spectrum. If the scattering is heterogeneous – waves with different spectra interact – then (21.22) is replaced by the more general conservation law of the total number of oscillations:

$$\frac{d}{dt}\left\{\int d\mathbf{k} N_{\mathbf{k}} + \int d\mathbf{k}' N_{\mathbf{k}'}\right\} = 0. \tag{21.23}$$

This result is in complete agreement with the physical picture we have described of the first scattering mechanism: the oscillations are only re-emitted but are neither absorbed nor created, as occurs in processes with the second mechanism described by the second square brackets in (21.21), in which the number of oscillations is accordingly not conserved. This

* An exception is the induced scattering of short-wave ion sound [see (9.77)] in an isotropic plasma investigated in Ch. III.

last assertion is a direct consequence of the symmetry of the contractions $V(\omega, k; \omega', k')$ and $S(\omega + \omega', k + k'; \omega, k)$.

These results confirm once more our deductions in the foregoing chapter when we studied the concrete processes of induced scattering of characteristic oscillations of an isotropic plasma with a small change in the frequency ($\omega'' = \omega - \omega' \ll \omega$) and establish the general features of waves scattering in a magnetoactive plasma. Let us now consider the change of entropy in a process of induced scattering in which the number of oscillations is conserved. In this case, the first term on the right-hand side of Eq. (21.21) can be written in the form

$$-\int d\mathbf{k}' N_{\mathbf{k}} N_{\mathbf{k}'} Q(\mathbf{k}, \mathbf{k}'), \tag{21.24}$$

where the kernel $Q(\mathbf{k}, \mathbf{k}')$ is antisymmetric:

$$Q(\mathbf{k}, \mathbf{k}') = -Q(\mathbf{k}', \mathbf{k}). \tag{21.25}$$

Using this notation, we can characterize the rate of change of the entropy of the equation

$$(2\pi)^3 \frac{dS}{dt} = \frac{1}{2}\int d\mathbf{k} d\mathbf{k}' (N_{\mathbf{k}} - N_{\mathbf{k}'}) Q(\mathbf{k}, \mathbf{k}'). \tag{21.26}$$

The equation for the change of the energy of oscillations of arbitrary polarization in the case of induced scattering can be obtained from the general equation (3.22b) in the same way as (21.20) if we use (17.50) instead of (17.60) for the spectral functions:

$$\begin{aligned}\frac{dW(\mathbf{k})}{dt} = &-\frac{\omega(\mathbf{k})}{\pi^2} W(\mathbf{k}) \int d\mathbf{k}' W(\mathbf{k}') \left\{\frac{\partial \omega \Delta'(\omega, \mathbf{k})}{\partial \omega}\right\}^{-1}_{\omega=\omega(\mathbf{k})} \left\{\frac{\partial \omega' \Delta'(\omega', \mathbf{k}')}{\partial \omega'}\right\}^{-1}_{\omega'=\omega'(\mathbf{k}')} \times \\ &\times \operatorname{Im}\{[V_{iajb}(\omega, \mathbf{k}; -\omega', -\mathbf{k}') e_{ji}(\omega, \mathbf{k}) e_{ba}(-\omega', -\mathbf{k}') - \\ &- S_{ijs}(\omega, \mathbf{k}; \omega', \mathbf{k}') S_{bca}(\omega - \omega', \mathbf{k} - \mathbf{k}'; \omega, \mathbf{k}) e_{ai}(\omega, \mathbf{k}) e_{sc}(\omega', \mathbf{k}') \times \\ &\times A_{jb}(\omega - \omega', \mathbf{k} - \mathbf{k}')] + [V_{iajb}(\omega, \mathbf{k}; \omega', \mathbf{k}') e_{ji}(\omega, \mathbf{k}) e_{ba}(\omega', \mathbf{k}') - \\ &- S_{ijs}(\omega, \mathbf{k}; -\omega', -\mathbf{k}') S_{bca}(\omega + \omega', \mathbf{k} + \mathbf{k}'; \omega, \mathbf{k}) e_{ai}(\omega, \mathbf{k}) e_{sc}(-\omega', -\mathbf{k}') A_{jb}(\omega + \omega', \mathbf{k} + \mathbf{k}')]\}.\end{aligned} \tag{21.27}$$

The transition to the number of oscillations is here made in exactly the same way as for Coulomb scattering of longitudinal waves:

$$\begin{aligned}\frac{dN_{\mathbf{k}}}{dt} = &-\frac{\hbar}{\pi^2} \int d\mathbf{k}' N_{\mathbf{k}} N_{\mathbf{k}'} \left\{\frac{\partial \Delta'(\omega, \mathbf{k})}{\partial \omega}\right\}^{-1}_{\omega=\omega(\mathbf{k})} \left\{\frac{\partial \Delta'(\omega', \mathbf{k}')}{\partial \omega'}\right\}^{-1}_{\omega'=\omega'(\mathbf{k})} \times \\ &\times \operatorname{Im}\{[V_{iajb}(\omega, \mathbf{k}; -\omega', -\mathbf{k}') e_{ji}(\omega, \mathbf{k}) e_{ba}(-\omega', -\mathbf{k}') - \\ &- S_{ijs}(\omega, \mathbf{k}; \omega', \mathbf{k}') S_{bca}(\omega - \omega', \mathbf{k} - \mathbf{k}'; \omega, \mathbf{k}) e_{ai}(\omega, \mathbf{k}) e_{sc}(\omega', \mathbf{k}') \times \\ &\times A_{jb}(\omega - \omega', \mathbf{k} - \mathbf{k}')] + [V_{iajb}(\omega, \mathbf{k}; \omega', \mathbf{k}') e_{ji}(\omega, \mathbf{k}) e_{ba}(\omega', \mathbf{k}') - \\ &- S_{ijs}(\omega, \mathbf{k}; -\omega', -\mathbf{k}') S_{bca}(\omega + \omega', \mathbf{k} + \mathbf{k}'; \omega, \mathbf{k}) e_{ai}(\omega, \mathbf{k}) e_{sc}(-\omega', -\mathbf{k}') A_{jb}(\omega + \omega', \mathbf{k} + \mathbf{k}')]\}.\end{aligned} \tag{21.28}$$

The structure of these equations is similar to the structure of the equations (21.20) and (21.21) for longitudinal waves. Therefore, all our conclusions concerning the conservation of the number of oscillations and the change of entropy drawn above for scattering of electrostatic oscillations remain true in the case (21.27) and (21.28) of the nonlinear interaction of oscillations of arbitrary polarization.

CHAPTER V

SPECIFIC NONLINEAR PROCESSES IN A MAGNETOACTIVE PLASMA

This chapter has five sections. The first three are devoted to the study of decay processes and the last two contain an analysis of induced Coulomb scattering of quasielectrostatic oscillations of a magnetoactive plasma on the plasma particles.

In § 22, we discuss decay of quasilongitudinal waves that are weakly damped in a cold plasma and in a plasma with approximately equal electron and ion temperatures (see [61] and [50], § 20). Here we obtain the kernels of more than ten nonlinear integrodifferential equations describing such processes. In particular, we consider decay processes in which there participate the low-frequency oscillations p [spectrum (18.4)], ion-cyclotron harmonics $\tilde{c}$ [spectrum [18.42)], electron-cyclotron harmonics c_i [spectrum (18.57)], and electron Langmuir oscillations modified by an external magnetic field [spectrum (18.24)].

In § 23 we also investigate decay processes but in this case with the possible participation of the quasilongitudinal waves that can exist (i.e., are weakly damped) only in a nonisothermal magnetoactive plasma with strongly differing ion and electron temperatures (see [28], § 7 and [64, 65]). These include above all oscillations of the type of ion sound $s_i^\prime$ (18.61), ion sound s_i (7.40a), and the form c_i of the ion-cyclotron harmonic (18.46). Here we give more than 20 kernels of nonlinear equations.

In § 24 we consider the induced combination scattering of high-frequency transverse electromagnetic waves in a cold magnetoactive plasma or, in other words, the coalescence of a high-frequency transverse electromagnetic wave ($\omega' \gg \Omega_e$) with a characteristic oscillation of the plasma leading to the formation of another high-frequency transverse electromagnetic wave ($\omega \gg \Omega_e$) (see [28], § 6 and [66]). This section is the only one in the chapter in which we discuss the nonlinear interaction of nonlongitudinal waves in a magnetoactive plasma.

Section 25 is based on the results of [67] (see also [28], § 8). Here we consider induced scattering on particles of a magnetoactive plasma of the low-frequency oscillation p with the spectrum (18.4), and of the electron oscillation s_e and the low-frequency ion sound $s_i^\prime$ through a slow (with phase velocity less than the ion thermal velocity) electrostatic virtual oscillation. We establish the relative importance of scattering of quasilongitudinal oscillations on free particles and on polarization clouds of the virtual wave. We consider only scattering in which there is no change (transformation) of the wave spectra as a result of the nonlinear interaction. The discussion of induced Coulomb scattering of ion-cyclotron and electron-cyclotron quasilongitudinal waves through a slow virtual oscillation $[(\omega''/|k_z''|) \ll v_{T_i}]$ in accordance with [68] is contained in § 26. As in the case of an isotropic plasma (see Ch. III), we here establish the direction of transfer of spectral energy density, and in concrete examples we find confirmation of the general assertion that there is conservation of the energy of oscillations with fixed frequency that is a multiple of (or equal to) the electron or ion gyroscopic frequency. It is assumed that the nonlinear interaction does not change the spectrum of the scattered cyclotron oscillations.

§ 22. Decay of Quasilongitudinal Characteristic Oscillations of an Isothermal Magnetoactive Plasma

We choose quasilongitudinal oscillations to illustrate the equations of nonlinear interaction in a magnetoactive plasma on account of their important role in the evolution of nonlinear effects in a plasma, which we have already noted in the linear theory of such oscillations [see formula (17.59) and the remarks on this formula in § 17, Ch. IV]. In addition, the analysis of the nonlinear interaction of longitudinal oscillations is as a rule much simpler* than in the

* When one studies decay processes with the participation of three oscillations of a magnetoactive plasma of arbitrary polarization, the finding of the kernel of the integrodifferential equation requires the calculation of five tensors (two three-index tensors S and three two-index tensors characterizing the spectral functions of the electric fields of the oscillations), each of which, in its turn, consists of six simpler tensor terms [see formulas (17.64) and (19.18) in a cold magnetoactive plasma]. In general, this procedure leads to many more terms than the calculation of the analogous contraction for waves of longitudinal polarization.

general case of arbitrary polarization and yields readily comprehensible equations for the spectral densities of the energy of oscillations belonging to different branches of the spectrum.

In this section we shall consider processes of coalescence of two longitudinal oscillations of a magnetoactive plasma into a third longitudinal oscillation. We shall take into account only such oscillations, for whose existence we do not need to assume that the plasma is nonisothermal, i.e., we consider oscillations that are weakly damped in a cold magnetoactive plasma ($v_T = 0$) and in a plasma with approximately equal ion and electron temperatures. The general properties of the equations that describe three-wave decay have already been discussed in detail (see § 21), and we shall therefore restrict ourselves here to studying only the kernels Q in (21.3) of the integrodifferential equation (21.5). To simplify the classification of the processes discussed below, we shall denote each of them by an abbreviated total symbol $\alpha\beta\gamma$ (compare [69], pp. 42-43), in which α denotes the oscillation with spectrum $\omega(\mathbf{k})$ that decays into two other oscillations β or γ or arises as a result of the coalescence of the oscillations β and γ with spectra $\omega'(\mathbf{k}')$ and $\omega''(\mathbf{k}'')$. In a decay process of such form

$$\omega(\mathbf{k}) = \omega'(\mathbf{k}') + \omega''(\mathbf{k}''), \qquad \mathbf{k} = \mathbf{k}' + \mathbf{k}'', \tag{22.1}$$

and the letters α, β, and γ take the values p, c_i, c_e, s introduced in § 18 to designate the corresponding characteristic oscillations of a magnetoactive plasma.

We begin with the coalescence of two p oscillations [Eq. (18.4)] into a third such oscillation.

1. ppp. The equality of the frequencies (22.1) reduces to the relation

$$\frac{|k_z|}{k}\omega_{L_e} = \frac{|k_z'|}{k'}\omega_{L_e} + \frac{|k_z''|}{k''}\omega_{L_e}, \tag{22.2}$$

which holds, for example, in the case of interaction of oscillations with the same wave numbers, $k \simeq k' \simeq k''$, and projections of their wave vectors onto the direction of $\mathbf{B}_0$ that satisfy the relation

$$|k_z| = |k_z'| + |k_z''|, \tag{22.3}$$

which follows, in its turn, from the condition of conservation of the wave vectors ($\mathbf{k} = \mathbf{k}' + \mathbf{k}''$), which have the same direction (along $\mathbf{B}_0$). In accordance with the definition (21.3), the kernel Q depends on the square of the modulus of the longitudinal contraction of the three-index tensor, which in this case is a variant of the expression (19.87) for $A_{0,0} = 1$ (because of the weak thermal motion):

$$S = -i\frac{e}{m}\frac{\omega_L^2}{\omega\omega'\omega''}\frac{1}{kk'k''}\left\{k_z k_z' k_z'' - i\frac{(\mathbf{k}[\mathbf{k}'\mathbf{h}])}{\Omega}(k_z\omega' - k_z'\omega)\right\}\left\{\frac{k_z}{\omega} + \frac{k_z'}{\omega'} + \frac{k_z''}{\omega''}\right\}. \tag{22.4}$$

Comparison of the electron and ion parts of the contraction shows that the kernel Q contains only the electron part, which exceeds the ion part by a factor $(M/m)^2$ if one allows for only the first term in the first curly brackets in (22.4) and by a factor M/m for the second term, which is "transverse" with respect to $\mathbf{h}$. As a result, we obtain for Q the expression

$$Q = \frac{1}{32\pi}\frac{1}{N_e mc^2}\left(\frac{c}{\omega_{L_e}}\right)^2\left(1 + \frac{\omega_{L_i}^2}{\Omega_i^2}\right)(k + k' + k'')^2\left\{1 + \frac{(\mathbf{k}[\mathbf{k}'\mathbf{h}])^2}{(k_z k_z' k_z'')^2}\frac{\omega_{L_e}^2}{\Omega_e^2}\left(1 + \frac{\omega_{L_i}^2}{\Omega_i^2}\right)^{-1}\left(k_z\frac{|k_z'|}{k'} - k_z'\frac{|k_z|}{k}\right)^2\right\}, \tag{22.5}$$

in which the second term in the curly brackets corresponds to our allowing for the part of the contraction (22.4) that is transverse relative to the direction of $\mathbf{B}_0$. In the special case of almost equal wave numbers of the interacting oscillations, $k \simeq k' \simeq k''$, and wave vectors with projections $k_z = |k_z|$, $k'_z = |k'_z|$, we can write down the following estimate for the kernel:

$$Q=\frac{9}{32\pi}\frac{1}{N_e mc^2}\left(\frac{ck}{\omega_{L_e}}\right)^2\left(1+\frac{\omega_{L_i}^2}{\Omega_i^2}\right). \tag{22.6}$$

2. Decay with the participation of these oscillations (18.26) l-l-l- for angles ϑ, ϑ', ϑ'' between the wave vectors **k**, **k'**, **k"** and the external magnetic field B_0 near $\pi/2$ ($\cos\vartheta \sim \cos\vartheta' \sim \cos\vartheta'' \ll 1$; see the inequalities (18.25)) can also be described by the kernel (22.5) if the sum $(1+\omega_{Li}^2/\Omega_i^2)$ in it is replaced by $(1+\omega_{Le}^2/\Omega_e^2)$. In this case one can also obtain a slightly different form of the expression for the kernel by using the fact that in the decay process l-l-l- the spectra of the interacting oscillations and the contraction S are determined solely by the electrons (the ions can be assumed fixed: $M=\infty$). Namely, on account of the relation (19.95) for the contraction S, the kernel Q can be represented in the form

$$Q=\frac{1}{32\pi N_e mc^2}\left(\frac{c^2}{\omega_{L_e}^2 k''^2}\right)\frac{[\mathbf{kk}']^4}{(kk')^2}\left(1+\frac{\omega_{L_e}^2}{\Omega_e^2}\right)^{-3}. \tag{22.7}$$

3. At angles ϑ, ϑ', ϑ'' not too near $\pi/2$ the magnetic-field-modified electron Langmuir oscillations $l_\pm$ are described by the spectra (18.24). The corresponding decay with the participation of only these oscillations is described by Eq. (21.5) with kernel

$$Q=\frac{1}{4\pi N_e mc^2}\left(\frac{c}{k''\omega_{L_e}}\right)^2\frac{[\mathbf{kk}']^4}{(kk')^2}\left[\frac{\partial\omega\varepsilon'(\omega,\mathbf{k})}{\partial\omega}\right]^{-1}_{\omega=\omega(\mathbf{k})}\left[\frac{\partial\omega'\varepsilon'(\omega',\mathbf{k}')}{\partial\omega'}\right]^{-1}_{\omega'=\omega'(\mathbf{k}')}\left[\frac{\partial\omega''\varepsilon'(\omega'',\mathbf{k}'')}{\partial\omega''}\right]^{-1}_{\omega''=\omega''(\mathbf{k}'')}, \tag{22.8}$$

in which the quantities $[\partial\omega\varepsilon'/\partial\omega]$ are given by

$$\left[\frac{\partial\omega\varepsilon'(\omega,\mathbf{k})}{\partial\omega}\right]=2\frac{\omega_{L_e}^2}{\omega^2}\left\{\frac{k_\perp^2}{k^2}\frac{\omega^4}{(\omega^2-\Omega_e^2)^2}+\frac{k_z^2}{k^2}\right\}, \tag{22.9}$$

which follows directly from (18.23) for the longitudinal permittivity of a cold magnetoactive plasma (here $\varepsilon' = \mathrm{Re}\,\varepsilon = \varepsilon$) and which we have already used to calculate the spectral function (18.28). Note a characteristic feature of the kernels (22.7) and (22.8) that arises from the form of the contraction (19.95): they vanish in the case of interaction of oscillations with parallel wave vectors $[\mathbf{kk}'] = 0$.

4. $\tilde{c}_i\tilde{c}_i\tilde{c}_i$. Considering the interaction of ion-cyclotron harmonics (18.42) with frequencies $\omega \simeq (l+n)\Omega_i$, $\omega' \simeq n\Omega_i$, $\omega'' \simeq l\Omega_i$, we must use the electron part of the contraction S_e (19.87), since the ion part S_i of the longitudinal contraction, which is determined by Eq. (19.90) and occurs in Q, is small compared with the electron part:

$$S_i \sim M^{-2}[\omega-(l+n)\Omega_i]^{-4} \ll M^{-2}(k_z v_{T_i})^{-4} \lesssim m^{-2}\Omega_i^{-4} \sim S_e. \tag{22.10}$$

Then the kernel Q can be written in the form

$$\begin{aligned}Q=&\frac{1}{4\pi N_e mc^2}\left(\frac{M}{m}\right)^3\left[\frac{k_z k'_z k''_z \rho_i^3}{n(n+l)l}\right]^2\frac{c^2}{\Omega_i^2}\left(\frac{k_z}{n+l}+\frac{k'_z}{n}+\frac{k''_z}{l}\right)^2 A_{l+n}(z_i)A_n(z'_i)A_l(z''_i)\times\\&\times\left\{1+(kr_{D_i})^2\left[1+\frac{k_\perp^2}{k^2}\frac{\omega_{L_i}^2}{\Omega_i^2}-\frac{k_z^2}{k^2}\frac{\omega_{L_e}^2}{(l\Omega_i+n\Omega_i)^2}\right]\right\}^{-2}\left\{1+(k'r_{D_i})^2\left[1+\left(\frac{k'_\perp}{k'}\frac{\omega_{L_i}}{\Omega_i}\right)^2-\right.\right.\\&\left.\left.-\left(\frac{k'_z\omega_{L_e}}{k'n\Omega_i}\right)^2\right]\right\}^{-2}\left\{1+(k''r_{D_i})^2\left[1+\left(\frac{k''_\perp}{k''}\frac{\omega_{L_i}}{\Omega_i}\right)^2-\left(\frac{k''_z\omega_{L_e}}{k''l\Omega_i}\right)^2\right]\right\}^{-2},\end{aligned} \tag{22.11}$$

if we retain in S_e only the first term (longitudinal with respect to **h**) of (19.87). Similarly, we obtain the kernels of the integrodifferential equations (21.5), which describe the decay processes $\tilde{c}_i c_i c_i$, $\tilde{c}_i\tilde{c}'_i\tilde{c}'_i$, and others in which only ion-cyclotron harmonics participate. They arise naturally from the contraction (19.87) and (19.90) and have a structure similar to (12.11).

5. For the decay process $c_e c_e c_e$ with the participation of three electron-cyclotron harmonics (18.57) with frequencies $\omega \approx (l+n)\Omega_e$, $\omega' \approx n\Omega_e$, $\omega'' \approx l\Omega_e$, we give the characteristic time of nonlinear interaction

$$\tau \sim 50 \frac{N_e mc^2}{W_l k^3} \frac{(n\Omega_e)^7}{(\omega_{L_e} c v_{T_e}^2 k^3)^2} (kr_{D_e})^{-8}, \tag{22.12}$$

which holds provided*

$$n \sim l \sim l+n, \quad A_n(z_e) \sim |A^e_{n,l}|^2 \sim 1, \quad (kr_{D_e})^2 \gg 1. \tag{22.13}$$

A rigorous expression for the kernel Q arises from the contraction (19.90) when the frequencies in this contraction are replaced by the corresponding spectra (18.57) of the electron-cyclotron oscillations.

6. As another example we can take the coalescence of an electron Langmuir oscillation l with a low-frequency oscillation p into an electron Langmuir oscillation: llp. A nonlinear interaction of this kind is characterized by the contraction

$$S \simeq S_e \simeq -i \frac{e}{m} \frac{\omega_{L_e}^2}{\omega^2} \left(\frac{k_z''}{\omega''}\right)^2 \frac{1}{k''} \left(\frac{\mathbf{kk}'}{kk'}\right), \tag{22.14}$$

which can be obtained from Eq. (19.23) [or the tensor (19.31)] if allowance is made for the electron gyroscopic frequency's being small compared with the frequencies of the Langmuir waves:

$$\omega, \ \omega' \gg \Omega_e \gg \omega''. \tag{22.15}$$

The corresponding kernel

$$Q = \frac{1}{32\pi N_e mc^2} \left(\frac{\mathbf{kk}'}{kk'}\right)^2 \left(\frac{ck''}{\omega_{L_e}}\right)^2 \left[1 + \frac{\omega_{L_i}^2}{\Omega_i^2}\right] \tag{22.16}$$

vanishes (in the given approximation) if the vectors of the Langmuir oscillations are perpendicular.

7. The coalescence $ll l_\pm$ is possible only if the frequency of the oscillations $l_\pm$ is appreciably lower than the electron plasma frequency. Since the existence of an "isotropic" electron Langmuir oscillation l requires the electron gyroscopic frequency to be small compared with the electron Langmuir frequency ($\omega_{Le} \gg \Omega_e$), and when this is the case the spectra (18.24) can be represented in the form

$$\omega^2 = (k_z^2/k^2)\,\Omega_e^2, \quad \omega^2 = \omega_{L_e}^2, \tag{22.17}$$

the decay $ll l_-$ is possible only if the first branch in (22.17) participates. Then all the oscillations in the decay are determined exclusively by electron variables, and the motion of the ions can be ignored:

$$Q = \frac{1}{32\pi N_e mc^2} \left(\frac{c}{\omega_{L_e} k''}\right)^2 \frac{[\mathbf{kk}']^4}{(kk')^2} \frac{\Omega_e^2}{\omega_{L_e}^2}. \tag{22.18}$$

This kernel Q is a special case of (22.8).

---

* The quantity $A^e_{n,l}$ in (22.13) is determined by the integral (19.67). The superscript e indicates that $A^e_{n,l}$ corresponds to the expression (19.67) for the electron values of the thermal velocity v_{Te} and Ω_e. This notation $A^e_{n,l}$ (or $A^i_{n,l}$ for ions) for the integral (19.67) for the given species of particle (electrons) will also be used in what follows [see, for example, formula (22.28) below].

8. The similar process of coalescence of electron Langmuir oscillations with an ion-cyclotron harmonic, $ll\tilde{c}_i$, is described by the contraction ($\omega'' \approx n\Omega_i$)

$$S = -i\frac{e}{m}\frac{\omega_{Le}^2}{\omega^2}\left(\frac{k_z''}{\omega''}\right)^2\frac{1}{k''}\left(\frac{\mathbf{kk}'}{kk'}\right) - i\frac{e_i}{M}\frac{\omega_{Li}^2}{\omega^2}\frac{k_z''^2}{(\omega''-n\Omega_i)^2}\frac{1}{k''}\left(\frac{\mathbf{kk}'}{kk'}\right)A_n(z_i''), \tag{22.19}$$

in which the ion part can be ignored if

$$1 > A_n(z_i''), \tag{22.20}$$

so that the kernel Q takes the form

$$Q = \frac{1}{16\pi N_e mc^2}\left(\frac{\mathbf{kk}'}{kk'}\right)^2\frac{M}{m}\left(\frac{ck_z''}{\Omega_i}\right)^2\frac{(k_z''\rho_i)^2}{n^4}A_n(z_i'')\left\{1+(k''r_{D_i})^2\left[1+\left(\frac{k_\perp''\omega_{Li}}{k''\Omega_i}\right)^2-\left(\frac{k_z''\omega_{Le}}{k''n\Omega_i}\right)^2\right]\right\}^{-2}. \tag{22.21}$$

Here the curly brackets are written down with allowance for the expression (18.42) for the $\tilde{c}_i$ spectrum. But if we take the modification (18.43) or (18.44) of this spectrum, the expression in the curly brackets on the right-hand side of (22.1) must be replaced by the reciprocal of the coefficient of $A_n(z_i'')$ in these spectra. Similarly, for the coalescence $ll\tilde{c}_i'$, when the spectrum of the ion-cyclotron oscillation is determined by Eq. (18.14) in a cold magnetoactive plasma, we have

$$Q = \frac{1}{16\pi N_e mc^2}\left(\frac{\mathbf{kk}'}{kk'}\right)^2\left(\frac{ck_z''}{\Omega_i}\right)^2. \tag{22.22}$$

9. When an electron Langmuir oscillation l interacts with an electron cyclotron harmonic llc_e leading to the formation of an electron Langmuir oscillation, we obtain, ignoring the motion of the ions, the following kernel for the equation that describes this coalescence [$\omega'' \simeq n\Omega_e$, spectrum (18.57)]:

$$Q = \frac{1}{16\pi N_e mc^2}\left(\frac{\mathbf{kk}'}{kk'}\right)^2\left(\frac{\sqrt{cv_{T_e}}\,k_z''}{n\Omega_e}\right)^4 A_n^{-3}(z_e'')\left\{(k''r_{D_e})^2-\frac{(k_\perp''\rho_e)^2}{n^2-1}-\left(\frac{k_z''\rho_e}{n}\right)^2\right\}^2. \tag{22.23}$$

10. As a generalization of the decay processes considered in § 22.6-22.9, we can consider processes in which the magnetic-field-modified Langmuir oscillation $l_\pm$ participates instead of the high-frequency Langmuir oscillation l. We then obtain kernels Q similar to those obtained above but with a slightly more complicated structure on account of the more complicated expression for this spectrum (18.24). For example, the coalescence $l_\pm l_\pm p$ is described by the kernel

$$Q = \frac{1}{8\pi N_e mc^2}\left(\frac{ck''}{\omega_{L_e}}\right)^2\left(\frac{\mathbf{kk}'}{kk'}\right)^2\left(1+\frac{\omega_{L_i}^2}{\Omega_i^2}\right)\left[\frac{\partial\omega\varepsilon'(\omega,k)}{\partial\omega}\right]^{-1}_{\omega=\omega(\mathbf{k})}\left[\frac{\partial\omega'\varepsilon'(\omega',\mathbf{k}')}{\partial\omega'}\right]^{-1}_{\omega'=\omega'(\mathbf{k}')}, \tag{22.24}$$

in whose derivation we have used the electron part of the contraction (19.22)

$$S = -i\frac{e}{m}\frac{\omega_{Le}^2}{\omega^2}\frac{k_i\Gamma_{is}(\omega)k_s'}{kk'}\frac{1}{k''}\left(\frac{k_z''}{\omega''}\right)^2 \tag{22.25}$$

and the simplifying equation*

$$i\omega_{L_e}^2\frac{k_i\Gamma_{ij}(\omega)k_j'}{\omega} = -\mathbf{kk}'. \tag{22.26}$$

---

* This equation is a direct consequence of Maxwell's equations for the longitudinal oscillation $l_\pm$ [cf. (19.92)], $\varepsilon_{ij}(\omega, 0)k_i = \{\delta_{ij} - i(\omega_{Le}^2/\omega)\Gamma_{ij}(\omega)\}k_j = 0$ and ther Hermiticity of the tensor $\Gamma_{ij} = \Gamma_{ji}^*$.

11. A further example of a decay process with different spectra of the participating oscillations is the process $c_e c_e p$ that gives rise to an electron-cyclotron harmonic c_e ($\omega \simeq n\Omega_e$) when it coalesces with low-frequency oscillations p:

$$Q = \frac{1}{8\pi N_e mc^2}\left(\frac{ck''}{\omega_{L_e}}\right)^2 \frac{\rho_e^2}{n^2} v_{T_e}^2 A_n^{-1}(z_e) A_n^{-1}(z_e')\left(1+\frac{\omega_{L_i}^2}{\Omega_i^2}\right)^2 \left[(k' r_{D_e})^2 - \frac{(k'_\perp \rho_e)^2}{n^2-1} - \left(\frac{k'_z \rho_e}{n}\right)^2\right]^2 |k_i \Gamma_{is}(\omega - n\Omega_e) k'_s|^2 I_n^2(k_\perp k'_\perp \rho_e^2) \exp\{-(k_\perp^2 + k_\perp'^2)\rho_e^2\} \tag{22.27}$$

and the process $c_e c_e \tilde{c}_i$ of coalescence of an electron-cyclotron harmonic with the ion-cyclotron (18.42) ($\omega'' = l\Omega_i$):

$$Q = \frac{1}{4\pi N_e mc^2}\frac{M}{m}\left(\frac{k''_z v_{T_e}}{\omega''}\right)^4 \left(\frac{e}{e_i}\right)^2 \left(\frac{c v_{T_e}}{n\Omega_e}\right)^2 A_n^{-1}(z_e') A_n^{-1}(z_e) A_l(z_i'') |k_i \Gamma_{is}(\omega - n\Omega_e) k'_s|^2 \times \left\{1 + (k'' r_{D_i})^2\left[1 + \left(\frac{k''_\perp \omega_{L_i}}{k''\Omega_i}\right)^2 - \left(\frac{k''_z \omega_{L_e}}{k'' n\Omega_i}\right)^2\right]\right\}^{-2} |A^e_{n,0}|^2. \tag{22.28}$$

The process $\tilde{c}_i\tilde{c}_i p$ is similar to the coalescence $c_e c_e p$, and the kernel of the equation that describes it has the form ($\omega \simeq n\Omega_i$)

$$Q = \frac{1}{8\pi N_e mc^2}\left(\frac{M}{m}\right)^2 \left(\frac{k_z k'_z k'' c v_{T_i}}{(n\Omega_i)^2 \omega_{L_e}}\right)^2 \left(1 + \frac{\omega_{L_i}^2}{\Omega_i^2}\right) A_n(z_i) A_n(z_i') \times \left\{1 + (k r_{D_i})^2\left[1 + \left(\frac{k_\perp \omega_{L_i}}{k\Omega_i}\right)^2 - \left(\frac{k_z \omega_{L_e}}{kn\Omega_i}\right)^2\right]\right\}^{-2} \left\{1 + (k' r_{D_i})^2\left[1 + \left(\frac{k'_\perp \omega_{L_i}}{k'\Omega_i}\right)^2 - \left(\frac{k'_z \omega_{L_e}}{k' n\Omega_i}\right)^2\right]\right\}^{-2} I_n^2(k_\perp k'_\perp \rho_e^2) \exp\{-(k_\perp^2 + k_\perp'^2)\rho_e^2\}. \tag{22.29}$$

Note that in deriving the kernels (22.27)-(22.29) we have used only the electron parts of the corresponding contractions.

12. To conclude this section we give a formula for the kernel of the integrodifferential equation (22.5), which corresponds to the coalescence $l_\pm c_e p$ of three oscillations with three different spectra ($\omega' \simeq \Omega_e$):

$$Q = \frac{1}{128 N_e mc^2}\left(\frac{k_z k'_z c v_{T_e}}{\Omega_e^2}\right)^2 \left(\frac{k''}{\omega_{L_e}}\right)^2 A_1^{-1}(z_e')\left(1 + \frac{\omega_{L_i}^2}{\Omega_i^2}\right) z_e' \exp\left(-\frac{1}{2}z_e'\right)\{I_0(z_e'/4) - I_1(z_e'/4)\}\left[\frac{\partial\omega\varepsilon'(\omega, \mathbf{k})}{\partial\omega}\right]^{-1}_{\omega=\omega(\mathbf{k})}. \tag{22.30}$$

In § 23, we shall describe decay processes with the participation of such oscillations; these processes can occur only in a nonisothermal magnetoactive plasma.

§ 23. Decay of Quasilongitudinal Characteristic Oscillations of a Nonisothermal Magnetoactive Plasma

Below, we give a list of formulas for the kernel Q [see the definition (21.3)] of the integrodifferential equation (21.5) that describes the decay processes with the participation of quasilongitudinal waves existing in a nonisothermal magnetoactive plasma with hot electrons and cold ions ($T_e \gg T_i$) (see [28], § 7 and [64, 65]). To facilitate the use of the table of formulas, let us recall the correspondence between the symbolic notation of the corresponding spectrum of the oscillation (for example, c_e) and the number of the relation that describes this spectrum [for c_e: (19.57)]:*

p (18.4); c_i (18.46); $\tilde{c}_i$ (18.43); c_e (18.57);
l_- (18.26); $l_\pm$ (18.24); s_i' (18.61); s_i (6.40a); l (6.34a).

* The letter l stands (as in § 22) for a high-frequency ($\omega \gg \Omega_e$) electron Langmuir oscillation (6.34a).

As in the foregoing section, we shall signify a decay process by the triple symbol $\alpha\beta\gamma$. The plasma is assumed to have two components: electrons and ions of one species. To shorten the formulas, we use this notation for the electron variables:

$$a_e \equiv (kr_{D_e})^2, \quad a'_e \equiv (k'r_{D_e})^2, \quad a''_e \equiv (k''r_{D_e})^2,$$
$$b_e \equiv \left(\frac{k_\perp \omega_{L_e}}{k\Omega_e}\right)^2, \quad b'_e \equiv \left(\frac{k'_\perp \omega_{L_e}}{k'\Omega_e}\right)^2, \quad b''_e \equiv \left(\frac{k''_\perp \omega_{L_e}}{k''\Omega_e}\right)^2 \tag{23.1}$$

and similarly for the ion variables:

$$a_i, \ a'_i, \ a''_i; \quad b_i, \ b'_i, \ b''_i. \tag{23.2}$$

The integrals $A_n(z)$ and $A_{0,0}(z, z', z'')$ in the expressions for the kernels Q are given by Eqs. (16.52) and (19.67), respectively [see also the definition (16.53) for z]. The quantity $A_n(z, z')$ is a special case of the integral (19.67):

$$A_n(z, z') \equiv \int\limits_0^\infty x dx e^{-x^2/2} J_n(k_\perp \rho x) J_n(k'_\perp \rho x) = I_n(k_\perp k'_\perp \rho^2) \exp\left\{-\frac{1}{2}(k_\perp^2 + k'^2_\perp)\rho^2\right\}, \quad z \equiv (k_\perp \rho)^2, \ z' \equiv (k'_\perp \rho)^2. \tag{23.3}$$

The order of succession of the formulas in the list corresponds to decrease in the frequencies of the oscillations participating in the decay process. To the right of the symbolic designation $\alpha\beta\gamma$ of the process we give the range of frequencies of the interacting waves in brackets. The form of the kernels Q depends on which term, electron or ion, is predominant in the longitudinal contraction of the three-index tensor S. Where this depends on the actual values of the plasma parameters and the wave vectors of the oscillations, formulas are given for both cases separately; the corresponding values of Q are donated by Q_e and Q_i. In a number of variants, the electron contribution and ion temperatures T_e and T_i. These conditions are specified after the equation for the kernel.*

1. $lls_i\,(\omega, \omega' \gg \Omega_e \gg \omega'' \gg \Omega_i)$:

$$Q_e = \frac{1}{32\pi}\frac{1}{N\varkappa T_e}\left(\frac{\mathbf{kk}'}{kk'}\right)^2,$$
$$Q_i = \frac{1}{32\pi}\frac{1}{NMc^2}\left(\frac{m}{M}\right)^2\left\{\frac{\mathbf{kk}'}{kk'}\frac{ck''}{\omega_{L_i}}\sqrt{1+\frac{1}{a''_e}} + \frac{\mathbf{kk}''}{kk''}\frac{ck'}{\omega_{L_e}} + \frac{\mathbf{k}'\mathbf{k}''}{k'k''}\frac{ck}{\omega_{L_e}}\right\}^2.$$

2. $lls'_i\,(\omega, \omega' \gg \Omega_e \gg \Omega_i \gg \omega'')$:

$$Q = \frac{1}{32\pi}\frac{1}{N\varkappa T_e}\left(\frac{\mathbf{kk}'}{kk'}\right)^2(1 + a''_e b''_i)^{-1}, \quad T_e \gg T_i \gtrsim T_e\frac{m}{M}.$$

3. $s_i s_i c_e\,(\omega, \omega' \gg \Omega_e;\ \omega'' \simeq n\Omega_e)$:

$$Q = \frac{1}{16\pi}\frac{1}{N\varkappa T_e}\left(\frac{k''_z}{k_z}\right)^4\left(\frac{\omega_{L_e}}{n\Omega_e}\right)^4\left(\frac{\mathbf{kk}'}{kk'}\right)^2 A_n^3(z''_e)\, a_e a'_e (1+a_e)^{-1}(1+a'_e)^{-1}\left\{a''_e - \frac{z''_e}{n^2-1} - \left(\frac{k''_z\rho_e}{n}\right)^2\right\}^{-2},$$
$$T_e \gg T_i \gtrsim T_e\frac{m}{M}.$$

4. $s_i s_i l_\pm\,(\omega, \omega' \gg \Omega_e;\ \omega'' \simeq \Omega_e)$:

$$Q = \frac{1}{32\pi}\frac{1}{N\varkappa T_e}\left(\frac{\mathbf{kk}'}{kk'}\right)^2(k_z r_{D_e})^{-2} a_e a'_e (1+a_e)^{-1}(1+a'_e)^{-1}\left(\frac{\omega_{L_e}}{k''_z}\right)^2\left\{\left(\frac{k''_z}{\omega''}\right)^2 + \frac{k''^2_\perp}{\omega''^2 - \Omega_e^2}\right\}^2\left\{\left(\frac{k''_z}{\omega''}\right)^2 + \left(\frac{k''_\perp\omega''}{\omega''^2-\Omega_e^2}\right)^2\right\},$$
$$T_e \gg T_i \gtrsim T_e\frac{m}{M}.$$

* For example, in seven of the 25 given cases the electron contribution to S is more important than the ion contribution in a plasma with temperatures $T_e \gg T_i \gtrsim (m/M)T_e$.

5. $s_i s_i l_- \, (\omega, \ \omega' \gg \Omega_e \gg \omega'')$:

$$Q = \frac{1}{32\pi} \frac{1}{N\varkappa T_e} \left(\frac{\mathbf{kk}'}{kk'}\right)^2 \left(\frac{k''}{k_z}\right)^2 (k_z r_{D_e})^{-2} \, a_e a'_e (1 + a_e)^{-1} (1 + a'_e)^{-1} (1 + b''_e).$$

6. $s_i s_i s'_i \, (\omega, \ \omega' \gg \Omega_e \gg \Omega_i \gg \omega'')$:

$$Q = \frac{1}{32\pi} \left(\frac{\mathbf{kk}'}{kk'}\right)^2 a_e a'_e a''_e \left\{ \left(\frac{a''_i}{N\varkappa T_i}\right)^{1/2} \left(1 + \frac{1}{a''_e} + b''_i\right) - \frac{4}{3} (N\varkappa T_e a''_e a'^2_e)^{-1/2} \right\}^2 (1 + a_e)^{-1} (1 + a'_e)^{-1} (1 + a''_e + a''_e b''_i)^{-1}.$$

7. $s_i s_i p \, (\omega, \ \omega' \gg \Omega_e \gg \Omega_i \gg \omega'')$:

$$Q = \frac{1}{32\pi} \left(\frac{\mathbf{kk}'}{kk'}\right)^2 \frac{m}{M} \left(\frac{k''}{\omega_{L_e}}\right)^2 \left\{ \frac{v_{T_i}}{\sqrt{N\varkappa T_i}} - \left(\frac{M}{m}\right)^{1/2} \left(\frac{\omega_{L_e}}{k_z}\right)^2 \frac{1}{v_{T_e}} \frac{1}{\sqrt{N\varkappa T_e}} \right\}^2 a_e a'_e (1 + a_e)^{-1} (1 + a'_e)^{-1} (1 + b''_i).$$

8. $l_\pm l_\pm s_i \, (\omega, \ \omega' \sim \Omega_e \gg \omega'')$:

$$Q = \frac{1}{32\pi} \frac{1}{N\varkappa T_e} \left(\frac{k_z k'_z}{kk'}\right)^2 \frac{1}{\omega^4} (1 + a''_e)^{-1} \left\{ \left(\frac{k_z}{k\omega}\right)^2 + \left(\frac{k_\perp}{k}\right)^2 \left(\frac{\omega}{\omega^2 - \Omega_e^2}\right)^2 \right\}^{-1} \left\{ \left(\frac{k'_z}{k'\omega'}\right)^2 + \left(\frac{k'_\perp}{k'}\right)^2 \left(\frac{\omega'}{\omega'^2 - \Omega_e^2}\right)^2 \right\}^{-1}.$$

9. $l_\pm l_\pm s'_i \, (\omega, \ \omega' \simeq \Omega_e \gg \Omega_i \gg \omega'')$:

$$Q = \frac{1}{32\pi} \frac{1}{N\varkappa T_e} \left(\frac{k_z k'_z}{kk'}\right)^2 \frac{1}{\omega^4} (1 + a''_e + a''_e b''_i)^{-1} \left\{ \left(\frac{k_z}{k\omega}\right)^2 + \left(\frac{k_\perp}{k}\right)^2 \left(\frac{\omega}{\omega^2 - \Omega_e^2}\right)^2 \right\}^{-1} \left\{ \left(\frac{k'_z}{k'\omega'}\right)^2 + \left(\frac{k'_\perp}{k'}\right)^2 \left(\frac{\omega'}{\omega'^2 - \Omega_e^2}\right)^2 \right\}^{-1}.$$

10. $l_- l_- s_i \, (\omega, \ \omega', \ \omega'' \ll \Omega_e)$:

$$Q = \frac{1}{32\pi} \frac{1}{N\varkappa T_e} (1 + a''_e)^{-1}, \quad T_e \gg T_i \gtrsim T_e \frac{m}{M}.$$

11. $s_i s_i l_- \, (\omega, \ \omega', \ \omega'' \ll \Omega_e)$:

$$Q = \frac{1}{64} \frac{1}{N\varkappa T_e} A_0^2 (z_e, \ z'_e) (1 + a_e)^{-1} (1 + a'_e)^{-1}, \quad T_e \gg T_i \gtrsim T_e \frac{m}{M}.$$

12. $l_- s_i s_i \, (\omega, \ \omega', \ \omega'' \ll \Omega_e)$:

$$Q_e = \frac{1}{64} \frac{1}{N\varkappa T_e} (1 + a'_e)^{-1} (1 + a''_e)^{-1},$$
$$Q_i = \frac{1}{32\pi} \frac{1}{NMc^2} \frac{m}{M} (a'_e a''_e)^2 (1 + a'_e)^{-2} (1 + a''_e)^{-2} (k_z k' k'')^{-2} \times$$
$$\times \left\{ \frac{k}{k_z} \frac{ck}{\omega_{L_e}} k (\mathbf{k}'\mathbf{k}'') (1 + b_e)^{1/2} + \frac{ck'}{\omega_{L_i}} k' (\mathbf{kk}'') \left(1 + \frac{1}{a'_e}\right)^{1/2} + \frac{ck''}{\omega_{L_i}} k'' (\mathbf{kk}') \left(1 + \frac{1}{a''_e}\right)^{1/2} \right\}^2.$$

13. $s_i s_i \tilde{c}_i \, (\omega, \ \omega' \gg \Omega_i; \ \omega'' \simeq n\Omega_i)$:

$$Q_e = \frac{1}{32\pi} \frac{1}{N\varkappa T_i} (kk' r_{D_e}^2)^{-2} (k''_z \rho_i)^{-2} n^2 A_0^2 (z_e, \ z'_e) A_n (z''_i) \left(1 + \frac{1}{a_e}\right) \left(1 + \frac{1}{a'_e}\right),$$
$$Q_i = \frac{1}{16\pi} \frac{1}{N\varkappa T_i} \left(\frac{\mathbf{kk}'}{kk'}\right)^2 A_n^{-1} (z''_i) \left(1 + \frac{1}{a_e}\right)^2 \left(1 + \frac{1}{a'_e}\right)^2.$$

14. $s_i s_i c_i \, (\omega, \ \omega' \gg \Omega_i; \ \omega'' \simeq \Omega_i)$:

$$Q_e = \frac{1}{16\pi} \frac{1}{N\varkappa T_i} A_1 (z''_i) (1 + a_e)^{-1} (1 + a'_e)^{-1},$$
$$Q_i = \frac{1}{16\pi} \frac{1}{N\varkappa T_i} \left(\frac{T_e}{T_i}\right)^2 \left(\frac{\mathbf{kk}'}{kk'}\right)^2 (k''_z \rho_i)^4 A_1^{-1} (z''_i) \, a_e a'_e (1 + a_e)^{-1} (1 + a'_e)^{-1}.$$

15. $s_i s_i s'_i \, (\omega, \ \omega' \gg \Omega_i \gg \omega'')$:

$$Q_e = \frac{1}{32\pi} \frac{1}{N\varkappa T_e} (1 + a_e)^{-1} (1 + a'_e)^{-1} (1 + a''_e + a''_e b''_i)^{-1};$$
$$Q_i = \frac{1}{32\pi} \frac{1}{N\varkappa T_i} \left(\frac{\mathbf{kk}'}{kk'}\right)^2 a_e a'_e a''_e a''_i (1 + a_e)^{-1} (1 + a'_e)^{-1} (1 + a''_e + a''_e b''_i)^{-1}.$$

16. $s_i s_i p\,(\omega,\ \omega' \gg \Omega_i \gg \omega'')$:

$$Q_e = \frac{1}{64}\frac{1}{N\varkappa T_e} A_0^2(z_e,\ z'_e)(1+a_e)^{-1}(1+a'_e)^{-1},$$
$$Q_i = \frac{1}{32\pi}\frac{1}{NMc^2}\left(\frac{\mathbf{kk}'}{kk'}\right)^2\left(\frac{ck''}{\omega_{L_e}}\right)^2 a_e a'_e (1+a_e)^{-1}(1+a'_e)^{-1}(1+b''_i).$$

17. $l_- s_i s'_i\,(\omega,\ \omega' \gg \Omega_i \gg \omega'')$:

$$Q_e = \frac{1}{64}\frac{1}{N\varkappa T_e}(1+a'_e)^{-1}(1+a''_e+a''_e b''_i)^{-1},$$
$$Q_i = \frac{1}{32\pi}\frac{1}{NMc^2}\left(\frac{ck''}{\omega_{L_e}}\right)^2\left(\frac{k}{k_z}\right)^2 a''_e(1+a''_e+a''_e b''_i)^{-1}.$$

18. $s_i l_- s'_i\,(\omega,\ \omega' \gg \Omega_i \gg \omega'')$:

$$Q_e = \frac{1}{64}\frac{1}{N\varkappa T_e}\frac{a''_e}{a'_e}(1+a_e)^{-1}(1+a''_e+a''_e b''_i)^{-1};\quad T_e \gg T_i \gtrsim T_e\frac{m}{M}.$$

19. $l_- l_- s'_i\,(\omega,\ \omega' \gg \Omega_i \gg \omega'')$:

$$Q = \frac{1}{32\pi}\frac{1}{N\varkappa T_e}(1+a''_e+a''_e b''_i)^{-1}.$$

20. $c_i c_i p\,(\omega,\ \omega' \simeq \Omega_i;\ \omega'' \ll \Omega_i)$;

$$Q_e = \frac{1}{16}\frac{1}{N\varkappa T_e}\left(\frac{T_e}{T_i}\right)^2 A_1(z_i)A_1(z'_i)A_0(z_e,\ z'_e),$$
$$Q_i = \frac{1}{4\pi}\frac{1}{NMc^2}\left(\frac{k''_z}{k''}\right)^2\left(\frac{T_e}{T_i}\right)^4 (k_z k''_z \rho_i^2)^2\omega_{L_i}^2\left\{\frac{ck'_z}{\Omega_i^2}A_1^{-2}(z'_i) + \right.$$
$$\left. + \frac{ck''_z}{\omega_{L_e}^2}\left(\frac{k''}{k''_z}\right)^2(1+b''_i) + \frac{ck_z}{\Omega_i}A_1^{-1}(z''_i)\left[\left(\frac{k''}{k''_z}\right)^2\frac{1}{\omega_{L_e}}(1+b''_i)^{1/2} + \frac{1}{\Omega_i}A_1^{-1}(z_i)\right]\right\}^2 (1+b''_i)^{-1}A_1^{-4}(z_i)A_1(z_i,\ z'_i).$$

21. $c_i c_i s'_i\,(\omega,\ \omega' \simeq \Omega_i;\ \omega'' \ll \Omega_i)$:

$$Q_e = \frac{1}{8\pi}\frac{1}{N\varkappa T_e}\left(\frac{T_e}{T_i}\right)^2(1+a''_e+a''_e b''_i)^{-1}A_1(z_i)A_1(z'_i)A_{0,\,0}^2(z_e,\ z'_e,\ z''_e),$$
$$Q_i = \frac{1}{8\pi}\frac{1}{NMc^2}\left(\frac{T_e}{T_i}\right)^4 (k_z k'_z \rho_i^2)^2\left(\frac{k''_z}{k''}\right)^2(1+a''_e+a''_e b''_i)^{-1}\left\{\frac{ck'_z\omega_{L_i}}{\Omega_i^2}A_1^{-2}(z'_i) + \right.$$
$$\left. + \frac{ck_z}{\Omega_i}A_1^{-1}(z_i)\left[\frac{k''}{k''_z}\left(1+\frac{1}{a''_e}+b''_i\right)^{1/2} + \frac{\omega_{L_i}}{\Omega_i}A_1^{-1}(z'_i)\right] + \frac{ck''_z}{\omega_{L_i}}\left(\frac{k''_z}{k}\right)^{-2}\left(1+\frac{1}{a''_e}+b''_i\right)\right\}^2 A_1^{-3}(z_i)A_1(z'_i)A_1^2(z_i,\ z'_i).$$

22. $p s'_i s'_i\,(\omega,\ \omega',\ \omega'' \ll \Omega_i)$:

$$Q = \frac{1}{64}\frac{1}{N\varkappa T_e}(1+a'_e+a'_e b'_i)^{-1}(1+a''_e+a''_e b''_i)^{-1},\quad T_e \gg T_i \gtrsim T_e\left(\frac{m}{M}\right)^{1/3}.$$

23. $s'_i p s'_i\,(\omega,\ \omega',\ \omega'' \ll \Omega_i)$:

$$Q = \frac{1}{64}\frac{1}{N\varkappa T_e}(1+a_e+a_e b_i)^{-1}(1+a''_e+a''_e b''_i)^{-1},\quad T_e \gg T_i \gtrsim T_e\left(\frac{m}{M}\right)^{1/3}.$$

24. $s'_i s'_i s'_i\,(\omega,\ \omega',\ \omega'' \ll \Omega_i)$:

$$Q_e = \frac{1}{32\pi}\frac{1}{N\varkappa T_e}(1+a_e+a_e b_i)^{-1}(1+a'_e+a'_e b'_i)^{-1}(1+a''_e+a''_e b''_i)^{-1},$$
$$Q_i = \frac{1}{32\pi}\frac{1}{NMc^2}\left(\frac{ck}{\omega_{L_i}}\right)^2\left\{\left(1+\frac{1}{a_e}+b_i\right)^{1/2} + \frac{k'}{k}\left(1+\frac{1}{a'_e}+b'_i\right)^{1/2} + \frac{k''}{k}\left(1+\frac{1}{a''_e}+b''_i\right)^{1/2}\right\}^2.$$

25. $p p s'_i\,(\omega,\ \omega',\ \omega'' \ll \Omega_i)$:

$$Q = \frac{1}{32\pi}\frac{1}{N\varkappa T_e}(1+a''_e+a''_e b''_i)^{-1},\quad T_e \gg T_i \gtrsim T_e\frac{m}{M}.$$

In deriving the above kernels Q we have in general considered the case of arbitrary mutual disposition of the wave vectors **k**, **k'**, **k"** of the interacting waves relative to each other and the direction **h** of the external magnetic field. However, we have assumed that the angles between the wave vectors and the external magnetic field are not too near 90° in the cases when this does not contradict the conditions for the existence of the spectra and the decay. Note also that in agreement with the general arguments formulated in § 12 the list of formulas for the kernels Q can be represented in the form of a table of cross sections of induced combination scattering of electrostatic oscillations in a magnetoactive plasma:

$$d\sigma(\mathbf{k}', \mathbf{k}) = \frac{d\mathbf{k}}{2\pi}\omega(\mathbf{k})\frac{1}{N}\left|\frac{\partial\omega'}{\partial\mathbf{k}'}\right|^{-1} Q(\mathbf{k}, \mathbf{k}')\{W_l(\mathbf{k}-\mathbf{k}')\omega(\mathbf{k})\,\mathrm{sgn}\,\varepsilon''(\omega, \mathbf{k}) - \\ - 2W_l(\mathbf{k})\omega''(\mathbf{k}-\mathbf{k}')\,\mathrm{sgn}\,\varepsilon''(\omega'', \mathbf{k}-\mathbf{k}')\}\,\delta(\omega(\mathbf{k}) - \omega'(\mathbf{k}') - \omega''(\mathbf{k}-\mathbf{k}')). \qquad (23.4)$$

In particular, for the process $s_i^l s_i^l s_i^l$ of coalescence of low-frequency ion-sound oscillations (see formula No. 24 in the list) with participation of only the electron part of the kernel Q_e and the first term in the curly brackets on the right-hand side of (23.4), we obtain

$$d\sigma(\mathbf{k}', \mathbf{k}) = \frac{d\mathbf{k}}{64\pi^2}\frac{W_l(\mathbf{k}-\mathbf{k}')}{N\varkappa T_e}\frac{k'}{N}(\varkappa\mathbf{h})^2\left\{a_e a_e' a_e''\left(1+\frac{1}{a_e'}+b_i'\right)^{1/2}\left(1+\frac{1}{a_e''}+b_i''\right)\times\right. \\ \times\left.\left(1+\frac{1}{a_e}+b_i\right)^2\right\}^{-1}\left\{[\varkappa'\mathbf{h}]^2+\left(1+\frac{1}{a_e'}+b_i'\right)^{-2}\left[\left(\frac{\varkappa'\mathbf{h}}{a_e'}\right)^2+\frac{b_i'^2(\varkappa'\mathbf{h})^4}{[\varkappa'\mathbf{h}]^2}\right]+2b_i'(\varkappa'\mathbf{h})^2\times \\ \times\left(1+\frac{1}{a_e'}+b_i'\right)^{-1}\right\}^{-1/2}\delta\left[(\varkappa\mathbf{h})\left(1+\frac{1}{a_e}+b_i\right)\right]^{-1/2}-(\varkappa'\mathbf{h})\left(1+\frac{1}{a_e'}+b_i'\right)^{-1/2}-(\varkappa''\mathbf{h})\left(1+\frac{1}{a_e''}+b_i''\right)^{-1/2}\Big]. \qquad (23.5)$$

Here $\varkappa$, $\varkappa'$, $\varkappa''$ are unit vectors in the directions **k**, **k'**, **k"**, respectively. In the limit of long waves of interacting acoustic oscillations s_i^l:

$$k,\ k',\ k'' \ll r_{D_e}^{-1} \qquad (23.6)$$

and strong magnetic fields $\mathbf{B}_0$:

$$b_i,\ b_i',\ b_i'' \ll 1 \qquad (23.7)$$

the cross section (23.5) simplifies considerably:

$$d\sigma(\mathbf{k}', \mathbf{k}) = \frac{d\mathbf{k}}{64\pi^2}\frac{W_l(\mathbf{k}-\mathbf{k}')}{N\varkappa T_e}\frac{k_z^2}{N}\delta(k_z - k_z' - k_z''). \qquad (23.8)$$

This limiting expression can be used, for example, in a plasma with density $N = 5 \cdot 10^9$ cm^{-3} in a magnetic field $B_0 = 5 \cdot 10^4$ G. For then

$$\frac{N}{B_0^2} \sim 1,\ b_i \sim b_i' \sim b_i'' \sim 4 \cdot 10^{-2} \ll 1, \qquad (23.9)$$

and the conditions for the existence of the spectrum of the low-frequency sound s_i^l reduce to a restriction on the wave numbers (in cm^{-1}) and the temperature (in °K):

$$5 \cdot 10^4 \ll k^2 T_e \ll 5 \cdot 10^{14}, \quad k^2 T_i \ll 10^8, \qquad (23.10)$$

so that at the electron temperature $T_e \simeq 10^6$ °K and wave numbers of order 1 cm^{-1} we have the inequalities (23.6)

$$a_e \sim a_e' \sim a_e'' \sim 10^{-8} k^2 T_e \ll 1. \qquad (23.11)$$

§ 24. Induced Combination Scattering of High-Frequency Transverse Waves in a Cold Magnetoactive Plasma†

In this section we consider (see [66] and [28], § 6) the decay interaction of two waves, one of which is a high-frequency transverse wave with frequency much greater than the electron gyroscopic ($\omega' \gg \Omega_e$) and the other is of arbitrary frequency ($\omega'' \lessgtr \Omega_e$). As a result of the nonlinear interaction there is formed a further high-frequency transverse electromagnetic wave ($\omega \gg \Omega_e$):

$$\omega,\ \omega' \gg \Omega_e \gtrless \omega''. \tag{24.1}$$

To calculate the kernel Q of the integrodifferential equation that describes the decay we make an additional simplifying assumption that puts an upper bound on the plasma temperature. We shall assume that the thermal motion of the particles is unimportant for all three waves that participate in this nonlinear process. In other words, the plasma is assumed to be cold with negligibly low velocities of the particle thermal motion:

$$v_T = 0. \tag{24.2}$$

The basis of our treatment is the generalized kinetic equation for the spectral function $(E_jE_i)_{\omega,\mathbf{k}}$ of the field of the high-frequency transverse wave (this equation allows for the possibility of polarization degeneracy):

$$\begin{aligned}\frac{d}{dt}(E_iE_a)_{\omega,\mathbf{k}} = i\frac{\omega}{2}\int d\omega' d\mathbf{k}' d\omega'' d\mathbf{k}''\delta(\omega-\omega'-\omega'')\,\delta(\mathbf{k}-\mathbf{k}'-\mathbf{k}'')(E_sE_c)_{\omega',\mathbf{k}'}\times\\ \times\{S^*_{abs}(\omega,\mathbf{k};\omega',\mathbf{k}')\,S_{rjc}(\omega,\mathbf{k};\omega',\mathbf{k}')\left[A^{\text{res}}_{ir}(\omega,\mathbf{k})(E_bE_j)_{\omega'',\mathbf{k}''}+\right.\\ \left.+2A^{\text{res}*}_{bj}(\omega'',\mathbf{k}'')(E_rE_i)_{\omega,\mathbf{k}}\right]-S_{ibc}(\omega,\mathbf{k};\omega',\mathbf{k}')\,S^*_{rjs}(\omega,\mathbf{k};\omega',\mathbf{k}')\times\\ \times\left[A^{\text{res}*}_{ar}(\omega,\mathbf{k})(E_jE_b)_{\omega'',\mathbf{k}''}+2A^{\text{res}}_{bj}(\omega'',\mathbf{k}'')(E_aE_r)_{\omega,\mathbf{k}}\right]\}.\end{aligned} \tag{24.3}$$

This equation describes the evolution of the four Stokes parameters [see the explicit expression (6.76) for the spectral function $(E_jE_i)_{\omega,\mathbf{k}}$] of the high-frequency transverse wave during the process of induced combination scattering on the characteristic oscillations of the magnetoactive plasma. The three-index tensors S on the right-hand side of Eq. (24.3) in the limit (24.1):

$$\omega,\ \omega' \gg \Omega_e, \qquad \omega,\ \omega' \gg \omega'' \tag{24.4}$$

have in accordance with (19.31) the form

$$S_{ijs}(\omega,\mathbf{k};\omega',\mathbf{k}') \simeq -i\frac{e}{m}\frac{\omega_L^2}{\omega^2}\left\{\delta_{is}\frac{k''_a}{\omega''}\Gamma_{aj}(\omega'')+\frac{k_s}{\omega}\Gamma_{ij}(\omega'')+\frac{k'_i}{\omega}\Gamma_{sj}(\omega'')\right\}. \tag{24.5}$$

The first term in the curly brackets on the right-hand side of (24.5) has the same order of magnitude as the other two if the phase velocities of the high-frequency waves and the characteristic oscillation of the magnetoactive plasma on which the scattering occurs are approximately the same:

$$\frac{\omega''}{k''} \sim \frac{\omega}{k}. \tag{24.6}$$

† The nonlinear interaction in a magnetoactive plasma of high-frequency transverse electromagnetic waves whose polarization is described by four Stokes parameters is studied in more detail in [105-107]. We are grateful to A. P. Kropotkin and N. V. Sholokhov for their assistance in establishing the importance of the polarization of the high-frequency transverse waves scattered by a plasma.

In this case, because of the small change in the frequency of the transverse waves as a result of the scattering ($\omega'' \ll \omega, \omega'$) the wavelength of the oscillation is appreciably greater than the wavelength of the high-frequency oscillations:

$$k'' \ll k, \; k'. \tag{24.7}$$

Calling one of the waves (ω, **k**) and (ω', **k'**) the incident and the other the scattered, we can say, in accordance with the inequalities (24.6) and (24.7), that scattering on fast oscillations in the plasma ($n'' \equiv ck''/\omega'' \sim 1$) occurs only through small angles. On the other hand, scattering can also occur on slow waves ($n'' \gg 1$), when the scattering angles are not small.

Experimentally, frequent use is made of almost monochromatic and fairly well collimated beams of high-frequency transverse electromagnetic waves (for example, laser beams). The effects that arise as a result of the combination scattering of such a beam can be split into two groups. First, there is a change in the spectral composition of the beam itself and the polarization of its spectral components. In describing such effects it is, in general, important to take into account scattering on fast waves, i.e., waves with refractive index n'' of order unity. However, if the angular width of the beam, ΔO, is large compared with the square of the relative width of the frequency spectrum $(\Delta\omega/\omega)^2$, then scattering on slow waves with large refractive index ($n'' \gg 1$) is decisive; for then

$$(n'')^2 = \left(\frac{ck''}{\omega''}\right)^2 \sim \frac{\Delta O c^2 k^2}{\omega''^2} \sim \Delta O \left(\frac{\omega}{\Delta\omega}\right)^2 \gg 1. \tag{24.8}$$

Secondly, there are formed scattered waves that propagate at appreciable angles to the direction of propagation of the incident waves; the evolution of the spectral density of the energy of these waves and the nature of their polarization are determined by scattering on slow oscillations with almost longitudinal polarization ($n'' \to \infty$).

Let us consider first the effects of the first group. We restrict ourselves to characteristic times over which the energy of the oscillations with frequency ω'' and wave vector $\mathbf{k}''$ formed in the decay remains much less than the energy of the incident wave. Allowing for this circumstance, we can represent Eq. (24.3) in the simpler form

$$\begin{aligned}\frac{d}{dt}(E_i E_a)_{\omega,\mathbf{k}} = i\omega \int d\omega' d\mathbf{k}' d\omega'' d\mathbf{k}'' \delta(\omega - \omega' - \omega'')\, \delta(\mathbf{k} - \mathbf{k}' - \mathbf{k}'')\, (E_s E_c)_{\omega',\mathbf{k}'} \times \\ \times \{ S^*_{abs}(\omega, \mathbf{k}; \omega', \mathbf{k}')\, S_{rjc}(\omega, \mathbf{k}; \omega', \mathbf{k}')\, A^{\text{res}*}_{bj}(\omega'', \mathbf{k}'')\, (E_r E_i)_{\omega,\mathbf{k}} - \\ - S_{ibc}(\omega, \mathbf{k}; \omega', \mathbf{k}')\, S^*_{rjs}(\omega, \mathbf{k}; \omega', \mathbf{k}')\, A^{\text{res}}_{bj}(\omega'', \mathbf{k}'')\, (E_a E_r)_{\omega,\mathbf{k}} \},\end{aligned} \tag{24.9}$$

which is equivalent to the following system of four equations for the Stokes parameters:

$$\begin{aligned}\frac{\partial W_{tr}(\mathbf{k})}{\partial t} &= \int d\mathbf{k}' Q(\mathbf{k}, \mathbf{k}') \{W_{tr} W'_{tr} + MM' + CC' + SS'\}, \\ \frac{\partial M(\mathbf{k})}{\partial t} &= \int d\mathbf{k}' Q(\mathbf{k}, \mathbf{k}') \{W_{tr} M' + W'_{tr} M\}, \\ \frac{\partial C(\mathbf{k})}{\partial t} &= \int d\mathbf{k}' Q(\mathbf{k}, \mathbf{k}') \{C W'_{tr} + C' W_{tr}\}, \\ \frac{\partial S(\mathbf{k})}{\partial t} &= \int d\mathbf{k}' Q(\mathbf{k}, \mathbf{k}') \{C' M - C M'\}.\end{aligned} \tag{24.10}$$

Here, the kernel $Q(\mathbf{k}, \mathbf{k}')$ is determined by the contraction of the three-index tensors S in accordance with the more general equation (24.3); the primes on the right-hand sides of (24.10) indicates the Stokes parameters that depend on **k'** [for example, $W_{tr} \equiv W_{tr}(\mathbf{k})$, $W'_{tr} \equiv W_{tr}(\mathbf{k}')$]. The system of equations (24.10) completely describes the evolution of a high-frequency beam that passes through the plasma in the case of combination scattering. It determines the change in the spectral energy density $W_{tr}(\mathbf{k})$ of the waves and the change of the polarization state of the individual spectral components.

Equations (24.10) enable us to draw certain general conclusions concerning the change of the polarization during the combination scattering process. If a naturally polarized high-frequency wave (M = C = S = 0) is incident on the plasma, then in accordance with (24.10) it will remain naturally polarized.

At the same time, the system of equations (24.10) reduces to a single equation for the spectral energy density:

$$\frac{\partial W_{tr}(\mathbf{k})}{\partial t} = W_{tr}(\mathbf{k}) \int d\mathbf{k}' Q(\mathbf{k}, \mathbf{k}') W_{tr}(\mathbf{k}'). \tag{24.11}$$

During the process of combination scattering, the high-frequency electromagnetic waves evolve in such a way that their spectral energy density is transferred from the shorter to the longer waves. This result is a natural consequence of our assuming that the energy density of the characteristic plasma oscillations (ω'', $\mathbf{k}''$) is small compared with the energy of the incident wave, since the processes of decay of high-frequency transverse waves allowed for in Eq. (24.9) obviously lead to a decrease in the frequencies of these waves, i.e., to a transfer of energy to the long-wavelength region.

The characteristic time τ of such spectral transfer is determined by the kernel $Q(\mathbf{k}, \mathbf{k}')$ of the integrodfifferential equations (24.10) and if the beam has a large angular width, $\Delta O \gg (\Delta\omega/\omega)^2$, the characteristic time is

$$\frac{1}{\tau} \sim \frac{\omega}{4} \frac{k^3 (\Delta\omega)^2}{N} \frac{W_{tr}(\mathbf{k})}{mc^2} \left(\frac{\Delta\omega}{\omega}\right)^2 \frac{[(\Delta\omega)^2 - \Omega_e^2]^2}{(\Delta\omega)^4 \sin^2\bar{\vartheta}'' + [(\Delta\omega)^2 - \Omega_e^2]^2 \cos^2\bar{\vartheta}''}. \tag{24.12}$$

Here $\bar{\vartheta}''$ is some mean value of the angle ϑ'' between the direction of propagation $\varkappa''$ of the oscillation of the plasma and the external magnetic field B_0, so that $\bar{\vartheta}'' = \pi/2$ for $[\varkappa \mathbf{h}] = 0$ and $\sin^2\bar{\vartheta}'' = \cos^2\bar{\vartheta}'' = 1/2$ for $(\varkappa\mathbf{h}) = 0$.

Turning to combination scattering on quasilongitudinal oscillations of the plasma (scattering through large angles), we shall restrict ourselves to the case when the incident wave interacts with thermal electrostatic fluctuations. This scattering can be described naturally by the appropriate effective cross section, an explicit expression for which also follows from Eq. (24.3) with allowance for the tensor (24.5):

$$d\sigma^{tt}(\mathbf{k}, \mathbf{k}') = \frac{1}{2} d\mathbf{k} \left(\frac{e^2}{mc^2}\right)^2 \left(\frac{\omega''}{\omega_{L_e}}\right)^2 (k'' r_{D_e}) \frac{(\omega''^2 - \Omega_e^2)^2}{(\varkappa''\mathbf{h})^2 (\omega''^2 - \Omega_e^2)^2 + \omega''^4 [\varkappa''\mathbf{h}]^2} \left\{1 + (\varkappa\varkappa')^2\right\} \delta(\omega - \omega' - \omega''). \tag{24.13}$$

The minimal thickness d of the plasma layer at which the polarization of a transverse electromagnetic wave propagating in the plasma can be assumed arbitrary and described by Stokes parameters [see the spectral function (6.76)] is determined by the condition that the advance of the phase difference between the ordinary and the extraordinary wave over this thickness be small

$$kd \ll \frac{\omega}{\Omega_e}. \tag{24.14}$$

In the opposite limit to the inequality (24.14), the polarization of transverse waves in the plasma is determined by the inverse tensor $A_{ij}(\omega, \mathbf{k})$, i.e., by the plasma parameters N, T, and B_0, and it cannot be specified arbitrarily.

§ 25. Induced Scattering of Low-Frequency Quasilongitudinal Characteristic Oscillations of a Magnetoactive Plasma on the Plasma Particles

Besides the decay-type processes (§ 22-24), an important role in the evolution of the spectral function of the electric fields of the characteristic oscillations of a magnetoactive plasma can be played by processes of induced scattering of oscillations on particles. In this section, we consider Coulomb scattering of quasilongitudinal waves in a magnetoactive plasma at a frequency appreciably lower than the ion gyroscopic ([67] and [28], § 8, p. 93):

$$\omega,\ \omega' \ll \Omega_i. \tag{25.1}$$

As in an isotropic plasma (see § 9), Coulomb scattering is here understood as induced scattering through an electrostatic (longitudinal) virtual oscillation. The external magnetic field is assumed to be so strong that, in addition to the inequalities (25.1), the electron Larmor radii are small compared with the wavelengths of the interacting oscillations. In accordance with our analysis in § 18 of the spectra of quasilongitudinal oscillations of a magnetoactive plasma, waves with the spectra (18.4), (18.61), and (18.69), i.e., the low-frequency oscillation p, ion sound, and electron sound, can exist under such conditions if the electron and ion temperatures bear the right ratio. The ion sound (18.61), whose Coulomb scattering is to be studied here, differs from ordinary ion sound in an isotropic plasma, (6.40a), by its frequency depending on the angle between the direction **h** of the magnetic field $\mathbf{B}_0$ and the wave vector of the oscillation:

$$\omega^2 = \frac{k_z^2}{k^2}\,\frac{\omega_{Li}^2}{1+\dfrac{\omega_{Le}^2}{k^2 v_{T_e}^2}}, \qquad \omega'^2 = \left(\frac{k'_z}{k'}\right)^2 \frac{\omega_{Li}^2}{1+\dfrac{\omega_{Le}^2}{k'^2 v_{T_e}^2}}. \tag{25.2}$$

This oscillation, as in the isotropic case, exists only in an nonisothermal plasma with hot electrons and cold ions (6.36). In contrast, the electron sound s_e [spectrum (18.69)] exists in a plasma with hot ions and cold electrons: (18.67). The specific feature of this oscillation is the large value of the ion Larmor radius ρ_i compared with the wavelength at right angles to the magnetic field: (18.70).

Of these three oscillations, only ion sound has a phase velocity that is small compared with the electron thermal velocity [compare the conditions (18.29), (18.62), and (18.68)]. The ions move at a velocity that is much lower than the phase velocity of all three oscillations. The condition for the phase velocity to be small compared with the ion thermal velocity limits the region of transparency and opacity of a homogeneous plasma, since from this condition we usually also obtain that the phase velocity is low compared with the electron thermal velocity provided the electron temperature is not too low compared with the ion temperature [$T \gtrsim (m/M)T_i$]. In the opacity region, the permittivity describes the Debye screening of the field of the charge in the plasma or, in other words, the virtual oscillation through which the scattering occurs (cf. § 9, the footnote after (9.21) on the role of ions and electrons in the scattering). In this section we shall assume that the frequency ω'' and the wave vector $\mathbf{k}''$ of the longitudinal virtual wave are equal to the difference of the frequencies ω and ω' of the wave vectors **k** and **k'** of the interacting oscillations (7.3) (compare the section on the scattering of ion sound in an isotropic plasma, § 9). In general, induced scattering of waves on particles can also occur when the phase velocity of the "virtual" oscillation is greater than the ion thermal velocity [in an isotropic plasma in the case of the scattering of longitudinal waves into transverse (§ 10) and transverse into transverse (§ 11) we have also considered the case (10.32), when the phase velocity of the virtual oscillation is appreciable higher than the electron thermal velocity]. However, in this case, the virtual wave is weakly damped, and if the dispersion relation is satisfied it becomes an actually existing characteristic oscillation of the magnetoactive plasma with spectrum $\omega''(\mathbf{k}'')$ and can participate in the corresponding decay process, so that induced scattering on particles competes with decay: induced combination scattering. This process of induced scattering of particles is independent of the decay process, since it occurs in a plasma in which the ions move with a thermal velocity appreciable greater than the phase velocity of the virtual wave along the external magnetic field (cf. [70]). Therefore, the exposition that follows below (see also § 26) augments naturally our earlier investigations of the decay of quasilongitudinal oscillations of a magnetoactive plasma (§ 22–24).

In accordance with the general properties of the equations of nonlinear interaction discussed in § 21, Ch. IV, [see in particular, (21.20)], the orginal equation that describes induced scattering of quasilongitudinal oscillations of a magnetoactive plasma with a small change in the frequency can be represented in the form

$$\frac{dW_l(\mathbf{k})}{dt} = -2\omega(\mathbf{k})\, W_l(\mathbf{k}) \int d\mathbf{k}' W_l(\mathbf{k}')\, Q(\mathbf{k}, \mathbf{k}') \left\{\frac{\partial \omega \varepsilon'(\omega, \mathbf{k})}{\partial \omega}\right\}^{-1}_{\omega=\omega(\mathbf{k})} \left\{\frac{\partial \omega' \varepsilon'(\omega', \mathbf{k}')}{\partial \omega'}\right\}^{-1}_{\omega'=\omega'(\mathbf{k}')} \quad (25.3)$$

with kernel

$$Q(\mathbf{k}, \mathbf{k}) \equiv \frac{1}{2\pi^2} \operatorname{Im}\{V(\omega(\mathbf{k}), \mathbf{k};\ -\omega'(\mathbf{k}'), -\mathbf{k}') -$$
$$- S(\omega(\mathbf{k}), \mathbf{k};\ \omega'(\mathbf{k}'), \mathbf{k}')\, S(\omega(\mathbf{k}) - \omega'(\mathbf{k}'), \mathbf{k} - \mathbf{k}';\ \omega(\mathbf{k}), \mathbf{k})\, \varepsilon^{-1}(\omega(\mathbf{k}) - \omega'(\mathbf{k}'), \mathbf{k} - \mathbf{k}')\}, \quad (25.4)$$

determined by the longitudinal contractions of the permittivity tensors ε, S, V. The longitudinal permittivity ε in the limit of low frequencies ($\omega'' \ll \Omega_i$) and Larmor radii that are small compared with the wavelength of the oscillations along the magnetic field has in accordance with (18.30) the form

$$\varepsilon(\omega'', \mathbf{k}'') = 1 + \frac{\omega_L^2}{k''^2 v_T^2}\left\{1 - A_0(z'')\, J_+\left(\frac{\omega''}{|k_z''|\, v_T}\right)\right\} \quad (25.5)$$

provided

$$k_z'' \gg \frac{\omega''}{\Omega_i} k_\perp''. \quad (25.6)$$

We shall assume that the condition (25.6) is always satisfied only for the virtual oscillation. We mention here that it is not too stringent a restriction on the angles, since it is satisfied for almost all angles ϑ'' between the magnetic field $\mathbf{B}_0$ and the wave vector $\mathbf{k}''$ of the virtual oscillation that are not too near $\pi/2$:

$$\frac{\omega''}{\Omega_i} \ll \left|\frac{\pi}{2} - \vartheta''\right|, \quad (25.7)$$

i.e., in general, for $k_\perp'' \gg k_z''$ as well. Therefore, the fact that the ion Larmor radius is small compared with the longitudinal wavelength does not mean that it is small compared with the transverse wavelength of the virtual oscillation. The quantities $\operatorname{Im} V(\omega, \mathbf{k}; -\omega', -\mathbf{k}') \equiv \operatorname{Im} V$ and $S(\omega, \mathbf{k}; \omega', \mathbf{k}') \equiv S$ are determined by (20.19) and (19.81), obtained earlier in § 19 and § 20 in our study of the three- and four-index permittivity tensors and their contractions in a magnetoactive plasma. For ease of reading we repeat these expressions:

$$S = i\frac{e}{m}\frac{\omega_L^2}{kk'k''}\frac{1}{v_T^2}\frac{A_{0,0}}{(\omega' k_z - \omega k_z')}\left\{\frac{k_z k_z' k_z''}{\omega' k_z - \omega k_z'} - i\frac{(\mathbf{k}[\mathbf{k}'\mathbf{h}])}{\Omega}\right\}\left\{k_z J_+\left(\frac{\omega}{|k_z|\, v_T}\right) - k_z' J_+\left(\frac{\omega'}{|k_z'|\, v_T}\right) - k_z'' J_+\left(\frac{\omega''}{|k_z''|\, v_T}\right)\right\}; \quad (25.8)$$

$$A_{0,0} \equiv A_{0,0}(z, z', z'') = \int_0^\infty x\,dx\, e^{-\frac{x^2}{2}} J_0(k_\perp \rho x)\, J_0(k_\perp' \rho x)\, J_0(k_\perp'' \rho x); \quad (25.9)$$

$$\operatorname{Im} V = \sqrt{\frac{\pi}{2}}\frac{e^2}{m^2}\frac{\omega_L^2}{v_T^2}\frac{k_z''^2}{(\omega' k_z - \omega k_z')^2}\left\{\left(\frac{k_z k_z'}{kk'}\right)^2 \frac{k_z''^2}{(\omega' k_z - \omega k_z')^2} + \frac{1}{\Omega^2}\frac{(\mathbf{k}[\mathbf{k}'\mathbf{h}])^2}{(kk')^2}\right\} B_{0,0,0}\frac{\omega''}{|k_z''|\, v_T}\exp\left\{-\frac{1}{2}\left(\frac{\omega''}{k_z'' v_T}\right)^2\right\}; \quad (25.10)$$

$$B_{0,0,0} \equiv B_{0,0,0}(z, z') = \int_0^\infty x\,dx\, e^{-\frac{x^2}{2}} J_0^2(k_\perp \rho x)\, J_0^2(k_\perp' \rho x). \quad (25.11)$$

The integrals (25.9) and (25.11) are bounded above by unity and tend to this maximal value for $z, z' \ll 1$. The condition

$$\omega' k_z - \omega k_z' \neq 0, \quad (25.12)$$

under which formulas (25.8) and (25.10) were obtained, is not an additional restriction on the processes but follows naturally from Eqs. (7.37) and the smallness of the virtual oscillation's phase velocity:

$$\frac{\omega''}{|k_z''|} \ll v_{T_i} \ll \frac{\omega}{|k_z|}, \quad \frac{\omega'}{|k_z'|}. \tag{25.13}$$

Equation (25.3) contains, in addition to the contraction (25.8), $S(\omega'', \mathbf{k}''; \omega, \mathbf{k})$, which, however, can obviously be reduced on account of the explicit expression (25.8) to (25.8):*

$$S(\omega'', \mathbf{k}''; \omega, \mathbf{k}) = -S(\omega, \mathbf{k}; \omega', \mathbf{k}') \doteq -S. \tag{25.14}$$

As a result, the kernel (25.4) takes the form

$$Q(\mathbf{k}, \mathbf{k}') = \frac{1}{2\pi^2} \operatorname{Im}\left\{V + \frac{S^2}{\varepsilon(\omega'', \mathbf{k}'')}\right\}. \tag{25.15}$$

As usual (cf. Ch. III), we have omitted on the right-hand side of (25.3) the linear term corresponding to damping of the scattered oscillations. This means that we consider waves of such large amplitude and with such high phase velocity that the linear effects can be ignored. It is this that enabled us in the derivation of the formula for Im V (see § 20) to ignore the contributions of poles of the form $(\omega - k_z v_z)^{-1}$, which correspond to the scattered oscillations. A consistent following through of this point of view entails ignoring the terms containing $\operatorname{Im} J_+ \times (\omega/|k_z| v_T)$ and $\operatorname{Im} J_+(\omega'/|k_z| v_T)$ in formula (25.8) for S.

To establish the physical meaning of the individual terms of the kernel (25.15), we use the terminology introduced in [71] and used systematically in the nonlinear theory of a magnetoactive plasma in [72]. Namely, we shall say that the part of Eq. (25.3) associated with the first term Im V of the kernel $Q(\mathbf{k}, \mathbf{k}')$ describes the induced scattering of longitudinal oscillations of the plasma on free particles. The particle species depends on which term in V – the electron or the ion term – determines the effect. This name can be justified to a certain extent by noting that in the generalized kinetic equation (3.22) of the field the first term, which corresponds to Im V, describes in the limit of a cold ($v_T = 0$) isotropic plasma induced Thomson scattering of short-wavelength transverse oscillations ($\omega \simeq ck$, $\omega' \simeq ck'$) on free electrons (cf. § 12). The second term in (25.3) and (25.4), which contains the permittivity $\varepsilon(\omega'', \mathbf{k}'')$ in the denominator, will be interpreted as scattering on the polarization clouds of the virtual wave. In this case we can speak of scattering on electrons or ions of the virtual oscillation if the electron or the ion term, respectively, predominates in $\varepsilon(\omega'', \mathbf{k}'')$ irrespective of the ratio of the electron and ion parts of S.† We emphasize that this division is arbitrary to a considerable degree, since both of the terms in (25.15) are proportional to

$$\sqrt{\frac{\pi}{2}} \frac{\omega''}{|k_z''| v_T} \exp\left\{-\frac{1}{2}\left(\frac{\omega''}{k_z'' v_T}\right)^2\right\}, \tag{25.16}$$

i.e., to the contribution of one and the same pole $\omega'' = k_z'' v_z$ in the complex plane of the variable ω'', and in a number of cases the total scattering effect is determined by the interference

---

* Equation (25.14) is a consequence of the more general symmetry properties of the tensor $S_{ijs}(\omega, \mathbf{k}; \omega', \mathbf{k}')$ in a magnetoactive plasma that follow from the symmetric expression (19.97) for S_{ijs} [see also (19.76)].

† Generally speaking, this terminology differs from that which we have used in an isotropic plasma (see § 9), although it sometimes overlaps. It is readily seen however that the verbal formulations we use, in both an isotropic and in a magnetoactive plasma, are merely an interpretation of the one or other mathematical procedure that describes one and the same process of induced scattering of waves on plasma particles.

of these physical mechanisms, this being formally expressed in the reduction of the leading terms of the first and the second expression in (25.15) (see, for example, the Coulomb scattering on electrons in an isotropic plasma, § 9-11). All the same, we feel that the introduction of this terminology simplifies the discussion and classification of the processes, since it enables us to give more concrete formulations to the different particular variants of the kernel (25.15). At the same time, recalling the methods and results of § 14 in Ch. III, we note that, as in an isotropic plasma, the approach to scattering theory used there enables us to obtain a formula for the scattering cross section (14.32) for a magnetoactive plasma too in which each term has a unique physical interpretation without the introduction of the special terminology of [71] (see the footnote in § 14 concerning allowance for $\mathbf{B}_0 = \text{const} \neq 0$ in the original equations).

1. Turning to the treatment of the actual expressions for the kernels Q, we point out that we are here only studying those nonlinear interactions in which waves belonging to one and the same spectral branch participate (i.e., the scattering does not change the form of the spectrum). We shall consider first the induced scattering of oscillations p with phase velocity along the magnetic field exceeding the electron thermal velocity and with wavelengths greater than the electron and ion Larmor radii. Such an oscillation p has the spectrum (18.4):

$$\omega^2 = \frac{k_z^2}{k^2}\,\omega_{L_e}^2 \left\{1 + \frac{k_\perp^2}{k^2}\,\frac{\omega_{L_i}^2}{\Omega_i^2}\right\}^{-1},$$

which simplifies considerably in a rarefied plasma ($\omega_{Li} \ll \Omega_i$)

$$\omega^2 = \frac{k_z^2}{k^2}\,\omega_{L_e}^2. \tag{25.17}$$

If we take into account the thermal correction on the right-hand side of (25.17) in the zeroth approximation in the parameters z_e, $z_i \ll 1$, we obtain

$$\omega^2 = \frac{k_z^2}{k^2}\left(\omega_{L_e}^2 + 3k^2 v_{T_e}^2\right), \tag{25.18}$$

which agrees with the spectrum of "Langmuir" waves whose nonlinear interaction was studied in [70] [see (4), p. 156]. In a dense plasma (18.9), we obtain from (18.4)

$$\omega^2 = \frac{k_z^2}{k^2}\,\frac{M}{m}\,\Omega_i^2. \tag{25.19}$$

In all the cases (18.4), (25.17), and (25.19) the condition (18.6) of a sufficient density of the plasma and the condition of smallness of the oscillation frequency compared with the ion gyroscopic frequency require almost perpendicular propagation of the oscillation relative to the external magnetic field $\mathbf{B}_0$ [see (18.8) and (18.8a)]. Oscillations with the spectrum (25.17) can exist not only in the range of frequencies below the ion gyroscopic frequency but also at frequencies ω appreciably greater than Ω

$$\Omega_i < \omega < \Omega_e. \tag{25.20}$$

In this case the condition $k_z \ll k$ is superfluous. The phase velocity of the virtual wave ($\omega''/|k_z''|$) is low compared with the ion thermal velocity only in a narrow interval of angles. For example, for the spectrum (25.17)

$$|\cos\vartheta - \cos\vartheta'| \ll \left(\frac{m}{M}\right)^{1/2} |k - k'|\, r_{D_i} \cos\vartheta. \tag{25.21}$$

With allowance for the conditions for the existence (i.e., weak damping) of the oscillation p and the inequalities (25.13), quantities (25.5), (25.8)-(25.11) in the kernel (25.15) simplify considerably:

$$\varepsilon(\omega'', \mathbf{k}'') = 1 + \frac{\omega_L^2}{(k''v_T)^2}\left\{1 - J_+\left(\frac{\omega''}{|k_z''|v_T}\right)\right\} \equiv 1 + \delta\varepsilon; \tag{25.22}$$

$$S = i\frac{e}{m}\frac{\omega_L^2}{v_T^2}\frac{1}{\omega k''}\left\{\frac{1}{\omega}\left(\frac{k_z k_z'}{kk'}\right) - \frac{i}{\Omega}\frac{(\mathbf{k}\,[\mathbf{k}'\mathbf{h}])}{kk'}\right\}\left[1 - J_+\left(\frac{\omega''}{|k_z''|v_T}\right)\right]; \tag{25.23}$$

$$\operatorname{Im} V = \sqrt{\frac{\pi}{2}}\frac{e^2}{m^2}\frac{\omega_L^2}{v_T^2}\frac{1}{\omega^2}\frac{\omega''}{|k_z''|v_T}\left\{\frac{1}{\omega^2}\left(\frac{k_z k_z'}{kk'}\right)^2 + \frac{1}{\Omega^2}\frac{(\mathbf{k}\,[\mathbf{k}'\mathbf{h}])^2}{(kk')^2}\right\}. \tag{25.24}$$

We draw attention to the second terms in the curly brackets in (25.23) and (25.24). Because they contain the gyroscopic frequency, their dependence on the particle masses is much weaker than that of the first terms, so that the ratio of the electron and ion parts of these (the second) terms is the same as the ratio of the electron and ion parts of the permittivity for the virtual wave (25.22)*

$$\frac{\operatorname{Re}\delta\varepsilon_e}{\operatorname{Re}\delta\varepsilon_i} = \frac{T_i}{T_e}, \qquad \frac{\operatorname{Im}\delta\varepsilon_e}{\operatorname{Im}\delta\varepsilon_i} = \left(\frac{T_i}{T_e}\right)^{3/2}\left(\frac{m}{M}\right)^{1/2}. \tag{25.25}$$

The first terms in the curly brackets in (25.23) and (25.24) depend more strongly on the ratio of the particle masses [the superscript (1) in (25.26) means that only the first terms of the right-hand sides have been taken from (25.23) and (25.24)]

$$\frac{\operatorname{Im} V_e^{(1)}}{\operatorname{Im} V_i^{(1)}} \sim \left(\frac{T_i}{T_e}\right)^{3/2}\left(\frac{M}{m}\right)^{3/2}, \qquad \frac{\operatorname{Im} S_e^{(1)}}{\operatorname{Im} S_i^{(1)}} \sim \frac{T_e}{T_i}\frac{M}{m},$$
$$\frac{\operatorname{Re} S_e^{(1)}}{\operatorname{Re} S_i^{(1)}} \sim \left(\frac{T_i}{T_e}\right)^{3/2}\left(\frac{M}{m}\right)^{1/2}. \tag{25.26}$$

It can be clearly seen from these relations that, in the framework of the posed problem, different variants arise depending on the ratio of the electron and ion temperatures. We could, of course, write out the general formula for the kernel Q, using the general expressions (25.5), (25.8), and (25.10). However, wishing to study in detail the features of the nonlinear interaction, we shall consider a number of special cases.

In an almost isothermal plasma, $T_e \sim T_i$, the contribution of scattering on free ions is less than the contribution of scattering on free electrons by a factor $(M/m)^{3/2}$ if allowance is made for only the first terms in (25.23) and (25.24). The nonlinear interaction is due in this case to scattering on the ion polarization clouds of the virtual wave, since the kernel that describes the process,

$$Q = \frac{1}{2\pi^2}\left(\frac{T_e}{T_i}\right)^{3/2}\left(1 + \frac{T_e}{T_i}\right)^{-2}\left(\frac{M}{m}\right)^{1/2}\sqrt{\frac{\pi}{2}}\frac{e^2}{m^2}\frac{\omega_{Le}^2}{v_{T_e}^2}\frac{1}{\omega^4}\left(\frac{k_z k_z'}{kk'}\right)^2\frac{\omega''}{|k_z''|v_{T_e}}, \tag{25.27}$$

is larger by a factor $(M/m)^{1/2}$ than the kernel corresponding to scattering on free electrons. The nonlinear interaction in the other range of electron and ion temperatures has the same mechanism. Namely,

$$Q = \frac{1}{2\pi^2}\left(\frac{T_i}{T_e}\right)^{1/2}\left(\frac{M}{m}\right)^{1/2}\operatorname{Im} V_e^{(1)}, \qquad \frac{M}{m} \gg \frac{T_e}{T_i} \gg 1; \tag{25.28}$$

$$Q = \frac{1}{2\pi^2}\left(\frac{T_e}{T_i}\right)^{3/2}\left(\frac{M}{m}\right)^{1/2}\operatorname{Im} V_e^{(1)}, \qquad \left(\frac{M}{m}\right)^{1/3} \gg \frac{T_i}{T_e} \gg 1, \tag{25.29}$$

* In deriving the first of Eqs. (25.25) we have allowed for the fact that the wavelength of the virtual oscillation is appreciably greater than the electron Debye radius ($k'' r_{D_e} \ll 1$) because this condition is satisfied for both the scattered oscillations.

where $\mathrm{Im}\, V_e^{(1)}$ is determined by the first term of (25.24):

$$\mathrm{Im}\, V_e^{(1)} \equiv \sqrt{\frac{\pi}{2}}\frac{e^2}{m^2}\frac{\omega_{Le}^2}{v_{T_e}^2}\frac{1}{\omega^4}\left(\frac{k_z k_z'}{kk'}\right)^2\frac{\omega''}{|k_z''|v_{T_e}}. \tag{25.30}$$

All these cases, (25.27)-(25.29), are combined by the equation

$$\frac{dW_p(\mathbf{k})}{dt} = -\frac{1}{4(2\pi)^{5/2}}\left(\frac{T_e}{T_i}\right)^{3/2}\left(1+\frac{T_e}{T_i}\right)^{-2}\left(\frac{M}{m}\right)^{1/2}\frac{\omega_{Le}}{N_e r_{D_e}^3} r_{D_e}^2 \frac{W_p(\mathbf{k})}{\varkappa T_e}\times$$
$$\times\int d\mathbf{k}' W_p(\mathbf{k}')\frac{|k_z|}{k}\frac{|k_z|k'-|k_z'|k}{|k_z-k_z'|}\frac{1}{kk'}\left(1+\frac{\omega_{Li}^2}{\Omega_i^2}\right)^{-1}, \tag{25.31}$$

from which (25.28) follows when $T_e \gg T_i$ and (25.29) when $T_i \gg T_e$. It is interesting that this equation also describes the scattering of a wave with the spectrum (25.17) at much higher frequencies (ω, $\omega' \gg \Omega_i$) provided the frequency of the virtual oscillation is low: $\omega'' \ll \Omega_i$. Note that since the frequencies of the scattered oscillations are low (25.1) compared with the ion gyroscopic frequency, the second terms in the curly brackets in (25.23) and (25.24) may become comparable in magnitude with the first terms only when $|k_z| \ll k_\perp$, $|k_z'| \ll k_\perp'$. However, one cannot, in general, ignore them by assuming $|k_z| \sim k_\perp$, $|k_z'| \sim k_\perp'$, since in the most interesting case, $\omega_{Le} \gg \Omega_i$, it turns out that waves of the spectrum (18.4) can propagate only almost perpendicularly to the external magnetic field. Before we take into account the influence of the second terms in the curly brackets (25.23) and (25.24) on the processes under consideration, we point out that the contributions to the kernel Q of the first and the second terms in (25.23) and (25.24) are additive. As a result of this they can be considered separately. The characteristic feature of the second terms is the reduction of the leading members of Im V and $\mathrm{Im}\,(S^2/\varepsilon)$. Allowance for the successive terms in order of magnitude in the zeroth approximation in the Larmor radius by means of the general formulas (25.8)-(25.11) gives the following expression for the total kernel:

$$Q = \frac{1}{2\pi^2}\left(\frac{T_e}{T_i}\right)^{3/2}\left(1+\frac{T_e}{T_i}\right)^{-2}\left(\frac{M}{m}\right)^{1/2}\sqrt{\frac{\pi}{2}}\frac{e^2}{m^2}\frac{\omega_{Le}^2}{v_{T_e}^2\omega^4}\left(\frac{k_z k_z'}{kk'}\right)^2\frac{\omega''}{|k_z''|v_{T_e}}\times$$
$$\times\left\{1+\frac{(\mathbf{k}\,[\mathbf{k}'\mathbf{h}])^2}{(k_z k_z')^2}\frac{\omega^2}{\Omega_e^2}\left[(k'' r_{D_e})^2 - \frac{v_{T_e}^2}{k_z''}\left(\frac{k_z^3}{\omega^2}-\frac{k_z'^3}{\omega'^2}\right)\right]^2\right\}, \tag{25.32}$$

which holds in accordance with what we have said above in the temperature range

$$\frac{M}{m} \gg \frac{T_e}{T_i} \gg \left(\frac{m}{M}\right)^{1/3}. \tag{25.33}$$

The second term in the curly brackets on the right-hand side of (25.32) corresponds to the contribution of the second terms in (25.23) and (25.24). In contrast to the first term (unity), the second term arises from the interference of two mechanisms: scattering on free electrons and ions and scattering on ions of the virtual oscillation. The kernel (25.32) is determined by the second term if the wavelength λ of the scattered oscillations is large compared with the electron Debye radius r_{De} and is bounded (in a dense plasma) above by $r_{De}(\omega_{Li}/\Omega_i)$:

$$r_{D_e} < \lambda < r_{D_e}\frac{\omega_{Li}}{\Omega_i}. \tag{25.34}$$

Outside the temperature range (25.33) in the case of very hot electrons,

$$\frac{T_e}{T_i} \gg \frac{M}{m}, \tag{25.35}$$

the complete kernel Q has the form

$$Q = \frac{1}{2\pi^2}\sqrt{\frac{\pi}{2}}\frac{e^2}{m^2}\frac{\omega_{Le}^2}{v_{T_e}^2}\frac{1}{\omega^4}\left(\frac{k_z k_z'}{kk'}\right)^2\frac{\omega''}{|k_z''|v_{T_e}}\left\{1+\left(\frac{T_i M}{T_e m}\right)^{1/2}\left(\frac{\mathbf{k}\,[\mathbf{k}'\mathbf{h}]}{k_z k_z'}\right)^2\frac{\omega^2}{\Omega_e^2}\left[(k'' r_{D_e})^2-\frac{v_{T_e}^2}{k_z''}\left(\frac{k_z^3}{\omega^2}-\frac{k_z'^3}{\omega'^2}\right)\right]^2\right\} \tag{25.36}$$

Here, both terms (the first and the second) are due to the reduction of the leading terms in (25.25). The first, which corresponds to scattering on free electrons, is due to the scattering on free ions and scattering on the ion polarization clouds of the virtual wave compensating each other. In a plasma with very hot ions,

$$T_i \gg \left(\frac{M}{m}\right)^{1/3} T_e, \tag{25.37}$$

the induced scattering of the oscillations p is characterized by the kernel

$$Q = \frac{1}{2\pi^2}\sqrt{\frac{\pi}{2}}\frac{e^2}{m^2}\frac{\omega_{Le}^2}{v_{T_e}^2}\frac{1}{\omega^4}\left(\frac{k_z k_z'}{kk'}\right)^2 \frac{\omega''}{|k_z''| v_{T_e}}\left\{\left(\frac{T_e}{T_i}\right)^{3/2}\left(\frac{M}{m}\right)^{1/2} + \left[\left(\frac{\mathbf{k}\,[\mathbf{k}'\mathbf{h}]}{k_z k_z'}\right)^2 \frac{\omega^2}{\Omega_e^2} + 1\right]\left[(k'' r_{D_e})^2 - \frac{v_{T_e}^2}{k_z''}\left(\frac{k_z^3}{\omega^2} - \frac{k_z'^3}{\omega^2}\right)\right]^2\right\}. \tag{25.38}$$

In contrast to all the processes considered above, the nonlinear interaction in this case is determined by scattering on free electrons and on the electron and not ion polarization clouds of the virtual oscillation. Summarizing our investigation of the scattering of the oscillations p with the spectrum (18.4),* we note that, using the concept of the characteristic time of nonlinear interaction, the fastest process is scattering in the temperature region $T_e \sim T_i$. More precisely, in a plasma with electron temperature T_e equal to thrice the ion temperature T_i,

$$T_e = 3T_i, \tag{25.39}$$

the characteristic time of nonlinear interaction is minimal, being

$$\tau \sim 10^3 \left(\frac{m}{M}\right)^{1/3} \frac{N_e r_{D_e}^3}{\omega_{L_e}} \frac{\varkappa T_e}{W_p} \frac{1}{r_{D_e}^2 |k_z k' - k_z' k|}. \tag{25.40}$$

For simplicity, we have restricted ourselves in this estimate to the first term in the kernel (25.32).

2. We now come to the scattering of electron sound (18.69) in a plasma with hot ions, (18.67). At a relatively low ion temperature,

$$\left(\frac{M}{m}\right)^{1/3} A_0^{2/3}(z_i'') T_e > T_i \gg T_e, \tag{25.41}$$

the permittivity $\varepsilon(\omega'', \mathbf{k}'')$, which characterizes the electrostatic virtual oscillation, has the form

$$\varepsilon(\omega'', \mathbf{k}'') = \frac{\omega_{Le}^2}{(k'' v_{T_e})^2} + i\sqrt{\frac{\pi}{2}} A_0(z_i'') \frac{\omega_{Li}^2}{(k'' v_{T_i})^2} \frac{\omega''}{|k_z''| v_{T_i}}, \tag{25.42}$$

and the values of Im V and S are determined by the electron terms on the right-hand sides of (25.23) and (25.24). In contrast to the induced Coulomb scattering of the oscillations p, considered in § 25.1, in the given case of scattering of the electron sound s_e there is no reduction of the leading terms in the kernel (25.15) on account of either the first or the second terms in the curly brackets in (25.23) and (25.24). Under the conditions (25.41), the Coulomb scattering of electron sound occurs on the ion polarization clouds of the virtual wave and is described by Eq. (25.3) with the kernel

$$Q = \frac{1}{2\pi^2} A_0(z_i'') \left(\frac{T_e}{T_i}\right)^{3/2} \left(\frac{M}{m}\right)^{1/2} \sqrt{\frac{\pi}{2}} \frac{e^2}{m^2} \frac{\omega_{Le}^2}{v_{T_e}^2} \frac{1}{\omega^4} \left(\frac{k_z k_z'}{kk'}\right)^2 \frac{\omega''}{|k_z''| v_{T_e}} \left\{1 + \frac{(\mathbf{k}\,[\mathbf{k}'\mathbf{h}])^2}{(k_z k_z')^2} \frac{\omega^2}{\Omega_e^2}\right\}. \tag{25.43}$$

* To avoid confusion we point out that the frequencies ω and ω' in the expressions for the kernels are to be understood everywhere as the expressions for the spectra $\omega(\mathbf{k})$ and $\omega'(\mathbf{k}')$ of the oscillations p in accordance with (18.4). We shall also use this abbreviated notation in the following sections.

Here, the first term [the unity in the curly brackets in (25.43)] is greater according to (25.41) than the kernel corresponding to scattering on free electrons, and the second term is greater than the first only in the case of almost perpendicular propagation of the scattered oscillations:

$$k_\perp^4 > k_z^4 \frac{\omega^2}{\Omega_e^2}. \tag{25.44}$$

Under conditions opposite to the left-hand member of the inequality (25.41),

$$\left(\frac{M}{m}\right)^{1/3} A_0^{2/3}(z_i'')\, T_e \ll T_i, \tag{25.45}$$

the kernel Q is reduced [compared with (25.43)]:

$$Q = \frac{1}{2\pi^2}\left\{A_0(z_i'')\left(\frac{T_e}{T_i}\right)^{3/2}\left(\frac{M}{m}\right)^{1/2}\left[1+\frac{(\mathbf{k}\,[\mathbf{k}'\mathbf{h}])^2}{(k_z k_z')^2}\frac{\omega^2}{\Omega_e^2}\right]+\left[\frac{v_{T_e}^2}{k_z''}\left(\frac{k_z^3}{\omega^2}-\frac{k_z'^3}{\omega'^2}\right)-\right.\right.$$
$$\left.\left.-\frac{T_e}{T_i}-\left(\frac{k''v_{T_e}}{\omega_{L_e}}\right)^2\right]^2\right\}\sqrt{\frac{\pi}{2}}\,\frac{e^2}{m^2}\frac{\omega_{L_e}^2}{v_{T_e}^2}\frac{1}{\omega^4}\left(\frac{k_z k_z'}{kk'}\right)^2\frac{\omega''}{|k_z''|\,v_{T_e}}. \tag{25.46}$$

The equation corresponding to the fastest process of Coulomb scattering of the electron sound (18.69) in a magnetoactive plasma with hot ions (18.67) can be represented by the equation* [the kernel (25.43)]:

$$\frac{dW_{s_e}(\mathbf{k})}{dt} = -\frac{1}{4(2\pi)^{5/2}}\frac{\omega_{L_e}}{N_e r_{D_e}^3}\frac{W_{s_e}(\mathbf{k})}{\varkappa T_e}\left[1+\left(\frac{k_\perp\omega_{L_e}}{k\Omega_e}\right)^2+\left(\frac{\omega_{L_i}}{kv_{T_i}}\right)^2\times\right.$$
$$\left.\times\left(1-A_0(k_\perp^2\rho_i^2)\right)\right]^{-1} r_{D_e}^2\int d\mathbf{k}' W_{s_e}(\mathbf{k}')\left[1+\left(\frac{k_\perp'\omega_{L_e}}{k'\Omega_e}\right)^2+\left(\frac{\omega_{L_i}}{k'v_{T_e}}\right)^2\times\right.$$
$$\left.\times\left(1-A_0(k_\perp'^2\rho_i^2)\right)\right]^{-1}\frac{k_z}{k}\frac{1}{|k_z-k_z'|}\left\{1+\frac{\omega^2}{\Omega_e^2}\left(\frac{\mathbf{k}\,[\mathbf{k}'\mathbf{h}]}{k_z k_z'}\right)^2\right\}\times$$
$$\times\left\{\frac{k_z}{k}\left[1+\left(\frac{k_\perp\omega_{L_e}}{k\Omega_e}\right)^2+\left(\frac{\omega_{L_i}}{kv_{T_i}}\right)^2\left(1-A_0(k_\perp^2\rho_i^2)\right)\right]-\frac{k_z'}{k'}\left[1+\left(\frac{k_\perp'\omega_{L_e}}{k'\Omega_e}\right)^2+\left(\frac{\omega_{L_i}}{k'v_{T_i}}\right)^2\left(1-A_0(k_\perp'^2\rho_i^2)\right)\right]\right\}. \tag{25.47}$$

3. The scattering of the ion sound (25.2) is investigated in exactly the same way as the above two examples. The permittivity $\varepsilon(\omega'', \mathbf{k}'')$ characterizing the virtual oscillation and the value of Im V are determined as before by (25.22) and (25.24). However, the expression (25.23) is suitable only for the ion term in S. The electron part of S is determined directly by means of the more general formula (25.8), since for the considered process the electron thermal velocity is higher than the phase velocities of both the virtual and the scattered oscillations. When calculating the kernels we shall take into account only the first terms on the right-hand sides of Eqs. (25.8) and (25.10), which do not contain the gyroscopic frequency. The point is that in the given case of Coulomb scattering of ion sound the conditions for the existence of the spectrum do not impose stringent restrictions on the ratio of k_z and $k_\perp$, in contrast to the two foregoing cases; therefore, we can neglect the small region in the wave-vector space in which $k_z \ll k_\perp$ and we need only allow for the contribution of the second terms to the right-hand sides of Eqs. (25.8) and (15.10). Then

$$S_e = -i\frac{e}{m}\frac{\omega_{L_e}^2}{kk'k''}\frac{1}{v_{T_e}^2} - \sqrt{\frac{\pi}{2}}\,\frac{e}{m}\frac{\omega_{L_e}^2}{v_{T_e}^2}\frac{1}{\omega^2}\left(\frac{k_z k_z'}{kk'}\right)\frac{\omega''}{|k_z''|\,v_{T_e}}. \tag{25.48}$$

The ratios of the electron and ion parts of Im V, S, and $\varepsilon(\omega'', \mathbf{k}'')$ are determined as before by Eqs. (25.25) and (25.26), the second of them in (25.26) being replaced by

$$\frac{\operatorname{Im} S_e^{(1)}}{\operatorname{Im} S_i^{(1)}} = \left(\frac{\omega\omega'}{k_z k_z'}\right)^2 \frac{1}{v_s^2}\frac{T_i}{T_e}. \tag{25.49}$$

* In (25.47) we assume $z_i'' \ll 1$, i.e., $A_0(z_i'') \simeq 1$, which does not, of course, exclude $(k_\perp\rho_i)^2 \sim 1$, $(k_\perp'\rho_i)^2 \sim 1$. The derivatives $(\partial\omega\varepsilon'/\partial\omega)$ in (25.47) can be readily found by means of (18.71).

Note that in the given case the imaginary part of the electron contraction (25.48) is always appreciably smaller than the imaginary part of the ion contraction S_i,

$$\operatorname{Im} S_e \ll \operatorname{Im} S_i, \tag{25.50}$$

since it follows from the plasma's being nonisothermal ($T_e \gg T_i$) and the low phase velocity of the ion sound,

$$\left(\frac{\omega}{k_z}\right)^2 < v_s^2, \quad v_s^2 \equiv \frac{\varkappa T_e}{M}, \tag{25.51}$$

that the right-hand side of (25.49) is always less than unity. In deriving the first of Eqs. (25.25) we have noted that the wavelength of the virtual oscillation is greater than the ion Debye radius, as can be seen by means of the inequalities

$$v_{T_i}^2 \ll \left(\frac{\omega}{k_z}\right)^2 < \left(\frac{\omega_{L_i}}{k}\right)^2, \quad v_{T_i}^2 \ll \left(\frac{\omega'}{k_z'}\right)^2 < \left(\frac{\omega_{L_i}}{k'}\right)^2, \tag{25.52}$$

which are included among the conditions for the existence of (weak damping) of the ion-sound spectrum. After these preliminary comments, we find that in a plasma with not too hot electrons,

$$\frac{M}{m} T_i \gg T_e \gg T_i \tag{25.53}$$

the kernel Q determining the Coulomb scattering of ion sound,

$$Q = \frac{1}{2\pi^2} \sqrt{\frac{\pi}{2}} \frac{e^2}{m^2} \frac{\omega_{L_e}^2}{v_{T_e}^2} \frac{1}{\omega^4} \left(\frac{k_z k_z'}{kk'}\right)^2 \frac{\omega''}{|k_z''| v_{T_e}}, \tag{25.54}$$

corresponds to scattering on free electrons. In a plasma with very hot electrons (25.35), the nonlinear interaction of ion-acoustic oscillations is determined in the first approximation solely by the ion terms. As in the case of the Coulomb scattering of the p oscillations [with spectrum (18.4)], there is a reduction of the principal contributions in this case. Allowance for the corrections is given by the kernel

$$Q = \frac{1}{2\pi^2} \left\{ 1 + \left(\frac{m}{M}\right)^{3/2} \left(\frac{T_e}{T_i}\right)^{3/2} \left[\frac{\omega\omega'}{k_z k_z'} \frac{1}{v_s^2} \frac{T_i}{T_e} + \frac{v_{T_i}^2}{k_z''} \left(\frac{k_z^3}{\omega^2} - \frac{k_z'^3}{\omega'^2}\right) - \right.\right.$$
$$\left.\left. - \frac{T_i}{T_e} - \left(\frac{k'' v_{T_i}}{\omega_{L_i}}\right)^2 \right]^2 \right\} \sqrt{\frac{\pi}{2}} \frac{e^2}{m^2} \frac{\omega_{L_e}^2}{v_{T_e}^2} \frac{1}{\omega^4} \left(\frac{k_z k_z'}{kk'}\right)^2 \frac{\omega''}{|k_z''| v_{T_e}}. \tag{25.55}$$

In both cases (25.54) and (25.55), the kernel contains a term that is exactly equal to $\operatorname{Im} V_e^{(1)}$ [the kernel (25.54) is entirely determined by it; see (25.30)]. Since $\operatorname{Im} V_e^{(1)}$ in accordance with (25.30) is proportional to $\omega''/|k_z''| v_{T_e}$, i.e., to the contribution of the electron pole corresponding to the virtual oscillation which is small in the given case $(\omega''/k_z'') \ll \omega/k_z$, ω'/k_z' compared with the contribution of the electron poles corresponding to the scattered ion-acoustic waves, we must ignore $\operatorname{Im} V_e^{(1)}$ as a small term. Thus, the nonlinear interaction of the ion-acoustic oscillations (25.2) is not small compared with the linear Landau damping effect only in the temperature region (25.35) and only in the case when the second term in (25.55) is greater than the first. If we restrict ourselves to a single species of particle, namely, we distinguish only the ion terms, we obtain the kernel

$$Q = \frac{1}{2\pi^2} (k'' r_{D_i})^4 \left\{ 1 - \frac{\omega_{L_i}^2}{k''^2} \left(\frac{k_z^2}{\omega^2} + \frac{k_z k_z'}{\omega\omega'} + \frac{k_z'^2}{\omega'^2} \right) \right\} \sqrt{\frac{\pi}{2}} \frac{e^2}{M^2} \frac{\omega_{L_i}^2}{v_{T_i}^2 \omega^4} \left(\frac{k_z k_z'}{kk'}\right)^2 \frac{\omega''}{|k_z''| v_{T_i}}, \tag{25.56}$$

which is proportional of the first power of the ion thermal velocity. In this, in particular, it differs significantly from the kernel (5.9) [with allowance for the first of Eqs. (5.3)] of Eq. (5.2) of [73]. The kernel (5.9) in the limit (25.13) of a low phase velocity of the virtual oscillation describes the induced scattering of ion sound in a isotropic plasma and arises because the nonlinear interaction is not one-dimensional ($[\mathbf{kk'}]^2 \neq 0$) in the lower order in v_{Ti}. Concluding our study of the Coulomb scattering of low-frequency ion sound in a magnetoactive nonisothermal plasma, we note that the nonlinear scattering on ions (25.55) is always large compared with the linear ion terms omitted in Eq. (25.3) due to the exponentially small linear damping on ions and that it can exceed the scattering on free electrons (25.54) only in the temperature region (25.35).

Summarizing, we emphasize that the variants of Coulomb induced scattering of quasi-longitudinal characteristic oscillations of a magnetoactive plasma considered in this section cover essentially all the cases of scattering of low-frequency longitudinal waves ($\omega \ll \Omega_i$) within any one branch of the spectrum (transformation of the spectrum as a result of scattering goes beyond the scope of this section). Knowledge of the kernels Q given above enables us when necessary to write the effective differential scattering cross sections for these processes if one uses the method of deriving such cross sections developed in Ch. III (§ 12). As we have already pointed out in § 25.1, the Coulomb scattering of waves with the spectrum (25.18) in the special case of a rarefied plasma ($\omega^2_{Li} \ll \Omega_i^2$) has been studied in [70], where the assumption is made that the phase velocity of the electrostatic virtual oscillation is high compared with the ion thermal velocity. Here, we have considered the opposite case of a low phase velocity of the virtual oscillation. The general expression (25.15) for the kernel Q, expressed here in terms of Im V, S, and ε, corresponds to formula (1.2) of [70]. In particular, ignoring the ion terms in Im V, S, and ε, and also the terms containing the gyroscopic frequency in Im V and S, we obtain from (25.15), (25.8)-(25.11) the kernel

$$Q = \frac{9}{2\pi^2} (kk')^2 r^4_{De} \sqrt{\frac{\pi}{2}} \frac{e^2}{m^2} \frac{\omega^2_{Le}}{v^2_{Te}} \left(\frac{k_z k'_z}{kk'}\right)^2 \frac{1}{\omega^4} \frac{\omega''}{|k''_z| v_{Te}} e^{-\frac{1}{2}\left(\frac{\omega''}{k''_z v_{Te}}\right)^2}, \tag{25.57}$$

which, after substitution of the spectrum (25.18), goes over into the "nonlinear growth rate" (13) of [70] (p. 161). If allowance is made for the ion terms in $\varepsilon(\omega'', \mathbf{k}'')$, we obtain in exactly the same way the relation

$$Q = \frac{1}{2\pi^3} \frac{16}{81} \sqrt{\frac{\pi}{2}} \frac{e^2}{m^2} \frac{\omega^2_{Le}}{v^2_{Te}} \frac{1}{\omega^4} \left(\frac{k_z k'_z}{kk'}\right)^2 \frac{\omega''}{|k''_z| v_{Te}} \left(\frac{m}{M}\right)^2 \frac{1}{(k+k')^4 r^4_{De}}, \tag{25.58}$$

from which we obtain formula (14) of [70], p. 162). We should point out that the kernels (25.57) and (25.58), which have the same order of magnitude, do not correspond to the fastest scattering processes. The point is that, for example, (25.57) is obtained because of the reduction of the principal terms of the electron parts of Im V, S, and ε. Therefore, the ratio of (25.57) to the kernel (25.27) arising from the leading term in S^2/ε is small and in a quasi-isothermal plasma is $(kk')^2 r^4_{De} (m/M)^{1/2}$, i.e., it is 10^{-3} in a hydrogen plasma with $(kr_{De})^2 \sim 10^{-1}$. Hence, the kernel (25.27) corresponds to a much faster process of induced Coulomb scattering.

Comparison of the results obtained in this section with the formulas for the induced Coulomb scattering of electrostatic oscillations in an isotropic plasma (§ 9) shows that there is a certain analogy between the process of scattering of the oscillations p with the spectrum (18.4) in a magnetoactive plasma and the scattering of electron Langmuir waves in an isotropic plasma. Indeed, Eq. (25.3) with the kernel (25.27) is identical with Eq. (9.27), which describes the Coulomb scattering of electron Langmuir waves on ions (i.e., on the ion polarization clouds of the virtual oscillation) if the factor $(\mathbf{kk'}/kk')^2$ in an isotropic plasma is replaced by $(k_z k'_z/kk')^2$ in a magnetoactive plasma. However, the difference is in reality increased by the fact that in the kernel (25.32), which generalizes (25.27), there is a second term, which depends explicitly on the magnetic field strength $\mathbf{B}_0$ and, in addition, the spectrum (18.4) itself depends explicitly on this field. This similarity with an isotropic plasma is explained by the fact that in the considered case of a very strong magnetic field (25.1) oscillations of the particles are possible only in the direction of this field and in this direction they occur in the same way as in an isotropic plasma. The actual fact of the existence of this similarity emphasizes once more the possibility of illustrating the general conservation laws and the direction of evolution of nonlinear processes by examples not only in an isotropic plasma, which were studied in detail in Ch. III (§ 9-11), but also in a magnetoactive plasma, in which the characteristic features of the nonlinear interaction remain the same as in an isotropic plasma, differing only by the greater variety of forms. This greater variety is manifested, for example, by the processes of induced scattering of cyclotron harmonics (which are, of course, absent in an isotropic plasma), to whose study the next paragraph is devoted.

§ 26. Induced Scattering of Quasilongitudinal Characteristic Cyclotron Oscillations of a Magnetoactive Plasma on the Plasma Particles

We now consider the nonlinear interaction of waves at frequencies near the ion and electron cyclotron resonances [68]:

$$\omega \simeq n\Omega_i, \qquad \omega \simeq n\Omega_e.$$

As everywhere in this chapter (except in § 24), we restrict our discussion to quasielectrostatic oscillations. They are of particular interest because, on the one hand, having a low phase velocity they can only be carried into a regime of kinetic (oscillatory) instability as a result of an anisotropy of the equilibrium velocity distribution function of the particles; on the other hand, waves with frequencies lying "within a line" of cyclotron resonance are rather efficiently damped. This damping allows one to use cyclotron resonance to "pump" electromagnetic energy into the plasma. And the mechanism of nonlinear interaction of waves responsible for transferring energy within the spectrum is in a number of cases decisive for the transformation of electromagnetic energy into thermal energy. As an example of such a mechanism, one can consider the nonlinear interaction of quasilongitudinal cyclotron waves with frequencies ω and ω' and wave vectors **k** and **k'** through a slow electrostatic virtual oscillation with frequency ω'' and wave vector **k"** equal to the differences of the frequencies, $\omega - \omega'$, and the wave vectors, **k** − **k'**, of the interacting waves. The assumption that the virtual oscillation is slow, i.e., that its phase velocity is low compared with the ion and electron thermal velocities, is made, as in the foregoing section, for the sake of greater simplicity and clarity of the resulting formulas. The nonlinear interaction effect also takes place, of course, when the phase velocity of the virtual oscillation is of the order of the thermal velocity of the particles.

1. We begin our systematic study with processes of induced scattering of ion-cyclotron oscillations $\tilde{c}_i'$ with the spectrum (18.14) at the gyroscopic frequency Ω_i and oscillations $\tilde{c}_i$ with the spectrum (18.44) at a frequency that is a multiple of the ion gyroscopic frequency: $n\Omega_i$ ($n > 1$). We shall assume that the plasma is fairly dense ($\omega_{Li}^2 \gg \Omega_i^2$). In accordance with the results of § 18 of the foregoing chapter, the first ion-cyclotron harmonic $\tilde{c}_i'$ can propagate in a wide interval of angles ϑ measured from the direction of the external magnetic field (18.15):

$$(\tilde{c}_i') \qquad \omega = \Omega_i\left\{1 - \frac{1}{2}\frac{m}{M}\frac{k_\perp^2}{k_z^2}\right\}, \qquad k_z \gg k_\perp\sqrt{\frac{m}{M}}. \tag{26.1}$$

In contrast, the oscillation (18.44) can only propagate almost at right angles to the magnetic field:*

$$(\tilde{c}_i) \qquad \omega = n\Omega_i\left\{1 - \frac{n^2-1}{2^n n!}z_i^{n-1}\right\}, \qquad k_z \ll k_\perp\sqrt{\frac{m}{M}}, \qquad n > 1. \tag{26.2}$$

* The spectrum (26.2) is a special case of (18.44) for oscillations with long wavelength at right angles to the external magnetic field compared with the ion Larmor radius: $z_i \equiv (k_\perp \rho_i)^2 \ll 1$ [see the asymptotic behavior (18.31) for the function $A_n(z)$].

We use some of the inequalities guaranteeing weak damping of the waves (26.1) and (26.2) and the other cyclotron harmonics during scattering,

$$|k_z|v_T \ll |\omega - n\Omega| \ll \Omega, \quad |k_z' v_T| \ll |\omega' - n\Omega| \ll \Omega, \quad n = 1, 2, 3\ldots$$

to simplify the longitudinal contraction S of the three-index tensor S_{ijs} and the imaginary part Im V of the longitudinal contraction of the four-index tensor $V_{i\alpha jb}$, which occur in the generalized kinetic equation:

$$S = -i\frac{e}{m}\frac{\omega_L^2}{v_T^2}\frac{1}{kk'k''}\sum_{n=-\infty}^{+\infty} e^{in(\theta'-\theta)}\int_0^\infty x dx e^{-\frac{x^2}{2}}\left\{\left[1 - J_+\left(\frac{\omega''}{|k_z''|v_T}\right)\right] \times \right.$$
$$\times\left[\frac{1}{\omega - n\Omega} J_n(k_\perp \rho x) J_n(k'_\perp \rho x) J_0(k''_\perp \rho x)\left(\frac{k_z k_z'}{\omega' - n\Omega} - i\frac{(\mathbf{k}'[\mathbf{kh}])}{\Omega}\right) + \right.$$
$$+ \frac{k_\perp k'_\perp}{2\Omega}\frac{1}{\omega - n\Omega} J_0(k''_\perp \rho x)\left(J_{n-1}(k_\perp \rho x) J_{n-1}(k'_\perp \rho x) - J_{n+1}(k_\perp \rho x) \times\right.$$
$$\left.\left.\times J_{n+1}(k'_\perp \rho x)\right)\right] - \frac{nJ_n(k_\perp \rho x) J_n(k'_\perp \rho x)}{(\omega - n\Omega)^2}\left[\frac{k''_\perp}{\rho}\frac{n}{x} J_1(k''_\perp \rho x) + J_0(k''_\perp \rho x)\left(k_z k_z' \frac{\Omega}{\omega - n\Omega} + i(\mathbf{k}'[\mathbf{kh}])\right)\right]\Bigg\}; \quad (26.3)$$

$$\mathrm{Im}\, V = \frac{e^2}{m^2}\frac{1}{(kk')^2}\frac{\omega_L^4}{v_T^4}\sqrt{\frac{\pi}{2}}\frac{\omega''}{|k_z''|v_T}\exp\left\{-\frac{1}{2}\left(\frac{\omega''}{k_z'' v_T}\right)^2\right\}\int_0^\infty x dx e^{-\frac{x^2}{2}} \times$$
$$\times\left|\sum_{n=-\infty}^{+\infty} e^{in(\theta'-\theta)}\left\{\frac{1}{\omega - n\Omega} J_n(k_\perp \rho x) J_n(k'_\perp \rho x)\left[\frac{k_z k_z'}{\omega - n\Omega} + i\frac{(\mathbf{k}'[\mathbf{kh}])}{\Omega}\right] + \right.\right.$$
$$\left.\left. + \frac{k_\perp k'_\perp}{2\Omega}\frac{1}{\omega - n\Omega}\left[J_{n-1}(k_\perp \rho x) J_{n-1}(k'_\perp \rho x) - J_{n+1}(k_\perp \rho x) J_{n+1}(k'_\perp \rho x)\right]\right\}\right|^2. \quad (26.4)$$

The notation employed here completely corresponds to that adopted in the present exposition. We recall, in particular, that θ and θ' are the angles of the components $\mathbf{k}_\perp$ and $\mathbf{k}'_\perp$ (at right angles to the magnetic field $\mathbf{B}_0$) of the wave vectors $\mathbf{k}$ and $\mathbf{k}'$ of the interacting oscillations with the x axis in the plane perpendicular to $\mathbf{B}_0$. The integrals with respect to the variable x of the products of three and four Bessel functions occur in formulas (26.3)-(26.4) because of the use of the explicit expressions for $A_{n,m}$ and $B_{n,m,l}$ [see (19.67) and (20.16) in the foregoing chapter]. For the spectra (26.1) and (26.2) considered in this section, the equations (26.3) and (26.4), on whose right-hand side summation over the particle species is assumed, determine both the electron and ion contributions to S and Im V. The general form of the nonlinear equation and its kernel is given by the relations (25.3), (25.4), and (25.15) of the foregoing section (§ 25). Taking into account the proximity of the frequencies of the scattered oscillations (26.1)-(26.2) to $n\Omega_i$ ($\omega \simeq n\Omega_i \ll \Omega_e$) and expanding all the quantities on the right-hand sides of (26.3)-(26.4) in powers of the small parameters

$$\frac{\omega''}{|k_z''|v_{T_{e,i}}} \ll 1, \quad z_{e,i} \ll 1, \quad z'_{e,i} \ll 1, \quad (26.5)$$

we obtain a simpler expression for the kernel Q in the case of the spectrum (26.1):

$$Q = \frac{1}{2\pi^2}\left(1 + \frac{T_i}{T_e}\right)^{-2}\left(1 + \sqrt{\frac{MT_i}{mT_e}}\right)\sin^2\left(\frac{\theta' - \theta}{2}\right)\mathrm{Im}\, V_e \quad (26.6)$$

After substitution of the explicit form of Im V_e from (26.6) we obtain the kernel

$$Q = \frac{1}{\pi^2}\sqrt{\frac{\pi}{2}}\frac{e^2}{mM}\frac{\sin(\vartheta' + \vartheta)\sin(\vartheta' - \vartheta)\sin^2\frac{\theta' - \theta}{2}}{\Omega_i^3 r_{D_e}^2 |k_z - k_z'| v_{T_e}}\left(1 + \frac{T_i}{T_e}\right)^{-2}\left(1 + \sqrt{\frac{MT_i}{mT_e}}\right). \quad (26.7)$$

In the temperature range $T_i \ll (m/M)T_e$ the scattering is determined by free electrons:

$$Q = \frac{1}{2\pi^2}\sin^2\left(\frac{\theta' - \theta}{2}\right)\mathrm{Im}\, V_e. \quad (26.8)$$

In an isothermal plasma, $T_i \simeq T_e$, the kernel Q as a function of T_i/T_e is maximal* and may exceed (26.8) by a factor $(M/m)^{1/2}$. In this case the scattering takes place on the polarization clouds of the virtual wave, so that the contributions of the ions and electrons are of the same order. Substituting the kernel (26.7) into Eq. (25.3), we arrive at the final equation, which characterizes the induced Coulomb scattering of the first ion-cyclotron harmonic (26.1):

$$\frac{dW_{\tilde{c}'_i}(\mathbf{k})}{dt} = -\frac{1}{4(2\pi)^{3/2}} \frac{m}{M} \frac{e^2}{M^2} \frac{\Omega_i^2}{\omega_{L_e}^2 v_{T_e}^3} W_{\tilde{c}'_i}(\mathbf{k}) \left(1 + \frac{T_i}{T_e}\right)^{-2} \left(1 + \sqrt{\frac{MT_i}{mT_e}}\right) \times$$

$$\times \int d\mathbf{k}' W_{\tilde{c}'_i}(\mathbf{k}') \left(\frac{k_\perp k'_\perp}{k_z k'_z}\right)^2 \frac{\sin\left(\frac{\theta' - \theta}{2}\right)}{|k_z - k'_z|} (\tan^2 \vartheta' - \tan^2 \vartheta). \tag{26.9}$$

From Eq. (26.9) it is easy to see that the total energy of the ion-cyclotron oscillations (26.1) is conserved [cf. (9.13)]

$$\int d\mathbf{k}' W_{\tilde{c}'_i}(\mathbf{k}) = \text{const}, \tag{26.10}$$

and that the energy is transferred within the spectrum during the process of nonlinear interaction from higher to lower frequencies.†

In the calculation of the kernel Q for the higher harmonics with the spectrum (26.2) we must distinguish two variants. If the number of harmonics n is higher than the second ($n > 2$), the main contribution to the kernel comes from the terms of the ion part of the contraction S, which is proportional to $(\omega - n\Omega_i)^{-2}$, i.e., from the terms from the second sum on the right-hand side of (26.3), which increase near the resonance $\omega \simeq n\Omega_i$ [in the same way as the components of the permittivity tensor of a cold magnetoactive plasma (16.37)-(16.38) increase as $\omega \to \Omega$]. In this case, the contribution of the electrons to S and Im V can be ignored. Then, at not too low electron temperatures,

$$\frac{z''_i}{(z_i z'_i)^{\frac{n-1}{2}}} \gg \left\{\sqrt{\frac{mT_e}{MT_i}} + \left(\frac{T_e}{T_i}\right)^2\right\}^{-1} \tag{26.11}$$

we obtain the following expression for the kernel Q:

$$Q = \frac{1}{(2\pi)^{3/2}} \frac{e^2}{M^2} \frac{\omega_{L_e}^2 [z_i'^{n-1} - z_i^{n-1}]}{\Omega_i^2 |k_z - k'_z| v_{T_e}^2} \left(1 + \frac{T_i}{T_e}\right)^{-2} \left[1 + \left(\frac{M}{m}\right)^{1/2} \left(\frac{T_e}{T_i}\right)^{3/2}\right] \times$$

$$\times \frac{(n-1)!\, 2^n}{(n^2-1)^3} (z_i z'_i)^{2-n} \left\{\left(\frac{n}{2} \frac{k''^2_\perp}{k_\perp k'_\perp}\right)^2 + \sin^2(\theta' - \theta)\right\}, \tag{26.12}$$

which is much larger in magnitude than the contribution due to scattering on free electrons:

$$\frac{Q}{\operatorname{Im} V_e} \simeq \frac{n^2}{2\pi^2} \left(1 + \frac{T_i}{T_e}\right)^{-2} \left[1 + \left(\frac{M}{m}\right)^{1/2} \left(\frac{T_e}{T_i}\right)^{3/2}\right] \left[\frac{(n-1)!\, 2^n}{(n^2-1)^2}\right]^2 (z_i z'_i)^{2-n}. \tag{26.13}$$

To the second variant ($n = 2$, $\omega \simeq 2\Omega_i$, $\omega' \simeq 2\Omega_i$) there corresponds the same order of the contributions of the resonance and nonresonance terms. Let us discuss the different special cases of the second variant that arise for different ratios of the ion and electron temperatures. In a

---

* Compare this with induced Coulomb scattering of the low-frequency oscillation p (considered in § 25.1).

† Compare this with the conclusion concerning the direction of energy transfer in the process of nonlinear interaction of transverse and longitudinal waves in an isotropic plasma (Ch. III, § 10).

nonisothermal plasma with hot electrons ($T_i \ll T_e$), the main contribution comes, as before (in the first variant), from the terms that contain "resonances of second order" ($\omega \simeq 2\Omega_i$). For this temperature ratio, the kernel Q takes the form

$$Q = \frac{\sqrt{2}}{27\pi^{3/2}} \frac{e^2}{M^2} \frac{\omega_{L_e}^2 (z_i' - z_i)}{\Omega_i^3 |k_z - k_z'| v_{T_e}^3} \left(\frac{M}{m}\right)^{1/2} \left(\frac{T_i}{T_e}\right)^{3/2} \left\{\frac{9}{16} + \left[\left(\frac{k_\perp''^2}{k_\perp k_\perp'}\right)^2 + \sin^2(\theta' - \theta)\right]\right\}. \quad (26.14)$$

In the opposite limiting case of hot ions ($T_i \gg T_e$),

$$Q = \frac{1}{12(2\pi)^{3/2}} \frac{e^2}{M^2} \frac{\omega_{L_e}^2 (z_i' - z_i)}{\Omega_i^3 |k_z - k_z'| v_{T_e}^3} \left(\frac{M}{m}\right)^{1/2} \left(\frac{T_e}{T_i}\right)^{3/2} \{3 - 2\cos 3(\theta' - \theta)\}. \quad (26.15)$$

As in the first variant (26.12), the scattering here occurs on the polarization clouds of the virtual oscillation and is much stronger than the scattering on the electrons. However, with increasing ion temperature, the kernel Q decreases as $(T_e/T_i)^{3/2}$.

2. Apart from what we have considered above, in a nonisothermal plasma ($T_e \gg T_i$) there also exist quasilongitudinal ion-cyclotron oscillations (18.46), which are weakly damped when there is a strong thermal motion of the electrons ($|k_z| v_{T_e} \gg \omega$) and go over in the long-wave limit (18.48) into the oscillations c_i with the spectrum (18.49). It turns out that oscillations with the spectrum (18.49) are almost always scattered on free electrons and are described by the nonlinear interaction equation with the kernel $Q = (1/2\pi^2) \operatorname{Im} V_e$. At the same time, for a Maxwell equilibrium distribution function (16.14) the linear damping on electrons (18.52) of the waves (18.49) is appreciably greater than the term describing the induced Coulomb scattering. In the short-wave region ($z_i \gtrsim 1$), the situation is different. For waves with the spectrum

$$\omega = \Omega_i \left\{1 + A_1(z_i)\left(1 + \frac{T_i}{T_e}\right)^{-1}\right\}, \quad z_i \gtrsim 3 \quad (26.16)$$

the ions make the main contribution to the contractions S and Im V. Retaining in the general expressions (26.3)-(26.4) the terms proportional to $(\omega - \Omega_i)^{-1}$, we can write down the following equation (25.3) of the nonlinear interaction of short-wave ($z_i \gtrsim 3$) ion-cyclotron harmonics [$\omega'' \equiv \omega(\mathbf{k}) - \omega'(\mathbf{k}')$]:

$$\frac{\partial W_{c_i}(\mathbf{k})}{\partial t} = -\frac{\Omega_i}{4\pi^3 N_e}\left(1 + \frac{T_i}{T_e}\right)^{-2} \frac{W_{c_i}(\mathbf{k})}{\varkappa T_e} \int d\mathbf{k}' W_{c_i}(\mathbf{k}') A_0(z_i'') \times$$
$$\times \sqrt{\frac{\pi}{2}} \frac{\omega''}{|k_z''| v_{T_i}} \exp\left\{-\frac{1}{2}\left(\frac{\omega''}{k_z'' v_{T_i}}\right)^2\right\} \left\{\sin^2(\theta' - \theta) z_i z_i' \int_0^\infty x dx e^{-\frac{x^2}{2}} \times\right.$$
$$\times J_1^2(k_\perp \rho_i x) J_1^2(k_\perp' \rho_i x) + \int_0^\infty \frac{dx}{x} e^{-\frac{x^2}{2}} \left[\frac{d}{dx}\left(J_1(k_\perp \rho_i x) J_1(k_\perp' \rho_i x)\right)\right]^2 +$$
$$+ A_1^{-1}(z_i) A_1^{-1}(z_i') \left|1 - \frac{T_e}{T_e + T_i} A_0(z_i'') J_+\left(\frac{\omega''}{|k_z''| v_{T_i}}\right)\right|^{-2} \times$$
$$\times \left[\sin^2(\theta' - \theta) z_i z_i' \left(\int_0^\infty x dx e^{-\frac{x^2}{2}} J_1(k_\perp \rho_i x) J_1(k_\perp' \rho_i x) J_0(k_\perp'' \rho_i x)\right)^2 + \right.$$
$$\left.\left. + z_i'' \left(\int_0^\infty dx e^{-\frac{x^2}{2}} J_1(k_\perp \rho_i x) J_1(k_\perp' \rho_i x) J_1(k_\perp'' \rho_i x)\right)^2\right]\right\}. \quad (26.16a)$$

Equation (26.16a) has the same form as the corresponding expression given [74] (after corrections of misprints in that paper). However, there is a difference. In Eq. (26.16a), the coefficient in front of the third term in the curly brackets of its kernel can be written in the form

$$\frac{A_0(z_i'')}{A_1(z_i) A_1(z_i')}. \quad (26.17)$$

In [74], this coefficient is given by the combination

$$\frac{A_0(z_i'') - A_1(z_i) - A_1(z_i')}{A_1(z_i) A_1(z_i')}, \tag{26.18}$$

which is very different from (26.17) in the considered short-wave region $z_i > 1$, $z_i' > 1$ for $z_i'' \gtrless 1$ in that it can change its sign. If we note that the first two terms in the kernel of Eq. (26.16a) under these conditions is small, we see that such a change of sign would mean that there exists a region in which energy is transferred in the opposite direction – from lower to higher frequencies. This change in the direction of transfer is impossible (in accordance with [74]) in the case of induced scattering of waves on particles with a Maxwell distribution.

3. We now turn to the study of induced Coulomb scattering of electron-cyclotron quasi-longitudinal oscillations. For the derivation of the spectra of such waves the ions are unimportant (their mass can be formally taken infinitely large: $M \to \infty$), and the electron quantities satisfy the conditions (18.56). The spectrum (18.59) of the first harmonic $c_e(\omega \simeq \Omega_e)$ can be conveniently considered in two limiting cases:

$$\omega \simeq \Omega_e\left\{1 + \frac{1}{2}\frac{k_\perp^2}{k^2}\frac{\omega_{Le}^2}{\Omega_e^2}\right\}, \qquad \frac{\omega_{Le}^2}{\Omega_e^2} \ll 1; \tag{26.19}$$

$$\omega \simeq \Omega_e\left\{1 - \frac{1}{2}\frac{k_\perp^2}{k_z^2}\right\}, \qquad \frac{\omega_{Le}^2}{\Omega_e^2} \gg 1, \quad k_z \gg k_\perp, \tag{26.20}$$

The spectrum of the higher harmonics (18.57) ($\omega \simeq n\Omega_e$, $n > 1$) can also be represented by two limiting expressions [with allowance for the asymptotic behavior (18.31) for $A_n(z_e)$ when $z_e \ll 1$]:

$$\omega(\mathbf{k}) = n\Omega_e\left\{1 + \frac{1}{(kr_{De})^2}\frac{z_e^n}{2^n n!}\right\}, \qquad \frac{\omega_{Le}^2}{\Omega_e^2} \ll 1, \quad k_z \ll k_\perp; \tag{26.21}$$

$$\omega(\mathbf{k}) = n\Omega_e\left\{1 - \frac{n^2-1}{2^n n!}z_e^{n-1}\right\}, \qquad \frac{\omega_{Le}^2}{\Omega_e^2} \gg 1, \quad k_z \ll k_\perp. \tag{26.22}$$

For the induced Coulomb scattering of electron-cyclotron quasilongitudinal waves (26.19)-(26.22) one can show that the contribution of the ions to the contractions S and V is negligibly small (compared with the electron contribution) if the phase velocities of the scattered oscillations appreciably exceed the ion thermal velocity:

$$\frac{\omega}{k}, \quad \frac{\omega'}{k'} \gg v_{T_i}. \tag{26.23}$$

Therefore, for the sake of greater generality of the formulas given below (which still remain fairly concrete), we go beyond the framework of the restrictions associated with a slow virtual oscillation ($\omega'' \ll |k_z''| v_{Ti}$) and small ion Larmor radius compared with the transverse wavelength of the virtual oscillation ($z_i'' \ll 1$). Namely, we shall assume that, in general, the phase velocity of the virtual wave is of the order of the ion thermal velocity:

$$\frac{\omega''}{|k_z''|} \lessgtr v_{T_i} \ll v_{T_e}, \tag{26.24}$$

and that the frequency difference of the scattered oscillations is small compared with the ion gyroscopic frequency:

$$\omega'' \ll \Omega_i. \tag{26.25}$$

As a result, we obtain from the contractions (26.3) and (26.4) and formula (25.15) the kernel Q ($\beta_i'' \equiv \omega''/|k_z''| v_{Ti}$):

$$Q = \frac{1}{2(2\pi)^{3/2}}\frac{e^2}{m^2}\left[\left(\frac{k_\perp}{k}\right)^2 - \left(\frac{k_\perp'}{k'}\right)^2\right]\frac{1}{\Omega_e |k_z - k_z'| v_{Te}^3}\frac{1}{(k'' r_{De})^4 |\varepsilon(\omega'', \mathbf{k}'')|^2} \times$$

$$\times\left\{\left(\frac{M}{m}\right)^{1/2}\left(\frac{T_e}{T_i}\right)^{3/2}\exp\left(-\frac{\beta_i''^2}{2}\right)\left\{\left[1-\frac{kk'}{k_\perp k_\perp'}(k_\perp'' r_{D_e})^2\right]^2+\right.\right.$$
$$\left.+4\sin^2(\theta'-\theta)(kk' r_{D_e}^2)^2\right\}+\left\{\left[(k'' r_{D_e})^2-\left(\frac{kk'}{k_\perp k_\perp'}\right)(k_\perp'' r_{D_e})^2+\frac{T_e}{T_i}-\right.\right.$$
$$\left.\left.\left.-\frac{T_e}{T_i}A_0(z_i'')\operatorname{Re}J_+(\beta_i'')\right]^2+4(kk' r_{D_e}^2)^2\sin^2(\theta'-\theta)\right\}\right\}, \tag{26.26}$$

which characterizes the scattering of waves with the spectrum (26.19). Here, $\varepsilon(\omega'', \mathbf{k}'')$ is the longitudinal permittivity, which depends on the frequency ω'' and the wave vector of the virtual oscillation [see (18.30)],

$$\varepsilon(\omega'', \mathbf{k}'')=1+(k'' r_{D_e})^{-2}\left\{1+\frac{T_e}{T_i}[1-A_0(z_i'')J_+(\beta_i'')]\right\}. \tag{26.27}$$

In a plasma of higher density ($\omega_{Le}^2 > \Omega_e^2$), the scattering of the first harmonic (26.20) of the electron-cyclotron waves is determined by Eq. (25.3) with the kernel

$$Q=\frac{4}{(2\pi)^{3/2}}\frac{e^2}{m^2}\left[\left(\frac{k_\perp}{k}\right)^2-\left(\frac{k_\perp'}{k'}\right)^2\right]\frac{\omega_{Le}^2\sin^2\left(\frac{\theta'-\theta}{2}\right)}{|k_z-k_z'|\,v_{T_e}^3\Omega_e^3}\frac{1}{(k'' r_{D_e})^4|\varepsilon(\omega'', \mathbf{k}'')|^2}\times$$
$$\times\left\{\left(\frac{M}{m}\right)^{1/2}\left(\frac{T_e}{T_i}\right)^{3/2}\exp\left(-\frac{\beta_i''^2}{2}\right)+\left[(k'' r_{D_e})^2+\frac{T_e}{T_i}-\frac{T_e}{T_i}A_0(z_i'')\operatorname{Re}J_+(\beta_i'')\right]^2\right\}. \tag{26.28}$$

In both the cases (26.27) and (26.28), the nonlinear interaction of the electron-cyclotron waves (26.19) and (26.20) represents scattering on the polarization clouds of the virtual oscillation.

In the scattering of the higher electron-cyclotron harmonics ($n > 1$), as in the case of scattering of ion-cyclotron waves (see § 26.1), one must distinguish two possibilities: $n = 2$ and $n > 2$. This distinction must be borne in mind only for the spectrum (26.22). The scattering of waves with the spectrum (26.21) is determined for all frequencies ($n > 1$) by one and the same formula for the kernel:

$$Q=\frac{1}{(2\pi)^{3/2}}\frac{e^2}{m^2}\left[\frac{z_e^n}{k^2}-\frac{z_e'^n}{k'^2}\right]\frac{\omega_{Le}^8}{\Omega_e^3|k_z-k_z'|\,v_{T_e}^9 k''^4}\frac{2^n(n-1)!}{(z_e z_e')^{n-1}}\times$$
$$\times\frac{1}{|\varepsilon(\omega'', \mathbf{k}'')|^2}\left[1+\left(\frac{M}{m}\right)^{1/2}\left(\frac{T_e}{T_i}\right)^{3/2}\exp\left(-\frac{\beta_i''^2}{2}\right)\right]\left[\left(\frac{n}{2}\frac{k_\perp''^2}{k_\perp k_\perp'}\right)^2+\sin^2(\theta'-\theta)\right]. \tag{26.29}$$

A characteristic feature of the expression (26.29) is the presence of the small parameter $(z_e z_e')^{n-1}$ in the denominator. This means that in a wide range of plasma parameters the induced Coulomb scattering on plasma particles determined by the kernel (26.29) is far stronger than scattering on free electrons. If the plasma density is high ($\omega_{Le}^2 \gg \Omega_e^2$) and the spectrum of the n-th electron-cyclotron harmonic ($n > 2$) is given by (26.22), the kernel of the nonlinear equation (25.3) has a form similar to (26.29). For the second ($\omega \simeq 2\Omega_e$, $\omega' \simeq 2\Omega_e$) electron-cyclotron harmonic (26.22), the nonlinear interaction through the longitudinal virtual oscillation depends strongly on the ratio of the ion and electron temperatures (in different temperature regions the kernel Q has completely different forms). Namely, in the wide interval (25.33) of ratios T_e/T_i of the electron and ion temperatures,

$$Q=\frac{1}{6}Q_0\left(\frac{M}{m}\right)^{1/2}\left(\frac{T_e}{T_i}\right)^{3/2}\exp\left(-\frac{\beta_i''^2}{2}\right)\left\{\left[\cos(\theta'-\theta)-\cos 2(\theta'-\theta)+\right.\right.$$
$$\left.+\frac{4}{3}\frac{k_\perp''^2}{k_\perp k_\perp'}\cos 2(\theta'-\theta)+\frac{8}{3}\sin(\theta'-\theta)\sin 2(\theta'-\theta)\right]^2+$$
$$\left.+\left[\frac{1}{2}\sin(\theta'-\theta)-\sin 2(\theta'-\theta)+\frac{4}{3}\frac{k_\perp''^2}{k_\perp k_\perp'}\sin 2(\theta'-\theta)-\frac{8}{3}\sin(\theta'-\theta)\cos 2(\theta'-\theta)\right]^2\right\}. \tag{26.30}$$

Here, we have sepatated out the factor Q_0,

$$Q_0 = \frac{1}{(2\pi)^3} \frac{e^2}{m^2} \frac{\omega_{Le}^2 (z_e' - z_e)}{\Omega_e^3 |k_z - k_z'| v_{T_e}^3} \frac{1}{(k'' r_{D_e})^4 |\varepsilon(\omega'', \mathbf{k}'')|^2}, \tag{26.31}$$

which also occurs in the kernel Q outside the interval (25.33). In a strongly nonisothermal plasma (25.35), the kernel Q is given by

$$Q = \frac{1}{6} Q_0 \left(\frac{T_e}{T_i}\right)^2 \Big\{ 1 + [\cos(\theta' - \theta) - \cos 2(\theta' - \theta)]^2 \, [1 - A_0(z_i'') \operatorname{Re} J_+(\beta_i'')]^2 + \\ + \left[\frac{1}{2} \sin(\theta' - \theta) - \sin 2(\theta' - \theta)\right]^2 [1 - A_0(z_i'') \operatorname{Re} J_+(\beta_i'')]^2 \Big\}. \tag{25.32}$$

Finally, in a plasma with cold electrons (25.37),

$$Q = \frac{8}{27} Q_0 \left\{ \frac{9}{16} + \left[\left(\frac{k_\perp''^2}{k_\perp k_\perp'} \right)^2 + \sin^2(\theta' - \theta) \right] \right\}. \tag{26.33}$$

In calculating these kernels we have borne in mind that, since the wavelength of the scattered electron-cyclotron harmonics (26.22) is fairly large ($k^2 r_{D_e}^2 \ll z_e \ll 1$), the virtual oscillation is also a long-wavelength one $[(k'' r_{D_e})^2 \ll 1]$. In a plasma with hot electrons and relatively cold ions, more precisely when

$$\frac{T_e}{T_i} \beta_i'' \exp\left(-\frac{\beta_i''^2}{2}\right) \gg 1, \tag{26.34}$$

the waves (26.22) are scattered in accordance with the nonlinear interaction equation (25.3) with the kernels (26.30)-(26.32) on the ion polarization clouds of the virtual oscillation, and in the opposite case ($T_e \ll T_i$) on the electron polarization clouds of the virtual wave [see the kernels (26.30) and (26.33)].

Summarizing our study of induced Coulomb scattering of quasilongitudinal cyclotron waves on particles of a magnetoactive plasma, let us point out the conclusions that can be drawn on the basis of the actual expressions given in this section for the kernels of the integrodifferential equation that describes such a nonlinear process. All these kernels are antisymmetric under the transposition of the wave vectors **k** and **k'** of the scattered waves. As in an isotropic plasma (Ch. III), this antisymmetry necessarily entails conservation of the total energy of the oscillation during the scattering [cf. (9.13)]:

$$\int d\mathbf{k} W_{c_i}(\mathbf{k}) = \text{const}, \qquad \int d\mathbf{k}' W_{c_e}(\mathbf{k}) = \text{const}. \tag{26.35}$$

This result is in complete agreement (in the first approximation) with the general conclusion (§ 21) concerning the conservation of the number of oscillations in induced scattering due to the approximate constancy ($\omega \simeq n\Omega_e$, $\omega \simeq n\Omega_i$) of their frequency. As we have already pointed out, the nonlinear interaction of cyclotron waves leads in all cases to not the dissipation of energy but its being transferred within the spectra from higher to lower frequencies. We must emphasize however that this transfer of energy from higher to lower frequencies gives rise to very different results in the different cases. If the oscillation frequency is less than the gyroscopic frequency or its multiple [the spectra (26.1), (26.2), (26.20), (26.22)], the maximum of the spectral energy density W(k) moves away with time from the region of strong absorption, which lies in the immediate proximity of the cyclotron harmonic ($|\omega - n\Omega_e| \lesssim |k_z| v_{T_e}$, $|\omega - n\Omega_i| \lesssim |k_z| v_{T_i}$). Conversely, if the oscillation frequency exceeds the gyroscopic frequency or its multiple [the spectra (26.16), (26.19), (26.21)], the nonlinear interaction shifts the spectral energy density of the oscillations into the region of strong cyclotron damping. This case is obviously of particular interest. Indeed, the two-stream instability enables one, for example, to effectively "excite" cyclotron oscillations and the nonlinear interaction considered here can rapidly transfer the energy received from the beam to the region of strong damping, this leading to efficient heating of the plasma. The characteristic time of the corresponding process can be estimated by means of the expressions given for the kernels Q. For example, the electron-cyclotron oscillations c_e with the spectrum (26.19) can be excited by a beam with low density and large thermal spread [the ratio of the electron densities of the beam, N_1, and the plasma, N_e, is small, $(N_1/N_e) \ll 1$, and the Debye radius r_{De1} of the beam electrons is large, $(kr_{De1})^2 \gg 1$]. Then the growth rate of such oscillations is (u is the beam velocity, the subscript 1 is appended to beam variables)

$$\gamma = \sqrt{\frac{\pi}{2}}\,\frac{\Omega_e}{2}\,\frac{(k_\perp \rho_{e1})^2}{(kr_{De1})^4} A_0(k_\perp^2 \rho_{e1}^2)\,\frac{\mathbf{ku}-\omega}{|k_z|\,v_{Te1}} \exp\left\{-\frac{1}{2}\left(\frac{\omega-\mathbf{ku}}{k_z v_{Te1}}\right)^2\right\} \tag{26.36}$$

and it is positive when

$$\frac{\mathbf{ku}-\Omega_e}{\Omega_e} > \frac{1}{2}\left(\frac{k_\perp \omega_{Le}}{k\Omega_e}\right)^2. \tag{26.37}$$

The characteristic time of the process of nonlinear transfer of the oscillation energy within the spectrum to lower frequencies is in this case approximately

$$\tau \sim \frac{10^3}{|\Omega_e|}\,\frac{T_e}{T_i}\,\frac{N_e \varkappa T_e}{W_0}, \qquad \frac{\omega''}{|k_z''|\,v_{Ti}} \sim 1 \tag{26.38}$$

[here, W_0 is the total energy of the electron-cyclotron oscillations (26.19)]. The exponentially small decay constant of these oscillations, which is proportional to $(\Omega_e^2/|k_z|v_{Te})\exp\left\{-\frac{1}{2}\left(\frac{\Omega_e}{k_z v_{Te}}\right)^2\right\}$, can be ignored compared with $1/\tau$, as can be seen for the example of a plasma with electron Larmor radius $\rho_e \approx 10^{-1}$ cm and the oscillations (26.19) with wavelength $2\pi/k_z \approx 10^{-1}$ cm and total energy $W_0 \approx 0.1 N_e \varkappa T_z$.

In conclusion, we note also that the kernels Q of the integrodifferential equations discussed above in § 25 and 26 enable one, as also in the case of an isotropic plasma, to calculate, using the method developed in § 12, the cross sections for the scattering of low-frequency and cyclotron waves on plasma particles. For example, the relation

$$d\sigma^{c_e c_e}(\mathbf{k}', \mathbf{k}) = \frac{1}{2\sqrt{2\pi}}\left(\frac{e^2}{mv_{Te}^2}\right)^2 \frac{k'}{|k_z - k_z'|\sin 2\vartheta'}\left(\frac{r_{De}}{\rho_e}\right)^3 \frac{T_i}{T_e}\{1+(T_i M/T_e m)^{1/2}\}\,d\mathbf{k} \tag{26.39}$$

determines the effective differential cross section for the scattering of the electron-cyclotron wave (26.19) with wave number k' into a similar oscillation with wave number k on electrostatic ion thermal fluctuations.

CONCLUSIONS

In § 4 of Ch. I and the first pages of Chs. II-V we have in fact already listed the results derived and discussed in this monograph. Therefore at this juncture we shall only emphasize the more general characteristic features of our review and note some possibilities for the further development of the theory of the nonlinear interaction of waves in a plasma and the application of this theory.

1. Our review has been based on the single basic generalized kinetic field equation in statistical nonlinear electrodynamics. This has made it possible to treat systematically numerous processes of nonlinear interaction of characteristic oscillations of isotropic and magnetoactive plasmas from a unified point of view.

2. Considerable attention has been devoted to the study of the structure of the three- and four-index complex permittivity tensors of collisionless isotropic and magnetoactive plasmas. The contractions of these tensors determine most directly the characteristic times of evolution of the nonlinear processes, and their general symmetry properties have enabled us to establish the conservation laws that these processes satisfy. In our opinion, the significance of the many-index complex permittivity tensors goes far beyond the framework of the statistical theory of nonlinear interaction of waves in a plasma developed here, since their study can also be of independent interest in other applications of the theory of weakly nonlinear plasma phenomena.

3. The general results of the theory of the nonlinear interaction of waves are supported by the admittedly completely exhaustive but very full treatment of the different concrete variants of the nonlinear processes: in an isotropic plasma the number of such variants is about 20-30, in a magnetoactive plasma we have obtained more than 50 kernels of nonlinear integrodifferential equations describing the evolution of the spectral energy density of the oscillations.

At the present time the further development of the theory of nonlinear interaction of waves in a plasma takes place in several different directions. On the one hand there is a still continuing flow of publications [75-79] containing calculations and discussions of cross sections of specific nonlinear processes in homogeneous and weakly nonhomogeneous magnetoactive plasmas; essentially, these do not go beyond the scope of the generalized kinetic field equation for three-wave interactions employed in this monograph. At the same time however studies also made of processes of higher order (more strongly nonlinear) with the participation of four oscillations [80-44], but these are also in the framework of statistical nonlinear electrodynamics, in which the nonlinear field equations are averaged over the statistical ensemble. Attemps are also made to solve (numerically and analytically) the nonlinear integrodifferential equations that arise in the theory of nonlinear interaction as special cases of the generalized kinetic field equation [41, 85, 86]. The finding of such solutions, for example, stationary spectral densities of the energy of superthermal oscillations, is important in connection with the question of the effect of weakly turbulent processes in a plasma on the distribution function of its particles, heating, etc., [87-89]. Recently, significant progress has been achieved in the theory of parametric resonance in a plasma [45, 90-92], some results of which for not too strong external electromagnetic fields applied to the plasma overlap with the results of the theory of nonlinear interaction. Considerable interest attaches to the application of the theory of nonlinear interaction of waves and parametric resonance in a plasma to the explanation of the phenomena resulting from the application of a powerful laser light beam to the very dense plasma formed by the "vaporization" of a solid [93-96]. The observed effects of light absorption [97] and ion acceleration [98] have a physical mechanism that overlaps evidently with the analogous effects in the interaction of an electromagnetic wave of the microwave range with a rarefied plasma in a wave guide [99-101]. This possibility of applying the theory is all the more important because at the present time the comparison of its results with experimental data is still in an initial stage (the qualitative agreement of the conclusions of the theory of the nonlinear interaction of waves with experiments has been obtained for a number of processes of induced combination scattering of high-frequency electromagnetic waves [102-103].

APPENDIX

Multi-Index Complex Permittivity Tensors of a Cold Plasma in the Framework of the Hydrodynamic Description

As in the main part of this review, we here consider a homogeneous unbounded collisionless plasma. The only difference from the main part of the exposition is in the method of describing such a plasma. Previously we used the system of self-consistent Vlasov equations; now we shall proceed from the hydrodynamic equations for the electron and ion variables separately, completely ignoring the thermal motion of the plasma particles ($v_T = 0$) but, in general, allowing for the effect of the external constant and homogeneous magnetic field $\mathbf{B}_0 = \text{const} \neq 0$.

We shall discuss first the simpler case of an isotropic plasma* (without external magnetic field: $\mathbf{B}_0 = 0$). The density $n(\mathbf{r}, t)$ and velocity $\mathbf{v}(\mathbf{r}, t)$ of the electrons and ions of such a plasma satisfy the continuity equation

$$\frac{\partial n}{\partial t} + \frac{\partial}{\partial \mathbf{r}}(n\mathbf{v}) = 0 \tag{A.1}$$

and the equation of motion

$$m\left\{\frac{\partial \mathbf{v}}{\partial t} + \left(\mathbf{v}\frac{\partial}{\partial \mathbf{r}}\right)\mathbf{v}\right\} = e\left\{\mathbf{E} + \frac{1}{c}[\mathbf{v}\mathbf{B}]\right\}. \tag{A.2}$$

As usual, we shall, of course, omit but understand the identifying indices of the particle species (electrons and ions) of the plasma for variables like the charge e, mass m, density $n(\mathbf{r}, t)$, and velocity $\mathbf{v}(\mathbf{r}, t)$. In connection with this, we emphasize that each of equations (A.1) and (A.2) corresponds in general to a system of equations whose number is equal to the number of particles species. The Lorentz force $e\{\mathbf{E} + 1/c[\mathbf{v}\mathbf{B}]\}$ on the right-hand side of (A.2) gives rise to a motion of the plasma particles under the influence of the electric $\mathbf{E}(\mathbf{r}, t)$ and magnetic $\mathbf{B}(\mathbf{r}, t)$ fields defined by the Maxwell equations

$$\left.\begin{aligned} &\operatorname{div}\mathbf{E} = 4\pi\rho, \qquad \operatorname{rot}\mathbf{E} = -\frac{1}{c}\frac{\partial \mathbf{B}}{\partial t}, \\ &\operatorname{rot}\mathbf{B} = \frac{1}{c}\frac{\partial \mathbf{E}}{\partial t} + \frac{4\pi}{c}\mathbf{j}, \qquad \operatorname{div}\mathbf{B} = 0, \end{aligned}\right\} \tag{A.3}$$

in which the charge density $\rho(\mathbf{r}, t)$,

$$\rho = en \tag{A.4}$$

and the current density $\mathbf{j}(\mathbf{r}, t)$

$$\mathbf{j} = en\mathbf{v} \tag{A.5}$$

are found in their turn from the continuity equation (A.1) and the equation of motion (A.2). On the right-hand sides of Eqs. (A.4) and (A.5), we understand summation over the particle species so that, for example, the continuity equation

$$\frac{\partial \rho}{\partial t} + \operatorname{div}\mathbf{j} = 0, \tag{A.6}$$

which relates the charge density ρ in the plasma to the current density $\mathbf{j}$ in the plasma, arises after summation over the particle species of the equation of continuity (A.1) for the charge density en of the given species. Assuming that the electric and magnetic fields are sufficiently weak,† we expand the density $n(\mathbf{r}, t)$ and the velocity $\mathbf{v}(\mathbf{r}, t)$ of the plasma particles in an inte-

* The nonlinear theory of an isotropic plasma that satisfies hydrodynamic equations is constructed in [85]. In the present exposition we shall follow the method by means of which Gorbunov obtained his results [104]. In particular, he derived the formulas given below, (A.32), (A.43), (A.48), and (A.49), for the three- and four-index tensors ε in an isotropic plasma.

† The explicit expression for the small parameter determining the applicability boundaries of the expansions used here can be obtained, for example, by means of the formulas (A.32) and (A.48) given below for the three- and four-index tensors S and V. Namely, comparison of

gropower series in the electric field [compare this with the expansion of the distribution function (5.4) in an isotropic plasma in the "kinetic" description in Ch. II):

$$n(\mathbf{r}, t) = N + \delta n(\mathbf{r}, t) + \delta n'(\mathbf{r}, t) + \delta n''(\mathbf{r}, t) + \ldots \tag{A.7}$$

$$\mathbf{v}(\mathbf{r}, t) = \delta\mathbf{v}(\mathbf{r}, t) + \delta\mathbf{v}'(\mathbf{r}, t) + \delta\mathbf{v}''(\mathbf{r}, t) + \ldots \tag{A.8}$$

Here, the equilibrium state, which does not depend on the electromagnetic field, is characterized by a constant and homogeneous density N and the absence of directed motion:

$$N = \text{const} \neq 0, \quad \mathbf{v}(\mathbf{r}, t) = 0, \quad \mathbf{E} = \mathbf{B} = 0, \tag{A.9}$$

and the nonequilibrium corrections to the density,

$$\delta n \gg \delta n' \gg \delta n'' \gg \ldots \tag{A.10}$$

and the nonequilibrium velocities,

$$\delta\mathbf{v} \gg \delta\mathbf{v}' \gg \delta\mathbf{v}'' \gg \ldots \tag{A.11}$$

are (in descending order) linear, bilinear, trilinear, etc., functionals of the electric field [cf. (5.5)]:

$$\delta\mathbf{v} \sim \delta n = O(E), \quad \delta\mathbf{v}' \sim \delta n' = O(E^2), \quad \delta\mathbf{v}'' \sim \delta n'' = O(E^3), \ldots. \tag{A.12}$$

In the linear approximation in the electric field, the system of equations (A.1)-(A.2) takes the form

$$\frac{\partial\delta n}{\partial t} + \operatorname{div}(N\delta\mathbf{v}) = 0, \quad \frac{\partial\delta\mathbf{v}}{\partial t} = \frac{e}{m}\mathbf{E} \tag{A.13}$$

and after the transition to Fourier variables [see, for example, the decomposition (2.1) into a Fourier integral in § 2] leads to the following expressions for the first nonequilibrium correction to the density δn and its first nonequilibrium velocity $\delta\mathbf{v}$:

$$\delta\mathbf{v}(\omega, \mathbf{k}) = i\frac{e}{m}\frac{1}{\omega}\mathbf{E}(\omega, \mathbf{k}); \tag{A.14}$$

$$\delta n(\omega, \mathbf{k}) = \frac{N}{\omega}\mathbf{k}\delta\mathbf{v}(\omega, \mathbf{k}) = i\frac{e}{m}\frac{N}{\omega^2}\mathbf{k}\mathbf{E}(\omega, \mathbf{k}). \tag{A.15}$$

Using (A.14) and the definition (A.5), we find the current density, which is linear in the field:

$$\delta\mathbf{j}(\omega, \mathbf{k}) = eN\delta\mathbf{v}(\omega, \mathbf{k}) = i\frac{e^2}{m}\frac{N}{\omega}\mathbf{E}(\omega, \mathbf{k}) \tag{A.16}$$

and the electric induction (1.4) [see § 1 and 6, formula (6.8)]

$$\mathbf{D}'(\omega, \mathbf{k}) = \mathbf{E}(\omega, \mathbf{k}) + \frac{4\pi i}{\omega}\mathbf{j}(\omega, \mathbf{k}), \tag{A.17}$$

which is related on account of (A.16) to the electric field $\mathbf{E}(\omega, \mathbf{k})$:

$$D'_i(\omega, \mathbf{k}) = \varepsilon_{ij}(\omega, \mathbf{k})E_j(\omega, \mathbf{k}) \tag{A.18}$$

---

these tensors with each other and with the permittivity tensor shows that the electric and magnetic fields discussed above must be sufficiently weak for the oscillation velocity that they impart to the plasma particles (electrons and ions) to be small compared with the phase velocities of the interacting oscillations and their "beats": $\frac{eE}{m\omega}\frac{k}{\omega} \ll 1$, $\frac{eE}{m\omega'}\frac{k'}{\omega'} \ll 1$, $\frac{eE}{m(\omega-\omega')} \times \frac{|\mathbf{k}-\mathbf{k}'|}{\omega-\omega'} \ll 1$. A more concrete expression for these small quantities arises in each special case with allowance for the specific conditions of the problem under consideration.

by the permittivity tensor of a cold isotropic collisionless plasma [cf. (7.25)]:

$$\varepsilon_{ij}(\omega, \mathbf{k}) = \delta_{ij}\left(1 - \frac{\omega_L^2}{\omega^2}\right), \quad \omega_L^2 \equiv \frac{4\pi N e^2}{m}. \tag{A.19}$$

The expression (A.19) is the simplest special case of the tensor (6.12) in the approximation of a cold plasma, when the equilibrium distribution function $f_0(\mathbf{v})$ can be assumed in the form

$$f_0 = N\delta(\mathbf{v}), \tag{A.20}$$

and the well-known derivation (A.13)-(A.19) of the permittivity tensor of a cold isotropic plasma given here clearly demonstrates the line of argument applied below to calculate the three- and four-index tensors ε with more complicated form. Indeed, in order to obtain an explicit expression for the three-index tensor ε_{ijs}, we must know the current density quadratic in the field:

$$\delta\mathbf{j}'(\mathbf{r}, t) = e\{N\delta\mathbf{v}'(\mathbf{r}, t) + \delta n(\mathbf{r}, t)\,\delta\mathbf{v}(\mathbf{r}, t)\}. \tag{A.21}$$

The velocity $\delta\mathbf{v}'(\mathbf{r}, t)$ on the right-hand side of (A.21) is found from the equation of motion (A.2) in the second approximation in the electric field:

$$\frac{\partial\delta\mathbf{v}'}{\partial t} = -\left(\delta\mathbf{v}\frac{\partial}{\partial\mathbf{r}}\right)\delta\mathbf{v} + \frac{e}{mc}[\delta\mathbf{v}\mathbf{B}]. \tag{A.22}$$

Namely, going over in Eq. (A.22) to Fourier transforms [see (5.21) for the Fourier transform of the product of two quantities] and eliminating the magnetic field $\mathbf{B}$ by means of one of the field equations (2.24), we obtain

$$\delta v_i'(\omega, \mathbf{k}) = \frac{1}{\omega}\int d\omega' d\mathbf{k}' d\omega'' d\mathbf{k}''\delta(\omega - \omega' - \omega'')\,\delta(\mathbf{k} - \mathbf{k}' - \mathbf{k}'')\times$$

$$\times\left\{\alpha_{ij,s}(\omega'', \mathbf{k}'')\left(i\frac{e}{m}\right)E_j(\omega'', \mathbf{k}'') + k_s''\delta v_i(\omega'', \mathbf{k}'')\right\}\delta v_s(\omega', \mathbf{k}'). \tag{A.23}$$

Here, the three-index tensor $\alpha_{ij,s}$ is the derivative with respect to the velocity v_s of the tensor α_{ij}, which we have already used on several occasions [see (5.16)]:

$$\alpha_{ij,s}(\omega, \mathbf{k}) \equiv \frac{\partial\alpha_{ij}(\omega, \mathbf{k}, \mathbf{v})}{\partial v_s} = \frac{1}{\omega}\{k_i\delta_{js} - \delta_{ij}k_s\}, \tag{A.24}$$

and the velocity $\delta\mathbf{v}_i(\omega, \mathbf{k})$ is determined by Eq. (A.14), allowance for which in conjunction with (A.24) somewhat simplifies the expression for $\delta v_i'(\omega, \mathbf{k})$:

$$\delta v_i'(\omega, \mathbf{k}) = \left(i\frac{e}{m}\right)^2\int d\omega' d\mathbf{k}'\,\frac{k_i - k_i'}{\omega - \omega'}\,\frac{\delta_{js}}{\omega\omega'}\,E_s(\omega', \mathbf{k}')\,E_j(\omega - \omega', \mathbf{k} - \mathbf{k}'). \tag{A.25}$$

Substituting (A.14), (A.15), and (A.25) into the Fourier transform of the current density (A.21),

$$\delta j_i'(\omega, \mathbf{k}) = e\left\{N\delta v_i'(\omega, \mathbf{k}) + \int d\omega' d\mathbf{k}'\delta n(\omega', \mathbf{k}')\,\delta v_i(\omega - \omega', \mathbf{k} - \mathbf{k}')\right\}, \tag{A.26}$$

we finally find the explicit form of $\delta\mathbf{j}'(\omega, \mathbf{k})$ as a function of the electric field $\mathbf{E}$:

$$\delta j_i'(\omega, \mathbf{k}) = -\frac{e^3N}{m^2}\int d\omega' d\mathbf{k}'\left\{\frac{k_i - k_i'}{\omega}\delta_{js} + \frac{k_s'}{\omega'}\delta_{ij}\right\}\frac{1}{\omega - \omega'}E_j(\omega - \omega', \mathbf{k} - \mathbf{k}')\frac{1}{\omega'}E_s(\omega', \mathbf{k}'). \tag{A.27}$$

This relation is conjunction with the "phenomenological" material equation (2.6) and (A.17) leads us to the desired equation for the three-index tensor ε_{ijs}:

$$\varepsilon_{ijs}(\omega, \mathbf{k}; \omega', \mathbf{k}') = -i\frac{e}{m}\,\frac{\omega_L^2}{\omega\omega'\omega''}\left\{\frac{k_i''}{\omega}\delta_{js} + \frac{k_s'}{\omega'}\delta_{ij}\right\}, \tag{A.28}$$

in which the frequencies ω, ω', ω'' and the wave vectors $\mathbf{k}$, $\mathbf{k}'$, $\mathbf{k}''$ are related because of the notation introduced for convenience of expression [cf. (7.3)]:

$$\omega'' = \omega - \omega', \quad \mathbf{k}'' = \mathbf{k} - \mathbf{k}'. \tag{A.29}$$

The tensor (A.28) is the hydrodynamic analog of the more general tensor (7.1), which allows for the thermal motion, with equilibrium distribution function in the form (A.20). Together with the tensor (A.28), one can introduce a further tensor that differs from it by transposition of the primes and indices ($\omega' \rightleftharpoons \omega''$, $\mathbf{k}' \rightleftharpoons \mathbf{k}''$, $s \rightleftharpoons j$):

$$\varepsilon_{isj}(\omega, \mathbf{k};\ \omega'', \mathbf{k}'') = -i\frac{e}{m}\frac{\omega_L^2}{\omega\omega'\omega''}\left\{\frac{k_i'}{\omega}\delta_{sj} + \frac{k_j''}{\omega''}\delta_{is}\right\}. \tag{A.30}$$

In accordance with the method by which the tensor ε_{ijs} is introduced [see (2.6)], as the kernel of an integral operator, physical meaning does not attach to each of the two possible expressions (A.28) and (A.30) but only to their symmetrized combination (7.4):

$$\varepsilon_{i(js)}(\omega, \mathbf{k};\ \omega', \mathbf{k}') = -\frac{i}{2}\frac{e}{m}\frac{\omega_L^2}{\omega\omega'\omega''}\left\{\frac{k_i}{\omega}\delta_{sj} + \frac{k_j''}{\omega''}\delta_{is} + \frac{k_s'}{\omega'}\delta_{ij}\right\}, \tag{A.31}$$

which we have already pointed out in the form of Eq. (2.7). Namely, such a sum of the two tensors ε occurs in the generalized kinetic field equation (3.22) in the form of the three-index tensor $S_{ijs}(\omega, \mathbf{k};\ \omega', \mathbf{k}')$ [see the definition (3.24)], which is twice the tensor (A.31):

$$S_{ijs}(\omega, \mathbf{k};\ \omega'\mathbf{k}') = -i\frac{e}{m}\frac{\omega_L^2}{\omega\omega'\omega''}\left\{\frac{k_i}{\omega}\delta_{js} + \frac{k_j''}{\omega''}\delta_{is} + \frac{k_s'}{\omega'}\delta_{ij}\right\}. \tag{A.32}$$

We emphasize that the expression (A.32) obtained here in the framework of hydrodynamic equations for the tensor S_{ijs} agrees with the limit (7.23) of the more general tensor (7.21) (obtained in a kinetic description) when thermal motion is ignored.

The velocity (A.25) quadratic in the field found above also determines in accordance with the continuity equation (A.1) (in the second approximation):

$$\frac{\partial \delta n'}{\partial t} + \operatorname{div}(N\delta\mathbf{v}' + \delta n\delta\mathbf{v}) = 0 \tag{A.33}$$

the second nonequilibrium correction $\delta n'$ to the density N of particles of given species:

$$\delta n'(\omega, \mathbf{k}) = \frac{N}{\omega}\mathbf{k}\delta\mathbf{v}'(\omega, \mathbf{k}) + \frac{\mathbf{k}}{\omega}\int d\omega' d\mathbf{k}' d\omega'' d\mathbf{k}''\delta(\omega - \omega' - \omega'')\,\delta(\mathbf{k} - \mathbf{k}' - \mathbf{k}'')\,\delta n(\omega', \mathbf{k}')\,\delta\mathbf{v}(\omega'', \mathbf{k}''), \tag{A.34}$$

which we require when calculating the current density cubic in the field:

$$\delta\mathbf{j}''(\mathbf{r}, t) = e\{N\delta\mathbf{v}''(\mathbf{r}, t) + \delta n(\mathbf{r}, t)\,\delta\mathbf{v}'(\mathbf{r}, t) + \delta n'(\mathbf{r}, t)\,\delta\mathbf{v}(\mathbf{r}, t)\}. \tag{A.35}$$

Here [in (A.35)] the velocity $\delta\mathbf{v}''$ is determined by the equation of motion (A.2):

$$\frac{\partial\delta\mathbf{v}''}{\partial t} = -\left(\delta\mathbf{v}\frac{\partial}{\partial\mathbf{r}}\right)\delta\mathbf{v}' - \left(\delta\mathbf{v}'\frac{\partial}{\partial\mathbf{r}}\right)\delta\mathbf{v} + \frac{e}{mc}[\delta\mathbf{v}'\mathbf{B}] \tag{A.36}$$

and can be represented in the form [cf. (A.23)]

$$\delta v_i''(\omega, \mathbf{k}) = \frac{1}{\omega}\int d\omega' d\mathbf{k}' d\omega'' d\mathbf{k}''\,\delta(\omega - \omega' - \omega'')\,\delta(\mathbf{k} - \mathbf{k}' - \mathbf{k}'')\times$$
$$\times\left\{a_{ij,s}(\omega'', \mathbf{k}'')\left(i\frac{e}{m}\right)E_j(\omega'', \mathbf{k}'') + k_s''\delta v_i(\omega'', \mathbf{k}'') + \delta_{is}k_j'\delta v_j(\omega'', \mathbf{k}'')\right\}\delta v_s'(\omega', \mathbf{k}'). \tag{A.37}$$

Substituting into the right-hand sides of Eqs. (A.34) and (A.36) the explicit expressions (A.14), (A.15), (A.24), and (A.25) for the quantities they contain, we obtain the desired relations for

$\delta n'(\omega, \mathbf{k})$ and $\delta v_i''(\omega, \mathbf{k})$ in terms of the electric field $\mathbf{E}(\omega, \mathbf{k})$:

$$\delta n'(\omega, \mathbf{k}) = \left(i\frac{e}{m}\right)^2 N \int d\omega' d\mathbf{k}' \frac{1}{\omega\omega'(\omega-\omega')} E_j(\omega-\omega', \mathbf{k}-\mathbf{k}') E_s(\omega', \mathbf{k}') \left\{\delta_{js}\frac{\mathbf{k}, \mathbf{k}-\mathbf{k}'}{\omega} + \frac{k_s' k_j}{\omega'}\right\}; \quad \text{(A.38)}$$

$$\delta v_i''(\omega, \mathbf{k}) = \left(i\frac{e}{m}\right)^3 \int d\omega' d\mathbf{k}' d\omega'' d\mathbf{k}'' \frac{1}{\omega\omega'(\omega-\omega')(\omega'-\omega'')\omega''} \times$$
$$\times E_j(\omega-\omega', \mathbf{k}-\mathbf{k}') E_s(\omega'-\omega'', \mathbf{k}'-\mathbf{k}'') E_r(\omega'', \mathbf{k}'') \{(k_i - k_i')(k_j' - k_j'')\delta_{sr} + k_j'(k_i' - k_i'')\delta_{sr}\}. \quad \text{(A.39)}$$

Formulas (A.14), (A.15), (A.38), and (A.39) enable us to write down the following expression for the Fourier transform of the current density (A.35):

$$\delta j_i''(\omega, \mathbf{k}) = -i\frac{e^4 N}{m^3} \int d\omega' d\mathbf{k}' d\omega'' d\mathbf{k}'' \frac{1}{\omega'(\omega-\omega')(\omega'-\omega'')\omega''} E_j(\omega-\omega', \mathbf{k}-\mathbf{k}') E_s(\omega'-\omega'', \mathbf{k}'-\mathbf{k}'') E_r(\omega'', \mathbf{k}'') \times$$
$$\times \left\{\delta_{sr}\left[\frac{(k_i - k_i')(k_j' - k_j'')}{\omega} + \frac{k_j'(k_i' - k_i'')}{\omega} + \frac{(k_j - k_j')(k_i' - k_i'')}{\omega-\omega'}\right] + \delta_{ij}\left[\delta_{sr}\frac{\mathbf{k}', \mathbf{k}'-\mathbf{k}''}{\omega'} + \frac{k_s' k_r''}{\omega''}\right]\right\}. \quad \text{(A.40)}$$

The contents of the curly brackets on the right-hand side of (A.40) can be slightly simplified (in its form) if the indices s and r and the corresponding variables are transposed:

$$\omega'' \rightleftarrows (\omega' - \omega''), \quad \mathbf{k}'' \rightleftarrows (\mathbf{k}' - \mathbf{k}''), \quad s \rightleftarrows r. \quad \text{(A.41)}$$

Then, namely,

$$\delta j_i''(\omega, \mathbf{k}) = -i\frac{e^4 N}{m^3} \int d\omega' d\mathbf{k}' d\omega'' d\mathbf{k}'' \frac{1}{\omega'(\omega-\omega')(\omega'-\omega'')\omega''} E_j(\omega-\omega', \mathbf{k}-\mathbf{k}') E_s(\omega'-\omega'', \mathbf{k}'-\mathbf{k}'') E_r(\omega'', \mathbf{k}'') \times$$
$$\times \left\{\delta_{sr}\left[\frac{(k_i - k_i')k_j''}{\omega} + \frac{k_j' k_i''}{\omega} + \frac{(k_j - k_j')k_i''}{\omega-\omega'}\right] + \delta_{ij}\left[\delta_{sr}\frac{\mathbf{k}'\mathbf{k}''}{\omega'} + \frac{k_r'(k_s' - k_s'')}{\omega'-\omega''}\right]\right\}. \quad \text{(A.42)}$$

Substituting the current density (A.42) into the right-hand side of (A.17) and taking into account the material equation (2.6), we find the explicit form of the four-index complex permittivity tensor ε_{ijsr}:

$$\varepsilon_{ijsr}(\omega, \mathbf{k}; \omega', \mathbf{k}'; \omega'', \mathbf{k}'') = \frac{e^2}{m^2} \frac{\omega_L^2}{\omega\omega'\omega''(\omega-\omega')(\omega'-\omega'')} \times$$
$$\times \left\{\delta_{sr}\left[\frac{(k_i - k_i')k_j''}{\omega} + \frac{k_j' k_i''}{\omega} + \frac{(k_j - k_j')k_i''}{\omega-\omega'}\right] + \delta_{ij}\left[\delta_{sr}\frac{\mathbf{k}_r'\mathbf{k}''}{\omega'} + \frac{k_2'(k_s' - k_s'')}{\omega'-\omega''}\right]\right\}. \quad \text{(A.43)}$$

Note that here, in contrast to formula (A.28) and (A.30) for the three-index tensors, the frequencies ω, ω', ω'' and the wave vectors $\mathbf{k}$, $\mathbf{k}'$, $\mathbf{k}''$ are independent [not related by Eqs. (A.29)]. The generalized kinetic field equation (3.22) contains in accordance with (3.23) the four-index tensor V, which depends only on two pairs of arguments: ω, $\mathbf{k}$ and ω', $\mathbf{k}'$. Its constituent terms arise from the tensor (A.43), which is more general in this sense, by an appropriate change of variables for $\varepsilon_{ijsr}(\omega, \mathbf{k}; \omega+\omega'; \mathbf{k}+\mathbf{k}'; \omega'\mathbf{k}')$:

$$(\omega, \mathbf{k}) \to (\omega', \mathbf{k}) \quad (\omega', \mathbf{k}') \to (\omega+\omega', \mathbf{k}+\mathbf{k}'), \quad (\omega'', \mathbf{k}'') \to (\omega', \mathbf{k}'),$$
$$(\omega-\omega', \mathbf{k}-\mathbf{k}') \to (-\omega', -\mathbf{k}'), \quad (\omega'-\omega'', \mathbf{k}'-\mathbf{k}'') \to (\omega, \mathbf{k}) \quad \text{(A.44)}$$

and similarly for $\varepsilon_{ijsr}(\omega, \mathbf{k}; \omega+\omega', \mathbf{k}+\mathbf{k}'; \omega, \mathbf{k})$:

$$(\omega, \mathbf{k}) \to (\omega, \mathbf{k}), \quad (\omega', \mathbf{k}') \to (\omega+\omega', \mathbf{k}+\mathbf{k}'), \quad (\omega'', \mathbf{k}'') \to (\omega, \mathbf{k}),$$
$$(\omega-\omega', \mathbf{k}-\mathbf{k}') \to (-\omega', -\mathbf{k}'), \quad (\omega'-\omega'', \mathbf{k}'-\mathbf{k}'') \to (\omega', \mathbf{k}'). \quad \text{(A.45)}$$

Namely, under such a transformation we obtain from the tensor (A.43)

$$\varepsilon_{ijsr}(\omega, \mathbf{k}; \omega+\omega', \mathbf{k}+\mathbf{k}'; \omega', \mathbf{k}') = -\frac{e^2}{m^2} \frac{\omega_L^2}{(\omega\omega')^2} \frac{1}{(\omega+\omega')} \times$$

$$\times\left\{\delta_{ij}\delta_{sr}\frac{\mathbf{k}',\ \mathbf{k}+\mathbf{k}'}{\omega+\omega'}+\delta_{ij}\frac{k_s}{\omega}(k_r+k'_r)+\delta_{sr}\left[\frac{k'_i k_j}{\omega}+\frac{k'_i k'_j}{\omega'}\right]\right\}; \tag{A.46}$$

$$\varepsilon_{ijrs}(\omega,\ \mathbf{k};\ \omega+\omega',\ \mathbf{k}+\mathbf{k}';\ \omega,\ \mathbf{k})=-\frac{e^2}{m^2}\frac{\omega_L^2}{(\omega\omega')^2}\frac{1}{(\omega+\omega')}\times$$

$$\times\left\{\delta_{ij}\delta_{sr}\frac{\mathbf{k},\ \mathbf{k}+\mathbf{k}'}{\omega+\omega'}+\delta_{ij}\frac{k'_r}{\omega'}(k_s+k'_s)+\delta_{rs}\left[-\frac{k'_i k_j}{\omega}+\frac{k_i}{\omega}(k_j+k'_j)+\frac{k'_j k_i}{\omega'}\right]\right\}. \tag{A.47}$$

Adding (A.46) and (A.47) in accordance with the definition (3.23), we obtain the desired expression for the four-index tensor V_{ijsr} in the generalized kinetic field equation (3.22) in a cold isotropic plasma:

$$V_{ijsr}(\omega,\ \mathbf{k};\ \omega',\ \mathbf{k}')=\varepsilon_{ijsr}(\omega,\ \mathbf{k};\ \omega+\omega',\ \mathbf{k}+\mathbf{k}';\ \omega',\ \mathbf{k}')+\varepsilon_{ijrs}(\omega,\ \mathbf{k};\ \omega+\omega',\ \mathbf{k}+\mathbf{k}';\ \omega,\ \mathbf{k})=$$

$$=-\frac{e^2}{m^2}\frac{\omega_L^2}{(\omega\omega')^2}\frac{1}{(\omega+\omega')}\left\{\delta_{ij}\delta_{sr}\frac{(\mathbf{k}+\mathbf{k}')^2}{\omega+\omega'}+\delta_{ij}\left[\frac{k_s}{\omega}(k_r+k'_r)+\frac{k'_r}{\omega'}(k_s+k'_s)\right]+\right.$$

$$\left.+\delta_{sr}\left[\frac{k_i}{\omega}(k_j+k'_j)+\frac{k'_j}{\omega'}(k_i+k'_i)\right]\right\}. \tag{A.48}$$

The tensor (A.48) is completely symmetric (remains unaltered) under the transposition of the frequencies and wave vectors $(\omega, \mathbf{k}) \rightleftharpoons (\omega', \mathbf{k}')$ with simultaneous transposition of the indices $i \rightleftharpoons j$, $s \rightleftharpoons r$. In particularly, the sum of the two tensors is completely symmetric:*

$$V_{iajb}(\omega,\ \mathbf{k};\ \omega',\ \mathbf{k}')+V_{iajb}(\omega,\ \mathbf{k};\ -\omega',\ -\mathbf{k}')=-\frac{e^2}{m^2}\frac{\omega_L^2}{(\omega\omega')^2}\times$$

$$\times\left\{\delta_{ia}\delta_{jb}\left[\left(\frac{\mathbf{k}+\mathbf{k}'}{\omega+\omega'}\right)^2+\left(\frac{\mathbf{k}-\mathbf{k}'}{\omega-\omega'}\right)^2\right]+\delta_{ia}\left[\frac{k_j}{\omega}\left(\frac{k_b+k'_b}{\omega+\omega'}+\frac{k_b-k'_b}{\omega-\omega'}\right)+\right.\right.$$

$$\left.\left.+\frac{k'_b}{\omega'}\left(\frac{k+k'_j}{\omega+\omega'}+\frac{k_j-k'_j}{\omega-\omega'}\right)\right]+\delta_{jb}\left[\frac{k_i}{\omega}\left(\frac{k_a+k'_a}{\omega+\omega'}+\frac{k_a-k'_a}{\omega-\omega'}\right)+\frac{k'_a}{\omega'}\left(\frac{k_i+k'_i}{\omega+\omega'}+\frac{k_i-k'_i}{\omega-\omega'}\right)\right]\right\}. \tag{A.49}$$

Four-index tensors depending on three pairs of independent arguments $(\omega, \mathbf{k})$, $(\omega', \mathbf{k}')$, $(\omega'', \mathbf{k}'')$ are used in the theory of the nonlinear interaction of four waves propagating in a plasma; the study of such interactions would, in general, go beyond the scope of the present exposition. Below, for reference purposes, we give an expression for the four-index tensor $\varepsilon_{i(jsr)}$, symmetrized with respect to the last three subscripts, as a function of three pairs of independent variables [see definition (20.5) in § 20 of Ch. IV)]:

$$\varepsilon_{i(jsr)}(\omega,\ \mathbf{k};\ \omega',\ \mathbf{k}';\ \omega'',\ \mathbf{k}'';\ \omega''',\ \mathbf{k}''')=\frac{1}{6}\frac{e^2}{m^2}\frac{\omega_L^2}{\omega\omega'\omega''\omega'''}\left\{\delta_{ij}\delta_{sr}\left(\frac{\mathbf{k}''+\mathbf{k}'''}{\omega''+\omega'''}\right)^2+\delta_{is}\delta_{jr}\left(\frac{\mathbf{k}'+\mathbf{k}'''}{\omega'+\omega'''}\right)^2+\delta_{ir}\delta_{js}\left(\frac{\mathbf{k}'+\mathbf{k}''}{\omega'+\omega''}\right)^2+\right.$$

$$+\frac{\delta_{ij}}{\omega''+\omega'''}\left[\frac{k''_s}{\omega''}(k''_r+k'''_r)+\frac{k'''_r}{\omega'''}(k''_s+k''_s)\right]+\frac{\delta_{is}}{\omega'+\omega'''}\left[\frac{k'_j}{\omega'}(k'_r+k'''_r)+\frac{k'''_r}{\omega'''}(k'_j+k'''_j)\right]+\frac{\delta_{ir}}{\omega'+\omega''}\left[\frac{k''_s}{\omega''}(k'_j+k''_j)+\frac{k'_j}{\omega'}(k'_s+k''_s)\right]+$$

$$\left.+\frac{\delta_{sr}}{\omega''+\omega'''}\left[\frac{k_i}{\omega}(k''_j+k'''_j)+\frac{k'_j}{\omega'}(k''_i+k'''_i)\right]+\frac{\delta_{jr}}{\omega'+\omega'''}\left[\frac{k_i}{\omega}(k'_s+k'''_s)+\frac{k''_s}{\omega''}(k'_i+k'''_i)\right]+\frac{\delta_{js}}{\omega'+\omega''}\left[\frac{k_i}{\omega}(k'_r+k''_r)+\frac{k'''_r}{\omega'''}(k'_i+k''_i)\right]\right\}. \tag{A.50}$$

We recall that the frequencies and wave vectors on the right-hand sides of Eq. (A.50) are related by the conditions (20.4). Note also that the tensor (A.50) arises as the symmetrized sum of the six terms (20.5), of which the five last are obtained by transposing the primes and the

---

* It should be noted that the right-hand sides of Eqs. (8.9) [see § 8 and (A.49)], which define one and the same sum of the tensors V in a cold isotropic plasma, are not equal. We recall that we derived Eq. (8.9) by integrating by parts, with respect to the velocity $d\mathbf{v}$, the general expression for the tensor V calculated by means of the Vlasov transport equation. This difference was first noted in [104]. In [104] the reasons for this discrepancy were also revealed; here, we shall not dwell on these, since the given discrepancy has no influence at all on the final results (the formulas for the kernels and the scattering cross sections) of the theory of the nonlinear interactions of waves given in the main exposition.

indices in the first tensor:*

$$\varepsilon_{ijsr}(\omega, \mathbf{k}; \omega'' + \omega''', \mathbf{k}'' + \mathbf{k}'''; \omega''', \mathbf{k}''') = \frac{e^2}{m^2}\frac{\omega_L^2}{\omega\omega'\omega''\omega'''}\frac{1}{\omega'' + \omega'''} \times$$
$$\times \left\{\delta_{ij}\delta_{sr}\frac{\mathbf{k}'', \mathbf{k}'' + \mathbf{k}'''}{\omega'' + \omega'''} + \delta_{ij}\frac{(k_s'' + k_s''')k_r'''}{\omega'''} + \delta_{sr}\left[\frac{k_i' k_j''}{\omega} + \frac{(k_j'' + k_j''')k_i''}{\omega} + \frac{k_j' k_i''}{\omega'}\right]\right\}. \tag{A.51}$$

The tensor (A.51) can be introduced by means of Eq. (A.40) for the current density $\delta j_i''$ if it is written in a slightly different form [compared with (A.40)]:

$$\delta j_i''(\omega, \mathbf{k}) = -i\frac{e^4 N}{m^3}\int d\omega' d\mathbf{k}' d\omega'' d\mathbf{k}'' d\omega''' d\mathbf{k}'''\delta(\omega - \omega' - \omega'' - \omega''')\times$$
$$\times\delta(\mathbf{k} - \mathbf{k}' - \mathbf{k}'' - \mathbf{k}''')\frac{1}{\omega'\omega''\omega'''(\omega'' + \omega''')}E_j(\omega', \mathbf{k}')E_s(\omega'', \mathbf{k}'')\times$$
$$\times E_r(\omega''', \mathbf{k}''')\left\{\delta_{ij}\delta_{sr}\frac{\mathbf{k}'', \mathbf{k} - \mathbf{k}'}{\omega - \omega'} + \delta_{ij}\frac{(k_s - k_s')k_r''}{\omega'''} + \delta_{sr}\left[\frac{k_i' k_j''}{\omega} + \frac{(k_j - k_j')k_i''}{\omega} + \frac{k_j' k_i''}{\omega'}\right]\right\}, \tag{A.52}$$

by redenoting the variables

$$\begin{gathered}(\omega', \mathbf{k}') \to (\omega - \omega', \mathbf{k} - \mathbf{k}'), \quad (\omega'', \mathbf{k}'') \to (\omega''', \mathbf{k}'''), \quad (\omega, \mathbf{k}) \to (\omega, \mathbf{k}),\\ (\omega - \omega', \mathbf{k} - \mathbf{k}') \to (\omega', \mathbf{k}'), \quad (\omega' - \omega'', \mathbf{k}' - \mathbf{k}'') \to (\omega'', \mathbf{k}'')\end{gathered} \tag{A.53}$$

and introducing δ functions by means of which Eqs. (20.4) are satisfied. In conclusion, let us point out the simplicity of the structure of the symmetrized tensor (A.50): the tensor (A.50) is composed of six vectors that have different dimensionalities in the inverse velocity [compare with (A.32), (A.48), and (A.49)]:

$$\frac{\mathbf{k}}{\omega}, \quad \frac{\mathbf{k}'}{\omega'}, \quad \frac{\mathbf{k}''}{\omega''}, \quad \frac{\mathbf{k}'' + \mathbf{k}'''}{\omega'' + \omega'''}, \quad \frac{\mathbf{k}' + \mathbf{k}''}{\omega' + \omega''}, \quad \frac{\mathbf{k}' + \mathbf{k}'''}{\omega' + \omega'''} \tag{A.54}$$

and the six Kronecker deltas

$$\delta_{ij}, \quad \delta_{is}, \quad \delta_{ir}, \quad \delta_{sr}, \quad \delta_{jr}, \quad \delta_{js}. \tag{A.55}$$

Concluding our treatment of an isotropic plasma, we write down the original equations (A.1)-(A.5) in Fourier components:

$$\frac{c^2k^2}{\omega^2}\left(\delta_{ij} - \frac{k_i k_j}{k^2}\right)E_j(\omega, \mathbf{k}) = E_i(\omega, \mathbf{k}) + \frac{4\pi i e}{\omega}\int d\omega' d\mathbf{k}' n(\omega', \mathbf{k}')v_i(\omega - \omega', \mathbf{k} - \mathbf{k}'); \tag{A.56}$$

$$v_i(\omega, \mathbf{k}) = i\frac{e}{m}\frac{1}{\omega}E_i(\omega, \mathbf{k}) + \frac{1}{\omega}\int d\omega' d\mathbf{k}' v_s(\omega - \omega', \mathbf{k} - \mathbf{k}')\times$$
$$\times\left\{(k_s - k_s')v_i(\omega', \mathbf{k}') + \left(i\frac{e}{m}\right)\left(\frac{k_i'}{\omega'}\delta_{js} - \delta_{ij}\frac{k_s'}{\omega'}\right)E_j(\omega', \mathbf{k}')\right\}; \tag{A.57}$$

$$n(\omega, \mathbf{k}) = \frac{k_i}{\omega}\int d\omega' d\mathbf{k}' n(\omega', \mathbf{k}')v_i(\omega - \omega', \mathbf{k} - \mathbf{k}'). \tag{A.58}$$

Essentially, by iterating this system with respect to the small electric field, we have calculated the three- and four-index tensors found above. Continuing the iteration, we obtain the complex permittivity tensors of even higher rank.

---

* For example, the second term in the curly brackets on the right-hand side of Eq. (20.5) [the tensor $\varepsilon_{irjs}(\omega, \mathbf{k}; \omega' + \omega'', \mathbf{k}' + \mathbf{k}''; \omega''\mathbf{k}'')$] arises from (A.51) by the index substitution $j \to r$, $s \to j$, $r \to s$ and argument substitution $(\omega', \mathbf{k}') \to (\omega''', \mathbf{k}''')$, $(\omega'', \mathbf{k}'') \to (\omega', \mathbf{k}')$, $(\omega''', \mathbf{k}''') \to (\omega'', \mathbf{k}'')$, i.e., by cyclic permutation of the indices and the corresponding frequencies and wave vectors.

If there is strong external constant and homogeneous magnetic field $\mathbf{B}_0$, the system of equations (A.1)-(A.5) is changed only due to a certain complication of the equation of motion (A.2):

$$m\left\{\frac{\partial \mathbf{v}}{\partial t}+\left(\mathbf{v}\frac{\partial}{\partial \mathbf{r}}\right)\mathbf{v}\right\}=e\left\{\mathbf{E}+\frac{1}{c}[\mathbf{v},\ \mathbf{B}_0+\mathbf{B}]\right\}. \tag{A.2a}$$

Solving this modified system of equations by expanding the density and velocity of the plasma particles, (A.7) and (A.8), in a series in powers of the electric field, we can readily obtain a generalization of the formulas obtained above to the case of a cold magnetoactive plasma. In particular, the solution of the linearized equation of motion (A.2a):

$$\frac{\partial \delta \mathbf{v}}{\partial t}=\frac{e}{mc}[\delta \mathbf{v} \mathbf{B}_0]+\frac{e}{m}\mathbf{E} \tag{A.59}$$

can be represented compactly by means of the previously introduced tensor $\mathbf{W}$ [see §16, formulas (16.12) and (16.13)]:

$$\delta v_i(\mathbf{r},\ t)=\frac{e}{m}\int\limits_{-\infty}^{0} d\tau W_{ij}(-\tau)E_j(\mathbf{r},\ t+\tau). \tag{A.60}$$

Going over in Eq. (A.60) to Fourier transforms after integrating with respect to τ by means of the relation used in (16.34)

$$\int\limits_{-\infty}^{0} d\tau W_{ij}(-\tau)e^{-i\omega\tau}=i\Gamma_{ij}(\omega) \tag{A.61}$$

we obtain an expression for the Fourier transform of the velocity linear in the electric field $\mathbf{E}(\omega, \mathbf{k})$:

$$\delta v_i(\omega,\ \mathbf{k})=i\frac{e}{m}\Gamma_{ij}(\omega)E_j(\omega,\ \mathbf{k}). \tag{A.62}$$

A similar form of $\delta\mathbf{v}(\omega, \mathbf{k})$ [we recall that the tensor $\Gamma_{ij}(\omega)$ is defined by formula (16.35) in §16] arises, naturally, from the solution of the algebraic equation

$$-i\omega\delta\mathbf{v}(\omega,\ \mathbf{k})=\frac{e}{mc}[\delta\mathbf{v}(\omega,\ \mathbf{k})\,\mathbf{B}_0]+\frac{e}{m}\mathbf{E}(\omega,\ \mathbf{k}), \tag{A.63}$$

corresponding to (A.59). The velocity (A.62) gives the current density

$$\delta j_i(\omega,\ \mathbf{k})=eN\delta v_i(\omega,\ \mathbf{k})=i\frac{e^2N}{m}\Gamma_{ij}(\omega)E_j(\omega,\ \mathbf{k}), \tag{A.64}$$

and hence the permittivity tensor of a cold magnetoactive plasma as well:

$$\varepsilon_{ij}(\omega,\ \mathbf{k})=\delta_{ij}-\frac{\omega_L^2}{\omega}\Gamma_{ij}(\omega), \tag{A.65}$$

which agrees with the tensor (16.34) already obtained in the kinetic description [see also the form of expression (16.37)-(16.38), which follows from (16.34)]. In accordance with (A.62), the first correction to the density [cf. (A.15)] has the form

$$\delta n(\omega,\ \mathbf{k})=\frac{N}{\omega}\mathbf{k}\delta\mathbf{v}(\omega,\ \mathbf{k})=i\frac{eN}{m}\frac{k_i}{\omega}\Gamma_{ij}(\omega)E_j(\omega,\ \mathbf{k}). \tag{A.66}$$

The velocity $\delta\mathbf{v}'$ next in order of magnitude is found from the equation of motion (A.2a) [cf. (A.22)]

$$\frac{\partial \delta\mathbf{v}'}{\partial t}-\frac{e}{mc}[\delta\mathbf{v}'\mathbf{B}_0]=-\left(\delta\mathbf{v}\frac{\partial}{\partial \mathbf{r}}\right)\delta\mathbf{v}+\frac{e}{mc}[\delta\mathbf{v}\mathbf{B}] \tag{A.67}$$

in exactly the same way as $\delta\mathbf{v}$:

$$\delta v_i'(\omega, \mathbf{k}) = \Gamma_{ia}(\omega) \int d\omega' d\mathbf{k}' d\omega'' d\mathbf{k}'' \delta(\omega - \omega' - \omega'') \delta(\mathbf{k} - \mathbf{k}' - \mathbf{k}'') \times$$
$$\times \left\{ a_{aj,s}(\omega'', \mathbf{k}'') \left(i \frac{e}{m} \right) E_j(\omega'', \mathbf{k}'') + k_s'' \delta v_a(\omega'', \mathbf{k}'') \right\} \delta v_s(\omega', \mathbf{k}'). \quad (A.68)$$

Substituting the velocity (A.62) into (A.68),

$$\delta v_i'(\omega, \mathbf{k}) = \left(i \frac{e}{m} \right)^2 \Gamma_{ia}(\omega) \int d\omega' d\mathbf{k}' d\omega'' d\mathbf{k}'' \delta(\omega - \omega' - \omega'') \delta(\mathbf{k} - \mathbf{k}' - \mathbf{k}'') \times$$
$$\times E_j(\omega'', \mathbf{k}'') E_s(\omega', \mathbf{k}') \Gamma_{bs}(\omega') \{ a_{aj,b}(\omega'', \mathbf{k}'') + k_b'' \Gamma_{aj}(\omega'') \} \quad (A.69)$$

and representing the resulting contraction of the two tensors Γ by their difference*

$$\omega' \Gamma_{ia}(\omega) \Gamma_{aj}(\omega'') = \Gamma_{ij}(\omega'') - \Gamma_{ij}(\omega), \quad \omega = \omega' + \omega'', \quad (A.70)$$

we obtain finally [with allowance for (A.24)]

$$\delta v_i'(\omega, \mathbf{k}) = \left(i \frac{e}{m} \right)^2 \int d\omega' d\mathbf{k}' d\omega'' d\mathbf{k}'' \delta(\omega - \omega' - \omega'') \delta(\mathbf{k} - \mathbf{k}' - \mathbf{k}'') \times$$
$$\times \frac{1}{\omega' \omega''} E_j(\omega'', \mathbf{k}'') E_s(\omega', \mathbf{k}') \{ -\omega \Gamma_{ij}(\omega) \Gamma_{as}(\omega') k_a'' + \omega' \Gamma_{js}(\omega') \Gamma_{ia}(\omega) k_a'' + \omega'' \Gamma_{ij}(\omega'') \Gamma_{as}(\omega') k_a'' \}. \quad (A.71)$$

In the limit of a vanishingly small external magnetic field ($\mathbf{B}_0 \to 0$), the expression (A.71) is identical with the expression (A.25) obtained previously for a cold isotropic plasma. Together with the equations (A.62) and (A.66), it gives the first nonlinear current density (A.26) in a cold magnetoactive plasma:

$$\delta j_i'(\omega, \mathbf{k}) = -\frac{e^3 N}{m} \int d\omega' d\mathbf{k}' d\omega'' d\mathbf{k}'' \delta(\omega - \omega' - \omega'') \delta(\mathbf{k} - \mathbf{k}' - \mathbf{k}'') \times$$
$$\times \frac{1}{\omega' \omega''} E_j(\omega'', \mathbf{k}'') E_s(\omega', \mathbf{k}') \{ \omega'' \Gamma_{ij}(\omega'') \Gamma_{as}(\omega') k_a + \omega' \Gamma_{js}(\omega') \Gamma_{ia}(\omega) k_a'' - \omega \Gamma_{ij}(\omega) \Gamma_{as}(\omega') k_a'' \}. \quad (A.72)$$

As in an isotropic plasma, we obtain directly from the current density (A.72) an explicit expression for the three-index tensors:

$$\varepsilon_{ijs}(\omega, \mathbf{k}; \omega', \mathbf{k}') = -i \frac{e}{m} \frac{\omega_L^2}{\omega \omega' \omega''} \{ \omega'' \Gamma_{ij}(\omega'') \Gamma_{as}(\omega') k_a + \omega' \Gamma_{js}(\omega') \Gamma_{ia}(\omega) k_a'' - \omega \Gamma_{ij}(\omega) \Gamma_{as}(\omega') k_a'' \}; \quad (A.73)$$

$$\varepsilon_{isj}(\omega, \mathbf{k}; \omega'', \mathbf{k}) = -i \frac{e}{m} \frac{\omega_L^2}{\omega \omega' \omega''} \{ \omega' \Gamma_{is}(\omega') \Gamma_{aj}(\omega'') k_a + \omega'' \Gamma_{sj}(\omega'') \Gamma_{ia}(\omega) k_a' - \omega \Gamma_{is}(\omega) \Gamma_{aj}(\omega'') k_a' \} \quad (A.74)$$

---

* The validity of Eq. (A.70) can be readily seen if one uses the explicit form (16.35) of the tensors Γ and the algebraic properties (16.20) of the tensor a_{ij}. In addition to Eq. (A.70), there are two others (for $\omega' + \omega'' = \omega$):

$$\omega'' \Gamma_{ia}(\omega) \Gamma_{aj}(\omega') = \Gamma_{ij}(\omega') - \Gamma_{ij}(\omega), \quad \omega \Gamma_{ia}(\omega') \Gamma_{ja}(\omega'') = \Gamma_{ij}(\omega') + \Gamma_{ji}(\omega'').$$

Note also that (A.70a)is a direct consequence of the property (16.21) of the contraction of the tensors W and arises by integrating (16.21) with respect to τ_0 and τ_1 with the $\exp(-i\omega\tau_0 - i\omega''\tau_1)$ in accordance with (A.61). In this sense, the entire derivation here of formulas for the three- and four-index tensors in a cold magnetoactive plasma can be carried through using only the tensor W (as was done in the kinetic description) and by going over to the tensor Γ only in the final expressions [see, for example, the transition from (19.11) to (19.16) in § 19].

and their sum S_{ijs}, which occurs in the generalized kinetic field equation (3.22):

$$S_{ijs}(\omega, \mathbf{k}; \omega', \mathbf{k}') = i\frac{e}{m}\frac{\omega_L^2}{\omega\omega'\omega''}\{\omega\Gamma_{is}(\omega)\Gamma_{aj}(\omega'')k'_a + \omega\Gamma_{ij}(\omega)\Gamma_{as}(\omega')\times$$
$$\times k''_a - \omega''\Gamma_{ij}(\omega'')\Gamma_{as}(\omega')k_a - \omega''\Gamma_{sj}(\omega'')\Gamma_{ia}(\omega)k'_a - \omega'\Gamma_{is}(\omega')\Gamma_{aj}(\omega'')k_a - \omega'\Gamma_{js}(\omega')\Gamma_{ia}(\omega)k''_a\}. \quad \text{(A.75)}$$

Here, the frequencies and wave vectors are related by Eqs. (A.29), and the tensors (A.73) and (A.74) go over into the tensors (A.28) and (A.30) in the limit of an isotropic plasma ($\mathbf{B}_0 = 0$). The tensor (A.75) is identical with that obtained previously in § 19 [see (19.16)] in the kinetic description of a magnetoactive plasma.

To calculate the current density (A.35) cubic in the field, we require, in addition to the already obtained quantities, the second correction to the density (A.34):*

$$\delta n'(\omega, \mathbf{k}) = -\frac{e^2N}{m^2}\int d\omega' d\mathbf{k}' d\omega'' d\mathbf{k}'' \delta(\omega - \omega' - \omega'')\delta(\mathbf{k} - \mathbf{k}' - \mathbf{k}'')\times$$
$$\times \frac{1}{\omega\omega'\omega''}E_j(\omega'', \mathbf{k}'')E_s(\omega', \mathbf{k}')\{\omega'' k_i\Gamma_{ij}(\omega'')\Gamma_{as}(\omega')k_a + \omega'\Gamma_{js}(\omega')k_i\Gamma_{ia}(\omega)k''_a - \omega k_i\Gamma_{ij}(\omega)\Gamma_{as}(\omega')k''_a\} \quad \text{(A.76)}$$

and the velocity $\delta\mathbf{v}''$, which satisfies the equation of motion [cf. (A.36)]:

$$\frac{\partial \delta\mathbf{v}''}{\partial t} - \frac{e}{mc}[\delta\mathbf{v}''\mathbf{B}_0] = -\left(\delta\mathbf{v}\frac{\partial}{\partial\mathbf{r}}\right)\delta\mathbf{v}' - \left(\delta\mathbf{v}'\frac{\partial}{\partial\mathbf{r}}\right)\delta\mathbf{v} + \frac{e}{mc}[\delta\mathbf{v}'\mathbf{B}]. \quad \text{(A.77)}$$

We find this last by using the Fourier transform of Eq. (A.77):

$$\delta v''_i(\omega, \mathbf{k}) = \Gamma_{ia}(\omega)\int d\omega' d\mathbf{k}' d\omega'' d\mathbf{k}'' \delta(\omega - \omega' - \omega'')\delta(\mathbf{k} - \mathbf{k}' - \mathbf{k}'')\times$$
$$\times\{\alpha_{aj,s}(\omega'', \mathbf{k}'')\left(i\frac{e}{m}\right)E_j(\omega'', \mathbf{k}'') + k'_b\delta v_b(\omega'', \mathbf{k}'')\delta_{as} + k''_s\delta v_a(\omega'', \mathbf{k}'')\}\delta v'_s(\omega', \mathbf{k}') \quad \text{(A.78)}$$

and substituting into the right-hand side of (A.78) the explicit expressions (A.24), (A.62), and (A.71) for the quantities it contains:

$$\delta v''_i(\omega, \mathbf{k}) = \left(i\frac{e}{m}\right)^3\Gamma_{ia}(\omega)\int d\omega' d\mathbf{k}' d\omega'' d\mathbf{k}'' \frac{1}{\omega''(\omega' - \omega'')}\times$$
$$\times E_j(\omega - \omega', \mathbf{k} - \mathbf{k}')E_s(\omega' - \omega'', \mathbf{k}' - \mathbf{k}'')E_r(\omega'', \mathbf{k}'')\left\{\frac{k_a - k'_a}{\omega - \omega'}\delta_{jc} - \right.$$
$$-\delta_{aj}\frac{k_c - k'_c}{\omega - \omega'} + k'_b\Gamma_{bj}(\omega - \omega')\delta_{ac} + (k_c - k'_c)\Gamma_{aj}(\omega - \omega')\}\times$$
$$\times\{(\omega' - \omega'')\Gamma_{rs}(\omega' - \omega'')\Gamma_{cd}(\omega')k''_d + \Gamma_{ds}(\omega' - \omega'')k''_d[\omega''\Gamma_{cr}(\omega'') - \omega'\Gamma_{cr}(\omega')]\}. \quad \text{(A.79)}$$

As a result, using (A.62), (A.66), (A.71), (A.76), and (A.79), we obtain the current density

$$\delta j''_i(\omega, \mathbf{k}) = -i\frac{e^4N}{m^3}\int d\omega' d\mathbf{k}' d\omega'' d\mathbf{k}'' \frac{1}{\omega'\omega''(\omega - \omega')(\omega' - \omega'')}\times$$
$$\times E_j(\omega - \omega', \mathbf{k} - \mathbf{k}')E_s(\omega' - \omega'', \mathbf{k}' - \mathbf{k}'')E_r(\omega'', \mathbf{k}'')\times$$
$$\times\{(\omega - \omega')\Gamma_{ij}(\omega - \omega')k_b\Gamma_{br}(\omega'')\omega''(k'_a - k''_a)\Gamma_{as}(\omega' - \omega'') +$$
$$+[\delta_{jb}(k_a - k'_a)\omega'\Gamma_{ia}(\omega) + \delta_{ib}(k_a - k'_a)\omega'\Gamma_{aj}(\omega - \omega') +$$
$$+(\omega - \omega')k_b\Gamma_{ij}(\omega - \omega') - \omega(k_b - k'_b)\Gamma_{ij}(\omega) + (\omega - \omega')k'_a\Gamma_{aj}(\omega - \omega')\times$$
$$\times\omega'\Gamma_{ib}(\omega)][(\omega' - \omega'')\Gamma_{rs}(\omega' - \omega'')\Gamma_{bc}(\omega')k''_c + \Gamma_{cs}(\omega' - \omega'')k''_c\,\omega'\Gamma_{br}(\omega'') - \Gamma_{cs}(\omega' - \omega'')k''_c\omega'\Gamma_{br}(\omega')]\} \quad \text{(A.80)}$$

---

* Note a helpful (for contraol purposes) consequence of the equation of continuity (A.6): $e\delta n'(\omega, \mathbf{k}) = \mathbf{k}\delta\mathbf{j}'(\omega, \mathbf{k})/\omega$, which relates formulas (A.72) and (A.76). In the limit $\mathbf{B}_0 \to 0$, the density (A.76) goes over into (A.38).

and its corresponding four-index complex permittivity tensor for a cold magnetoactive plasma

$$\begin{aligned}
&\varepsilon_{ijsr}(\omega, \mathbf{k}; \omega', \mathbf{k}'; \omega'', \mathbf{k}'') = \frac{e^2}{m^2} \frac{\omega_L^2}{\omega\omega'\omega''(\omega-\omega')(\omega'-\omega'')} \times \\
&\times \{(\omega-\omega')\Gamma_{ij}(\omega-\omega') k'_b \Gamma_{br}(\omega'')\omega''(k'_a - k''_a)\Gamma_{as}(\omega'-\omega'') + \\
&+ [\delta_{jb}(k_a - k'_a)\omega'\Gamma_{ia}(\omega) + \delta_{ib}(k_a - k'_a)\omega'\Gamma_{aj}(\omega-\omega') + \\
&+ (\omega-\omega') k_b \Gamma_{ij}(\omega-\omega') - \omega(k_b - k'_b)\Gamma_{ij}(\omega) + \\
&+ (\omega-\omega') k'_a \Gamma_{aj}(\omega-\omega')\omega'\Gamma_{ib}(\omega)][(\omega'-\omega'')\Gamma_{rs}(\omega'-\omega'') \times \\
&\times \Gamma_{bc}(\omega') k''_c + \Gamma_{cs}(\omega'-\omega'') k''_c \omega'' \Gamma_{br}(\omega'') - \Gamma_{cs}(\omega'-\omega'') k''_c \omega' \Gamma_{br}(\omega')]\}.
\end{aligned} \quad \text{(A.81)}$$

In the limit of a vanishingly small external magnetic field, this tensor is equal to the expression (A.43) calculated in an isotropic plasma.

LITERATURE CITED

1. V. P. Silin and A. A. Rukhadze, Electromagnetic Properties of Plasmas and Similar Media [in Russian], Gosatomizdat (1961).
2. A. A. Vedenov, in: Reviews of Plasma Physics, Vol. 3, Consultants Bureau, New York (1967), p. 229.
3. A. A. Vedenov, Theory of a Turbulent Plasma [in Russian], VINITI, Moscow (1965).
4. B. B. Kadomtsev, Plasma Turbulence, Academic Press, London (1965).
5. L. M. Kovrizhnykh, Tr. Fiz. Inst. Akad. Nauk SSSR, 32:173 (1965); Rapport interne EUR-CEA-FC, Fontenay-aux-Roses (Seine), France (1964), p. 258.
6. V. N. Tsytovich, Nonlinear Effects in Plasmas, New York (1970).
7. L. M. Kovizhnykh, Preprint [in Russian], P. N. Lebedev Physics Institute, No. 117, Moscow' (1968).
8. A. I. Akhiezer et al., Collective Oscillations in a Plasma, Oxford (1967) [the page number(s) given in the text refers to the Russian original: Atominzdat (1964)].
9. A. G. Sitenko, Electromagnetic Fluctuations in a Plasma, New York (1967).
10. A. G. Sitenko and Yu. A. Kirochkin, Usp. Fiz. Nauk, 89:227 (1966).
11. R. Z. Sagdeev and A. A. Galeev, Lectures on the Nonlinear Theory of Plasma, International Atomic Energy Report N IC 66 64, Trieste (1966).
12. L. M. Gorbunov, V. V. Pustovalov, and V. P. Silin, Zh. Eksp. Teor. Fiz., 47:1437 (1964).
13. Yu. L. Klimontovich, The Statistical Theory of Non-Equilibrium Processes in a Plasma, Pergamon Press, Oxford (1967) [the page number(s) given in the text refers to the Russian original: Izd-vo MGU (1964)].
14. S. A. Akhmanov and R. V. Khokhlov, Problems of Nonlinear Optics (Electromagnetic Waves in Nonlinear Dispersive Media) [in Russian], VINITI, Moscow (1964).
15. N. Bloembergen, Nonlinear Optics, Benjamin, New York (1965); [the page number(s) given in the text refers to the Russian translation: Mir (1966)].
16. L. D. Landau and E. M. Lifshitz, Electrodynamics of Continuous Media, Pergamon Press, Oxford (1960) [the page number(s) given in the text refers to the Russian original: Fizmatgiz (1959)].
17. V. L. Ginzburg, The Propagation of Electromagnetic Waves in Plasmas, Pergamon Press, Oxford (1964) [the page number(s) given in the text refers to the Russian original: Nauka (1967)].
18. R. Becker, Electron Theory [Russian translation], ONTI (1936).
19. V. P. Silin, Prikl. Mekh. Tekh. Fiz., No. 1, 32 (1964).
20. N. Rostoker and K. Matsuda, Plasma Physics and Controlled Nuclear Fusion Research, Vol. 1, Vienna (1966), p. 747.

21. N. N. Bogolyubov, "Problems of a dynamical theory in statistical physics," in: Studies in Statistical Mechanics, Vol. 1 (ed. J. de Boer and G. E. Uhlenbeck, North-Holland, Amsterdam (1962).
22. K. P. Turov, Fundamentals of Kinetic Theory [in Russian], Nauka, Moscow (1966).
23. V. I. Petviashvili, Candidate's Dissertation [in Russian], Tbilisi (1963).
24. Yu. L. Klimontovich and V. P. Silin, Zh. Eksp. Teor. Fiz., 42:286 (1962).
25. Yu. L. Klimontovich and V. P. Silin, Dokl. Akad. Nauk SSSR, 145:764 (1962).
26. W. A. Shurcliff, Polarized Light, Harvard University Press (1962).
27. L. D. Landau and E. M. Lifshitz, The Classical Theory of Fields, Addison-Wesley, Cambridge, Mass., (1951) [the page number(s) given in the text refers to a later Russian edition: Nauka (1967)].
28. A. P. Kropotkin, Candidate's Dissertation [in Russian], Fiz. Inst. Akad. Nauk, Moscow (1967).
29. L. M. Gorbunov and V. P. Silin, Zh. Eksp. Teor. Fiz., 47:203 (1964).
30. W. E. Drummond and D. Pines, Ann. Physics, 28:478 (1964).
31. L. M. Gorbunov, Candidate's Dissertation [in Russian], Fiz, Inst. Akad, Nauk, Moscow (1964).
32. A. Gailitis, L. M. Gorbunov, L. M. Kovrizhnykh, V. V. Pustovalov, V. P. Silin, and V. N. Tsytovich, Plasma Physics and Controlled Nuclear Fusion Research, Vol. 1, Vienna (1966), p. 673.
33. L. M. Gorbunov, V. V. Pustovalov, and V. P. Silin, Izv. Vyssh. Ucheb. Zaved., Radiofiz., 8:461 (1965).
34. L. M. Gorbunov and V. P. Silin, Zh. Eksp. Teor. Fiz., 50:1095 (1966).
35. L. M. Gorbunov and A. M. Timerbulatov, Zh. Eksp. Teor. Fiz., 53:1492 (1967).
36. D. A. Tidman and G. H. Weiss, Phys. Fluids, 4:866 (1961).
37. V. P. Silin, Zh. Eksp. Teor. Fiz., 45:866 (1963).
38. N. V. Sholokhov, Izv. Vyssh. Ucheb. Zaved., Radiofiz., 7:452 (1964).
39. O. P. Pogutse, Izv. Vyssh. Ucheb. Zaved., Radiofiz., 7:280 (1964).
40. F. G. Bass and A. Ya. Blank, Zh. Eksp. Teor. Fiz., 43:1479 (1962).
41. V. I. Petviashvili, Dokl. Akad. Nauk SSSR, 153:1295 (1963).
42. P. A. Sturrock, Proc. Roy. Soc., 242A:277 (1957).
43. L. I. Rudakov, Doctoral Dissertation [in Russian], IAÉ, Moscow (1965).
44. V. A. Golovko, Zh. Eksp. Teor. Fiz., 47:1765 (1964).
45. V. P. Silin, Zh. Eksp. Teor. Fiz., 48:1679 (1965).
46. V. D. Shafranov, in: Reviews of Plasma Physics, Vol. 3, Consultants Bureau, New York (1967), p. 1.
47. M. N. Rosenbluth and N. Rostoker, Phys. Fluids, 5:776 (1962).
48. F. Villars and V. F. Weisskopf, Phys. Rev., 94:232 (1954).
49. O. Buneman, J. Geophys. Res., 67:2050 (1962).
50. V. V. Pustovalov, Candidate's Dissertation [in Russian], Fiz. Inst. Akad. Nauk, Moscow (1966).
51. T. H. Stix, The Theory of Plasma Waves, McGraw-Hill, New York (1962) [the page number(s) given in the text refers to the Russian translation: Atomizdat (1965)].
52. A. P. Kropotkin and V. V. Pustovalov, Plasma Physics and Controlled Nuclear Fusion Research, Vol. 1, Vienna (1966), p. 695.
53. V. I. Pakhomov, V. F. Aleksin, and K. N. Stepanov, Zh. Tekh. Fiz., 31:1170 (1961).
54. K. N. Stepanov, Doctoral Dissertation [in Russian], Khar'kov (1964).
55. V. F. Aleksin and K. N. Stepanov, in: Plasma Physics and the Problem of Controlled Thermonuclear Fusion, Vol. 3 [in Russian], Izd-vo Akad. Nauk Ukr. SSR, Kiev (1963), p. 64.
56. A. J. McConnell, Introduction to Tensor Analysis with Applications to Geometry, Mechanics, and Physics [Russian translation], Fizmatgiz (1963).

57. W. P. Allis, S. J. Buchsbaum, and A. Bers, Waves in Anisotropic Plasmas, MIT Press, Cambridge, Mass. (1963).
58. M. Bocher, Introduction to Higher Algebra, New York (1907).
59. A. A. Rukhadze, Doctoral Dissertation [in Russian], Fiz. Inst. Akad. Nauk, Moscow (1964).
60. V. L. Ginzburg and A. A. Rukhadze, Waves in Magnetoactive Plasmas [in Russian], Izd-vo Nauka, Moscow (1970) [the page number(s) given in the text refers to the Russian original: Izd-vo, Nauka, Moscow (1970)].
61. A. P. Kropotkin and V. V. Pustovalov, Zh. Eksp. Teor. Fiz., 49:1345 (1965).
62. L. D. Landau and E. M. Lifshitz, Statistical Physics, London (1958) [the page number(s) given in the text refers to the Russian original: Izd-vo Nauka, Moscow (1964)].
63. L. I. Rudakov, Zh. Eksp. Teor. Fiz., 48:1372 (1965).
64. A. P. Kropotkin and V. V. Pustovalov, Preprint [in Russian], P. N. Lebedev Physics Institute, No. 13, Moscow (1966).
65. A. P. Kropotkin and V. V. Pustovalov, Phys. Fluids, 10:241 (1967).
66. A. P. Kropotkin and V. V. Pustovalov, Izv. Vyssh, Ucheb. Zaved. Radiofiz., 8:886 (1965).
67. A. A. Kropotkin, V. V. Pustovalov, and N. V. Sholokhov, Zh. Tekh. Fiz., 38:240 (1968).
68. A. P. Kropotkin and N. V. Sholokhov, Zh. Tekh. Fiz., 39:628 (1968); Preprint [in Russian], P. N. Lebedev Physics Institute, No. 58, Moscow (1967).
69. Yu. A. Kirochkin, Candidate's Dissertation [in Russian], Khar'kov, KhGU (1964).
70. V. D. Shapiro and V. I. Shevchenko, in: High-Frequency Properties of Plasmas [in Russian], Naukova Dumka (1965), p. 156.
71. A. Gailitis and V. N. Tsytovich, Zh. Eksp. Teor. Fiz., 47:1468 (1964).
72. A. B. Shvartsburg, Candidate's Dissertation [in Russian], Fiz. Inst. Akad. Nauk, Moscow (1966).
73. L. M. Kovrizhnykh, Zh. Eksp. Teor. Fiz., 48:1114 (1965).
74. A. A. Galeev, V. I. Karpman, and R. Z. Sagdeev, Nuclear Fusion, 5:20 (1965).
75. N. P. Giorgadze, E. M. Khirseli, and N. L. Tsintsadze, Izv. Vyssh. Ucheb. Zaved., Radiofiz., 9:489 (1966).
76. R. N. Sudan, A. Cavaliere, and M. N. Rosenbluth, Phys. Rev., 158:387 (1967).
77. P. M. Platzman, P. A. Wolf, and N. Tzoar, Phys. Rev., 174:489 (1967).
78. R. A. Ellis and M. Porcolab, Phys. Rev. Lett., 21:529 (1968).
79. R. Cano, C. Etievant, I. Fidone, and G. Granata, Nucl. Fusion, 9:223 (1969).
80. G. I. Suramlishvili, Dokl. Akad. Nauk SSSR, 157:83 (1964).
81. G. I. Suramlishvili, Candidate's Dissertation [in Russian], IAÉ, Moscow (1965).
82. L. M. Kovrizhnykh, Zh. Eksp. Teor. Fiz., 49:237 (1965).
83. S. Krishan, Wave-Wave Interaction between Transverse Plasma Oscillations (Preprint), Department of Physics, University of Alberta, Edmonton, Canada, March (1967).
84. V. A. Liperovskii and V. N. Tsytovich, Izv. Vyssh. Ucheb. Zaved., Radiofiz., 12:823 (1969).
85. V. E. Zakharov, Zh. Eksp. Teor. Fiz., 51:688 (1966).
86. V. N. Tsytovich, Preprint No. 157 [in Russian], P. N. Lebedev Physics Institute, Moscow (1969).
87. L. M. Kovrizhnykh, Zh. Eksp. Teor. Fiz., 52:1406 (1967).
88. A. P. Kropotkin, Zh. Eksp. Teor. Fiz., 53:1765 (1967).
89. V. L. Sizonenko and K. N. Stepanov, Preprint FTI Akad. Nauk, UkrSSr, No. 218, Khar'kov (1968); ZhETF Pis. Red., 9:468 (1963).
90. V. P. Silin, Survey of Phenomena in Ionized Gases, Vienna (1968), p. 205.
91. V. P. Silin, Zh. Eksp. Teor. Fiz., 57:183 (1969).
92. N. E. Andreev, A. Yu. Kirii, and V. P. Silin, Zh. Eksp. Teor. Fiz., 57:1024 (1969).
93. Yu. V. Afanas'ev, N. G. Basov, O. N. Krokhin, N. V. Morachevskii, and G. V. Sklizkov, Zh. Tekh. Fiz., 39:894 (1969).

94. U. Ascoli-Bartoli, B. Brunelli, A. Caruso, A. de Angels, G. Gatti, R. Gratton, F. Parlange, and H. Salzmann, Plasma Physics and Controlled Nuclear Fusion Research, Vol. 1, Vienna (1969), p. 917.
95. A. F. Haught, D. H. Polk, and W. J. Fader, Plasma Physics and Controlled Nuclear Fusion Research Vol. 1, Vienna (1969), p. 925.
96. M. J. Lubin, H. S. Dunn, and W. Friedman, Plasma Physics and Controlled Nuclear Fusion Research, Vol. 1, Vienna (1969), p. 945.
97. N. G. Basov, S. D. Zakharov, P. G. Kryukov, Yu. V. Senatskii, and S. V. Chekalin, ZhETF Pis. Red., 8:26 (1968).
98. W. I. Linlor, Appl. Phys. Lett., 3:210 (1963).
99. I. R. Gekker and O. V. Sizukhin, ZhETF Pis. Red., 9:408 (1969).
100. G. M. Batanov and K. A. Sarksyan, ZhETF Pis. Red., 13:539 (1971).
101. K. F. Sergeichev and V. E. Trofimov, ZhETF Pis. Red., 13:236 (1971).
102. B. A. Demidov and S. D. Fanchenko, ZhETF Pis. Red., 3:533 (1966).
103. N. F. Perepelkin, ZhETF Pis. Red., 3:258 (1966).
104. L. M. Gorbunov, Zh. Eksp. Teor. Fiz., 55:2298 (1968).
105. N. V. Sholokhov, Candidate's Dissertation [in Russian], Fiz. Inst. Akad. Nauk, Moscow (1970).
106. N. V. Sholokhov, Izv. Vyssh. Ucheb. Zaved., Radiofiz., 13, 857 (1970).
107. N. V. Sholokhov, Teor. Mat. Fiz., 2:117 (1970).